# FUNDAMENT



The great advances made in VLSI technology in recent years have been underpinned by rapid developments in the design and fabrication of CMOS and bipolar devices, particularly at the deep submicron level. This book examines in detail the basic properties and design of these devices, including chip integration, and discusses the various factors that affect their performance.

The authors begin with a thorough review of the relevant aspects of semiconductor physics and proceed to a description of the design of CMOS and bipolar devices. The optimization of these devices for VLSI applications is also covered. The authors highlight the intricate interdependencies and subtle tradeoffs between those device parameters, such as power consumption and packing density, that affect circuit performance and manufacturability. They also discuss in detail the scaling, and physical limits to the scaling, of CMOS and bipolar devices.

The book contains many exercises, and can be used as a textbook for senior undergraduate or first-year graduate courses on microelectronics or VLSI devices. It will also be a valuable reference volume for practicing engineers involved in research and development in the electronics industry.

Yuan Taur received his Ph.D. from the University of California, Berkeley, in 1974. From 1975 to 1979 he worked at NASA, Goddard Institute for Space Studies, and from 1979 to 1981 at Rockwell International Science Center. Since 1981 he has been with the Silicon Technology Department at the IBM T. J. Watson Research Center. He has authored or coauthored over 100 technical papers and holds 7 U.S. patents. He is a Fellow of the IEEE and has been an editor of *IEEE Electron Device Letters.*

Tak Ning received his Ph.D. from the University of Illinois, Urbana-Champaign, in 1971. He has been with the IBM T. J. Watson Research Center since 1973, serving as manager of the Advanced Silicon Device Technology Department from 1982 to 1991. He has authored or coauthored about 100 technical papers and holds 19 U.S. patents. He has served as associate editor of *IEEE Transactions on Electron Devices*, and is a Fellow of the IEEE. He is also an IBM Fellow and a member of the U.S. National Academy of Engineering.

FU[illegible]NTALS OF MODERN
VLSI DEVICES

The great advances [illegible] technology in recent years have been underpinned by rapid development [illegible] and fabrication of CMOS and bipolar devices, particularly in the [illegible]. This book examines in detail the basic properties and design of these devices, including chip integration, and discusses the various factors that affect their performance.

The authors begin with a thorough review of the relevant aspects of semiconductor physics and proceed to a description of the design of CMOS and bipolar devices. The optimization of these devices for VLSI applications is also covered. The authors highlight the intricate interdependencies and subtle tradeoffs between those device parameters, such as power consumption and packing density, that affect circuit performance and manufacturability. They also discuss in detail the scaling and physical limits to the scaling of CMOS and bipolar devices.

The book contains many exercises, and can be used as a textbook for senior undergraduate or first-year graduate courses on microelectronics or VLSI devices. It will also be a valuable reference volume for practicing engineers involved in research and development in the electronics industry.

Yuan Taur received a Ph.D. from the University of California, Berkeley, in 1974. From 1975 to 1979 he worked at NASA Goddard Institute for Space Studies, and from 1979 to 1981 at Rockwell International Science Center. Since 1981 he has been with the Silicon Technology Department at the IBM T. J. Watson Research Center. He has authored or coauthored over 100 technical papers and holds 7 U.S. patents. He is a Fellow of the IEEE and has been an editor of *IEEE Electron Device Letters*.

Tak H. Ning received his Ph.D. from the University of Illinois, Urbana-Champaign, in 1971. He has been with the IBM T. J. Watson Research Center since 1973, serving as manager of the Advanced Silicon Device Technology Department from 1982 to 1991. He has authored or coauthored about 100 technical papers and holds 15 U.S. patents. He has served as associate editor of *IEEE Transactions on Electron Devices*, and is a Fellow of the IEEE. He is also an IBM Fellow and a member of the U.S. National Academy of Engineering.

# FUNDAMENTALS OF MODERN VLSI DEVICES

YUAN TAUR    TAK H. NING

CAMBRIDGE
UNIVERSITY PRESS

PUBLISHED BY THE PRESS SYNDICATE OF THE UNIVERSITY OF CAMBRIDGE
The Pitt Building, Trumpington Street, Cambridge, CB2 1RP, United Kingdom

CAMBRIDGE UNIVERSITY PRESS
The Edinburgh Building, Cambridge CB2 2RU, UK
40 West 20th Street, New York, NY 10011-4211, USA
477 Williamstown Road, Port Melbourne, VIC 3207, Australia
Ruiz de Alarcón 13, 28014 Madrid, Spain
Dock House, The Waterfront, Cape Town 8001, South Africa

http://www.cambridge.org



First published 1998
Reprinted 1999, 2000, 2001, 2002
First South Asian edition 2003
Reprinted 2004, 2006

Printed in India at Replika Press Pvt. Ltd.

Typeset in Times Roman 11/14 pt. and Eurostile in LaTeX 2ε [TB]

*A catalogue record for this book is available from the British Library*

*Library of Congress Cataloging-in-Publication Data*

Taur, Yuan, 1946–
Fundamentals of modern VSLI devices / Yuan Taur, Tak H. Ning.
p. cm.
ISBN 0-521-55056-4. – ISBN 0-521-55959-6 (pbk.)
1. Metal oxide semiconductors, Complementary. 2. Bipolar transistors. 3. Integrated circuits – Very large scale integration.
I. Ning. Tak H. 1943– . II. Title.
TK7871.99.M44T38 1998
621.39'5–dc21 98-16162
CIP

ISBN 0-521-54085-2 paperback

Special edition for sale in South Asia only, not for export elsewhere.

# CONTENTS

# PHYSICAL CONSTANTS AND UNIT CONVERSIONS

| Description | Symbol | Value and unit |
|---|---|---|
| Electronic charge | $q$ | $1.6 \times 10^{-19}$ C |
| Boltzmann's constant | $k$ | $1.38 \times 10^{-23}$ J/K |
| Vacuum permittivity | $\varepsilon_0$ | $8.85 \times 10^{-14}$ F/cm |
| Silicon permittivity | $\varepsilon_{si}$ | $1.04 \times 10^{-12}$ F/cm |
| Oxide permittivity | $\varepsilon_{ox}$ | $3.45 \times 10^{-13}$ F/cm |
| Velocity of light in vacuum | $c$ | $3 \times 10^{10}$ cm/s |
| Planck's constant | $h$ | $6.63 \times 10^{-34}$ J-s |
| Free-electron mass | $m_0$ | $9.1 \times 10^{-31}$ kg |
| Thermal voltage ($T = 300$ K) | $kT/q$ | 0.0259 V |
| Angstrom | Å | 1 Å = $10^{-8}$ cm |
| Nanometer | nm | 1 nm = $10^{-7}$ cm |
| Micrometer (micron) | μm | 1 μm = $10^{-4}$ cm |
| Millimeter | mm | 1 mm = 0.1 cm |
| Meter | m | 1 m = $10^2$ cm |
| Electron-volt | eV | 1 eV = $1.6 \times 10^{-19}$ J |
| Energy = charge × voltage | $E = qV$ | Joule = Coulomb × Volt |
| Charge = capacitance × voltage | $Q = CV$ | Coulomb = Farad × Volt |
| Power = current × voltage | $P = IV$ | Watt = Ampere × Volt |
| Time = resistance × capacitance | $t = RC$ | second = Ω (ohm) × Farad |
| Current = charge/time | $I = Q/t$ | Ampere = Coulomb/second |
| Resistance = voltage/current | $R = V/I$ | Ω (ohm) = Volt/Ampere |

# LIST OF SYMBOLS

| Symbol | Description | Unit |
|---|---|---|
| $A$ | Area | $cm^2$ |
| $A_E$ | Emitter area | $cm^2$ |
| $\alpha$ | Common-base current gain | None |
| $\alpha_0$ | Static common-base current gain | None |
| $\alpha_F$ | Forward common-base current gain in the Ebers–Moll model | None |
| $\alpha_R$ | Reverse common-base current gain in the Ebers–Moll model | None |
| $\alpha_T$ | Base transport factor | None |
| $\alpha_n$ | Electron-initiated rate of electron–hole pair generation per unit distance | $cm^{-1}$ |
| $\alpha_p$ | Hole-initiated rate of electron–hole pair generation per unit distance | $cm^{-1}$ |
| $BV$ | Breakdown voltage | V |
| $BV_{CBO}$ | Collector–base junction breakdown voltage with emitter open circuit | V |
| $BV_{CEO}$ | Collector–emitter breakdown voltage with base open circuit | V |
| $BV_{EBO}$ | Emitter–base junction breakdown voltage with collector open circuit | V |
| $\beta$ | Common-emitter current gain | None |
| $\beta_0$ | Static common-emitter current gain | None |
| $\beta_F$ | Forward common-emitter current gain in the Ebers–Moll model | None |
| $\beta_R$ | Reverse common-emitter current gain in the Ebers–Moll model | None |
| $c$ | Velocity of light in vacuum ($= 3 \times 10^{10}$ cm/s) | cm/s |
| $C$ | Capacitance | F |
| $C_d$ | Depletion-layer capacitance (per unit area) | F ($F/cm^2$) |

| Symbol | Description | Unit |
|---|---|---|
| $C_{dm}$ | Maximum depletion-layer capacitance (per unit area) | F (F/cm$^2$) |
| $C_{dBE}$ | Base–emitter diode depletion-layer capacitance (per unit area) | F (F/cm$^2$) |
| $C_{dBC}$ | Base–collector diode depletion-layer capacitance (per unit area) | F (F/cm$^2$) |
| $C_D$ | Diffusion capacitance | F |
| $C_{Dn}$ | Diffusion capacitance due to excess electrons | F |
| $C_{Dp}$ | Diffusion capacitance due to excess holes | F |
| $C_{DE}$ | Emitter diffusion capacitance | F |
| $C_g$ | Intrinsic gate capacitance | F |
| $C_i$ | Inversion-layer capacitance (per unit area) | F (F/cm$^2$) |
| $C_{it}$ | Interface trap capacitance per unit area | F/cm$^2$ |
| $C_j$ | Junction capacitance per unit area | F/cm$^2$ |
| $C_J$ | Junction capacitance | F |
| $C_L$ | Load capacitance | F |
| $C_{in}$ | Equivalent input capacitance of a logic gate | F |
| $C_{out}$ | Equivalent output capacitance of a logic gate | F |
| $C_{ov}$ | Gate-to-source (-drain) overlap capacitance (per edge) | F |
| $C_{ox}$ | Oxide capacitance per unit area | F/cm$^2$ |
| $C_{si}$ | Silicon capacitance per unit area | F/cm$^2$ |
| $C_w$ | Wire capacitance per unit length | F/cm |
| $C_\pi$ | Base–emitter capacitance in the small-signal hybrid-$\pi$ equivalent-circuit model | F |
| $C_\mu$ | Base–collector capacitance in the small-signal hybrid-$\pi$ equivalent-circuit model | F |
| $d$ | Width of diffusion region in a MOSFET | cm |
| $D_n$ | Electron diffusion coefficient | cm$^2$/s |
| $D_{nB}$ | Electron diffusion coefficient in the base of an n–p–n transistor | cm$^2$/s |
| $D_p$ | Hole diffusion coefficient | cm$^2$/s |
| $D_{pE}$ | Hole diffusion coefficient in the emitter of an n–p–n transistor | cm$^2$/s |
| $\Delta V_t$ | Threshold voltage roll-off due to short-channel effect | V |
| $\Delta E_g$ | Apparent bandgap narrowing | J |
| $\Delta E_{gB}$ | Bandgap-narrowing parameter in the base region | J |
| $\Delta E_{g,SiGe}$ | Total base bandgap narrowing due to Ge in the SiGe-base bipolar transistor | J |
| $E$ | Energy | J |
| $E_c$ | Conduction-band edge | J |
| $E_v$ | Valence-band edge | J |

| Symbol | Description | Unit |
|---|---|---|
| $E_a$ | Ionized-acceptor energy level | J |
| $E_d$ | Ionized-donor energy level | J |
| $E_f$ | Fermi energy level | J |
| $E_g$ | Energy gap of silicon | J |
| $E_i$ | Intrinsic Fermi level | J |
| $\mathscr{E}$ | Electric field | V/cm |
| $\mathscr{E}_c$ | Critical field for velocity saturation | V/cm |
| $\mathscr{E}_{eff}$ | Effective vertical field in MOSFET | V/cm |
| $\mathscr{E}_{ox}$ | Oxide electric field | V/cm |
| $\mathscr{E}_s$ | Electric field at silicon surface | V/cm |
| $\mathscr{E}_x$ | Vertical field in silicon | V/cm |
| $\mathscr{E}_y$ | Lateral field in silicon | V/cm |
| $\varepsilon_0$ | Vacuum permittivity ($= 8.85 \times 10^{-14}$ F/cm) | F/cm |
| $\varepsilon_{si}$ | Silicon permittivity ($= 1.04 \times 10^{-12}$ F/cm) | F/cm |
| $\varepsilon_{ox}$ | Oxide permittivity ($= 3.45 \times 10^{-13}$ F/cm) | F/cm |
| $f$ | Probability that an electronic state is filled | None |
| $f$ | Frequency, clock frequency | Hz |
| $f_{\max}$ | Maximum frequency of oscillation | Hz |
| $f_T$ | Cutoff frequency | Hz |
| FI | Fan-in | None |
| FO | Fan-out | None |
| $\phi$ | Barrier height | V |
| $\phi_{ox}$ | Silicon–silicon dioxide interface potential barrier for electrons | V |
| $\phi_{ms}$ | Work-function difference between metal and silicon | V |
| $\phi_n$ | Electron quasi-Fermi potential | V |
| $\phi_p$ | Hole quasi-Fermi potential | V |
| $g$ | Number of degeneracy | None |
| $g_{ds}$ | Small-signal output conductance | A/V |
| $g_m$ | Small-signal transconductance | A/V |
| $G_E$ | Emitter Gummel number | s/cm$^4$ |
| $G_B$ | Base Gummel number | s/cm$^4$ |
| $\gamma$ | Emitter injection efficiency | None |
| $h$ | Planck's constant ($= 6.63 \times 10^{-34}$J-s) | J-s |
| $i$ | Time-dependent current | A |
| $i_B$ | Time-dependent base current in a bipolar transistor | A |
| $i_b$ | Time-dependent small-signal base current | A |
| $i_C$ | Time-dependent collector current in a bipolar transistor | A |

| Symbol | Description | Unit |
|---|---|---|
| $i_c$ | Time-dependent small-signal collector current | A |
| $i_E$ | Time-dependent emitter current in a bipolar transistor | A |
| $I$ | Current | A |
| $I_B$ | Static base current in a bipolar transistor | A |
| $I_C$ | Static collector current in a bipolar transistor | A |
| $I_E$ | Static emitter current in a bipolar transistor | A |
| $I_S$ | Switch current in an ECL circuit | A |
| $I_g$ | Gate current in a MOSFET | A |
| $I_0$ | MOSFET current per unit width at threshold | A/cm |
| $I_{dsat}$ | MOSFET saturation current | A |
| $I_{on}$ | MOSFET on current (per unit width) | A (A/cm) |
| $I_{off}$ | MOSFET off current (per unit width) | A (A/cm) |
| $I_n$ | nMOSFET current per unit width | A/cm |
| $I_p$ | pMOSFET current per unit width | A/cm |
| $I_N$ | nMOSFET current | A |
| $I_P$ | pMOSFET current | A |
| $I_{ds}$ | Drain-to-source current in a MOSFET | A |
| $I_{sx}$ | Substrate current in a MOSFET | A |
| $J$ | Current density | A/cm$^2$ |
| $J_B$ | Base current density | A/cm$^2$ |
| $J_C$ | Collector current density | A/cm$^2$ |
| $J_n$ | Electron current density | A/cm$^2$ |
| $J_p$ | Hole current density | A/cm$^2$ |
| $k$ | Boltzmann's constant ($= 1.38 \times 10^{-23}$J/K) | J/K |
| $\kappa$ | Scaling factor (>1) | None |
| $l$ | Mean free path | cm |
| $L$ | Length, MOSFET channel length | cm |
| $L_D$ | Debye length | cm |
| $L_n$ | Electron diffusion length | cm |
| $L_p$ | Hole diffusion length | cm |
| $L_{met}$ | Metallurgical channel length of MOSFET | cm |
| $L_{eff}$ | Effective channel length of MOSFET | cm |
| $L_w$ | Wire length | cm |
| $m$ | MOSFET body-effect coefficient | None |
| $m_0$ | Free-electron mass ($= 9.1 \times 10^{-31}$kg) | kg |
| $m^*$ | Electron effective mass | kg |
| $M$ | Avalanche multiplication factor | None |
| $\mu$ | Carrier mobility | cm$^2$/V-s |

| Symbol | Description | Unit |
|---|---|---|
| $\mu_{eff}$ | Effective mobility | $cm^2/V\text{-}s$ |
| $\mu_n$ | Electron mobility | $cm^2/V\text{-}s$ |
| $\mu_p$ | Hole mobility | $cm^2/V\text{-}s$ |
| $n$ | Density of free electrons | $cm^{-3}$ |
| $n_0$ | Density of free electrons at thermal equilibrium | $cm^{-3}$ |
| $n_i$ | Intrinsic carrier density | $cm^{-3}$ |
| $n_{ie}$ | Effective intrinsic carrier density | $cm^{-3}$ |
| $n_{ieB}$ | Effective intrinsic carrier density in base of bipolar transistor | $cm^{-3}$ |
| $n_{ieE}$ | Effective intrinsic carrier density in emitter of bipolar transistor | $cm^{-3}$ |
| $n_n$ | Density of electrons in n-region | $cm^{-3}$ |
| $n_p$ | Density of electrons in p-region | $cm^{-3}$ |
| $N_a$ | Acceptor impurity density | $cm^{-3}$ |
| $N_d$ | Donor impurity density | $cm^{-3}$ |
| $N_b$ | Impurity concentration in bulk silicon | $cm^{-3}$ |
| $N_c$ | Effective density of states of conduction band | $cm^{-3}$ |
| $N_v$ | Effective density of states of valence band | $cm^{-3}$ |
| $N_B$ | Base doping concentration | $cm^{-3}$ |
| $N_C$ | Collector doping concentration | $cm^{-3}$ |
| $N_E$ | Emitter doping concentration | $cm^{-3}$ |
| $p$ | Density of free holes | $cm^{-3}$ |
| $p_0$ | Density of free holes at thermal equilibrium | $cm^{-3}$ |
| $p_n$ | Density of holes in n-region | $cm^{-3}$ |
| $p_p$ | Density of holes in p-region | $cm^{-3}$ |
| $P$ | Power dissipation | W |
| $P_{ac}$ | Active power dissipation | W |
| $P_{off}$ | Standby power dissipation | W |
| $q$ | Electronic charge ($= 1.6 \times 10^{-19}$ C) | C |
| $Q$ | Charge | C |
| $Q_B$ | Excess minority charge in the base | C |
| $Q_E$ | Excess minority charge in the emitter | C |
| $Q_{BE}$ | Excess minority charge in the base–emitter space-charge region | C |
| $Q_{BC}$ | Excess minority charge in the base–collector space-charge region | C |
| $Q_{DE}$ | Total stored minority-carrier charge in a bipolar transistor biased in the forward-active mode | C |
| $Q_{pB}$ | Hole charge per unit area in base of n–p–n transistor | $C/cm^2$ |

| Symbol | Description | Unit |
|---|---|---|
| $Q_s$ | Total charge per unit area in silicon | $C/cm^2$ |
| $Q_d$ | Depletion charge per unit area | $C/cm^2$ |
| $Q_i$ | Inversion charge per unit area | $C/cm^2$ |
| $Q_f$ | Fixed oxide charge per unit area | $C/cm^2$ |
| $Q_m$ | Mobile charge per unit area | $C/cm^2$ |
| $Q_{it}$ | Interface trapped charge per unit area | $C/cm^2$ |
| $Q_{ot}$ | Oxide trapped charge per unit area | $C/cm^2$ |
| $Q_{ox}$ | Equivalent oxide charge density per unit area | $C/cm^2$ |
| $r$, $R$ | Resistance | Ω |
| $r_b$ | Base resistance | Ω |
| $r_{bi}$ | Intrinsic base resistance | Ω |
| $r_{bx}$ | Extrinsic base resistance | Ω |
| $r_c$ | Collector series resistance | Ω |
| $r_e$ | Emitter series resistance | Ω |
| $r_0$ | Output resistance in small-signal hybrid-$\pi$ equivalent-circuit model | Ω |
| $r_\pi$ | Input resistance in small-signal hybrid-$\pi$ equivalent-circuit model | Ω |
| $R_L$ | Load resistance in a circuit | Ω |
| $R_s$ | Source series resistance | Ω |
| $R_d$ | Drain series resistance | Ω |
| $R_{sd}$ | Source–drain series resistance | Ω |
| $R_{ch}$ | MOSFET channel resistance | Ω |
| $R_w$ | Wire resistance per unit length | Ω/cm |
| $R_{Sbi}$ | Sheet resistance of intrinsic-base layer | Ω/□ |
| $R_{sw}$ | Equivalent switching resistance of a CMOS gate | Ω |
| $R_{swn}$ | Equivalent switching resistance of nMOSFET pulldown | Ω |
| $R_{swp}$ | Equivalent switching resistance of pMOSFET pullup | Ω |
| $\rho$ | Resistivity | Ω-cm |
| $\rho$ | Charge density | $C/cm^3$ |
| $\rho_{sh}$ | Sheet resistivity | Ω/□ |
| $\rho_{ch}$ | Sheet resistivity of MOSFET channel | Ω/□ |
| $\rho_{sd}$ | Sheet resistivity of source or drain region | Ω/□ |
| $\rho_c$ | Specific contact resistivity | Ω-$cm^2$ |
| $S$ | MOSFET subthreshold slope | V/decade |
| $S_p$ | Surface recombination velocity for holes | cm/s |
| $\sigma_V$ | Vertical straggle of Gaussian doping profile | cm |
| $\sigma_L$ | Lateral straggle of Gaussian doping profile | cm |
| $t$ | Time | s |

| Symbol | Description | Unit |
|---|---|---|
| $t_B$ | Base transit time | s |
| $t_E$ | Emitter delay time | s |
| $t_{BE}$ | Base–emitter depletion-layer transit time | s |
| $t_{BC}$ | Base–collector depletion-layer transit time | s |
| $t_{ox}$ | Oxide thickness | cm |
| $t_w$ | Thickness of wire | cm |
| $T$ | Absolute temperature | K |
| $\tau$ | Lifetime | s |
| $\tau$ | Circuit delay | s |
| $\tau_b$ | Buffered delay | s |
| $\tau_{int}$ | Intrinsic, unloaded delay | s |
| $\tau_F$ | Forward transit time of bipolar transistor | s |
| $\tau_n$ | Electron lifetime | s |
| $\tau_n$ | nMOSFET pulldown delay | s |
| $\tau_{nB}$ | Electron lifetime in base of n–p–n transistor | s |
| $\tau_p$ | Hole lifetime | s |
| $\tau_p$ | pMOSFET pullup delay | s |
| $\tau_{pE}$ | Hole lifetime in emitter of n–p–n transistor | s |
| $\tau_R$ | Reverse transit time of bipolar transistor | s |
| $\tau_w$ | Wire *RC* delay | s |
| $v$ | Velocity | cm/s |
| $v_{th}$ | Thermal velocity | cm/s |
| $v_d$ | Carrier drift velocity | cm/s |
| $v_{sat}$ | Saturation velocity of carriers | cm/s |
| $V$ | Voltage | V |
| $V$ | Quasi-Fermi level along MOSFET channel | V |
| $V_A$ | Early voltage | V |
| $V_{app}$ | Applied voltage across p–n diode | V |
| $V'_{app}$ | Applied voltage appearing immediately across p–n junction (smaller than $V_{app}$ by *IR* drops in series resistances) | V |
| $V_{BE}$ | Base–emitter bias voltage | V |
| $V_{BC}$ | Base–collector bias voltage | V |
| $V_{CE}$ | Collector-to-emitter voltage | V |
| $V_{dd}$ | Power-supply voltage | V |
| $V_{ds}$ | Source-to-drain voltage | V |
| $V_{dsat}$ | MOSFET drain saturation voltage | V |
| $V_{fb}$ | Flat-band voltage | V |
| $V_{ox}$ | Potential drop across oxide | V |
| $V_g$ | Gate voltage | V |

| Symbol | Description | Unit |
|---|---|---|
| $V_{bs}$ | Substrate reverse-bias voltage | V |
| $V_t$ | Threshold voltage ($2\psi_B$ definition) | V |
| $V_{on}$ | Linearly extrapolated threshold voltage | V |
| $V_{in}$ | Input node voltage of a logic gate | V |
| $V_{out}$ | Output node voltage of a logic gate | V |
| $V_x$ | Node voltage between stacked nMOSFETs of a NAND gate | V |
| $W$ | Width, MOSFET width | cm |
| $W_n$ | nMOSFET width | cm |
| $W_p$ | pMOSFET width | cm |
| $W_B$ | Intrinsic-base width | cm |
| $W_d$ | Depletion-layer width | cm |
| $W_{dBE}$ | Base–emitter junction depletion-layer width | cm |
| $W_{dBC}$ | Base–collector junction depletion-layer width | cm |
| $W_{dm}$ | Maximum depletion-layer width | cm |
| $W_{dm}^0$ | Maximum depletion-layer width in long-channel device | cm |
| $W_E$ | Emitter-layer width (thickness) | cm |
| $W_S$ | Source junction depletion-layer width | cm |
| $W_D$ | Drain junction depletion-layer width | cm |
| $\omega$ | Angular frequency | rad/s |
| $x_j$ | Junction depth | cm |
| $x_c$, $x_i$ | Depth of inversion channel | cm |
| $\psi$ | Potential | V |
| $\psi_B$ | Difference between Fermi level and intrinsic level | V |
| $\psi_{bi}$ | Built-in potential | V |
| $\psi_f$ | Fermi potential | V |
| $\psi_i$ | Intrinsic potential | V |
| $\psi_s$ | Surface potential | V |

# PREFACE

It has been fifty years since the invention of the bipolar transistor, more than forty years since the invention of the integrated-circuit (IC) technology, and more than thirty-five years since the invention of the MOSFET. During this time, there has been a tremendous and steady progress in the development of the IC technology with a rapid expansion of the IC industry. One distinct characteristic in the evolution of the IC technology is that the physical feature sizes of the transistors are reduced continually over time as the lithography technologies used to define these features become available. For almost thirty years now, the minimum lithography feature size used in IC manufacturing has been reduced at a rate of 0.7× every three years. In 1997, the leading-edge IC products have a minimum feature size of 0.25 μm.

The basic operating principles of large and small transistors are the same. However, the relative importance of the various device parameters and performance factors for transistors of the 1-μm and smaller generations is quite different from those for transistors of larger-dimension generations. For example, in the case of CMOS, the power-supply voltage was lowered from the standard 5 V, starting with the 0.6- to 0.8-μm generation. Since then CMOS power supply voltage has been lowered in steps once every few years as the device physical dimensions are reduced. At the same time, many physical phenomena, such as short-channel effect and velocity saturation, which are negligible in large-dimension MOSFETs, are becoming more and more important in determining the behavior of MOSFETs of deep-submicron dimensions. In the case of bipolar devices, breakdown voltage and base-widening effects are limiting their performance, and power dissipation is limiting their level of integration on a chip. Also, the advent of SiGe-base bipolar technology has extended the frequency capability of small-dimension bipolar transistors into the range previously reserved for GaAs and other compound-semiconductor devices.

The purpose of this book is to bring together the device fundamentals that govern the behavior of CMOS and bipolar transistors into a single text, with emphasis on those parameters and performance factors that are particularly important for VLSI (very-large-scale-integration) devices of deep-submicron dimensions. The book starts with a comprehensive review of the properties of the silicon material, and the basic physics of p–n junctions and MOS capacitors, as they relate to the fundamental principles of MOSFET and bipolar transistors. From there, the basic operation of MOSFET and bipolar devices, and their design and optimization for VLSI applications are developed. A great deal of the volume is devoted to in-depth discussions of the intricate interdependence and subtle tradeoffs of the various device parameters pertaining to circuit performance and manufacturability. The effects which are particularly important in small-dimension devices, e.g., quantization of the two-dimensional surface inversion layer in a MOSFET device and the heavy-doping effect in the intrinsic base of a bipolar transistor, are covered in detail. Also included in this book are extensive discussions on scaling and limitations to scaling of MOSFET and bipolar devices.

This book is suitable for use as a textbook by senior undergraduate or graduate students in electrical engineering and microelectronics. The necessary background assumed is an introductory understanding of solid-state physics and semiconductor physics. For practicing engineers and scientists actively involved in research and development in the IC industry, this book serves as a reference in providing a body of knowledge in modern VLSI devices for them to stay up to date in this field.

VLSI devices are too huge a subject area to cover thoroughly in one book. We have chosen to cover only the fundamentals necessary for discussing the design and optimization of the state-of-the-art CMOS and bipolar devices in the sub-0.5-μm regime. Even then, the specific topics covered in this book are based on our own experience of what the most important device parameters and performance factors are in modern VLSI devices.

Many people have contributed directly and indirectly to the topics covered in this book. We have benefited enormously from the years of collaboration and interaction we had with our colleagues at IBM, particularly in the areas of advanced silicon-device research and development. These include Douglas Buchanan, Hu Chao, T. C. Chen, Wei Chen, Kent Chuang, Peter Cook, Emmanuel Crabbé, John Cressler, Bijan Davari, Robert Dennard, Max Fischetti, David Frank, Charles Hsu, Genda Hu, Randall Isaac, Khalid Ismail, G. P. Li, Shih-Hsien Lo, Yuh-Jier Mii, Edward Nowak, George Sai-Halasz, Stanley Schuster, Paul Solomon, Hans Stork, Jack Sun, Denny Tang, Lewis Terman, Clement Wann, James Warnock, Siegfried Wiedmann, Philip Wong, Matthew Wordeman, Ben Wu, and Hwa Yu.

We would like to acknowledge the secretarial support of Barbara Grady and the support of our management at IBM Thomas J. Watson Research Center where this book was written. Finally, we would like to give special thanks to our families – Teresa, Adrienne, and Brenda Ning and Betty, Ying, and Hsuan Taur – for their support and understanding during this seemingly endless task.

Yuan Taur
Tak H. Ning
Yorktown Heights, New York, October, 1997

# 1 INTRODUCTION

Since the invention of the bipolar transistor in 1947, there has been an unprecedented growth of the semiconductor industry, with an enormous impact on the way people work and live. In the last twenty years or so, by far, the strongest growth area of the semiconductor industry has been in silicon very-large-scale-integration (VLSI) technology. The sustained growth in VLSI technology is fueled by the continued shrinking of transistors to ever smaller dimensions. The benefits of miniaturization – higher packing densities, higher circuit speeds, and lower power dissipation – have been key in the evolutionary progress leading to today's computers and communication systems that offer superior performance, dramatically reduced cost per function, and much reduced physical size, in comparison with their predecessors. On the economic side, the integrated-circuit (IC) business has grown worldwide in sales from $1 billion in 1970 to $20 billion in 1984 and is projected to reach $185 billion in 1997. The electronics industry is now among the largest industries in terms of output as well as employment in many nations. The importance of microelectronics in the economic, social, and even political development throughout the world will no doubt continue to ascend. The large worldwide investment in VLSI technology constitutes a formidable driving force that will all but guarantee the continued progress in IC integration density and speed, for as long as physical principles will allow.

## 1.1 EVOLUTION OF VLSI DEVICE TECHNOLOGY

An excellent account of the evolution of the metal–oxide–semiconductor field-effect transistor (MOSFET), from its initial concept to VLSI applications in the mid-1980s, can be found in the paper by Sah (Sah, 1988). Figure 1.1 gives a chronology of the major milestone events in the development of VLSI technology. The bipolar transistor technology was developed early on and was applied to the first integrated-circuit memory in mainframe computers in the 1960s. Bipolar transistors have been used all along where raw circuit speed is most important, for bipolar circuits remain the fastest at the individual-circuit level. However, the large power dissipation of

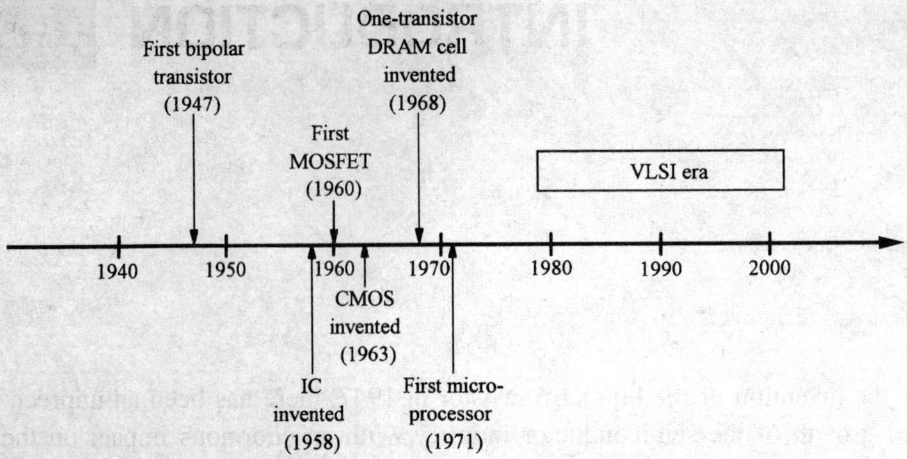


**FIGURE 1.1.** A brief chronology of the major milestones in the development of VLSI.

bipolar circuits has severely limited their integration level, to about $10^4$ circuits per chip. This integration level is quite low by today's VLSI standard.

The idea of modulating the surface conductance of a semiconductor by the application of an electric field was first reported in 1930. However, early attempts to fabricate a surface-field-controlled device were not successful because of the presence of large densities of surface states which effectively shielded the surface potential from the influence of an external field. The first MOSFET on a silicon substrate using $SiO_2$ as the gate insulator was fabricated in 1960 (Kahng and Atalla, 1960). During the 1960s and 1970s, n-channel and p-channel MOSFETs were widely used, along with bipolar transistors, for implementing circuit functions on a silicon chip. Although the MOSFET devices were slow compared to the bipolar devices, they had a higher layout density and were relatively simple to fabricate; the simplest MOSFET chip could be made using only four masks and a single doping step. However, just like bipolar circuits, single-polarity MOSFET circuits suffered from large standby power dissipation, and hence were limited in the level of integration on a chip.

The major breakthrough in the level of integration came in 1963 with the invention of CMOS (complementary MOS) (Wanlass and Sah, 1963), in which both n-channel and p-channel MOSFETs are constructed simultaneously on the same substrate. A CMOS circuit typically consists of an n-channel MOSFET and a p-channel MOSFET connected in series between the power-supply terminals, so that there is negligible standby power dissipation. Significant power is dissipated only during switching of the circuit (i.e., only when the circuits are active.) By cleverly designing the "switch activities" of the circuits on a chip to minimize active power dissipation, engineers have been able to integrate hundreds of millions of CMOS transistors on a single chip and still have the chip readily air-coolable. Until recently, the integration level of CMOS was not limited by chip-level power dissipation, but by chip fabrication technology. Another advantage of CMOS circuits comes from the ratioless, full rail-to-rail logic swing, which improves the noise margin and makes a CMOS chip easier to design.

As linear dimensions reached the 0.5-μm level in the early 1990s, the performance advantage of bipolar transistors was outweighed by the significantly greater circuit density of CMOS devices. The system performance benefit of integrated functionality superseded that of raw transistor performance, and practically all the VLSI chips in production today are based on CMOS technology. Bipolar transistors are used only where raw circuit speed makes an important difference. Consequently, bipolar transistors are usually used in small-size bipolar-only chips, or in so-called BiCMOS chips where most of the functions are implemented using CMOS transistors and only a relatively small number are implemented using bipolar transistors.

Advances in lithography and etching technologies have enabled the industry to scale down transistors in physical dimensions, and to pack more transistors in the same chip area. Such progress, combined with a steady growth in chip size, resulted in an exponential growth in the number of transistors and memory bits per chip. The recent trends and future projections in these areas are illustrated in Fig. 1.2. Dynamic random-access memories (DRAMs) have characteristically contained the highest component count of any IC chips. This has been so because of the small size of the one-transistor memory cell (Dennard, 1968) and because of the large and often insatiable demand for more memory in computing systems. It is interesting to note that the entire content of this book can be stored in one 64-Mb DRAM chip, which is in volume production in 1997 and has an area equivalent to a square of about $1.2 \times 1.2\ \text{cm}^2$.

One remarkable feature of silicon devices that fuels the rapid growth of the information technology industry is that their speed increases and their cost decreases

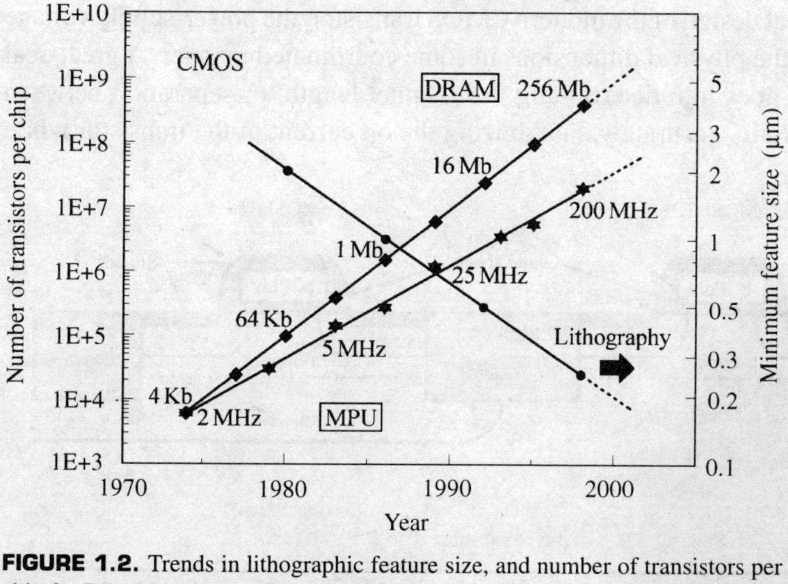


**FIGURE 1.2.** Trends in lithographic feature size, and number of transistors per chip for DRAM and microprocessor chips.

as their size is reduced. The transistors manufactured today are 20 times faster and occupy less than 1% of the area of those built 20 years ago. This is illustrated in the trend of microprocessor units (MPUs) in Fig. 1.2. The increase in the clock frequency of microprocessors is the result of a combination of improvements in microprocessor architecture and improvements in transistor speed.

## 1.2 MODERN VLSI DEVICES

It is clear from Fig. 1.2 that modern transistors of practical interest have feature sizes of 0.5 μm and smaller. Although the basic operation principles of large and small transistors are the same, the relative importance of the various device parameters and performance factors for the small-dimension modern transistors is quite different from that for the transistors of the early 1980s or earlier. It is our intention to focus our discussion in this book on the fundamentals of silicon devices of sub-0.5-μm generations.

### 1.2.1 MODERN CMOS TRANSISTORS

A schematic cross section of modern CMOS transistors, consisting of an n-channel MOSFET and a p-channel MOSFET integrated on the same chip, is shown in Fig. 1.3. A generic process flow for fabricating the CMOS transistors is outlined in Appendix 1. The key physical features of the modern CMOS technology, as illustrated in Fig. 1.3, include: p-type polysilicon gate for the p-channel MOSFET and n-type polysilicon gate for the n-channel MOSFET, refractory metal silicide on the polysilicon gate as well as on the source and drain diffusion regions, and shallow-trench oxide isolation.

In the electrical design of the modern CMOS transistor, the power-supply voltage is reduced with the physical dimensions in some coordinated manner. A great deal of design detail goes into determining the channel length, or separation between the source and drain, accurately, maximizing the on current of the transistor while

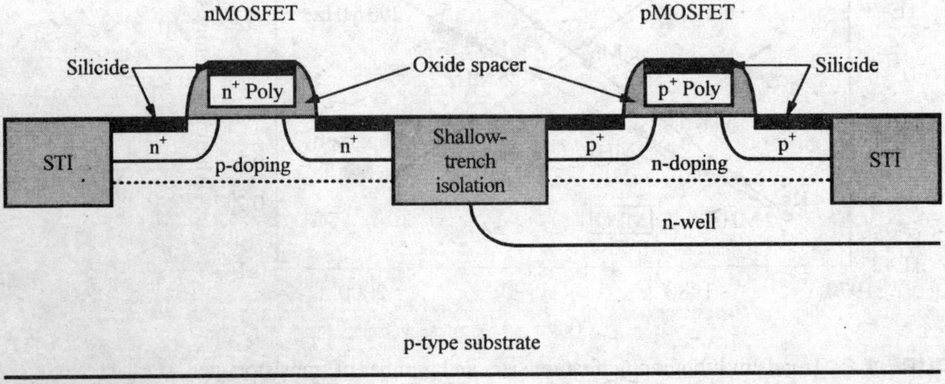


**FIGURE 1.3.** Schematic device cross section for an advanced CMOS technology.

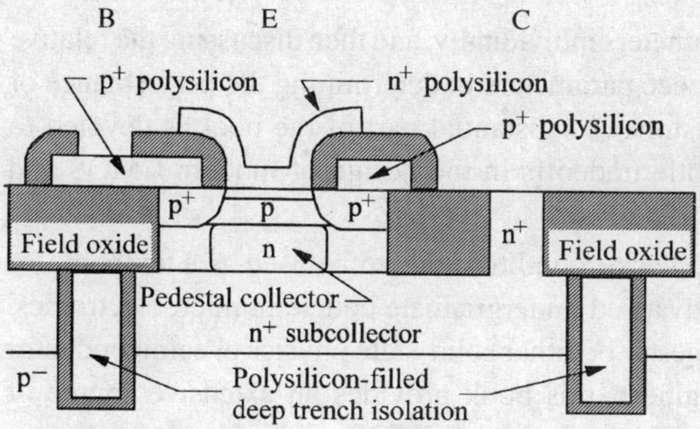


**FIGURE 1.4.** Schematic cross section of a modern n–p–n bipolar transistor.

maintaining an adequately low off current, minimizing variation of the transistor characteristics with process tolerances, and minimizing the parasitic resistances and parasitic capacitances.

### 1.2.2 MODERN BIPOLAR TRANSISTORS

The schematic cross section of a modern silicon bipolar transistor is shown in Fig. 1.4. The process outline for fabricating this transistor is shown in Appendix 2. The salient features of this transistor include: shallow-trench field oxide and deep-trench isolation, polysilicon emitter, polysilicon base contact which is self-aligned to the emitter contact, and a pedestal collector which is doped to the desired level only directly underneath the emitter.

Unlike CMOS, the power-supply voltage for a bipolar transistor is usually kept constant as the transistor physical dimensions are reduced. Without the ability to reduce the operating voltage, electrical breakdown is a severe concern in the design of modern bipolar transistors. In designing a modern bipolar transistor, a lot of effort is spent tailoring the doping profile of the various device regions in order to maintain adequate breakdown-voltage margins while maximizing the device performance. At the same time, unlike the bipolar transistors before the early 1980s when the device performance was mostly limited by the device physical dimensions practical at the time, a modern bipolar transistor often has its performance limited by its current-density capability and not by its physical dimensions. Attempts to improve the current-density capability of a transistor usually lead to reduced breakdown voltages.

## 1.3 SCOPE AND BRIEF DESCRIPTION OF THE BOOK

In writing this book, it is our goal to address the factors governing the performance of modern VLSI devices in depth. This is carried out by first discussing the design of the various device parameters individually, and then discussing the relative importance of the individual device parameters in determining the performance of small-dimension modern transistors. A substantial part of the book is devoted to in-depth discussions on the subtle tradeoffs in the design of modern CMOS and bipolar transistors.

This book contains sufficient background tutorials to be used as a textbook for students taking a graduate or advanced undergraduate course in microelectronics. The prerequisite will be one semester of either solid-state physics or semiconductor physics. For the practicing engineer, this book provides an extensive source of reference material that covers the fundamentals of CMOS and bipolar technologies, devices, and circuits. It should be useful to VLSI process engineers interested in learning basic device principles, and to device design or characterization engineers who desire more in-depth knowledge in their specialized areas. Below is a brief description of each chapter.

## CHAPTER 2: BASIC DEVICE PHYSICS

Chapter 2 is devoted to introducing just the right level of basic device physics to make the book self-contained, and to prepare the reader with the necessary background on device operation and material physics to follow the discussion in the rest of the book. It is assumed that the reader has the basic knowledge equivalent to one semester of upper-level undergraduate or graduate-level solid-state physics or semiconductor physics. From there, the device concepts or device physics needed to understand the subsequent chapters are introduced.

Starting with the energy bands in silicon, Chapter 2 first introduces the basic concepts of Fermi level, carrier concentration, drift and diffusion current transport, and Poisson's equation. The next two sections focus on the most elementary building blocks of silicon devices: the p–n junction and the MOS capacitor. Basic knowledge of their characteristics is a prerequisite to further understand the operation of the VLSI devices they lead into: bipolar and MOSFET transistors. The rest of Chapter 2 covers high-field effects, Si–$SiO_2$ systems, hot carriers, and the physics of tunneling and breakdown relevant to VLSI device reliability. The purpose here is to introduce to the reader the basic physical concepts needed to follow the discussion in this book. The details on the recent advances in each of these areas are beyond the scope of this book.

## CHAPTER 3: MOSFET DEVICES

Chapter 3 describes the basic characteristics of MOSFET devices, using the n-channel MOSFET as an example for most of the discussions. It is divided into two parts. The first part deals with the more elementary long-channel MOSFETs, including subsections on drain current models, $I$–$V$ characteristics, subthreshold currents, channel mobility, and intrinsic capacitances. These serve as a foundation for understanding the more important but more complex short-channel MOSFETs, which have lower capacitances and carry higher currents per gate voltage swing. The second part of Chapter 3 covers the specific features of short-channel MOSFETs important for device design purposes. The subsections include short-channel effects, velocity saturation and overshoot, channel-length modulation, and source–drain series resistance.

### CHAPTER 4: CMOS DEVICE DESIGN

Chapter 4 considers the major device design issues in a CMOS technology. It begins with the concept of MOSFET scaling – the most important guiding principle for achieving density, speed, and power improvements in VLSI evolution. Several non-scaling factors are addressed, notably, the thermal voltage and the silicon bandgap, which have significant implications on the deviation of the CMOS evolution path from ideal scaling. Two key CMOS device design parameters – threshold voltage and channel length – are then discussed in detail. Subsections on threshold voltage include off-current requirement, nonuniform channel doping, gate work-function effects, channel profile design, and quantum-mechanical and discrete dopant effects on threshold voltage. Subsections on channel length include the definition of effective channel length, its extraction by the conventional method and the shift-and-ratio method, and the physical interpretation of effective channel length.

### CHAPTER 5: CMOS PERFORMANCE FACTORS

Chapter 5 examines the key factors that govern the switching performance and power dissipation of basic digital CMOS circuits which form the building blocks of a VLSI chip. Starting with a brief description of static CMOS logic gates and their layout, we examine the parasitic resistances and capacitances that may adversely affect the delay of a CMOS circuit. These include source and drain series resistance, junction capacitance, overlap capacitance, gate resistance, and interconnect capacitance and resistance. Next, we formulate a delay equation and use it to study the sensitivity of CMOS delay performance to a variety of device and circuit parameters such as wire loading, device width and length, gate oxide thickness, power-supply voltage, threshold voltage, parasitic components, and substrate sensitivity in stacked circuits. The last section of Chapter 5 further extends the delay equation to project the performance factors of several advanced CMOS materials and device structures. These include SOI CMOS, high-mobility Si–SiGe CMOS, and low-temperature CMOS. The unique advantage of each approach is discussed in depth.

### CHAPTER 6: BIPOLAR DEVICES

The basic components of a bipolar transistor are described in Chapter 6. The discussion is based entirely on the vertical n–p–n transistor, since practically all high-speed bipolar transistors used in digital circuits are of the vertical n–p–n type. However, the basic device operation concept and device physics can be readily extended to other types of bipolar transistors, such as p–n–p bipolar transistors and lateral bipolar transistors.

The basic operation of a bipolar transistor is described in terms of two p–n diodes connected back to back. The basic theory of a p–n diode is modified and applied to derive the current equations for a bipolar transistor. From these current equations, other important device parameters and phenomena, such as current gain, Early voltage, base–collector junction avalanche, emitter–collector punch-through, base widening, and diffusion capacitance, are examined. Finally, the basic equivalent-circuit models relating the device parameters to circuit parameters are developed. These equivalent-circuit models form the starting point for discussing the performance of a bipolar transistor in circuit applications.

## CHAPTER 7: BIPOLAR DEVICE DESIGN

Chapter 7 covers the basic design of a bipolar transistor. The design of the individual device regions, namely the emitter, the base, and the collector, are discussed separately. Since the detailed characteristics of a bipolar transistor depend on its operating point, the focus of this chapter is on optimizing the device design according to its intended operating condition and environment, and on the trade-offs that must be made in the optimization process. The sections include an examination of the effect of grading the base doping profile to enhance the drift field in the intrinsic base, and a derivation of the collector-current equations when there is significant heavy doping effect in the base and when Ge is used to engineer the intrinsic-base bandgap. The chapter concludes with a discussion of the salient features of the most commonly used modern bipolar device structure.

## CHAPTER 8: BIPOLAR PERFORMANCE FACTORS

The major factors governing the performance of bipolar transistors in circuit applications are discussed in Chapter 8. Several of the commonly used figures of merit, namely, cutoff frequency, maximum oscillation frequency, and logic gate delay, are examined, and how a bipolar transistor can be optimized for a given figure of merit is discussed. Sections are devoted to examining the important delay components of a logic gate, and how these components can be minimized. The power–delay trade-offs in the design of a bipolar transistor under various circuit loading conditions are also examined. Finally, the scaling properties of bipolar transistors, and how the large standby power dissipation of bipolar circuits limits the integration level of bipolar circuit chips, are discussed.

# BASIC DEVICE PHYSICS 2

This chapter reviews the basic concepts of semiconductor device physics. Starting with electrons and holes and their transport in silicon, we focus on the most elementary types of devices in VLSI technology: p–n junction and metal-oxide-semiconductor (MOS) capacitor. The rest of the chapter deals with subjects of importance to VLSI device reliability: high-field effects, the Si–$SiO_2$ system, and dielectric breakdown.

## 2.1 ELECTRONS AND HOLES IN SILICON

The first section covers energy bands in silicon, Fermi level, n-type and p-type silicon, electrostatic potential, drift and diffusion current transport, and basic equations governing VLSI device operation. These will serve as the basis for understanding the more advanced device concepts discussed in the rest of the book.

### 2.1.1 ENERGY BANDS IN SILICON

The starting material used in the fabrication of VLSI devices is silicon in the crystalline form. The silicon wafers are cut parallel to either the ⟨111⟩ or ⟨100⟩ planes (Sze, 1981), with ⟨100⟩ material being the most commonly used. This is largely due to the fact that ⟨100⟩ wafers, during processing, produce the lowest charges at the oxide–silicon interface as well as higher mobility (Balk *et al.*, 1965). In a silicon crystal each atom has four valence electrons to share with its four nearest neighboring atoms. The valence electrons are shared in a paired configuration called a *covalent bond*. ***The most important result of the application of quantum mechanics to the description of electrons in a solid is that the allowed energy levels of electrons will be grouped into bands*** (Kittel, 1976). ***The bands are separated by regions of energy that the electrons in the solid cannot possess: forbidden gaps***. The highest energy band that is completely filled by electrons at 0 K is called the *valence band*. The next higher energy band, separated by a forbidden gap from the valence band, is called the *conduction band*, as shown in Fig. 2.1.

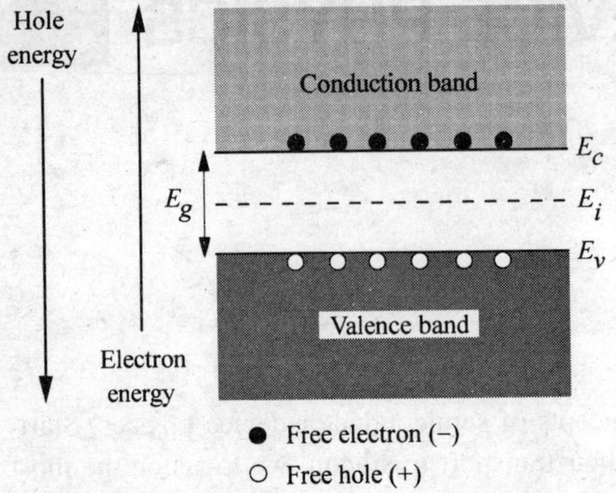


**FIGURE 2.1.** Energy-band diagram of silicon.

#### 2.1.1.1 BANDGAP OF SILICON

What sets a semiconductor such as silicon apart from a metal or an insulator is that at absolute zero temperature, the valence band is completely filled with electrons, while the conduction band is completely empty, and that the separation between the conduction band and valence band, or the *bandgap*, is on the order of 1 eV. On one hand, no electrical conduction is possible at 0 K, since there are no current carriers in the conduction band, whereas the electrons in the completely filled valence band cannot be accelerated by an electric field and gain energy. On the other hand, the bandgap is small enough that at room temperature a small fraction of the electrons are excited into the conduction band, leaving behind vacancies, or *holes*, in the valence band. This allows limited conduction to take place from the motion of both the electrons in the conduction band and the holes in the valence band. In contrast, an insulator has a much larger forbidden gap of at least several electron volts, making room-temperature conduction virtually impossible. Metals, on the contrary, have partially filled conduction band even at absolute zero temperature, so that the electrons can gain an infinitesimal amount of energy from the applied electric field. This makes them good conductors at any temperature.

As shown in Fig. 2.1, the energy of the electrons in the conduction band increases upward, while the energy of the holes in the valence band increases downward. The bottom of the conduction band is designated $E_c$, and the top of the valence band $E_v$. Their separation, or the bandgap, is $E_g = E_c - E_v$. For silicon, $E_g$ is 1.12 eV at room temperature or 300 K. The bandgap decreases slightly as the temperature increases, with a temperature coefficient of $dE_g/dT \approx -2.73 \times 10^{-4}$ eV/K for silicon near 300 K (Arora, 1993). Other important physical parameters of silicon and silicon dioxide are listed in Table 2.1.

#### 2.1.1.2 STATISTICAL DISTRIBUTION FUNCTION

The energy distribution of electrons in a solid is governed by the laws of Fermi–Dirac statistics. The principal result of these statistics is the *Fermi–Dirac distribution*

**TABLE 2.1** Physical Properties of Si and $SiO_2$ at Room Temperature (300 K)

| Property | Si | $SiO_2$ |
|---|---|---|
| Atomic/molecular weight | 28.09 | 60.08 |
| Atoms or molecules/cm$^3$ | $5.0 \times 10^{22}$ | $2.3 \times 10^{22}$ |
| Density (g/cm$^3$) | 2.33 | 2.27 |
| Crystal structure | Diamond | Amorphous |
| Lattice constant (Å) | 5.43 | — |
| Energy gap (eV) | 1.12 | 8–9 |
| Dielectric constant | 11.7 | 3.9 |
| Intrinsic carrier concentration (cm$^{-3}$) | $1.4 \times 10^{10}$ | — |
| Carrier mobility (cm$^2$/V-s) | Electron: 1430 | — |
| | Hole: 470 | — |
| Effective density of states (cm$^{-3}$) | Conduction band, $N_c$: $3.2 \times 10^{19}$ | — |
| | Valence band, $N_v$: $1.8 \times 10^{19}$ | — |
| Breakdown field (V/cm) | $3 \times 10^5$ | $>10^7$ |
| Melting point (°C) | 1415 | 1600–1700 |
| Thermal conductivity (W/cm-°C) | 1.5 | 0.014 |
| Specific heat (J/g-°C) | 0.7 | 1.0 |
| Thermal diffusivity (cm$^2$/s) | 0.9 | 0.006 |
| Thermal expansion coefficient (°C$^{-1}$) | $2.5 \times 10^{-6}$ | $0.5 \times 10^{-6}$ |

*function*, which gives the probability that an electronic state at energy $E$ is occupied by an electron,

$$f(E) = \frac{1}{1 + e^{(E-E_f)/kT}}. \tag{2.1}$$

Here $k = 1.38 \times 10^{-23}$ J/K is Boltzmann's constant, and $T$ is the absolute temperature. This function contains a parameter, $E_f$, called the *Fermi level*. The Fermi level is the energy at which the probability of occupation of an energy state by an electron is exactly one-half. At absolute zero temperature, $T = 0$ K, all the states below the Fermi level are filled ($f = 1$ for $E < E_f$), and all the states above the Fermi level are empty ($f = 0$ for $E > E_f$). At finite temperatures, some states above the Fermi level are filled as some states below become empty. In other words, the probability distribution $f(E)$ makes a smooth transition from unity to zero as the energy increases across the Fermi level. The width of the transition is governed by the thermal energy, $kT$. It is important to keep in mind that the thermal energy at room temperature is 0.026 eV, or roughly $\frac{1}{40}$ of the silicon bandgap. In most cases when the energy is at least several $kT$ above or below the Fermi level, Eq. (2.1) can be approximated by the simple formulas

$$f(E) \approx e^{-(E-E_f)/kT} \quad \text{for} \quad E > E_f \tag{2.2}$$

and

$$f(E) \approx 1 - e^{-(E_f-E)/kT} \quad \text{for} \quad E < E_f. \tag{2.3}$$

Equation (2.3) should be interpreted as stating that the probability of finding a hole (i.e., an empty state *not* occupied by an electron) at an energy $E < E_f$ is $e^{-(E_f-E)/kT}$. The last two equations follow directly from the Maxwell–Boltzmann statistics for classical particles, which is a good approximation to the Fermi–Dirac statistics when the energy is at least several $kT$ away from $E_f$.

#### 2.1.1.3 INTRINSIC CARRIER CONCENTRATION

Based on Eqs. (2.2) and (2.3), one can write down the concentration of electrons in the conduction band as

$$n = N_c e^{-(E_c-E_f)/kT}, \tag{2.4}$$

and the concentration of holes in the valence band as

$$p = N_v e^{-(E_f-E_v)/kT}, \tag{2.5}$$

where $N_c$ and $N_v$ are the *effective densities of states* in the conduction and valence bands, respectively. Their expressions are derived in Appendix 3. Both $N_c$ and $N_v$ are proportional to $T^{3/2}$. Their values at room temperature are listed in Table 2.1.

For an intrinsic silicon, $n = p$, since for every electron excited into the conduction band, a vacancy or hole is left behind in the valence band. The Fermi level for intrinsic silicon, or the *intrinsic Fermi level*, $E_i$, is then obtained by equating Eq. (2.4) and Eq. (2.5) and solving for $E_f$:

$$E_i = E_f = \frac{E_c + E_v}{2} - \frac{kT}{2} \ln\left(\frac{N_c}{N_v}\right). \tag{2.6}$$

By substituting Eq. (2.6) for $E_f$ in Eq. (2.4) or Eq. (2.5), one obtains the intrinsic carrier concentration, $n_i = n = p$:

$$n_i = \sqrt{N_c N_v} e^{-(E_c-E_v)/2kT} = \sqrt{N_c N_v} e^{-E_g/2kT}. \tag{2.7}$$

Since the thermal energy, $kT$, is much smaller than the silicon bandgap $E_g$, ***the intrinsic Fermi level is very close to the midpoint between the conduction band and the valence band***. In fact, $E_i$ is sometimes referred to as the midgap energy level, since the error in assuming $E_i$ to be $(E_c + E_v)/2$ is only about $0.3kT$. The intrinsic carrier concentration $n_i$ at room temperature is $1.4 \times 10^{10}$ cm$^{-3}$, as given in Table 2.1, which is very small compared with the atomic density of silicon.

### 2.1.2 n-TYPE AND p-TYPE SILICON

Intrinsic silicon at room temperature has an extremely low free-carrier concentration; therefore, its resistivity is very high. In practice, intrinsic silicon hardly exists at room temperature, since it would require materials with an unobtainably

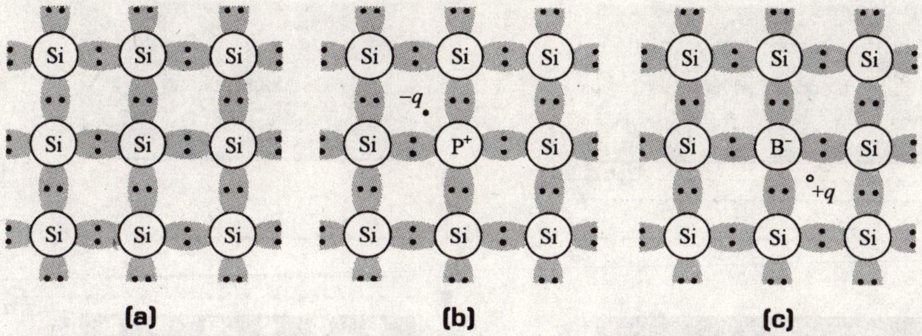


**FIGURE 2.2.** Three basic bond pictures of silicon: (a) intrinsic Si with no impurities, (b) n-type silicon with donor (phosphorus), (c) p-type silicon with acceptor (boron). (After Sze, 1981.)

high purity. Most impurities in silicon introduce additional energy levels in the forbidden gap and can be easily ionized to add either electrons to the conduction band or holes to the valence band, depending on where the impurity level is (Kittel, 1976). ***The electrical conductivity of silicon is then dominated by the type and concentration of the impurity atoms, or dopants***, and the silicon is called *extrinsic*.

#### 2.1.2.1 DONORS AND ACCEPTORS

Silicon is a column-IV element with four valence electrons per atom. There are two types of impurities in silicon that are electrically active: those from column V such as arsenic or phosphorus, and those from column III such as boron. As is shown in Fig. 2.2, a column-V atom in a silicon lattice tends to have one extra electron loosely bonded after forming covalent bonds with other silicon atoms. In most cases, the thermal energy at room temperature is sufficient to ionize the impurity atom and free the extra electron to the conduction band. Such types of impurities are called *donors*; they become positively charged when ionized. Silicon material doped with column-V impurities or donors is called *n-type* silicon, and its electrical conductivity is dominated by electrons in the conduction band. On the other hand, a column-III impurity atom in a silicon lattice tends to be deficient by one electron when forming covalent bonds with other silicon atoms (Fig. 2.2). Such an impurity atom can also be ionized by accepting an electron from the valence band, which leaves a free-moving hole that contributes to electrical conduction. These impurities are called *acceptors*; they become negatively charged when ionized. Silicon material doped with column-III impurities or acceptors is called *p-type* silicon, and its electrical conductivity is dominated by holes in the valence band. It should be noted that impurity atoms must be in a *substitutional* site (as opposed to *interstitial*) in silicon in order to be electrically active.

In terms of the energy-band diagrams in Fig. 2.3, donors add allowed electron states in the bandgap close to the conduction-band edge, while acceptors

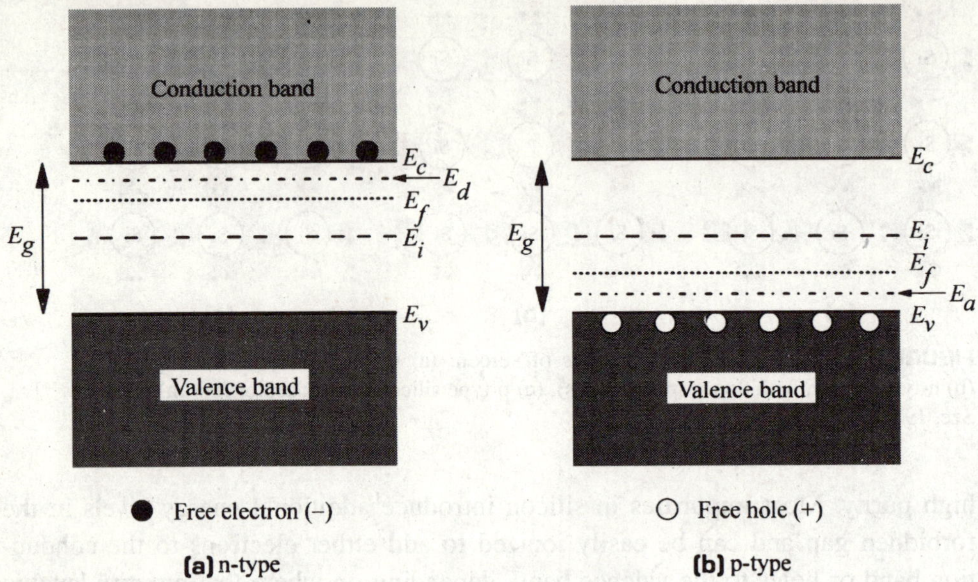


**FIGURE 2.3.** Energy-band diagram representation of (a) donor level $E_d$ and Fermi level $E_f$ in n-type silicon, (b) acceptor level $E_a$ and Fermi level $E_f$ in p-type silicon.

add allowed states just above the valence-band edge. Donor levels contain positive charge when ionized (emptied). Acceptor levels contain negative charge when ionized (filled). The ionization energies are denoted by $E_c - E_d$ for donors and $E_a - E_v$ for acceptors, respectively. Figure 2.4 shows the donor and acceptor levels of common impurities in silicon and their ionization energies (Sze, 1981). Phosphorus and arsenic are commonly used donors, or n-type dopants, with low ionization energies on the order of $2kT$, while boron is a commonly used acceptor or p-type dopant with a comparable ionization energy. Figure 2.5 shows the *solid solubility* of important impurities in silicon as a function of annealing temperature (Trumbore, 1960). Arsenic, boron, and phosphorus have the highest solid solubility among all the impurities, which makes them the most important doping species in VLSI technology.

#### 2.1.2.2 FERMI LEVEL IN EXTRINSIC SILICON

In contrast to intrinsic silicon, the Fermi level in an extrinsic silicon is not located at the midgap. The Fermi level in n-type silicon moves up towards the conduction band, consistent with the increase in electron density as described by Eq. (2.4). On the other hand, the Fermi level in p-type silicon moves down towards the valence band, consistent with the increase in hole density as described by Eq. (2.5). These cases are depicted in Fig. 2.3. The exact position of the Fermi level depends on both the ionization energy and the concentration of dopants. For example, for an n-type material with a donor impurity concentration $N_d$, the charge neutrality condition

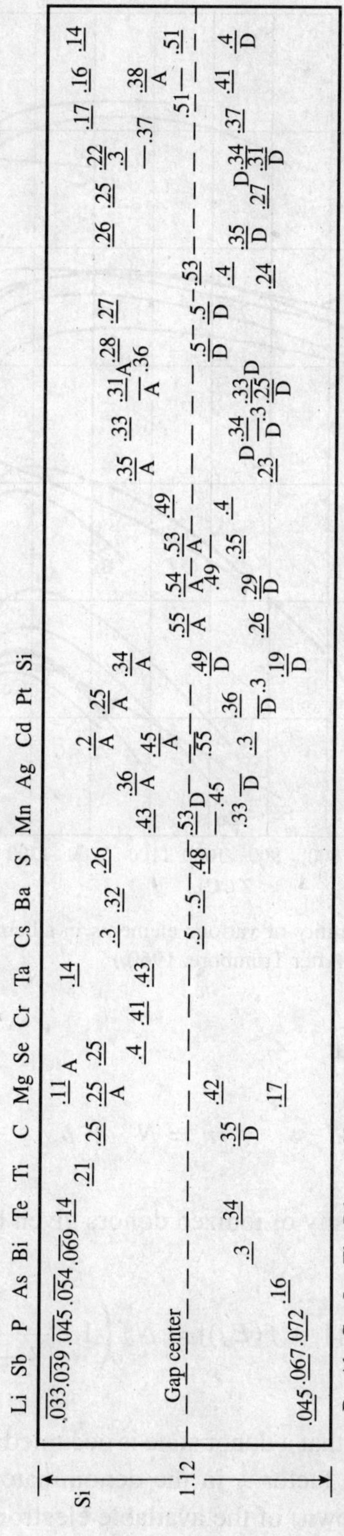

**FIGURE 2.4.** Donor and acceptor levels of various impurities in silicon. Numbers next to the level indicate ionization energies $E_c - E_d$ (donors) or $E_a - E_v$ (acceptors) in electron volts. (After Sze, 1981.)

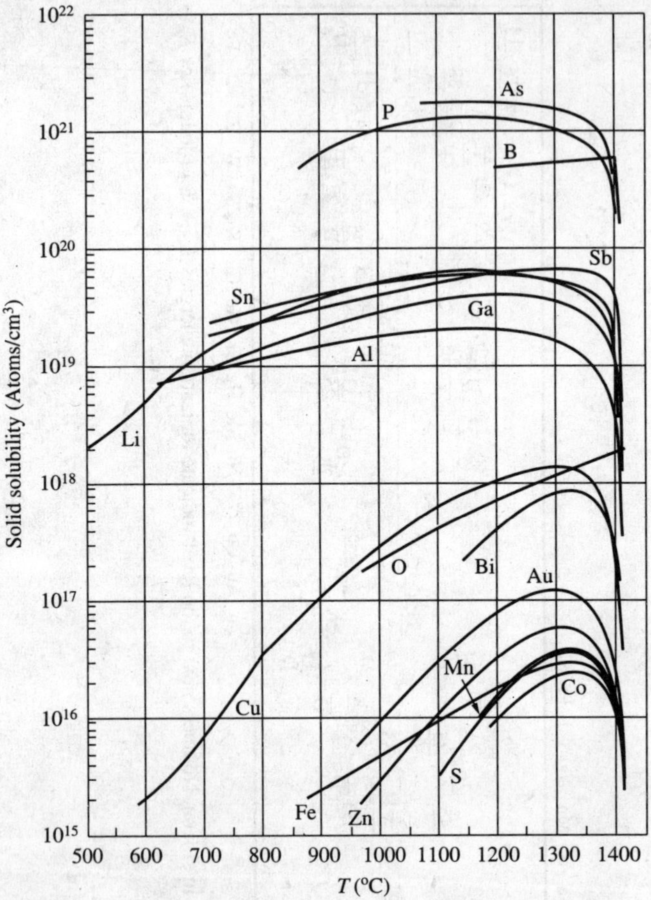


**FIGURE 2.5.** Solid solubility of various elements in silicon as a function of temperature. (After Trumbore, 1960.)

in silicon requires that

$$n = N_d^+ + p, \tag{2.8}$$

where $N_d^+$ is the density of ionized donors given by

$$N_d^+ = N_d[1 - f(E_d)] = N_d\left(1 - \frac{1}{1 + \frac{1}{2}e^{(E_d - E_f)/kT}}\right), \tag{2.9}$$

since the probability that a donor state is occupied by an electron (i.e., in the neutral state) is $f(E_d)$. The factor $\frac{1}{2}$ in the denominator of $f(E_d)$ arises from the spin degeneracy (up or down) of the available electronic states associated with an ionized donor level (Ghandhi, 1968). Substituting Eq. (2.4) and Eq. (2.5) for $n$ and $p$

in Eq. (2.8), one obtains

$$N_c e^{-(E_c-E_f)/kT} = \frac{N_d}{1+2e^{-(E_d-E_f)/kT}} + N_v e^{-(E_f-E_v)/kT}, \tag{2.10}$$

which is an algebraic equation that can be solved for $E_f$. In n-type silicon, electrons are the majority current carriers, while holes are the minority current carriers, which means that the second term on the right-hand side (RHS) of Eq. (2.10) can be neglected. For shallow donor impurities with low to moderate concentration at room temperature, $(N_d/N_c)\exp[(E_c-E_d)/kT] \ll 1$, a good approximate solution for $E_f$ is

$$E_c - E_f = kT \ln\left(\frac{N_c}{N_d}\right). \tag{2.11}$$

In this case, the Fermi level is at least a few $kT$ below $E_d$ and essentially all the donor levels are empty (ionized), i.e., $n = N_d^+ = N_d$.

A useful, general relationship between majority and minority carriers can be found by multiplying Eq. (2.4) and Eq. (2.5) together:

$$np = n_i^2 = N_c N_v e^{-(E_c-E_v)/kT} = N_c N_v e^{-E_g/kT}. \tag{2.12}$$

In other words, the product $np$ in equilibrium is a constant, independent of the dopant type and Fermi level position. The hole density in n-type silicon is then given by

$$p = n_i^2/N_d. \tag{2.13}$$

Likewise, for p-type silicon with a shallow acceptor concentration $N_a$, the Fermi level is given by

$$E_f - E_v = kT \ln\left(\frac{N_v}{N_a}\right), \tag{2.14}$$

the hole density is $p = N_a^- = N_a$, and the electron density is

$$n = n_i^2/N_a. \tag{2.15}$$

Figure 2.6 plots the Fermi-level position in the energy gap versus temperature for a wide range of impurity concentration (Grove, 1967). The slight variation of the silicon bandgap with temperature is also incorporated in the figure. It is seen that as the temperature increases, the Fermi level approaches the intrinsic value near midgap. When the intrinsic carrier concentration becomes larger than the doping concentration, the silicon is intrinsic. In an intermediate range of temperature including room temperature, all the donors or acceptors are ionized. The majority

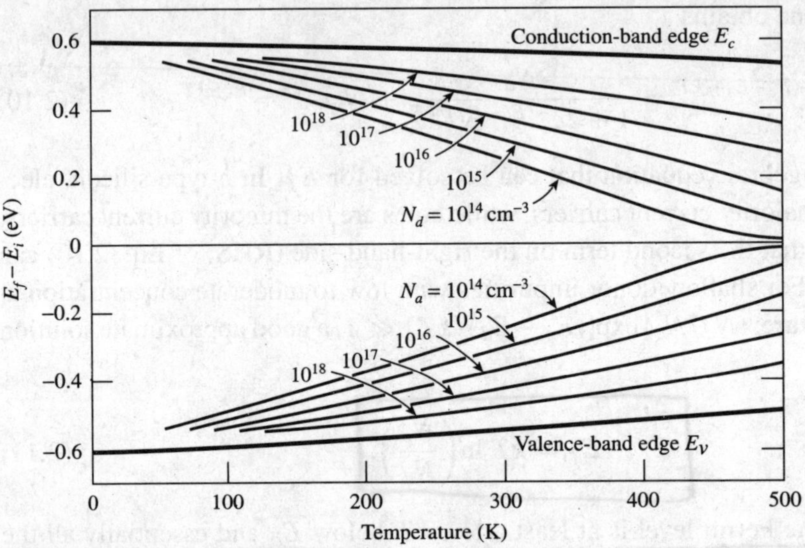


**FIGURE 2.6.** The Fermi level in silicon as a function of temperature for various impurity concentrations. (After Grove, 1967.)

carrier concentration is then given by the doping concentration, independent of temperature. For temperatures below this range, *freeze-out* occurs, i.e., the thermal energy is no longer sufficient to ionize all the impurity atoms even with their shallow levels (Sze, 1981). In this case, the majority-carrier concentration is less than the doping concentration, and one would have to solve Eq. (2.10) numerically to find $E_f$, $n$, and $p$ (Shockley, 1950).

Instead of using $N_c$, $N_v$ and referring to $E_c$ and $E_v$, Eq. (2.11) and Eq. (2.14) can be written in a more useful form in terms of $n_i$ and $E_i$ defined by Eq. (2.6) and Eq. (2.7):

$$E_f - E_i = kT \ln\left(\frac{N_d}{n_i}\right) \tag{2.16}$$

for n-type silicon, and

$$E_i - E_f = kT \ln\left(\frac{N_a}{n_i}\right). \tag{2.17}$$

for p-type silicon. In other words, ***the distance between the Fermi level and the intrinsic Fermi level near the midgap is a logarithmic function of doping concentration***. These expressions will be used extensively throughout the book.

#### 2.1.2.3 FERMI LEVEL IN DEGENERATELY DOPED SILICON

For heavily doped silicon, the impurity concentration $N_d$ or $N_a$ can exceed the effective density of states $N_c$ or $N_v$, so that $E_f > E_c$ or $E_f < E_v$ according to

Eq. (2.11) and Eq. (2.14). In other words, the Fermi level moves into the conduction band for $n^+$ silicon, and into the valence band for $p^+$ silicon. In addition, when the impurity concentration is higher than $10^{18}$–$10^{19}$ cm$^{-3}$, the donor (or acceptor) levels broaden into bands. This results in an effective decrease in the ionization energy until finally the impurity band merges with the conduction (or valence) band and the ionization energy becomes zero. Under these circumstances, the silicon is said to be *degenerate*. Strictly speaking, Fermi statistics should be used for the electron concentration in calculation of the Fermi level when $E_c - E_f \leq kT$ (Ghandhi, 1968). For practical purposes, it is a good approximation [within $(1\text{–}2)kT$] to assume that the Fermi level of the degenerate $n^+$ silicon is at the conduction-band edge, and that of the degenerate $p^+$ silicon is at the valence-band edge.

### 2.1.3 CARRIER TRANSPORT IN SILICON

Carrier transport or current flow in silicon is driven by two different mechanisms: (a) the *drift* of carriers, which is caused by the presence of an electric field, and (b) the *diffusion* of carriers, which is caused by an electron or hole concentration gradient in silicon. The drift current will be discussed first.

#### 2.1.3.1 DRIFT CURRENT AND MOBILITY

When an electric field is applied to a conducting medium containing free carriers, the carriers are accelerated and acquire a drift velocity superimposed upon their random thermal motion. This is described in more detail in Appendix 4. The drift velocity of holes is in the direction of the applied field, and the drift velocity of electrons is opposite to the field. The velocity of the carriers does not increase indefinitely under field acceleration, since they are scattered frequently and lose their acquired momentum after each collision. At low electric fields, the drift velocity $v_d$ is proportional to the electric field strength $\mathscr{E}$ with a proportionality constant $\mu$, defined as the *mobility*, in units of cm$^2$/V-s, i.e.,

$$v_d = \mu\mathscr{E}. \tag{2.18}$$

The mobility is proportional to the time interval between collisions and is inversely proportional to the effective mass of the carriers (Appendix 4). Electron and hole mobilities in silicon at low impurity concentrations are listed in Table 2.1. The electron mobility is approximately three times the hole mobility, since the effective mass of electrons in the conduction band is much lighter than that of holes in the valence band.

Figure 2.7 plots the electron and hole mobilities at room temperature versus n-type or p-type doping concentration. At low impurity levels, the mobilities are mainly limited by carrier collisions with the silicon lattice or acoustic phonons (Kittel, 1976). As the doping concentration increases beyond $10^{15}$–$10^{16}$/cm$^3$,

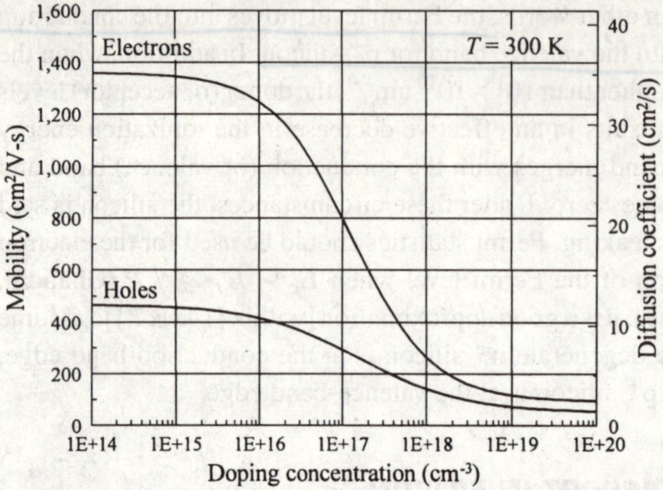


**FIGURE 2.7.** Electron and hole mobilities in bulk silicon at 300 K as a function of doping concentration.

collisions with the charged (ionized) impurity atoms through Coulomb interaction become more and more important and the mobilities decrease. In general, one can use *Matthiessen's rule* to include different contributions to the mobility:

$$\frac{1}{\mu} = \frac{1}{\mu_L} + \frac{1}{\mu_I} + \cdots, \tag{2.19}$$

where $\mu_L$ and $\mu_I$ correspond to the lattice- and impurity-scattering-limited components of mobility, respectively. At high temperatures, the mobility tends to be limited by lattice scattering and is proportional to $T^{-3/2}$, relatively insensitive to the doping concentration (Sze, 1981). At low temperatures, the mobility is higher, but is a strong function of doping concentration as it becomes more limited by impurity scattering. What is shown in Fig. 2.7 is the *bulk mobility* applicable to conduction in silicon substrates far from the surface. ***In the inversion layer of a MOSFET device, the current flow is governed by the surface mobility, which is much lower than the bulk mobility***. This is mainly due to additional scattering mechanisms between the carriers and the Si–$SiO_2$ interface in the presence of high electric fields normal to the surface. Surface scattering adds another term to Eq. (2.19). Carrier mobility in the surface inversion channel of a MOSFET will be discussed in more detail in Section 3.1.6.

#### 2.1.3.2 RESISTIVITY

For a homogeneous n-type silicon with a free-electron density $n$, the drift current density under an electric field $\mathscr{E}$ is

$$J_{n,drift} = qnv_d = qn\mu_n\mathscr{E}, \tag{2.20}$$

where $q = 1.6 \times 10^{-19}$ C is the electronic charge and $\mu_n$ is the electron mobility. The resistivity, $\rho_n$, of n-type silicon defined by $J_{n,drift} = \mathscr{E}/\rho_n$ (a form of Ohm's law) is then given by

$$\rho_n = \frac{1}{qn\mu_n}. \tag{2.21}$$

Similarly, for p-type silicon,

$$J_{p,drift} = qp\mu_p\mathscr{E} \tag{2.22}$$

and

$$\rho_p = \frac{1}{qp\mu_p}, \tag{2.23}$$

where $\mu_p$ is the hole mobility. In general, the total resistivity should include both the majority and the minority carrier components:

$$\rho = \frac{1}{qn\mu_n + qp\mu_p}, \tag{2.24}$$

since both electrons and holes contribute to electrical conduction. Figure 2.8 shows the measured resistivity of n-type (phosphorus-doped) and p-type (boron-doped) silicon versus impurity concentration at room temperature.

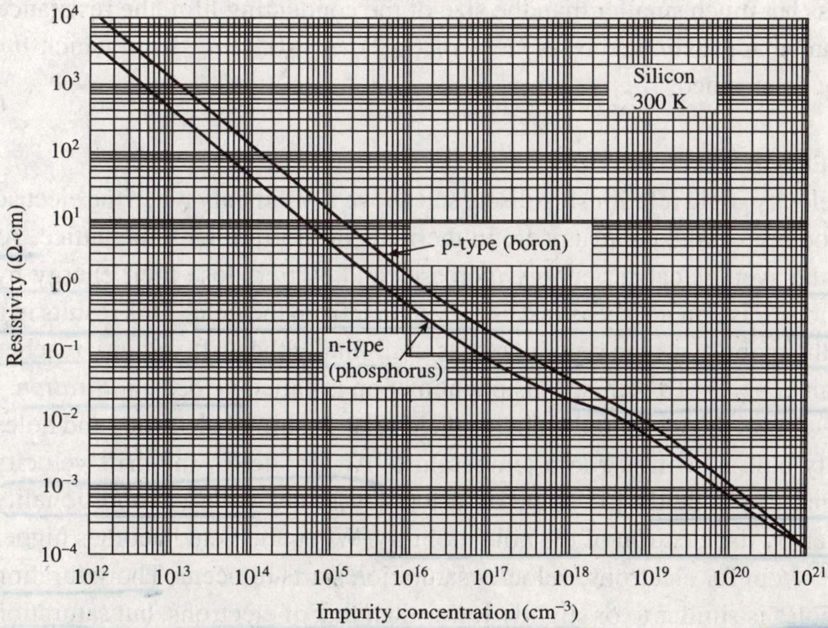


**FIGURE 2.8.** Resistivity versus impurity concentration for n-type and p-type silicon at 300 K. (After Sze, 1981.)

#### 2.1.3.3 SHEET RESISTIVITY

The resistance of a uniform conductor of length $L$, width $W$, and thickness $t$ is given by

$$R = \rho \frac{L}{Wt}, \tag{2.25}$$

where $\rho$ is the resistivity in ohm-centimeters. In a planar IC technology, the thickness of conducting regions is uniform and normally much less than both the length and the width of the regions. It is then useful to define a quantity, called the *sheet resistivity*, as

$$\rho_{sh} = \frac{\rho}{t} \tag{2.26}$$

in units of Ω/□ (ohms per square). Then

$$R = \frac{L}{W}\rho_{sh}, \tag{2.27}$$

i.e., the total resistance is equal to the number of squares ($L/W = 1$ is one square) of the line times the sheet resistivity. Note that sheet resistivity does not depend on the size of the square. The most common technique of measuring the sheet resistivity of a thin film is the four-point method, in which a small current is passed through the two outer probes and the voltage is measured between the two inner probes (Sze, 1981). If the spacing between the probes is much greater than the film thickness but much smaller than the size of the conducting film, the resistance measured can be approximated by $V/I = \rho_{sh}(\ln 2)/\pi \approx 0.22\rho_{sh}$, from which $\rho_{sh}$ can be easily determined.

#### 2.1.3.4 VELOCITY SATURATION

The linear velocity–field relationship discussed above is valid only when the electric field is not too high and the carriers are in thermal equilibrium with the lattice. At high fields, the average carrier energy increases and carriers lose their energy by optical-phonon emission nearly as fast as they gain it from the field. This results in a decrease of the mobility as the field increases until finally the drift velocity reaches a limiting value, $v_{sat} \approx 10^7$ cm/s. This phenomenon is called *velocity saturation.*

Figure 2.9 shows the measured velocity–field relationship of electrons and holes in high-purity bulk silicon at room temperature. At low fields, the drift velocity is proportional to the field (45° slope on a log–log scale) with a proportionality constant given by the electron or the hole mobility. When the field becomes higher than $3 \times 10^3$ V/cm for electrons, velocity saturation starts to occur. The saturation velocity of holes is similar to or slightly lower than that of electrons, but saturation for holes takes place at a much higher field because of their lower mobility. For more highly doped material, low-field mobilities are lower because of impurity

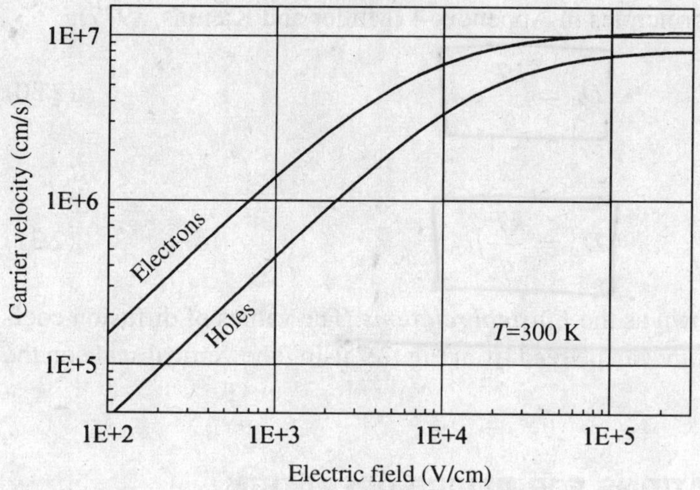


**FIGURE 2.9.** Velocity–field relationship of electrons and holes in silicon at 300 K.

scattering (Fig. 2.7). However, the saturation velocity remains essentially the same, independent of impurity concentration. There is a weak dependence of $v_{sat}$ on temperature. It decreases slightly as the temperature increases (Arora, 1993).

### 2.1.3.5 DIFFUSION CURRENT

The discussion in this section so far has dealt only with the case when the carrier concentration within the silicon is uniform and the carriers move under the influence of an electric field. ***If the carrier concentration is not uniform, carriers will also diffuse under the influence of the concentration gradient***. This leads to an additional current contribution in proportion to the concentration gradient:

$$J_{n,diff} = qD_n \frac{dn}{dx} \tag{2.28}$$

for electrons, and

$$J_{p,diff} = -qD_p \frac{dp}{dx} \tag{2.29}$$

for holes. The proportionality constants $D_n$ and $D_p$ are called the electron and hole *diffusion coefficients* and have units of cm$^2$/s. There is a negative sign in Eq. (2.29), since diffusion current flows in the direction of decreasing hole (positive charge) concentration.

Physically, both drift and diffusion are closely associated with the random thermal motion of carriers and their collisions with the silicon lattice in thermal equilibrium. A simple relationship between the diffusion coefficient and the mobility is derived from the basic principles in Appendix 4 (Muller and Kamins, 1977):

$$D_n = \frac{kT}{q}\mu_n \tag{2.30}$$

for electrons, and

$$D_p = \frac{kT}{q}\mu_p \tag{2.31}$$

for holes. These are known as the *Einstein relations*. The values of diffusion coefficient at room temperature can be read from Fig. 2.7 using the vertical scale on the right-hand side.

### 2.1.4 BASIC EQUATIONS FOR DEVICE OPERATION

#### 2.1.4.1 POISSON'S EQUATION

One of the key equations governing the operation of VLSI devices is *Poisson's equation.* It comes from Maxwell's first equation, which in turn is based on Coulomb's law for electrostatic force of a charge distribution. Poisson's equation is expressed in terms of the electrostatic potential, which is defined as the carrier energy divided by the electronic charge $q$. Carrier energies are either at the conduction-band edge or at the valence-band edge, as discussed before in connection with the energy-band diagram, Fig. 2.1. Since one is only interested in the spatial variation of the electrostatic potential, it can be defined with an arbitrary additive constant. It makes no difference whether $E_c$, $E_v$, or any other quantity displaced from the band edges by a fixed amount is used to represent the potential. Conventionally, the electrostatic potential is defined in terms of the intrinsic Fermi level,

$$\psi_i = -\frac{E_i}{q}. \tag{2.32}$$

There is a negative sign because $E_i$ is defined as electron energy while $\psi_i$ is defined for a positive charge (i.e., increases downward in a band diagram).

The electric field $\mathscr{E}$, which is defined as the electrostatic force per unit charge, is equal to the negative gradient of $\psi_i$,

$$\mathscr{E} = -\frac{d\psi_i}{dx}. \tag{2.33}$$

Now we can write Poisson's equation as

$$\frac{d^2\psi_i}{dx^2} = -\frac{d\mathscr{E}}{dx} = -\frac{\rho(x)}{\varepsilon_{si}}, \tag{2.34}$$

where $\rho(x)$ is the charge density per unit volume at $x$, and $\varepsilon_{si}$ is the permittivity of silicon equal to $11.7\varepsilon_0$. Here $\varepsilon_0 = 8.85 \times 10^{-14}$ F/cm is the vacuum permittivity.

Equation (2.34) is a one-dimensional equation, which is adequate for describing most of the basic device operations. In some cases (for example, in the short-channel effect of MOSFETs or in the base resistance calculation of bipolar devices), the two-dimensional or even the three-dimensional Poisson's equation must be used.

Another form of Poisson's equation is *Gauss's law*, which is obtained by integrating Eq. (2.34):

$$\mathscr{E} = \frac{1}{\varepsilon_{si}} \int \rho(x)\, dx = \frac{Q_s}{\varepsilon_{si}}, \tag{2.35}$$

where $Q_s$ is the integrated charge density per unit area.

There are two sources of charge in silicon: *mobile charge* and *fixed charge*. Mobile charges are electrons and holes, whose densities are represented by $n$ and $p$. Fixed charges are ionized donor (positively charged) and acceptor (negatively charged) atoms whose densities are represented by $N_d^+$ and $N_a^-$, respectively. Equation (2.34) can then be written as

$$\frac{d^2\psi_i}{dx^2} = -\frac{d\mathscr{E}}{dx} = -\frac{q}{\varepsilon_{si}}[p(x) - n(x) + N_d^+(x) - N_a^-(x)]. \tag{2.36}$$

For a homogeneous n-type or p-type silicon with no applied field, the RHS of Eq. (2.36) is zero and the potential is constant throughout the sample.

#### 2.1.4.2 CARRIER CONCENTRATION AS A FUNCTION OF ELECTROSTATIC POTENTIAL

Many of the parameters discussed before can be expressed in terms of the electrostatic potential $\psi_i$. For example, Eq. (2.16) and Eq. (2.17) can be combined into one equation for both n-type and p-type silicon:

$$\boxed{\psi_B \equiv |\psi_f - \psi_i| = \frac{kT}{q} \ln\left(\frac{N_b}{n_i}\right),} \tag{2.37}$$

where $\psi_f = -E_f/q$ is the Fermi potential and $N_b$ is either the donor or the acceptor concentration. Equation (2.37) is a very useful expression relating the separation of the Fermi potential from the midgap, $\psi_B$, to the doping concentration. Also, Eq. (2.4) and Eq. (2.5) can be expressed in a more useful form:

$$n = n_i e^{(E_f - E_i)/kT} = n_i e^{q(\psi_i - \psi_f)/kT} \tag{2.38}$$

and

$$p = n_i e^{(E_i - E_f)/kT} = n_i e^{q(\psi_f - \psi_i)/kT} \tag{2.39}$$

The last two equations are often referred to as *Boltzmann's relations* and are valid for either n-type or p-type silicon. Note that Eq. (2.37) is based on charge neutrality and is valid only if the local mobile carrier concentration equals the doping concentration. We shall see later that in the presence of band bending, there is a net charge due to an imbalance between the mobile and the fixed charge densities. In that case, $|\psi_f - \psi_i|$ is no longer given by Eq. (2.37), but Eqs. (2.38) and (2.39) remain valid.

### 2.1.4.3 DEBYE LENGTH

In an inhomogeneous silicon, the doping concentration varies spatially. Both the intrinsic Fermi level and the bands generally follow the variation according to Eq. (2.37). However, if the doping concentration changes abruptly on a very short length scale, the bands do not respond immediately, since both $\psi_i$ and its first-order spatial derivative are continuous owing to the thermal diffusion effects. The length scale in an n-type silicon, for example, can be estimated by substituting Eq. (2.38) into Eq. (2.36):

$$\frac{d^2\psi_i}{dx^2} = -\frac{q}{\varepsilon_{si}}\left[N_d(x) - n_i e^{q(\psi_i-\psi_f)/kT}\right]. \tag{2.40}$$

For an incremental change of doping concentration $\Delta N_d(x)$ with respect to a uniformly doped background, the corresponding change in the intrinsic potential $\Delta\psi_i(x)$ can be found by expanding the exponential term in Eq. (2.40) and keeping only the first-order term (zeroth-order terms have no spatial dependence):

$$\frac{d^2(\Delta\psi_i)}{dx^2} - \frac{q^2 N_d}{\varepsilon_{si}kT}\Delta\psi_i = -\frac{q}{\varepsilon_{si}}\Delta N_d(x). \tag{2.41}$$

This is a second-order differential equation whose solution $\Delta\psi_i$ takes the form $\exp(-x/L_D)$, where

$$L_D \equiv \sqrt{\frac{\varepsilon_{si}kT}{q^2 N_d}} \tag{2.42}$$

is called the *Debye length*. Physically, this means that ***it takes a distance on the order of $L_D$ for the silicon bands to respond to an abrupt change in $N_d$***. A small electric field is set up in this region due to the charge imbalance. The Debye length is usually much smaller than the lateral device dimension. For example, $L_D = 0.04$ μm for $N_d = 10^{16}$ cm$^{-3}$.

### 2.1.4.4 CURRENT-DENSITY EQUATIONS

The next set of equations are current-density equations. The total current density is the sum of the drift current density given by Eq. (2.20) and Eq. (2.22) and the diffusion current density given by Eq. (2.28) and Eq. (2.29). In other words,

$$J_n = qn\mu_n\mathscr{E} + qD_n\frac{dn}{dx} \tag{2.43}$$

for the electron current density, and

$$J_p = qp\mu_p\mathscr{E} - qD_p\frac{dp}{dx} \tag{2.44}$$

for the hole current density. The total conduction current density is $J = J_n + J_p$.

Using Eq. (2.33) and the Einstein relations, Eq. (2.30) and Eq. (2.31), one can write the current densities as

$$J_n = -qn\mu_n\left(\frac{d\psi_i}{dx} - \frac{kT}{qn}\frac{dn}{dx}\right) = -qn\mu_n\frac{d\phi_n}{dx} \tag{2.45}$$

and

$$J_p = -qp\mu_p\left(\frac{d\psi_i}{dx} + \frac{kT}{qp}\frac{dp}{dx}\right) = -qp\mu_p\frac{d\phi_p}{dx}, \tag{2.46}$$

where the *quasi-Fermi potentials* $\phi_n$ and $\phi_p$ are defined by

$$\phi_n \equiv \psi_i - \frac{kT}{q}\ln\left(\frac{n}{n_i}\right) \tag{2.47}$$

and

$$\phi_p \equiv \psi_i + \frac{kT}{q}\ln\left(\frac{p}{n_i}\right). \tag{2.48}$$

The electron and the hole quasi-Fermi potentials can be displaced from each other under nonequilibrium conditions.

### 2.1.4.5 CONTINUITY EQUATIONS

The next set of equations are *continuity equations* based on the conservation of mobile charge:

$$\frac{\partial n}{\partial t} = \frac{1}{q}\frac{\partial J_n}{\partial x} - R_n + G_n \tag{2.49}$$

and

$$\frac{\partial p}{\partial t} = -\frac{1}{q}\frac{\partial J_p}{\partial x} - R_p + G_p, \tag{2.50}$$

where $G_n$ and $G_p$ are the electron and hole generation rates, $R_n$ and $R_p$ are the electron and hole recombination rates, and $\partial J_n/\partial x$ and $\partial J_p/\partial x$ are the net flux of mobile charges in and out of $x$. At thermal equilibrium, the generation rate is equal to the recombination rate and $np = n_i^2$. When excess minority carriers are injected by light or other means, the recombination rate exceeds the generation rate, which establishes a tendency to return to equilibrium.

In silicon, the probability of direct band-to-band recombination by a *radiative* (transfer of energy to a photon) or *Auger* (transfer of energy to another carrier) process is very low due to its indirect bandgap. Most of the recombination processes take place indirectly via a trap or a deep impurity level near the middle of the forbidden gap. This is often referred to as the *Shockley–Read* recombination (Shockley and Read, 1952). Under low-injection conditions, the recombination rate is inversely proportional to the *minority-carrier lifetime*, $\tau$, which is in the range of $10^{-4}$ to $10^{-9}$ s, depending on the quality of the silicon crystal. The *minority-carrier diffusion length*, which is the average distance a minority carrier travels before it recombines with a majority carrier, is given by $L = (D\tau)^{1/2}$, where $D$ is the diffusion coefficient. The diffusion length is typically a few microns to a few millimeters in silicon. (A discussion of the minority-carrier diffusion process can be found in Section 2.2.4.) ***Since L is much larger than the active dimensions of a VLSI device, generation–recombination in general plays very little role in device operation***. Only in a few special circumstances, such as CMOS latch-up, the SOI floating-body effect, junction leakage current, and radiation-induced soft error, must the generation–recombination mechanism be taken into account.

In the steady state, $\partial n/\partial t = \partial p/\partial t = 0$. If the generation and recombination rates are also negligible, the continuity equations are reduced to $dJ_n/dx = dJ_p/dx = 0$, which simply means conservation of electron and hole currents.

#### 2.1.4.6 DIELECTRIC RELAXATION TIME

In contrast to the minority-carrier lifetime discussed above, the majority-carrier response time is very short in a semiconductor. It can be estimated for a one-dimensional (1-D) homogeneous n-type silicon as follows. Neglecting $R_n$ and $G_n$ in the continuity equation (2.49), one obtains

$$\frac{\partial n}{\partial t} = \frac{1}{q}\frac{\partial J_n}{\partial x}. \tag{2.51}$$

Substituting $J_n = \mathscr{E}/\rho_n$ (Ohm's law) and $\partial\mathscr{E}/\partial x = -qn/\varepsilon_{si}$ (Poisson's equation with only the majority-carrier density) into Eq. (2.51) yields

$$\frac{\partial n}{\partial t} = -\frac{n}{\rho_n \varepsilon_{si}}. \tag{2.52}$$

The solution to this equation takes the form of $n(t) \propto \exp(-t/\rho_n\varepsilon_{si})$, where $\rho_n\varepsilon_{si}$ is the majority-carrier response time, or *dielectric relaxation time*. ***The majority-carrier response time in silicon is typically on the order of $10^{-12}$ s, which is shorter than most device switching times.***

Note that $\rho_n\varepsilon_{si}$ is the minimum response time for an ideal 1-D case without any parasitic capacitances. In practice, the majority carrier response time is limited by the *RC* delay of the specific silicon structure and contacts.

## 2.2 p–n JUNCTIONS

p–n junctions, also called *p–n diodes*, are important devices as well as important components of all MOSFET and bipolar devices. The characteristics of p–n diodes are therefore important in determining the characteristics of VLSI devices and circuits. A p–n diode is formed when one region of a semiconductor substrate is doped n-type and an immediately adjacent region is doped p-type. In practice, a silicon p–n diode is usually formed by counterdoping a local region of a larger region of doped silicon. For instance, a region of a p-type silicon substrate or "well" can be counterdoped with n-type impurities to form the n-type region of a p–n diode. The n-type region thus formed has a donor concentration higher than its acceptor concentration.

A doped semiconductor region is called *compensated* if it contains both donor and acceptor impurities such that neither impurity concentration is negligible compared to the other. For a compensated semiconductor region, it is the *net* doping concentration, i.e., $N_d - N_a$ if it is n-type and $N_a - N_d$ if it is p-type, that determines its Fermi level and its mobile carrier concentration. However, for simplicity, we shall derive the characteristics and behavior of p–n diodes assuming none of the doped regions are compensated, i.e., the n-sides of the diodes have a net donor concentration of $N_d$ and the p-sides have a net acceptor concentration of $N_a$. The resultant equations can be extended to diodes with compensated doped regions simply by replacing $N_d$ by $N_d - N_a$ for the n-regions and replacing $N_a$ by $N_a - N_d$ for the p-regions.

### 2.2.1 BUILT-IN POTENTIAL AND APPLIED POTENTIAL

Figure 2.10(a) shows the energy-band diagrams of a p-type silicon and an n-type silicon physically separated from each other. As discussed in Section 2.1.2, the Fermi level for an n-type silicon lies close to its conduction band, and that for a p-type silicon lies close to its valence band. Also, as we shall show below, the

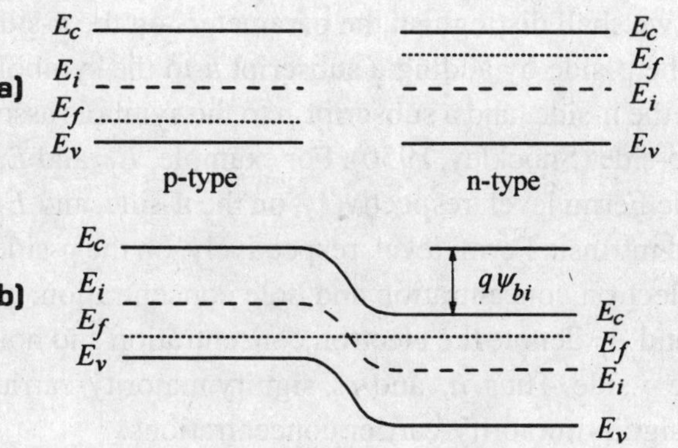


**FIGURE 2.10.** Energy-band diagrams of a p–n junction at thermal equilibrium: (a) when the p-region and the n-region are physically separated; (b) when the p-region and the n-region are joined to form a p–n junction.

Fermi level of a semiconductor is flat, i.e., spatially constant, when there is no current flow in it. Therefore, as the p-type region and the n-type region are brought together to form a p–n junction, the Fermi level must remain flat across the entire structure if there is no current flow in and across the junction. This causes the energy bands of the p-type region to lie higher than the corresponding energy bands of the n-type region, as illustrated in Fig. 2.10(b). The potential difference between the corresponding energy bands on the p- and n-sides is called the *built-in potential* $\psi_{bi}$ of the p–n junction. This is indicated in Fig. 2.10(b).

At thermal equilibrium, there is no net electron or hole current flow. Therefore, Eqs. (2.43) and (2.44) give

$$J_n = q\mu_n\left(n\mathscr{E} + \frac{kT}{q}\frac{dn}{dx}\right) = 0, \tag{2.53}$$

$$J_p = q\mu_p\left(p\mathscr{E} - \frac{kT}{q}\frac{dp}{dx}\right) = 0. \tag{2.54}$$

Substituting Eqs. (2.38) for $n$, (2.39) for $p$, and (2.33) for $\mathscr{E}$, we have

$$\frac{dE_f}{dx} = 0, \tag{2.55}$$

which simply states that the Fermi level is spatially constant. Thus ***the condition for zero electron or hole current is that the Fermi level is spatially constant.***

The potential variation, or band bending, across the junction implies an electric field in the junction region given by Eq. (2.33), namely $\mathscr{E} = -d\psi_i/dx$. This electric field causes a drift component of electron and hole currents to flow. At thermal equilibrium, this drift-current component is exactly balanced by a diffusion component of electron and hole currents flowing in the opposite direction caused by the large electron and hole concentration gradients across the junction. The net result is zero electron and hole currents across the p–n junction at thermal equilibrium. The n- and p-type regions sufficiently far away from the junction, where the bands are essentially flat, i.e., the electric field is negligible, are referred to as the *bulk quasineutral regions.*

To facilitate description of both the n-side and the p-side of a diode simultaneously, when necessary for clarity, we shall distinguish the parameters on the n-side from the corresponding ones on the p-side by adding a subscript $n$ to the symbols associated with the parameters on the n-side, and a subscript $p$ to the symbols associated with the parameters on the p-side (Shockley, 1950). For example, $E_{fn}$ and $E_{in}$ denote the Fermi level and intrinsic Fermi level, respectively, on the n-side, and $E_{fp}$ and $E_{ip}$ denote the Fermi level and intrinsic Fermi level, respectively, on the p-side. Similarly, $n_n$ and $p_n$ denote the electron concentration and hole concentration, respectively, on the n-side, and $n_p$ and $p_p$ denote the electron concentration and hole concentration, respectively, on the p-side. Thus, $n_n$ and $p_p$ signify majority-carrier concentrations, while $n_p$ and $p_n$ signify minority-carrier concentrations.

Consider the n-side of a p–n diode at thermal equilibrium. If the n-side is non-degenerately doped to a concentration of $N_d$, then the separation between its Fermi level, which is flat across the diode, and its intrinsic Fermi level is given by Eq. (2.16), namely

$$E_{fn} - E_{in} = kT \ln\left(\frac{n_{n0}}{n_i}\right) = kT \ln\left(\frac{N_d}{n_i}\right), \tag{2.56}$$

where $n_{n0}$ denotes the n-side electron concentration at thermal equilibrium.

Similarly, for the nondegenerately doped p-side of a p–n diode at thermal equilibrium, with a doping concentration of $N_a$, we have

$$E_{ip} - E_{fp} = kT \ln\left(\frac{p_{p0}}{n_i}\right) = kT \ln\left(\frac{N_a}{n_i}\right), \tag{2.57}$$

where $p_{p0}$ is the p-side hole concentration at thermal equilibrium.

The built-in potential across the p–n diode is

$$q\psi_{bi} = E_{ip} - E_{in} = kT \ln\left(\frac{n_{n0}p_{p0}}{n_i^2}\right). \tag{2.58}$$

Since Eq. (2.12) gives $n_{n0}p_{n0} = n_{p0}p_{p0} = n_i^2$, Eq. (2.58) can also be written as

$$q\psi_{bi} = kT \ln\left(\frac{p_{p0}}{p_{n0}}\right) = kT \ln\left(\frac{n_{n0}}{n_{p0}}\right), \tag{2.59}$$

which relates the built-in potential to the electron and hole densities on the two sides of the p–n diode.

## 2.2.2 ABRUPT JUNCTIONS

Analysis of a p–n diode is much simpler if the junction is assumed to be abrupt, i.e., the doping impurities are assumed to change abruptly from p-type on one side to n-type on the other side of the junction. The abrupt-junction approximation is reasonable for modern VLSI devices, where the use of ion implantation for doping the junctions, followed by low-thermal-cycle diffusion and/or annealing, results in junctions that are fairly abrupt. Besides, the abrupt-junction approximation often leads to closed-form solutions which render the device physics much easier to understand.

### 2.2.2.1 DEPLETION APPROXIMATION

Analysis of an abrupt junction becomes even simpler in the depletion approximation in which the p–n diode is approximated by three regions as illustrated in Fig. 2.11(a). Both the bulk p-region, i.e., the region with $x < -x_p$, and the bulk

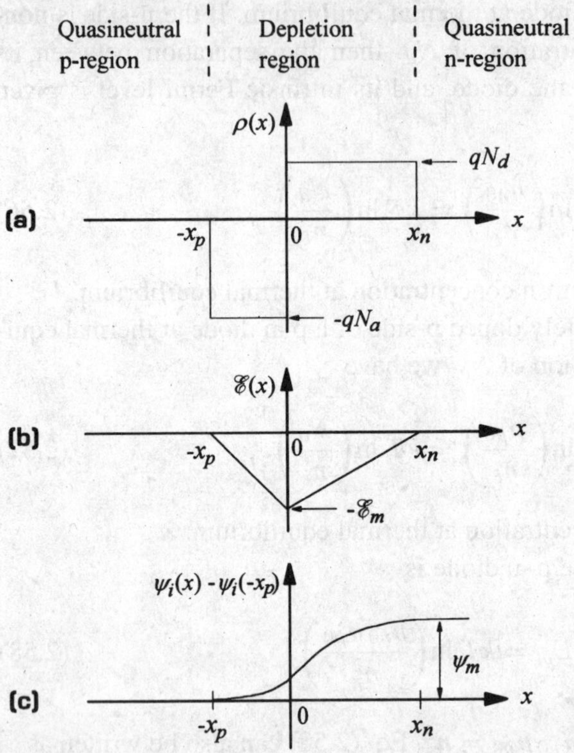


**FIGURE 2.11.** Depletion approximation of a p–n junction: (a) charge distribution, (b) electric field, and (c) electrostatic potential.

n-region, i.e., the region with $x > x_n$, are assumed to be charge-neutral, while the transition region, i.e., the region with $-x_p < x < x_n$, is assumed to be depleted of mobile electrons and holes. As we shall show later, the depletion-layer widths, $x_p$ and $x_n$, are dependent on the donor concentration $N_d$ on the n-side and the acceptor concentration $N_a$ on the p-side, as well as on the applied voltage $V_{app}$ across the junction. The depletion approximation is quite accurate for all applied voltages except at large forward biases, where the mobile-charge densities are not negligible compared to the ionized impurity concentrations in the transition region. (The behavior of p–n diodes at large forward biases will be discussed in Chapter 6 in connection with bipolar devices.) The transition region is often referred to as the *depletion region or depletion layer*. Since the transition region is not charge-neutral, it is also referred to as the *space-charge region* or *space-charge layer*.

The Poisson's equation, i.e., Eq. (2.36), for the depletion region is

$$-\frac{d^2\psi_i}{dx^2} = \frac{d\mathscr{E}}{dx} = \frac{q}{\varepsilon_{si}}\left[p(x) - n(x) + N_d^+(x) - N_a^-(x)\right] \tag{2.60}$$

$$= \frac{q}{\varepsilon_{si}}\left[N_d^+(x) - N_a^-(x)\right],$$

where $N_d^+$ is the ionized-donor concentration and $N_a^+$ is the ionized-acceptor concentration, and where the mobile-electron and -hole concentrations have been set to zero, consistent with the depletion approximation. For simplicity, we shall assume that all the donors and acceptors within the depletion region are ionized, and that the junction is abrupt and not compensated, i.e., there are no donor impurities on the p-side and no acceptor impurities on the n-side. With these assumptions, Eq. (2.60) becomes

$$-\frac{d^2\psi_i}{dx^2} = \frac{qN_d}{\varepsilon_{si}} \quad \text{for} \quad 0 \le x \le x_n \tag{2.61}$$

and

$$-\frac{d^2\psi_i}{dx^2} = -\frac{qN_a}{\varepsilon_{si}} \quad \text{for} \quad -x_p \le x \le 0. \tag{2.62}$$

Integrating Eq. (2.61) once from $x = 0$ to $x = x_n$, and Eq. (2.62) once from $x = -x_p$ to $x = 0$, subject to the boundary conditions of $d\psi_i/dx = 0$ at $x = -x_p$ and at $x = x_n$, we obtain the maximum electric field, $\mathscr{E}_m$, which is located at $x = 0$. That is,

$$\mathscr{E}_m \equiv \left|\frac{-d\psi_i}{dx}\right|_{x=0} = \frac{qN_d x_n}{\varepsilon_{si}} = \frac{qN_a x_p}{\varepsilon_{si}}. \tag{2.63}$$

It is clear from Eq. (2.63) that the total space charge inside the n-side of the depletion region is equal (but opposite in sign) to the total space charge inside the p-side of the depletion region. Thus, in Fig. 2.11(a), the two charge distribution plots have the same area. Equation (2.63) could have been obtained directly from Gauss's law, i.e., Eq. (2.35).

Let $\psi_m$ be the total potential drop across the p–n junction, i.e., $\psi_m = [\psi_i(x_n) - \psi_i(-x_p)]$. The total potential drop can be obtained by integrating Eqs. (2.61) and (2.62) twice, the second time from $x = -x_p$ to $x = x_n$. That is,

$$\psi_m = \int_{-x_p}^{x_n} d\psi_i(x) = -\int_{-x_p}^{x_n} \mathscr{E}(x)\,dx \tag{2.64}$$
$$= \frac{\mathscr{E}_m(x_n + x_p)}{2} = \frac{\mathscr{E}_m W_d}{2},$$

where $W_d = x_n + x_p$ is the total width of the depletion layer. It can be seen from Eq. (2.64) that $\psi_m$ is equal to the area in the $\mathscr{E}(x)$–$x$ plot, i.e., Fig. 2.11(b). Eliminating $\mathscr{E}_m$ from Eqs. (2.63) and (2.64) gives

$$W_d = \sqrt{\frac{2\varepsilon_{si}(N_a + N_d)\psi_m}{qN_aN_d}}. \tag{2.65}$$

This equation relates the total width of the depletion layer to the total potential drop across the junction and to the doping concentrations of the two sides of the diode.

As we shall show in Section 2.2.3, an externally applied voltage, $V_{app}$, across a p–n diode has the effect of shifting the Fermi level of the bulk neutral n-region relative to that of the bulk neutral p-region. That is, the total potential drop is the sum of the built-in potential and the externally applied potential, namely

$$\psi_m = \psi_{bi} \pm V_{app}, \tag{2.66}$$

where the + sign is for the case where the junction is reverse-biased and $\psi_m > \psi_{bi}$, and the − sign is for the case where the junction is forward-biased and $\psi_m < \psi_{bi}$. As we shall show in Section 2.2.4, large currents can flow in a diode when it is forward biased, while there is relatively little current flow when it is reverse biased. If Eq. (2.66) is used in Eq. (2.65), it gives the total depletion-layer width of a forward- or reverse-biased diode.

The charge per unit area on either side of the depletion region is

$$|Q_d| = qN_d x_n = qN_a x_p = \varepsilon_{si}\mathscr{E}_m, \tag{2.67}$$

and the depletion-layer capacitance per unit area is

$$C_d \equiv \frac{d|Q_d|}{d\psi_m} = \frac{\varepsilon_{si}}{W_d}. \tag{2.68}$$

That is, the depletion-layer capacitance of a diode is equivalent to a parallel-plate capacitor of separation $W_d$ and dielectric constant $\varepsilon_{si}$. Physically, this is due to the fact that only the mobile charge at the edges of the depletion layer, not the space charge within the depletion region, responds to changes of the applied voltage.

#### 2.2.2.2 ONE-SIDED JUNCTIONS

In many applications, such as the source or drain junction of a MOSFET or the emitter–base diode of a bipolar transistor, one side of the p–n diode is degenerately doped while the other side is lightly to moderately doped. In this case, practically all the voltage drop and the depletion layer occur across the lightly doped side of the diode. That this is the case can be inferred readily from Eq. (2.63), which implies that $x_n = N_a W_d/(N_a + N_d)$ and $x_p = N_d W_d/(N_a + N_d)$. The characteristics of a one-sided p–n diode are therefore determined primarily by the properties of the lightly doped side alone. In this sub-subsection, we shall derive the equations for an $n^+$–p diode where the characteristics are determined by the p-side. The results can be extended straightforwardly to a $p^+$–n diode.

As discussed in Section 2.1.2, for a lightly to moderately doped p-type silicon, the Fermi level is given by Eq. (2.16), and for a heavily or degenerately doped n-type silicon, it is a good approximation to assume its Fermi level to be at the conduction-band edge. Therefore, the built-in potential for an $n^+$–p diode, from Eqs. (2.57) and (2.58), is given by

$$q\psi_{bi} = E_{fn} - E_{in} + kT\ln\left(\frac{N_a}{n_i}\right) \tag{2.69}$$
$$\approx E_{cn} - E_{in} + kT\ln\left(\frac{N_a}{n_i}\right)$$
$$\approx \frac{E_g}{2} + kT\ln\left(\frac{N_a}{n_i}\right),$$

where we have made a further approximation that the intrinsic Fermi level is located half way between the conduction- and valence-band edges, $E_{cn}$ and $E_{vn}$, on the n-side. [See Eq. (2.6) and the discussion that follows.] Figure 2.12 is a plot of $\psi_{bi}$, as approximated by Eq. (2.69), as a function of the doping concentration of the lightly doped side.

The depletion-layer width, from Eqs. (2.65) and (2.66), is

$$W_d = \sqrt{\frac{2\varepsilon_{si}(\psi_{bi} \pm V_{app})}{qN_a}}. \tag{2.70}$$

The depletion-layer capacitance is given by Eq. (2.68). Figure 2.13 is a plot of the depletion-layer width and capacitance as a function of doping concentration for $V_{app} = 0$.

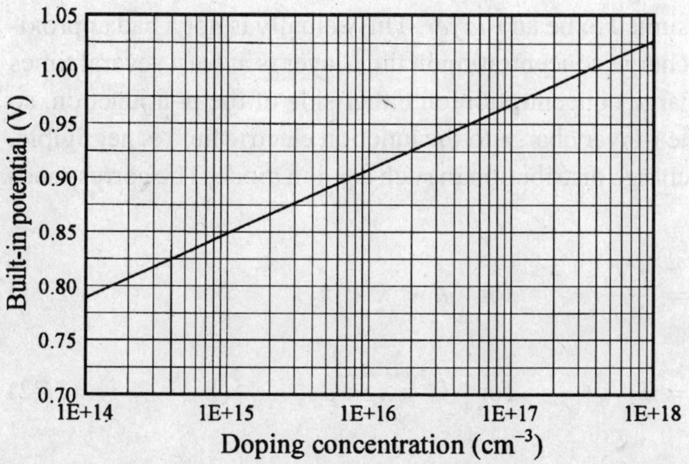


**FIGURE 2.12.** Built-in potential for a one-sided p–n junction versus the doping concentration of the lightly doped side.

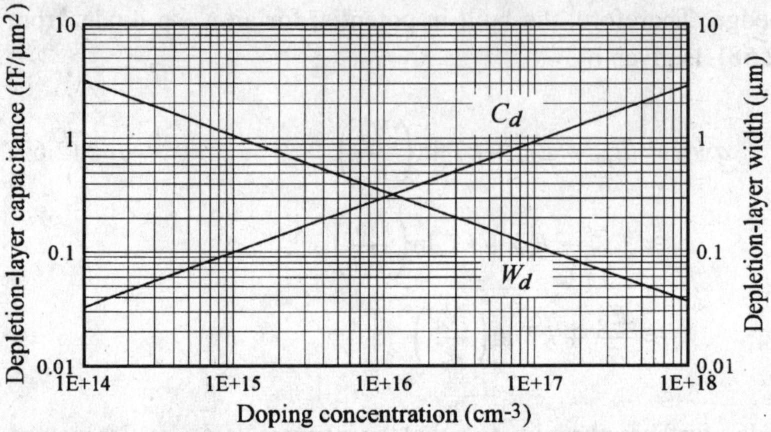


**FIGURE 2.13.** Depletion-layer width and depletion-layer capacitance, at zero bias, as a function of doping concentration of the lightly doped side of a one-sided p–n junction.

### 2.2.2.3 THIN-i-LAYER p–i–n DIODES

Many modern VLSI devices operate at very high electric fields within the depletion regions of some of their p–n diodes. In fact, the junction fields are often so high that detrimental high-field effects, such as avalanche multiplication and hot-carrier effects, limit the attainable device and circuit performance. To overcome the constraints imposed by high fields in a diode, device designers often introduce a thin but lightly doped region between the n- and the p-side. In practice, this can be accomplished by sandwiching a lightly doped layer during epitaxial growth of the doped layers, or by grading the doping concentrations at or near the junction by ion implantation and/or diffusion. Analyses of such a diode structure become very simple if the lightly doped region is assumed to be intrinsic or undoped, i.e., if the lightly doped region is assumed to be an *i-layer*. This actually is not a bad approximation as long as the net charge concentration in the i-layer is at least several times smaller than the space-charge concentration on either side of the p–n junction, so that the contribution by the i-layer charge to the junction electric field is negligible.

Figure 2.14 shows the charge distribution in such a p–i–n diode. The corresponding Poisson's equation is

$$-\frac{d^2\psi_i}{dx^2} = \frac{qN_d}{\varepsilon_{si}} \qquad \text{for} \quad d < x < x_n, \tag{2.71}$$

$$-\frac{d^2\psi_i}{dx^2} = 0 \qquad \text{for} \quad 0 < x < d, \tag{2.72}$$

$$-\frac{d^2\psi_i}{dx^2} = -\frac{qN_a}{\varepsilon_{si}} \qquad \text{for} \quad -x_p < x < 0. \tag{2.73}$$

These equations can be solved in the same way as Eqs. (2.61) and (2.62). Thus,

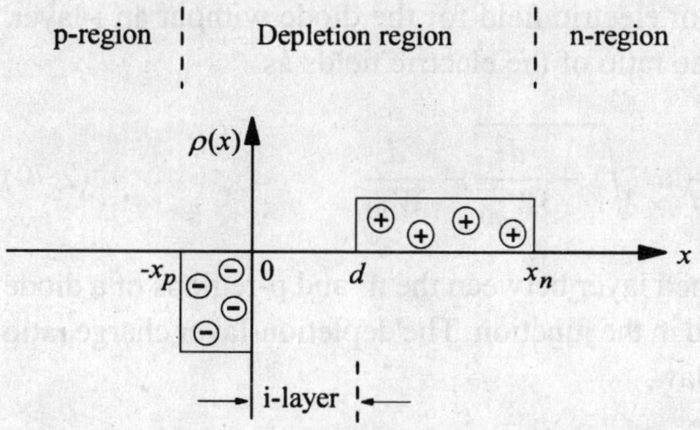


**FIGURE 2.14.** Charge distribution in a p–i–n diode.

integrating the equations once, subject to the boundary conditions that the electric field is zero at $x = -x_p$ and at $x = x_n$, gives

$$\mathscr{E}_m = \frac{qN_a x_p}{\varepsilon_{si}} = \frac{qN_d(x_n - d)}{\varepsilon_{si}}, \tag{2.74}$$

where $\mathscr{E}_m$ is the maximum electric field which exists in the region $0 \le x \le d$. Integrating the equations twice gives the total potential drop $\psi_m$ across the junction as

$$\psi_m = \frac{\mathscr{E}_m(W_d + d)}{2}, \tag{2.75}$$

where $W_d = x_n + x_p$ is the total depletion-layer width. Eliminating $\mathscr{E}_m$ from Eqs. (2.74) and (2.75) gives

$$W_d = \sqrt{\frac{2\varepsilon_{si}(N_a + N_d)\psi_m}{qN_aN_d} + d^2}. \tag{2.76}$$

It is interesting to compare two diodes with the same externally applied voltage and the same p-side and n-side doping concentrations, one with an i-layer and one without. These two diodes have the same $\psi_m$. From Eq. (2.76), we can write

$$W_d = \sqrt{W_{d0}^2 + d^2} \tag{2.77}$$

for the diode with an i-layer, where $W_{d0}$ is the depletion-layer width, given by Eq. (2.65), for the diode without an i-layer. Therefore,

$$\frac{W_d}{W_{d0}} = \sqrt{1 + \frac{d^2}{W_{d0}^2}}. \tag{2.78}$$

If we denote by $\mathscr{E}_{m0}$ the maximum electric field for the diode without an i-layer, then Eqs. (2.64) and (2.75) give the ratio of the electric fields as

$$\frac{\mathscr{E}_m}{\mathscr{E}_{m0}} = \frac{W_{d0}}{W_d + d} = \sqrt{1 + \frac{d^2}{W_{d0}^2}} - \frac{d}{W_{d0}}. \tag{2.79}$$

Thus, introduction of a lightly doped layer between the n- and p-regions of a diode reduces the maximum electric field in the junction. The depletion-layer charge ratio for the two diodes is, by Gauss's law,

$$\frac{Q_d}{Q_{d0}} = \frac{\mathscr{E}_m}{\mathscr{E}_{m0}} = \sqrt{1 + \frac{d^2}{W_{d0}^2}} - \frac{d}{W_{d0}}, \tag{2.80}$$

where $Q_{d0}$ is the depletion-layer charge for the diode without an i-layer.

The depletion-layer capacitance can be calculated from $d|Q_d|/d\psi_m$, and the result is

$$C_d = \frac{\varepsilon_{si}}{W_d}. \tag{2.81}$$

The junction depletion-layer capacitance is related to the depletion-layer width in exactly the same way with or without an i-layer. This is expected from the physical picture of a parallel-plate capacitor where the capacitance is determined by the separation of the plates and not by any fixed charge distribution between the plates. The ratio of the capacitance with an i-layer to that without an i-layer is

$$\frac{C_d}{C_{d0}} = \frac{W_{d0}}{W_d} = \frac{1}{\sqrt{1 + d^2/W_{d0}^2}}, \tag{2.82}$$

where $C_{d0} = \varepsilon_{si}/W_{d0}$ is the depletion-layer capacitance for the diode without an i-layer.

### 2.2.3 THE DIODE EQUATION

In considering the current–voltage characteristics of a p–n diode, it is much more convenient to work with the quasi-Fermi potentials, instead of the intrinsic potential. The current densities and the quasi-Fermi potentials are given by Eqs. (2.45), (2.46), (2.47), and (2.48). These are repeated here:

$$J_n = -qn\mu_n \frac{d\phi_n}{dx}, \tag{2.83}$$

$$J_p = -qp\mu_p \frac{d\phi_p}{dx}, \tag{2.84}$$

where

$$\phi_n \equiv \psi_i - \frac{kT}{q}\ln\left(\frac{n}{n_i}\right) \tag{2.85}$$

is the quasi-Fermi potential for electrons and

$$\phi_p \equiv \psi_i + \frac{kT}{q}\ln\left(\frac{p}{n_i}\right) \tag{2.86}$$

is the quasi-Fermi potential for holes, and $\psi_i$ is the intrinsic potential. In terms of the quasi-Fermi potentials, the $pn$ product is

$$pn = n_i^2 \exp[q(\phi_p - \phi_n)/kT]. \tag{2.87}$$

#### 2.2.3.1 SPATIAL VARIATION OF $\phi_n$ AND $\phi_p$ AT LOW TO INTERMEDIATELY HIGH CURRENTS

Let us examine the spatial variation of the quasi-Fermi potentials in a diode. First let us examine the diode contacts and regions far enough away from the junction where the electron and hole concentrations are equal to their equilibrium values, i.e., where $n = n_0$ and $p = p_0$. In all these regions, we have $-q\phi_n = -q\phi_p = E_f$. This can be seen by using $n = n_0$ and $p = p_0$ in Eqs. (2.85) and (2.86) and then comparing them with Eqs. (2.16) or (2.17).

Now, let us consider the spatial variation of $\phi_n$ in regions of nonequilibrium. We first consider the magnitude of $d\phi_n/dx$ inside the quasineutral n-region of the diode. If the n-region has a doping concentration of $N_d = 1 \times 10^{17}$ cm$^{-3}$, then $n = 1 \times 10^{17}$ cm$^{-3}$. From Fig. 2.7, $\mu_n$ is on the order of 1000 cm$^2$/V-s. If $J_n$ is zero, $\phi_n$ is flat as required by Eq. (2.83). Even for an intermediately high current density $J_n = 1 \times 10^3$ A/cm$^2$, Eq. (2.83) gives $q(d\phi_n/dx) \approx 60$ eV/cm. Now, as can be seen from Figs. 2.12 and 2.13, the built-in potential is on the order of 1 V, and the depletion-layer thickness is on the order of 0.1 μm. The average rate of change of the built-in potential energy across the space-charge region is therefore $\langle q d\psi_{bi}/dx\rangle_{\text{avg}} \approx 10^5$ eV/cm. That is, even for $J_n = 1 \times 10^3$ A/cm$^2$, $q(d\phi_n/dx)$ is much smaller than $\langle q d\psi_{bi}/dx\rangle_{\text{avg}}$, and $\phi_n$ is essentially flat within the quasineutral n-region. [We shall return later to discuss the case where $J_n$ is large enough so that $q(d\phi_n/dx)$ is not negligible compared to $\langle q\, d\psi_{bi}/dx\rangle_{\text{avg}}$.]

Let us next compare the magnitudes of $d\phi_n/dx$ at the boundaries of the space-charge region of the diode (Shockley, 1950). Within the space-charge region, $J_n$ is constant if generation–recombination current is negligible. Therefore, referring to Fig. 2.11, this implies that $n\mu_n d\phi_n/dx$ at $x_n$ is equal to $n\mu_n\, d\phi_n/dx$ at $-x_p$.

Since $n(x_n)$, which is the n-region electron concentration, is much larger than $n(-x_p)$, which is the p-region electron concentration, while the electron mobilities at $x_n$ and at $-x_p$ are comparable, $d\phi_n/dx$ at $x_n$ must be much smaller than $d\phi_n/dx$ at $-x_p$. Also, as we shall show immediately below, $d\phi_n/dx$ can be appreciable inside the quasineutral p-region. Therefore, $\phi_n$ remains essentially flat across the space-charge region, until at or close to the depletion-layer edge on the p-side.

Finally, let us consider $d\phi_n/dx$ in the quasineutral p-region. As we shall show in the next sub-subsection, the concentration of excess electrons and $J_n$ on the p-side of the diode are nonzero only within some distance from the depletion-layer edge. This distance is determined by the electron diffusion length. As the concentration of excess electrons falls off with distance from the depletion-layer edge, $\phi_n$ varies with distance according to Eq. (2.85), and $d\phi_n/dx$ varies with $J_n$ according to Eq. (2.83). Where the concentration of excess electrons reaches zero, $-q\phi_n$ is equal to the Fermi level, as discussed earlier.

In summary, at low to intermediately high electron current densities, $\phi_n$ is essentially flat inside the n-region and across the space-charge region of a p–n diode. ***Practically all the spatial variation of $\phi_n$ occurs in the p-region close to the depletion-layer edge***. Similarly, $\phi_p$ is essentially flat on the p-side and within the space-charge region. ***Practically all the spatial variation of $\phi_p$ occurs on the n-side close to the depletion-layer edge***. The variation of $\phi_n$ and $\phi_p$ with distance is illustrated in Fig. 2.15.

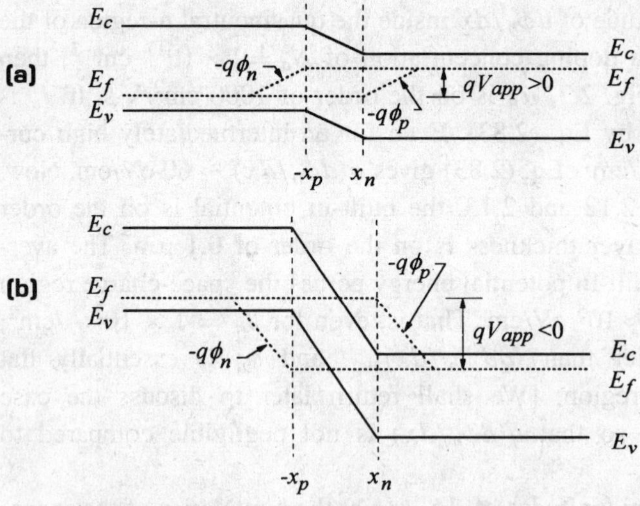


**FIGURE 2.15.** Variation of the quasi-Fermi potentials, $\phi_p$ for holes and $\phi_n$ for electrons, as a function of distance in a p–n diode: (a) forward-biased junction and (b) reverse-biased junction.

#### 2.2.3.2 RELATIONSHIP BETWEEN MINORITY-CARRIER DENSITY AND JUNCTION VOLTAGE

As can be seen in Fig. 2.15, *within the depletion region and at its edges*, the difference $\phi_p - \phi_n$ is equal to the applied voltage across the p–n junction, namely

$$V_{app} = \phi_p - \phi_n \quad \text{for} \quad -x_p \leq x \leq x_n. \tag{2.88}$$

For a forward-biased diode, $V_{app} > 0$ and $-q\phi_n$ lies above $-q\phi_p$ in an energy-band diagram, as illustrated in Fig. 2.15(a). For a reverse-biased diode, $V_{app} < 0$ and $-q\phi_p$ lies above $-q\phi_n$, as illustrated in Fig. 2.15(b).

Equations (2.87) and (2.88) can be combined to give the electron density at the depletion-layer edge on the p-side as

$$\begin{aligned} n_p(x=-x_p) &= \frac{n_i^2}{p_p(x=-x_p)} \exp(qV_{app}/kT) \qquad (2.89) \\ &= \frac{n_{p0}(x=-x_p)p_{p0}(x=-x_p)}{p_p(x=-x_p)} \exp(qV_{app}/kT) \\ &\approx n_{p0}(x=-x_p)\exp(qV_{app}/kT), \end{aligned}$$

where we have made the ***assumption that the majority-carrier concentration is about the same as its equilibrium value***. This assumption is a good one for all biases where the injected minority-carrier concentration is small compared to the majority-carrier concentration. (The case of large forward bias, where the injected minority-carrier concentration is comparable to or larger than the majority-carrier concentration, will be discussed later in this subsection.) Similarly,

$$p_n(x=x_n) \approx p_{n0}(x=x_n)\exp(qV_{app}/kT) \tag{2.90}$$

is the hole density at the depletion-layer edge on the n-side. Equations (2.89) and (2.90) are ***the most important boundary conditions governing a p–n junction***. They relate the minority-carrier concentrations at the boundaries of the depletion layer to their thermal-equilibrium values and to the applied voltage across the junction. They apply to both a forward-biased ($V_{app} > 0$) junction, resulting in $n_p \gg n_{p0}$ at $x = -x_p$ and $p_n \gg p_{n0}$ at $x = x_n$, and to a reverse-biased ($V_{app} < 0$) junction, resulting in $n_p \ll n_{p0}$ at $x = -x_p$ and $p_n \ll p_{n0}$ at $x = x_n$.

#### 2.2.3.3 EFFECT OF FINITE RESISTIVITY

In the previous discussion, by assuming $\phi_n$ to be flat inside the n-region and $\phi_p$ to be flat inside the p-region, we have implicitly assumed that the voltage drops due to the finite resistivity of the quasineutral n- and p-regions are negligible. However, as a current flows in these resistive regions, $\phi_n$ and $\phi_p$ vary with distance according to

Eqs. (2.83) and (2.84), respectively. The variation in $\phi_n$ inside the n-region means that the voltage appearing immediately across the junction, $V'_{app}$, is smaller than the voltage applied to the diode terminal, $V_{app}$, by an amount equal to $\psi_i(x_n) - \psi_i$(n-side contact) $\approx \phi_n(x_n) - \phi_n$(n-side contact). Here $\psi_i(x_n) - \psi_i$(n-side contact) is simply the voltage drop associated with the current flow inside the n-region. Similarly, if the hole current in the p-region also causes an appreciable variation in $\phi_p$, the internal junction voltage is less than the diode terminal voltage by a corresponding additional amount.

However, even when there is significant voltage drop inside the n- or p-region, our argument for $\phi_n$ and $\phi_p$ being essentially flat across the space-charge layer still holds. Thus, it should be understood that in Eqs. (2.88) to (2.90), ***$V_{app}$ should be replaced by $V'_{app}$ whenever the voltage drop in the n-region or in the p-region is not negligible***. This distinction is certainly important in the forward-biased emitter–base diode of a bipolar transistor, which will be discussed in Section 6.3. It should also be noted that for a reverse-biased diode, there is negligible current flow, and hence there is no distinction between $V_{app}$ and $V'_{app}$.

#### 2.2.3.4 DIODE EQUATION AT LARGE FORWARD BIASES

As stated in the derivation of Eqs. (2.89) and (2.90), these equations are valid when the majority-carrier concentrations are about the same as their equilibrium values. If this condition is not met, these equations are not valid and Eq. (2.87) should be used instead. At sufficiently large forward biases, the injected minority-carrier concentration, particularly on the lightly doped side of the diode, could be so large that, in order to maintain quasineutrality, the electron and hole concentrations become approximately equal. In this case, Eq. (2.87) gives $n \approx p \approx n_i \exp(qV'_{app}/2kT)$. At such high levels of minority-carrier injection, the concept of a well-defined transition region is no longer valid, and the quasi-Fermi potentials do not have simple behavior in any regions of the diode (Gummel, 1967). An example of how the "boundary" of a p–n junction can be "relocated" at high minority-carrier injections can be found in Section 6.3.3 in connection with the discussion of base-widening effects in a bipolar transistor.

### 2.2.4 CURRENT-VOLTAGE CHARACTERISTICS

As discussed in Section 2.2.1, at thermal equilibrium the drift component of the current caused by the electric field in the depletion region is exactly balanced out by the diffusion component of the current caused by the electron and hole concentration gradient across the junction, resulting in zero current flow in the diode. When an external voltage is applied, this current component balance is upset, and current will flow in the diode. If carriers are generated by light or some other means, thermal equilibrium is disturbed, and current can also flow in the diode. Here the current flow in a diode as a result of an external applied voltage is described.

Consider a forward-biased p–n diode. Electrons are injected from the n-side into the p-side, and holes are injected from the p-side into the n-side. If generation and recombination in the depletion region are negligible, then the hole current leaving the p-side is the same as the hole current entering the n-side. Similarly, the electron current leaving the n-side is equal to the electron current entering the p-side. To determine the total current flowing in the diode, all we need to do is to determine the hole current entering the n-side and the electron current entering the p-side.

The starting point for describing the current–voltage characteristics is the continuity equations. For electrons, it is given by Eq. (2.49), which is repeated here:

$$\frac{\partial n}{\partial t} = \frac{1}{q}\frac{\partial J_n}{\partial x} - R_n + G_n, \tag{2.91}$$

where $R_n$ and $G_n$ are the electron recombination and generation rates, respectively. Equation (2.91) can be rewritten as

$$\frac{\partial n}{\partial t} = \frac{1}{q}\frac{\partial J_n}{\partial x} - \frac{n - n_0}{\tau_n}, \tag{2.92}$$

where

$$\tau_n \equiv \frac{n - n_0}{R_n - G_n} \tag{2.93}$$

is the *electron lifetime*, and $n_0$ is the electron concentration at thermal equilibrium. Substituting Eq. (2.43) for $J_n$ into Eq. (2.92) gives

$$\frac{\partial n}{\partial t} = n\mu_n\frac{\partial \mathscr{E}}{\partial x} + \mu_n\mathscr{E}\frac{\partial n}{\partial x} + D_n\frac{\partial^2 n}{\partial x^2} - \frac{n - n_0}{\tau_n}. \tag{2.94}$$

#### 2.2.4.1 DIODES WITH UNIFORMLY DOPED REGIONS

Let us consider electrons in the p-region of a p–n diode. For simplicity, we shall assume the p-region to be uniformly doped, so that at low electron injection currents the hole density is uniform in the p-region. As will be shown in Section 6.1.1, the electric field is zero for a region where the majority-carrier concentration is uniform. Thus, for the p-region under discussion, $\partial\mathscr{E}/\partial x = 0$ and $\mathscr{E} = 0$. For electrons in this p-region, Eq. (2.94) reduces to

$$\frac{\partial n_p}{\partial t} = D_n\frac{\partial^2 n_p}{\partial x^2} - \frac{n_p - n_{p0}}{\tau_n}. \tag{2.95}$$

At steady state, Eq. (2.95) becomes

$$\frac{\partial n_p}{\partial t} = 0 = D_n\frac{\partial^2 n_p}{\partial x^2} - \frac{n_p - n_{p0}}{\tau_n}, \tag{2.96}$$

which can be rewritten as

$$\frac{d^2 n_p}{dx^2} - \frac{n_p - n_{p0}}{L_n^2} = 0, \tag{2.97}$$

where

$$L_n \equiv \sqrt{\tau_n D_n} = \sqrt{\frac{kT \mu_n \tau_n}{q}} \tag{2.98}$$

is the *electron diffusion length* in the p-region. It should be noted that the quantities in Eq. (2.98) are all for minority carriers, not majority carriers.

A note on the coordinate system to be followed here. Earlier in this chapter, the physical junction of a p–n diode was assumed to be located at $x = 0$, and the p-side depletion-layer edge to be located at $x = -x_p$. The excess electrons in the p-region of the diode were injected from the n-side. If we assume negligible generation and recombination in the depletion layer, then all the electrons leaving the n-side will enter the quasineutral p-region at the p-side depletion-layer boundary. These excess electrons will then move further into the p-region, contributing to electron current and becoming recombined along the way. The p-side depletion-layer boundary is therefore really the starting location for considering the distribution and transport of the excess electrons in the p-region. For simplicity, we shall change our coordinate system and let $x = 0$ be the location of the depletion-region boundary, and $x = W$ the location of the ohmic contact, of the p-side of the diode.

The electron concentrations are given by Eq. (2.89) at $x = 0$, and equal to $n_{p0}$ at $x = W$, i.e.,

$$n_p = n_{p0} \exp(qV_{app}/kT) \quad \text{at} \quad x = 0, \tag{2.99}$$

$$n_p = n_{p0} \quad \text{at} \quad x = W. \tag{2.100}$$

Solving Eq. (2.97) subject to these boundary conditions gives

$$n_p - n_{p0} = n_{p0}\left[\exp\left(\frac{qV_{app}}{kT}\right) - 1\right]\frac{\sinh[(W - x)/L_n]}{\sinh(W/L_n)}. \tag{2.101}$$

Since there is no electric field in the p-region, there is no electron drift-current component, only an electron diffusion-current component. The electron current density entering the p-region is

$$\begin{aligned} J_n(x = 0) &= qD_n\left(\frac{dn_p}{dx}\right)_{x=0} \\ &= -\frac{qD_n n_{p0}[\exp(qV_{app}/kT) - 1]}{L_n \tanh(W/L_n)} \\ &= -\frac{qD_n n_i^2[\exp(qV_{app}/kT) - 1]}{p_{p0} L_n \tanh(W/L_n)}, \end{aligned} \tag{2.102}$$

where in writing the last equation we have used the fact that $n_{p0}p_{p0} = n_i^2$. Equations (2.101) and (2.102) are valid for a p-region of arbitrary width $W$. Note that $J_n$ is negative, since electrons have a charge of $-q$ and are flowing in the $+x$ direction.

The hole density in the n-region and the hole current density entering n-side have the same forms as Eq. (2.101) and Eq. (2.102), respectively, and can be derived in an analogous manner (cf. Exercise 2.16). ***The total current flowing through a p–n diode is the sum of the electron current and hole current***. Since it is straightforward to sum the two currents, the total current will not be shown explicitly here.

#### 2.2.4.2 EMITTER AND BASE OF A DIODE

Since $p_{p0} \approx N_a$, Eq. (2.102) shows that the minority-carrier current is inversely proportional to the doping concentration. Thus, in a one-sided diode, the minority-carrier current in the lightly doped side is much larger than that in the heavily doped side. The diode current is dominated by the flow of minority carriers in the lightly doped side of the diode, while minority-carrier current in the heavily doped side usually can be neglected in comparison. (The effect of heavy doping can increase the minority-current flowing in the heavily doped region substantially. The heavy-doping effect is particularly important in bipolar devices and will be covered in Chapter 6. The effect of heavy doping on the magnitudes of the currents in a diode will be discussed as exercises.) The lightly doped side is often referred to as the *base* of the diode. The heavily doped side is often referred to as the *emitter* of the diode, since the minority carriers entering the base are emitted from it.

In discussing the current–voltage characteristics of a diode, often only the minority-carrier current flow in the base is considered, since the minority-carrier current flow in the emitter is small in comparison. As a result, unless stated explicitly, the region of the diode under discussion is assumed to be the base. If needed, the minority-carrier current in the emitter can be obtained from the base-current equations simply by substituting the appropriate doping and width parameters.

#### 2.2.4.3 FORWARD-BIASED n$^+$–p DIODES

We first consider the case where the n$^+$–p diode is forward biased, i.e., $V_{app} > 0$, and $qV_{app}/kT \gg 1$. In this case, Eqs. (2.101) and (2.102) become

$$n_p - n_{p0} = n_{p0} \exp\left(\frac{qV_{app}}{kT}\right) \frac{\sinh[(W - x)/L_n]}{\sinh(W/L_n)} \quad \text{(forward biased)} \tag{2.103}$$

and

$$J_n(x = 0) = -\frac{qD_n n_i^2 \exp(qV_{app}/kT)}{p_{p0}L_n \tanh(W/L_n)} \quad \text{(forward biased).} \tag{2.104}$$

That is, both ***the excess minority-carrier concentration and the minority-carrier current increase exponentially with the applied voltage.***

#### 2.2.4.4 REVERSE-BIASED n$^+$–p DIODES

Next we consider the case where the n$^+$–p diode is reverse-biased, i.e., $V_{app} < 0$, and $|qV_{app}| \gg kT$. In this case, Eqs. (2.101) and (2.102) become

$$n_p - n_{p0} = -n_{p0}\frac{\sinh[(W-x)/L_n]}{\sinh(W/L_n)} \quad \text{(reverse biased)} \qquad (2.105)$$

and

$$J_n(x=0) = \frac{qD_n n_i^2}{p_{p0}L_n \tanh(W/L_n)} \quad \text{(reverse biased).} \qquad (2.106)$$

Notice that $n_p - n_{p0}$ is negative, and $J_n$ is positive. The reverse bias causes a gradual depletion of electrons in the p-region near the depletion-region boundary, and this electron concentration gradient causes an electron current to flow from the neutral p-region towards the depletion region. This is the electron diffusion component of the leakage current in a reverse-biased diode. It is also referred to as the electron *saturation current* of a diode.

The hole saturation current has an analogous form. The total diffusion leakage current in a diode is the sum of the electron and hole saturation currents.

#### 2.2.4.5 WIDE-BASE n$^+$–p DIODES

A diode is *wide-base* if its base width $W$ is large compared to the minority-carrier diffusion length in the base. In this case, this means $W/L_n \gg 1$. For a forward-biased wide-base diode, Eqs. (2.103) and (2.104) reduce to

$$\begin{aligned} &n_p - n_{p0} \qquad (2.107)\\ &= n_{p0}\exp(qV_{app}/kT)\exp(-x/L_n) \quad \text{(forward, wide base)} \end{aligned}$$

and

$$J_n(x=0) = -\frac{qD_n n_i^2}{p_{p0}L_n}\exp(qV_{app}/kT) \quad \text{(forward, wide base).} \qquad (2.108)$$

Thus, for a forward-biased wide-base diode, the excess minority-carrier concentration decreases exponentially with increasing distance from the depletion-region boundary, and the minority-carrier current is independent of the base width.

For a reverse-biased wide-base diode, Eqs. (2.105) and (2.106) reduce to

$$n_p - n_{p0} = -n_{p0}\exp(-x/L_n) \quad \text{(reverse, wide base)} \qquad (2.109)$$

and

$$J_n(x = 0) = \frac{qD_n n_i^2}{p_{p0} L_n} \qquad \text{(reverse, wide base).} \tag{2.110}$$

That is, the minority-carrier electrons in the base within a diffusion length of the depletion-region boundary diffuse towards the depletion region, with a saturation current density given by Eq. (2.110) which is independent of the base width.

#### 2.2.4.6 NARROW-BASE n$^+$–p DIODES

A diode is called *narrow-base* if its base width $W$ is small compared to the minority-carrier diffusion length in the base. In the case, this means $W/L_n \ll 1$. For a forward-biased narrow-base diode, Eqs. (2.103) and (2.104) reduce to

$$\begin{aligned} n_p - n_{p0} & \\ &= n_{p0} \exp\left(\frac{qV_{app}}{kT}\right)\left(1 - \frac{x}{W}\right) \qquad \text{(forward, narrow base)} \end{aligned} \tag{2.111}$$

and

$$J_n(x = 0) = -\frac{qD_n n_i^2}{p_{p0} W} \exp\left(\frac{qV_{app}}{kT}\right) \qquad \text{(forward, narrow base).} \tag{2.112}$$

For a reverse-biased narrow-base diode, the corresponding equations are

$$n_p - n_{p0} = -n_{p0}\left(1 - \frac{x}{W}\right) \qquad \text{(reverse, narrow base)} \tag{2.113}$$

and

$$J_n(x = 0) = \frac{qD_n n_i^2}{p_{p0} W} \qquad \text{(reverse, narrow base).} \tag{2.114}$$

For both forward and reverse biases, the minority-carrier current density in a narrow-base diode increases rapidly as $W$ is decreased.

#### 2.2.4.7 TEMPERATURE DEPENDENCE AND MAGNITUDE OF THE DIODE LEAKAGE CURRENTS

The temperature dependence of the electron and hole diffusion currents is dominated by the temperature dependence of the factor $n_i^2$, which, as shown in Eq. (2.12), is

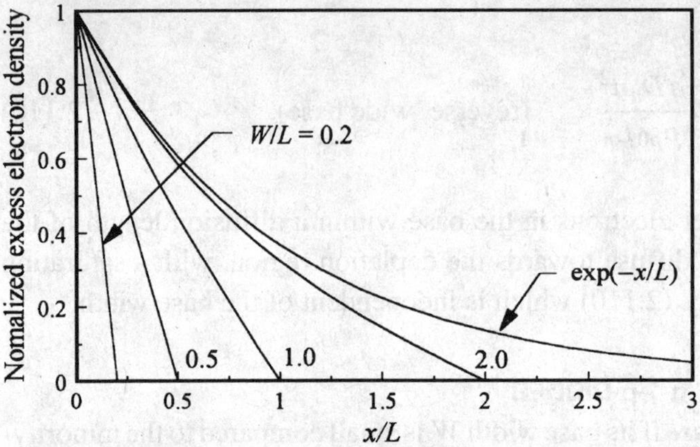


**FIGURE 2.16.** Relative magnitude of the excess minority-carrier concentration in the base of a diode as a function of distance from the base depletion-layer edge, with $W/L$ as a parameter, where $L$ is the minority-carrier diffusion length in the base and $W$ is the base-region width. The case of $W/L = \infty$ is given by $\exp(-x/L)$.

proportional to $\exp(-E_g/kT)$, where $E_g$ is the bandgap energy. This temperature dependence, corresponding to an activation energy of about 1.1 eV, is often used to distinguish the diffusion currents from the generation currents coming from within the depletion region of a diode. Generation currents are proportional to $n_i$, and hence are proportional to $\exp(-E_g/2kT)$, corresponding to an activation energy of about 0.5 eV (Grove and Fitzgerald, 1966).

For an n$^+$–p diode with a base doping concentration of $1 \times 10^{17}$ cm$^{-3}$, using the values of $D_n/L_n = L_n/\tau_n$ in Fig. 2.18 (to be derived later), Eq. (2.110) gives an electron diffusion current of about $8 \times 10^{-13}$ A/cm$^2$ at room temperature. For a typical n$^+$–p diode, the observed diffusion current (which is the sum of the electron and the hole diffusion currents) is comparable to the generation current, both being on the order of $10^{-13}$ A/cm$^2$, at room temperature (Kircher, 1975). However, the diffusion current, due to its larger activation energy, is usually larger than the generation current at elevated temperatures.

#### 2.2.4.8 DISTRIBUTION OF EXCESS MINORITY CARRIERS

It can be seen from Eqs. (2.103) and (2.105) that both a forward-biased diode and a reverse-biased diode have the same $\sinh[(W - x)/L]$ spatial dependence for the distribution of excess minority carriers (actually depletion of minority carriers in a reverse-biased diode). Figure 2.16 is a plot of the relative magnitude of the excess minority-carrier density as a function of $x/L$ with $W/L$ as a parameter. The $\exp(-x/L)$ distribution is for the case of $W/L = \infty$. It shows that a diode behaves like a wide-base diode for $W/L > 2$. For $W/L < 2$, the diode behavior depends strongly on $W$. For $W/L < 1$, the distribution can be approximated by the $1 - x/W$ dependence of a narrow-base diode.

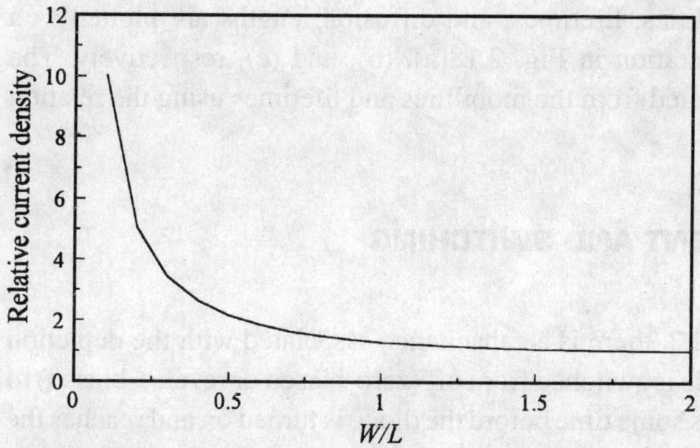


**FIGURE 2.17.** Relative magnitude of the minority-carrier current density in the base region of a diode as a function of $W/L$, normalized to the current at $W/L = \infty$. Here $L$ is the minority-carrier diffusion length in the base, and $W$ is the width of the base region.

### 2.2.4.9 DEPENDENCE OF MINORITY-CARRIER CURRENT ON BASE WIDTH

Figure 2.17 is a plot of the minority-carrier current density given by Eq. (2.102), normalized to its wide-base value. It shows that when $W/L < 1$, the minority-carrier current increases very rapidly as the diode base width decreases.

### 2.2.4.10 MINORITY-CARRIER MOBILITY, LIFETIME, AND DIFFUSION LENGTH

There have been many attempts to measure the minority-carrier lifetimes, mobilities, and diffusion lengths. For doping concentrations greater than about $1 \times 10^{19}$ cm$^{-3}$, the experiments are quite difficult, since the minority-carrier concentrations are too small, and as a result there is quite a bit of spread in the reported data (Dziewior and Silber, 1979; Dziewior and Schmid, 1977; del Alamo *et al.*, 1985a, b). For purposes of device modeling, the following empirical equations have been proposed for minority-carrier electrons (Swirhun *et al.*, 1986) and minority-carrier holes (del Alamo *et al.*, 1985a, b):

$$\mu_n = 232 + \frac{1180}{1 + (N_a/8 \times 10^{16})^{0.9}} \text{ cm}^2\text{-V-s}^{-1} \tag{2.115}$$

$$\mu_p = 130 + \frac{370}{1 + (N_d/8 \times 10^{17})^{1.25}} \text{ cm}^2\text{-V-s}^{-1} \tag{2.116}$$

$$\frac{1}{\tau_n} = 3.45 \times 10^{-12} N_a + 0.95 \times 10^{-31} N_a^2 \text{ s}^{-1} \tag{2.117}$$

$$\frac{1}{\tau_p} = 7.8 \times 10^{-13} N_d + 1.8 \times 10^{-31} N_d^2 \text{ s}^{-1}. \tag{2.118}$$

The minority-carrier mobilities, lifetimes, and diffusion lengths are plotted as a function of doping concentration in Fig. 2.18(a), (b), and (c), respectively. The diffusion lengths are calculated from the mobilities and lifetimes using the relation $L = (kT\mu\tau/q)^{1/2}$.

### 2.2.5 TIME-DEPENDENT AND SWITCHING CHARACTERISTICS

As discussed in Section 2.2.2, there is a capacitance associated with the depletion layer of a diode. As the diode is switched from off (zero-biased or reverse-biased) to on (forward-biased), it takes some time before the diode is turned on and reaches the steady state. This time is associated with charging up the depletion-layer capacitor and filling up the p- and n-regions with excess minority carriers. Similarly, when a diode is switched from the on state to the off state, it takes some time before the diode is turned off. This time is associated with discharging the depletion-layer capacitor and discharging the excess minority carriers stored in the p- and n-regions. The majority-carrier response time, or dielectric relaxation time, is negligibly short, on the order of $10^{-12}$ s, as shown in Section 2.1.4.

Consider the time needed to charge and discharge the depletion-layer capacitor. From Fig. 2.13, the depletion-layer capacitance $C_d$ is typically on the order of 1 fF/μm$^2$. To turn a diode from off to on, and from on to off, the voltage swing $V$ is typically about 1 V. If the diode is connected so that it carries a current density $J$ of 1 mA/μm$^2$, then the time associated with charging and discharging the depletion-layer capacitor is on the order of $C_d V/J$, which is on the order of $10^{-12}$ s. Of course, this time changes in proportion to the current density $J$. However, as we shall show below, the time needed to charge and discharge the depletion-layer capacitor is usually very short compared with the time associated with charging and discharging the p- and n-regions of their minority carriers.

#### 2.2.5.1 EXCESS MINORITY CARRIERS IN THE BASE AND BASE CHARGING TIME

Consider an n$^+$–p diode with a p-region base width $W$. When a forward bias is applied to it, minority-carrier electrons are injected into the base. As discussed in Section 2.2.4, for a wide-base diode, the minority-carrier density decreases exponentially with increasing distance, and practically all the minority carriers recombine before they reach the minority-carrier sink at $x = W$. For a narrow-base diode, on the other hand, practically all the minority carriers can travel across the base region without recombining.

The total *excess minority-carrier charge* (electrons) in the p-type base region is

$$Q_B = -q \int_0^W (n_p - n_{p0})\,dx. \tag{2.119}$$

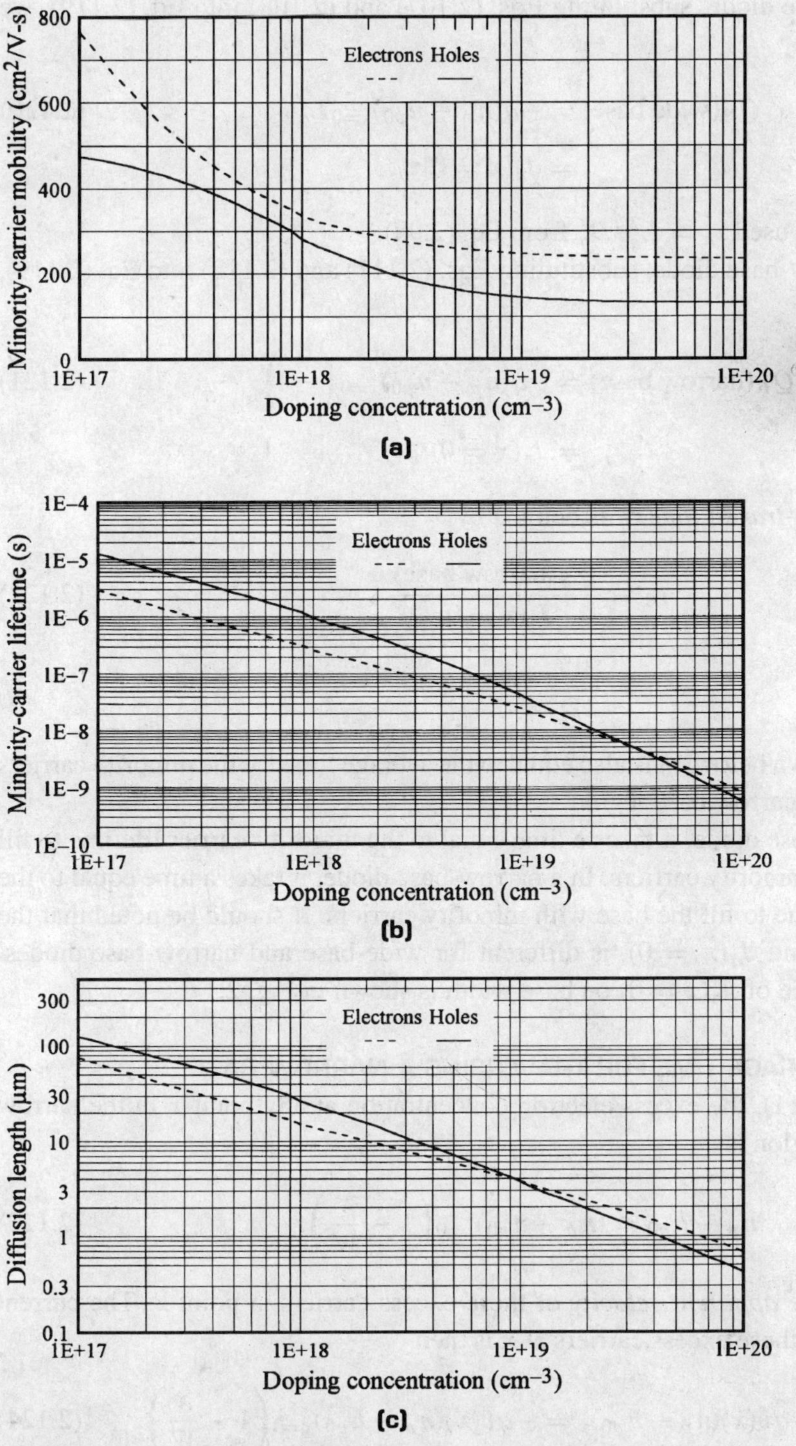


**FIGURE 2.18.** Minority-carrier (a) mobilities, (b) lifetimes, and (c) diffusion lengths as a function of doping concentration, calculated using the empirical equations (2.115) to (2.118).

For a wide-base diode, substituting Eqs. (2.107) and (2.108) into Eq. (2.119), we obtain

$$\begin{aligned} Q_B(\text{wide base}) &= -q(n_p - n_{p0})_{x=0} L_n \\ &= J_n(x=0)\tau_n, \end{aligned} \tag{2.120}$$

where we have used $\tau_n = L_n^2/D_n$ from Eq. (2.98).

For a narrow-base diode, substituting Eqs. (2.111) and (2.112) into Eq. (2.119), we obtain

$$\begin{aligned} Q_B(\text{narrow base}) &= -q(n_p - n_{p0})_{x=0}\left(\frac{W}{2}\right) \\ &= J_n(x=0)t_B, \end{aligned} \tag{2.121}$$

where the *base-transit time* $t_B$ is defined by

$$\begin{aligned} t_B &\equiv \frac{Q_B(\text{narrow base})}{J_n(x=0)} \\ &= \frac{W^2}{2D_n}. \end{aligned} \tag{2.122}$$

As will be shown below, $t_B$ is also equal to the average time for the minority carriers to traverse the narrow base region.

In a wide-base diode, it takes a time equal to the minority-carrier lifetime to fill the base with minority carriers. In a narrow-base diode, it takes a time equal to the base-transit time to fill the base with minority carriers. It should be noted that the charging current, $J_n(x=0)$, is different for wide-base and narrow-base diodes. The dependence of $J_n(x=0)$ on base width is shown in Fig. 2.17.

#### 2.2.5.2 AVERAGE TIME FOR TRAVERSING A NARROW BASE

From Eq. (2.111), the excess electron concentration at any point $x$ in the narrow p-type base region is

$$n_p - n_{p0} = (n_p - n_{p0})_{x=0}\left(1 - \frac{x}{W}\right). \tag{2.123}$$

Let $v(x)$ be the *apparent velocity* of these excess carriers at point $x$. The current density due to these excess carriers at $x$ is then

$$J_n(x) = -qv(x)(n_p - n_{p0})_x = -qv(x)(n_p - n_{p0})_{x=0}\left(1 - \frac{x}{W}\right). \tag{2.124}$$

The electron current density at $x = 0$ is given by Eq. (2.112), i.e.,

$$J_n(x=0) = -\frac{qD_n}{W}(n_p - n_{p0})_{x=0}. \tag{2.125}$$

Assuming negligible recombination in the narrow base region, then current continuity requires $J_n(x)$ to be independent of $x$, i.e.,

$$v(x) = \frac{D_n}{W - x}. \tag{2.126}$$

The average time for traversing the base is thus given by

$$t_{avg} \equiv \int_0^W \frac{dx}{v(x)} = \frac{W^2}{2D_n}. \tag{2.127}$$

Comparison of Eqs. (2.122) and (2.127) shows that the base-transit time is equal to the average time for the minority carriers to traverse the narrow base.

It is instructive to estimate the magnitude of $t_B$. Modern n–p–n bipolar transistors typically have base widths of about 0.1 μm, and a peak base doping concentration of about $2 \times 10^{18}$ cm$^{-3}$ (Nakamura and Nishizawa, 1995). The corresponding minority electron mobility, from Fig. 2.18, is about 300 cm$^2$/V-s. The base-transit time is therefore less than $1 \times 10^{-11}$ s, which is extremely short compared with the corresponding minority-carrier lifetime on the order of $1 \times 10^{-7}$ s. ***Recombination is negligible in the base layers of modern bipolar transistors.***

#### 2.2.5.3 DISCHARGE TIME OF A FORWARD-BIASED DIODE

Consider an n$^+$–p diode in a circuit configuration shown in Fig. 2.19(a). For simplicity, let us assume the diode to have a unit cross-sectional area, so that current and current density can be used interchangeably. Furthermore, for simplicity, let us assume the external voltage, $V_F$ or $V_R$, driving the circuit to be large compared to the internal junction voltage, i.e., the voltage immediately across the diode depletion layer, which is typically less than 1.0 V. At $t < 0$, there is a forward current of $I_F \approx V_F/R$ as illustrated in Fig. 2.19(b), and an excess electron distribution in the base region as illustrated in Fig. 2.19(c).

At time $t = 0$, the external bias is switched to a reverse voltage of $V_R$. The excess electrons in the base start to diffuse back towards the depletion region of the diode. Those electrons at the edge of the depletion region are swept away by the electric field in the depletion region towards the n$^+$ emitter at a saturated velocity of about $10^7$ cm/s. As shown in Fig. 2.13, the depletion-layer width is typically on the order of 0.1 μm. The transit time across the depletion region is therefore on the order of $10^{-12}$ s. As we shall see later, except for diodes of very narrow base widths, this time is extremely short compared to the total time for emptying the excess electrons out of the base region. Thus, as long as there are sufficient excess electrons in the base region, the reverse current is limited not by the diffusion of excess electrons but by the external resistor and has a value of $I_R \approx V_R/R$, and the slope $(dn_p/dx)_{x=0}$, being proportional to $I_R$, is approximately constant.

As the excess electrons are discharged, part of the external voltage starts to appear across the p–n junction, and the junction becomes less forward biased. However, as

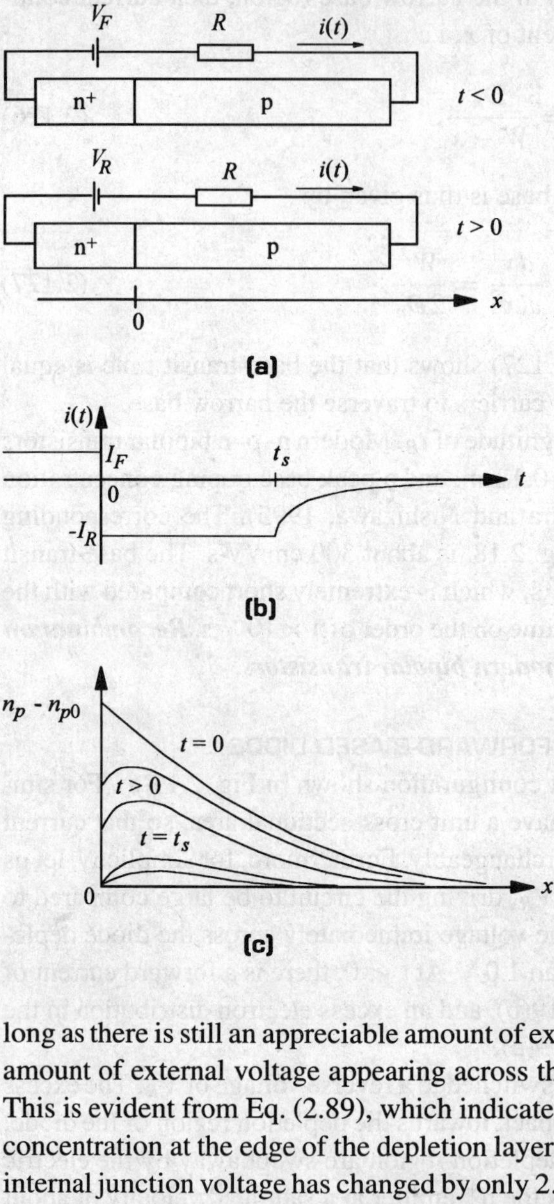


**FIGURE 2.19.** Schematics showing the switching of a $n^+$–p diode from forward bias to reverse bias: (a) the circuit schematics, (b) the diode current as a function of time, and (c) the excess-electron distribution in the base for different times.

long as there is still an appreciable amount of excess electrons stored in the base, the amount of external voltage appearing across the p–n junction remains very small. This is evident from Eq. (2.89), which indicates that even after the excess-electron concentration at the edge of the depletion layer has decreased by a factor of 10, the internal junction voltage has changed by only $2.3kT/q$, or 60 mV. This is consistent with our assumptions that the reverse current remains essentially constant. During this time, the diode remains in the on condition.

At time $t = t_s$, the excess electrons have been depleted to the point that the reverse current is limited by the diffusion of electrons instead of by the external resistor. The rate of voltage change across the junction increases. Finally, when all the excess electrons are removed, the p–n diode is completely off. The external reverse-bias voltage appears entirely across the junction, and the reverse current is limited by the diode leakage current.

The time needed to switch off a forward-biased diode can be estimated from a charge-control analysis (Kuno, 1964). For simplicity, we shall estimate only the time during which the reverse current is approximately constant, and during which the diode remains in the on condition. Since the junction voltage remains approximately constant during this time, charging and discharging of the depletion-layer capacitance of the junction can be ignored. From Eq. (2.92), the continuity equation for electrons in the p-type base region is

$$\frac{\partial(n_p - n_{p0})}{\partial t} = \frac{1}{q}\frac{\partial i_n(t)}{\partial x} - \frac{n_p - n_{p0}}{\tau_n}, \tag{2.128}$$

where $i_n(t)$ is the non-steady-state electron current in the base region. (Current and current density have been used interchangeably for our diode of unit cross section.) Multiplying Eq. (2.128) by $-q$ and integrating over the base region, we have

$$-q\frac{\partial}{\partial t}\int_0^W (n_p - n_{p0})\, dx = -\int_0^W \frac{\partial i_n(t)}{\partial x}\, dx + \frac{q}{\tau_n}\int_0^W (n_p - n_{p0})\, dx, \tag{2.129}$$

or

$$\frac{dQ_B(t)}{dt} = -i_n(W, t) + i_n(0, t) - \frac{Q_B(t)}{\tau_n}, \tag{2.130}$$

where $Q_B(t)$ is the excess minority charge stored in the base region, given by Eq. (2.119). The current difference $i_n(0, t) - i_n(W, t)$ is simply the current flowing through the external resistor $R$. That is, Eq. (2.130) can also be written as

$$\frac{dQ_B(t)}{dt} + \frac{Q_B(t)}{\tau_n} = I_R. \tag{2.131}$$

These equations are simply the continuity equation in the base region stated in the charge-control form. The solution for Eq. (2.131) is

$$Q_B(t) = I_R\tau_n + [Q_B(0) - I_R\tau_n]\, \exp(-t/\tau_n), \tag{2.132}$$

or

$$\frac{Q_B(t)}{Q_B(0)} = \frac{I_R\tau_n}{Q_B(0)}\left[1 - \exp\left(\frac{-t}{\tau_n}\right)\right] + \exp\left(\frac{-t}{\tau_n}\right), \tag{2.133}$$

where $Q_B(0)$ is the excess minority charge just after the diode is switched from forward bias to reverse bias.

- *Discharge time for a wide-base diode.* For a wide-base diode, Eq. (2.120) gives $Q_B(0) = -\tau_n I_F$. (Note the negative sign in the expression for $Q_B$, since $Q_B$ is negative for electrons.) Therefore, Eq. (2.133) gives

$$\frac{Q_B(t)}{Q_B(0)} = \left(1 + \frac{I_R}{I_F}\right)\exp\left(\frac{-t}{\tau_n}\right) - \frac{I_R}{I_F}. \tag{2.134}$$

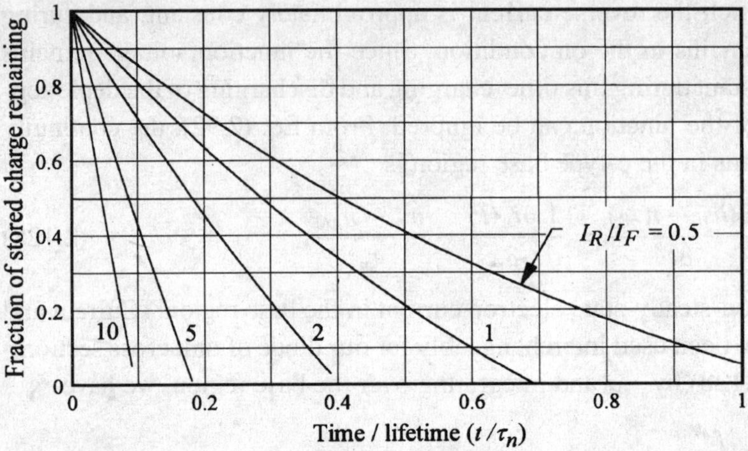


**FIGURE 2.20.** Plot of charge ratio $Q_B(t)/Q_B(0)$ as a function of $t/\tau_n$ during the discharge of a forward-biased wide-base diode, with the ratio of discharge current to charging current, $I_R/I_F$, as a parameter.

Figure 2.20 is a plot of the charge ratio $Q_B(t)/Q_B(0)$ as a function of $t/\tau_n$ with the current ratio $I_R/I_F$ as a parameter. It shows that a forward-biased diode discharges with a time constant approximately equal to the minority-carrier lifetime, unless the reverse discharge current is much larger than the forward charging current. Even for $I_R/I_F = 10$, the diode discharges in a time of approximately $\tau_n/10$, which, as can be seen from Fig. 2.18(b), is larger than $10^{-8}$ s for most diodes of practical doping concentrations. This time is very long compared to the typical switching delays of VLSI circuits. The important point is that ***it takes a long time to drain off the excess minority carriers stored in a wide-base diode and turn it off***. It is important to minimize excess minority carriers stored in forward-biased diodes if these diodes are to be switched off fast.

- *Discharge time for a narrow-base diode.* For a narrow-base diode, Eq. (2.121) gives $Q_B(0) = -t_B I_F$, and Eq. (2.133) gives

$$\frac{Q_B(t)}{Q_B(0)} = \frac{I_R \tau_n}{-I_F t_B}\left[1 - \exp\left(\frac{-t}{\tau_n}\right)\right] + \exp\left(\frac{-t}{\tau_n}\right). \tag{2.135}$$

As we shall show below, a narrow-base diode discharges in a time small compared to $\tau_n$. Therefore, Eq. (2.135) is physically valid for $t/\tau_n \ll 1$, and can be approximated by

$$\frac{Q_B(t)}{Q_B(0)} = 1 - \frac{I_R t}{I_F t_B}. \tag{2.136}$$

It shows that the discharge time for a narrow-base diode lasts approximately $(I_F/I_R)t_B$, which, for a large $I_R/I_F$ ratio, can be much shorter than the base transit time. Even for $I_R/I_F = 1$, the discharge time lasts about a time equal to $t_B$, which, as discussed following Eq. (2.127), is very small compared to $\tau_n$. The important point is that ***forward-biased narrow-base diodes can be switched off fast.***

### 2.2.6 DIFFUSION CAPACITANCE

For a forward-biased diode, in addition to the capacitance associated with the depletion layer, there is an important capacitance component associated with the rearrangement of the excess minority carriers in response to a change in the applied voltage. This minority-carrier capacitance is called *diffusion capacitance*, $C_D$.

Again consider an n$^+$–p diode of unit cross section so that current and current density can be used interchangeably for simplicity. We shall assume the n$^+$-side (emitter $E$) to be wide compared to its hole diffusion length, and the p-side (base $B$) to have a width $W_B$ small compared to its electron diffusion length. At a forward bias $V_{app}$, the excess electron charge in the base is given by Eq. (2.121) as

$$Q_n(V_{app}) = -I_n(V_{app})t_B, \tag{2.137}$$

where $-I_n$ is the electron current injected from the emitter into the base, and $t_B$ is the base transit time. Note that $Q_n$ is negative for the electron charge, and $I_n$ is a positive quantity. The diffusion capacitance due to these stored electrons is

$$C_{Dn} \equiv \frac{d|Q_n|}{dV_{app}} = \frac{dI_n}{dV_{app}}t_B. \tag{2.138}$$

As shown in Section 2.2.4, the minority-carrier injection current for a forward-biased diode is proportional to $\exp(qV_{app}/kT)$. Therefore, Eq. (2.138) gives

$$C_{Dn} = \frac{q}{kT}I_n t_B. \tag{2.139}$$

Similarly, from Eq. (2.120), the diffusion capacitance due to the excess holes in the emitter is

$$C_{Dp} = \frac{q}{kT}I_p \tau_{pE}, \tag{2.140}$$

where $-I_p$ is the hole current injected from the base into the emitter (holes flowing in the $-x$ direction give rise to a negative current), and $\tau_{pE}$ is the hole lifetime in the emitter region. The total diffusion capacitance is

$$C_D = C_{Dn} + C_{Dp} = \frac{q}{kT}\left(I_n t_B + I_p \tau_{pE}\right). \tag{2.141}$$

It is instructive to examine the relative magnitude of the two diffusion-capacitance components. Using the hole equivalent of Eq. (2.108) for hole current and Eq. (2.112) for electron current, we have

$$\frac{C_{Dn}}{C_{Dp}} = \frac{I_n t_B}{I_p \tau_{pE}} \tag{2.142}$$

$$= \frac{D_{nB}}{W_B N_B} \frac{L_{pE} N_E}{D_{pE}} \frac{W_B^2}{2D_{nB}\tau_{pE}},$$

where $N_B$ is the base-region doping concentration, and $N_E$ is the doping concentration of the emitter region. Using Eq. (2.98), Eq. (2.142) can be simplified to

$$\frac{C_{Dn}}{C_{Dp}} = \frac{N_E}{N_B} \frac{W_B}{2L_{pE}}. \tag{2.143}$$

Since the ratio $N_E/N_B$ is typically about 100, the ratio $C_{Dn}/C_{Dp}$ is much larger than unity for most n$^+$–p diodes. That is, the diffusion capacitance of a one-sided p–n diode is dominated by the minority charge stored in the base region of the diode. The diffusion capacitance due to the minority charge stored in the emitter region is usually small in comparison. [The effect of heavy doping, when included, will increase the amount of stored charge and hence the diffusion capacitance. Since the heavy-doping effect is larger in the more heavily doped emitter than in the base, it will make the ratio $C_{Dn}/C_{Dp}$ smaller than that given by Eq. (2.143). Evaluation of the diffusion capacitance including the heavy-doping effect is left to Exercise 2.18.]

## 2.3 MOS CAPACITORS

The metal–oxide–semiconductor (MOS) structure is the basis of CMOS technology. The Si–$SiO_2$ MOS system has been studied extensively (Nicollian and Brews, 1982) because it is directly related to most planar devices and integrated circuits. In this section, we review the fundamental properties of MOS capacitors and the basic equations that govern their operation. The effects of charges in the oxide layer and at the oxide–silicon interface are discussed in Section 2.3.6.

### 2.3.1 SURFACE POTENTIAL: ACCUMULATION, DEPLETION, AND INVERSION

#### 2.3.1.1 ENERGY-BAND DIAGRAM OF AN MOS SYSTEM

The cross section of an MOS capacitor is shown in Fig. 2.21. It consists of a conducting gate electrode (metal or heavily doped polysilicon) on top of a thin layer of silicon dioxide grown on a silicon substrate. The energy-band diagrams of the three components are shown in Fig. 2.22. Silicon dioxide is an insulator with a large energy gap in the range of 8–9 eV. It is convenient to relate the band structures of all three materials to a common reference potential, the *vacuum level*. The vacuum

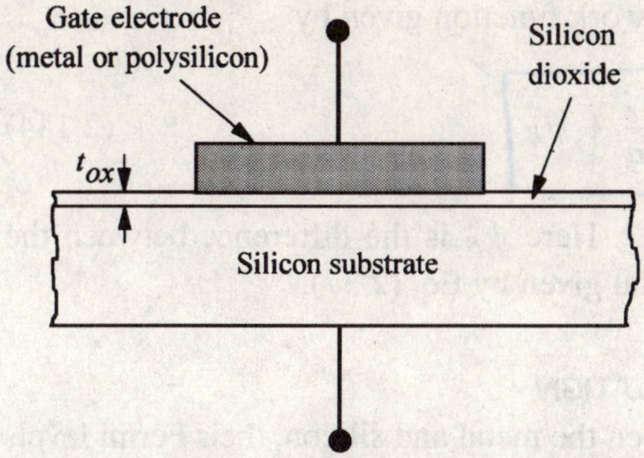


**FIGURE 2.21.** Schematic cross section of an MOS capacitor.

level is defined as the energy level at which the electron is free, i.e., no longer bonded to the lattice. In silicon, the vacuum level is 4.05 eV above the conduction band, as shown in Fig. 2.22. In other words, an electron at the conduction-band edge must gain a kinetic energy of 4.05 eV (called the *electron affinity*, $\chi$) in order to break loose from the crystal field of silicon. In silicon dioxide, the vacuum level is 0.95 eV above its conduction band, which means that ***the potential barrier is 4.05* eV *− 0.95* eV *= 3.1* eV *between the conduction bands of silicon and silicon dioxide***. This figure has important significance when discussing the reliability of Si–$SiO_2$ systems.

In metals, the energy difference between the vacuum level and the Fermi level (remember that the conduction band is half filled in metals) is called the *work function* of the metal. Different metals have different work functions. For simplicity, we start the discussion assuming that the silicon is p-type and that the metal work

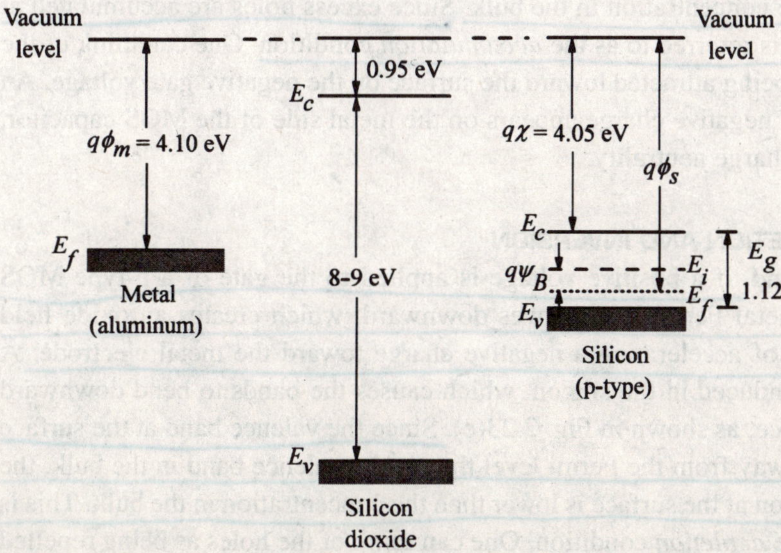


**FIGURE 2.22.** Energy-band diagram of the three components of an MOS capacitor: metal (aluminum), silicon dioxide, and p-type silicon.

function, $\phi_m$, is the same as the silicon work function given by

$$\phi_s = \chi + \frac{E_g}{2q} + \psi_B \tag{2.144}$$

for p-type silicon as shown in Fig. 2.22. Here $\psi_B$ is the difference between the Fermi potential and the intrinsic potential given by Eq. (2.37).

### 2.3.1.2 FLAT BAND AND ACCUMULATION

When there is no applied voltage between the metal and silicon, their Fermi levels line up. Since the work functions are equal, their vacuum levels line up as well, and the bands in both the silicon and the oxide are flat, as shown in Fig. 2.23(a). This is called the *flat-band condition.* There is no charge, no field, and the carrier concentration is at the equilibrium value throughout the silicon. Now consider the case when a negative voltage is applied to the gate of a p-type MOS capacitor, as shown in Fig. 2.23(b). This raises the metal Fermi level (i.e., electron energy) with respect to the silicon Fermi level and creates an electric field in the oxide that would accelerate a negative charge toward the silicon substrate. A field is also induced at the silicon surface in the same direction as the oxide field. Because of the low carrier concentration in silicon (compared with metal), the bands bend upward toward the oxide interface. ***The Fermi level stays flat within the silicon, since there is no net flow of conduction current,*** as was discussed in Section 2.2.1. The potential at the silicon surface is called the *surface potential.* Due to the band bending, the Fermi level at the surface is much closer to the valence band than is the Fermi level in the bulk silicon. This results in a hole concentration much higher at the surface than the equilibrium hole concentration in the bulk. Since excess holes are accumulated at the surface, this is referred to as the *accumulation* condition. One can think of the excess holes as being attracted toward the surface by the negative gate voltage. An equal amount of negative charge appears on the metal side of the MOS capacitor, as required for charge neutrality.

### 2.3.1.3 DEPLETION AND INVERSION

On the other hand, if a positive voltage is applied to the gate of a p-type MOS capacitor, the metal Fermi level moves downward, which creates an oxide field in the direction of accelerating a negative charge toward the metal electrode. A similar field is induced in the silicon, which causes the bands to bend downward toward the surface, as shown in Fig. 2.23(c). Since the valence band at the surface is now farther away from the Fermi level than is the valence band in the bulk, the hole concentration at the surface is lower than the concentration in the bulk. This is referred to as the *depletion* condition. One can think of the holes as being repelled away from the surface by the positive gate voltage. The situation is similar to the depletion layer in a p–n junction discussed in Section 2.2.2. The depletion of holes

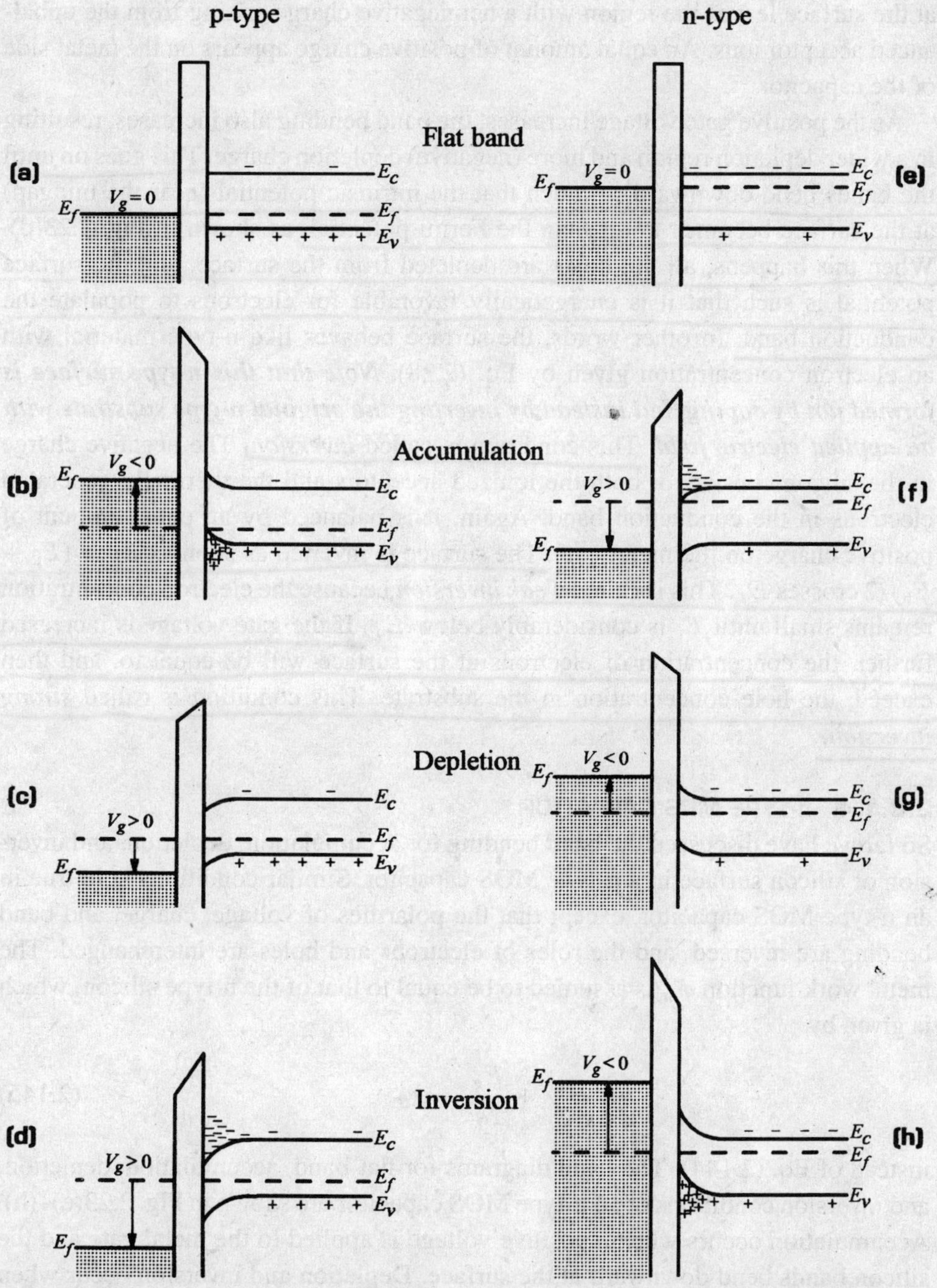


**FIGURE 2.23.** Energy-band diagrams for ideal (a)–(d) p-type and (e)–(h) n-type MOS capacitors under different bias conditions: (a), (e), flat band; (b), (f), accumulation; (c), (g), depletion; (d), (h), inversion. (After Sze, 1981.)

at the surface leaves the region with a net negative charge arising from the unbalanced acceptor ions. An equal amount of positive charge appears on the metal side of the capacitor.

As the positive gate voltage increases, the band bending also increases, resulting in a wider depletion region and more (negative) depletion charge. This goes on until the bands bend downward so much that the intrinsic potential (near the midgap) at the surface becomes lower than the Fermi potential, as shown in Fig. 2.23(d). When this happens, all the holes are depleted from the surface, and the surface potential is such that it is energetically favorable for electrons to populate the conduction band. In other words, the surface behaves like n-type material with an electron concentration given by Eq. (2.38). ***Note that this n-type surface is formed not by doping, but instead by inverting the original p-type substrate with an applied electric field***. This condition is called *inversion*. The negative charge in the silicon consists of both the ionized acceptors and the thermally generated electrons in the conduction band. Again, it is balanced by an equal amount of positive charge on the metal gate. The surface is inverted as soon as $E_i = (E_c + E_v)/2$ crosses $E_f$. This is called *weak inversion* because the electron concentration remains small until $E_i$ is considerably below $E_f$. If the gate voltage is increased further, the concentration of electrons at the surface will be equal to, and then exceed, the hole concentration in the substrate. This condition is called *strong inversion.*

### 2.3.1.4 n-TYPE MOS CAPACITOR

So far we have discussed the band bending for accumulation, depletion, and inversion of silicon surface in a p-type MOS capacitor. Similar conditions hold true in an n-type MOS capacitor, except that the polarities of voltage, charge, and band bending are reversed, and the roles of electrons and holes are interchanged. The metal work function $\phi_m$ is assumed to be equal to that of the n-type silicon, which is given by

$$\phi_s = \chi + \frac{E_g}{2q} - \psi_B, \tag{2.145}$$

instead of Eq. (2.144). The band diagrams for flat-band, accumulation, depletion, and inversion conditions of an n-type MOS capacitor are shown in Fig. 2.23(e)–(h). Accumulation occurs when a positive voltage is applied to the metal gate and the silicon bands bend downward at the surface. Depletion and inversion occur when the gate voltage is negative and the bands bend upward toward the surface.

### 2.3.1.5 FIELD RELATIONSHIP AT THE SILICON–OXIDE INTERFACE

In most cases, the field in the oxide is constant, since there is usually no net charge in the oxide, and Poisson's equation becomes $d\mathscr{E}/dx = 0$. ***A simple relationship***

***exists between the field in the oxide, $\mathscr{E}_{ox}$, and the field at the silicon surface, $\mathscr{E}_s$, based on the boundary condition that the normal component of electrical displacement is continuous across the interface***, that is,

$$\varepsilon_{ox}\mathscr{E}_{ox} = \varepsilon_{si}\mathscr{E}_s, \tag{2.146}$$

or $\mathscr{E}_{ox} \approx 3\,\mathscr{E}_s$. This relationship is valid even in the presence of oxide charge, provided that $\mathscr{E}_{ox}$ is defined as the electric field in the oxide at the oxide–silicon interface. Oxide charge will be discussed in Section 2.3.6.

## 2.3.2 ELECTROSTATIC POTENTIAL AND CHARGE DISTRIBUTION IN SILICON

### 2.3.2.1 SOLVING POISSON'S EQUATION

In this sub-subsection, the relations among the surface potential, charge, and electric field are derived by solving Poisson's equation in the surface region of silicon. A more detailed band diagram at the surface of a p-type silicon is shown in Fig. 2.24. The potential $\psi(x) = \psi_i(x) - \psi_i(x = \infty)$ is defined as the amount of band bending at position $x$, where $x = 0$ is at the silicon surface and $\psi_i(x = \infty)$ is the intrinsic potential in the bulk silicon. Remember that $\psi(x)$ is positive when the bands bend downward. The boundary conditions are $\psi = 0$ in the bulk silicon, and $\psi = \psi(0) = \psi_s$ at the surface. The surface potential $\psi_s$ depends on the applied gate voltage, as will be discussed in Section 2.3.3. Poisson's equation, Eq. (2.36), is

$$\frac{d^2\psi}{dx^2} = -\frac{d\mathscr{E}}{dx} = -\frac{q}{\varepsilon_{si}}\left[p(x) - n(x) + N_d^+(x) - N_a^-(x)\right]. \tag{2.147}$$

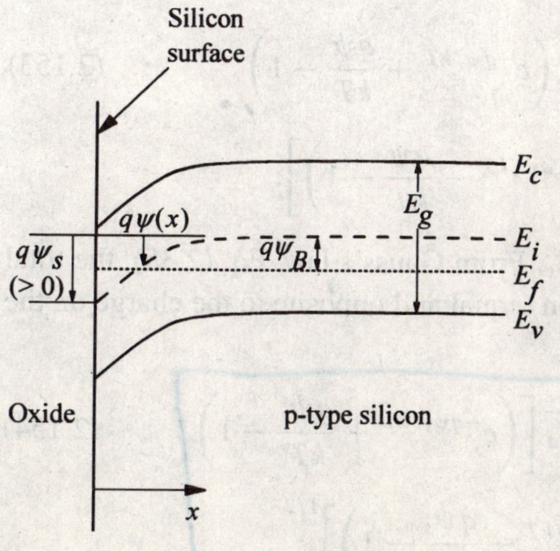


**FIGURE 2.24.** Energy-band diagram near the silicon surface of a p-type MOS device. The band bending $\psi$ is defined as positive when the bands bend downward with respect to the bulk. Accumulation occurs when $\psi_s < 0$. Depletion and inversion occur when $\psi_s > 0$.

In the bulk silicon, $p = N_a$, $n = n_i^2/N_a$. Charge neutrality condition for a uniformly doped p-type silicon requires

$$N_d^+(x) - N_a^-(x) = -N_a + \frac{n_i^2}{N_a}. \tag{2.148}$$

In the surface region, $p(x)$ and $n(x)$ are given by Eq. (2.39) and Eq. (2.38), which can be expressed in terms of $\psi$ as

$$p(x) = n_i e^{q(\psi_f - \psi_i)/kT} = n_i e^{q(\psi_B - \psi)/kT} = N_a e^{-q\psi/kT} \tag{2.149}$$

and

$$n(x) = n_i e^{q(\psi_i - \psi_f)/kT} = n_i e^{q(\psi - \psi_B)/kT} = \frac{n_i^2}{N_a} e^{q\psi/kT} \tag{2.150}$$

In the above steps, $\psi_B = \psi_f - \psi_i(x = \infty)$ and $\exp(q\psi_B/kT) = N_a/n_i$ from Eq. (2.37) were used.

Substituting the last three equations into Eq. (2.147) yields

$$\frac{d^2\psi}{dx^2} = -\frac{q}{\varepsilon_{si}}\left[N_a(e^{-q\psi/kT} - 1) - \frac{n_i^2}{N_a}(e^{q\psi/kT} - 1)\right]. \tag{2.151}$$

Multiplying $(d\psi/dx)\,dx$ on both sides of Eq. (2.151) and integrating from the bulk ($\psi = 0$, $d\psi/dx = 0$) toward the surface, one obtains

$$\int_0^{d\psi/dx} \frac{d\psi}{dx}\, d\left(\frac{d\psi}{dx}\right) \tag{2.152}$$

$$= -\frac{q}{\varepsilon_{si}} \int_0^{\psi} \left[N_a(e^{-q\psi/kT} - 1) - \frac{n_i^2}{N_a}(e^{q\psi/kT} - 1)\right] d\psi,$$

which gives the electric field at $x$, $\mathscr{E} = -d\psi/dx$, in terms of $\psi$:

$$\mathscr{E}^2(x) = \left(\frac{d\psi}{dx}\right)^2 = \frac{2kTN_a}{\varepsilon_{si}}\left[\left(e^{-q\psi/kT} + \frac{q\psi}{kT} - 1\right) + \frac{n_i^2}{N_a^2}\left(e^{q\psi/kT} - \frac{q\psi}{kT} - 1\right)\right]. \tag{2.153}$$

At $x = 0$, we let $\psi = \psi_s$ and $\mathscr{E} = \mathscr{E}_s$. From Gauss's law, Eq. (2.35), the total charge per unit area induced in the silicon (equal and opposite to the charge on the metal gate) is

$$Q_s = -\varepsilon_{si}\mathscr{E}_s = \pm\sqrt{2\varepsilon_{si}kTN_a}\left[\left(e^{-q\psi_s/kT} + \frac{q\psi_s}{kT} - 1\right) + \frac{n_i^2}{N_a^2}\left(e^{q\psi_s/kT} - \frac{q\psi_s}{kT} - 1\right)\right]^{1/2} \tag{2.154}$$

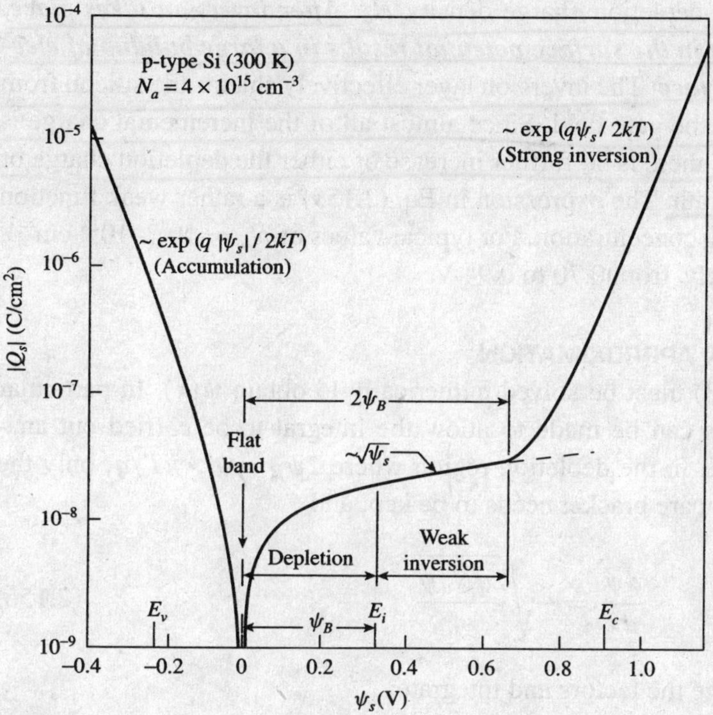


**FIGURE 2.25.** Variation of total charge density (fixed plus mobile) in silicon as a function of surface potential $\psi_s$ for a p-type MOS device. (After Sze, 1981.)

This function is plotted in Fig. 2.25. At the flat-band condition, $\psi_s = 0$ and $Q_s = 0$. In accumulation, $\psi_s < 0$ (bands bending upward) and the first term in the square bracket dominates once $-q\psi_s/kT > 1$. The accumulation charge density is then proportional to $\exp(-q\psi_s/2kT)$ as indicated in the figure. In depletion, $\psi_s > 0$ and $q\psi_s/kT > 1$, but $\exp(q\psi_s/kT)$ is not large enough to make the $n_i^2/N_a^2$ term appreciable. Therefore, the $q\psi_s/kT$ term in the square bracket dominates and the negative depletion charge density (from ionized acceptor atoms) is proportional to $\psi_s^{1/2}$. When $\psi_s$ increases further, the $(n_i^2/N_a^2)\exp(q\psi_s/kT)$ term eventually becomes larger than the $q\psi_s/kT$ term and dominates the square bracket. This is when inversion occurs. The negative inversion charge density is proportional to $\exp(q\psi_s/2kT)$ as indicated in Fig. 2.25.

*A popular criterion for the onset of strong inversion is for the surface potential to reach a value such that* $(n_i^2/N_a^2)\exp(q\psi_s/kT) = 1$, i.e.,

$$\psi_s(\text{inv}) = 2\psi_B = 2\frac{kT}{q}\ln\left(\frac{N_a}{n_i}\right). \tag{2.155}$$

Under this condition, the electron concentration given by Eq. (2.150) at the surface becomes equal to the depletion charge density $N_a$. ***After inversion takes place, even a slight increase in the surface potential results in a large buildup of electron density at the surface.*** The inversion layer effectively shields the silicon from further penetration of the gate field. Since almost all of the incremental charge is taken up by electrons, there is no further increase of either the depletion charge or the depletion-layer width. The expression in Eq. (2.155) is a rather weak function of the substrate doping concentration. For typical values of $N_a = 10^{16}$–$10^{18}$ cm$^{-3}$, $2\psi_B$ varies only slightly, from 0.70 to 0.94 V.

### 2.3.2.2 DEPLETION APPROXIMATION

In general, Eq. (2.153) must be solved numerically to obtain $\psi(x)$. In particular cases, approximations can be made to allow the integral to be carried out analytically. For example, in the depletion region where $2\psi_B > \psi > kT/q$, only the $q\psi/kT$ term in the square bracket needs to be kept and

$$\frac{d\psi}{dx} = -\sqrt{\frac{2qN_a\psi}{\varepsilon_{si}}}. \tag{2.156}$$

One can then rearrange the factors and integrate:

$$\int_{\psi_s}^{\psi} \frac{d\psi}{\sqrt{\psi}} = -\int_0^x \sqrt{\frac{2qN_a}{\varepsilon_{si}}}\, dx, \tag{2.157}$$

where $\psi_s$ is the surface potential at $x = 0$ as assumed before. Therefore,

$$\psi = \psi_s\left(1 - \sqrt{\frac{qN_a}{2\varepsilon_{si}\psi_s}}\,x\right)^2, \tag{2.158}$$

which can be written as

$$\psi = \psi_s\left(1 - \frac{x}{W_d}\right)^2. \tag{2.159}$$

This is a parabolic equation with the vertex at $\psi = 0$, $x = W_d$, where

$$W_d = \sqrt{\frac{2\varepsilon_{si}\psi_s}{qN_a}} \tag{2.160}$$

is the depletion-layer width defined as the distance to which the band bending extends. The total depletion charge density in silicon, $Q_d$, is equal to the charge per unit area of ionized acceptors in the depletion region:

$$Q_d = -qN_aW_d = -\sqrt{2\varepsilon_{si}qN_a\psi_s}. \tag{2.161}$$

These results are very similar to those of the one-sided abrupt p–n junction under the depletion approximation, discussed in Section 2.2.2. ***In the MOS case, however,*** $W_d$ ***reaches a maximum value*** $W_{dm}$ ***at the onset of strong inversion when*** $\psi_s = 2\psi_B$. Substituting Eq. (2.155) into Eq. (2.160) gives the maximum depletion width:

$$W_{dm} = \sqrt{\frac{4\varepsilon_{si}kT\ln(N_a/n_i)}{q^2N_a}}. \tag{2.162}$$

### 2.3.2.3 STRONG INVERSION

Beyond strong inversion, the $(n_i^2/N_a^2)\exp(q\psi/kT)$ term representing the inversion charge in Eq. (2.153) becomes appreciable and must be kept, together with the depletion charge term:

$$\frac{d\psi}{dx} = -\sqrt{\frac{2kTN_a}{\varepsilon_{si}}\left(\frac{q\psi}{kT} + \frac{n_i^2}{N_a^2}e^{q\psi/kT}\right)}. \tag{2.163}$$

This equation can only be integrated numerically. The boundary condition is $\psi = \psi_s$ at $x = 0$. After $\psi(x)$ is solved, the electron distribution $n(x)$ in the inversion layer can be calculated from Eq. (2.150). Examples of numerically calculated $n(x)$ are plotted in Fig. 2.26 for two values of $\psi_s$ with $N_a = 10^{16}$ cm$^{-3}$. ***The electrons are distributed extremely close to the surface with an inversion-layer width less than 50 Å.*** A higher surface potential or field tends to confine the electrons even closer to the surface. In general, electrons in the inversion layer must be

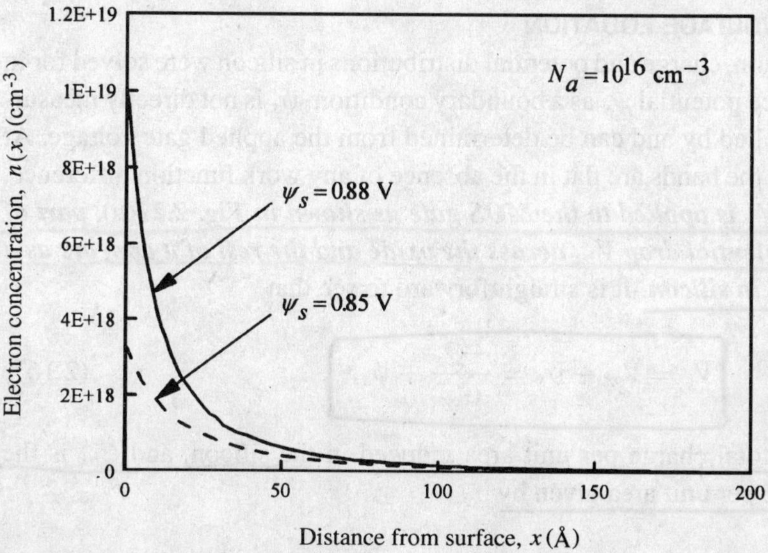


**FIGURE 2.26.** Electron concentration versus distance in the inversion layer of a p-type MOS device.

treated quantum-mechanically as a 2-D gas (Stern and Howard, 1967). According to the quantum-mechanical model, inversion-layer electrons occupy discrete energy bands and have a peak distribution 10–20 Å away from the surface. More details will be discussed in Section 4.2.4.

When the inversion charge density per unit area, $Q_i$, is much greater than the depletion charge density, Eq. (2.154) can be approximated by

$$Q_i = -\sqrt{\frac{2\varepsilon_{si}kTn_i^2}{N_a}e^{q\psi_s/2kT}}. \tag{2.164}$$

Since the electron concentration at the surface is

$$n(0) = \frac{n_i^2}{N_a}e^{q\psi_s/kT}, \tag{2.165}$$

one can write

$$|Q_i| = \sqrt{2\varepsilon_{si}kTn(0)}. \tag{2.166}$$

The effective inversion-layer thickness (classical model) can be estimated from $Q_i/qn(0) = 2\varepsilon_{si}kT/qQ_i$, which is inversely proportional to $Q_i$. Similar expressions also hold true for the surface charge density of extra holes under accumulation, except that the factor $n_i^2/N_a$ is replaced by $N_a$.

## 2.3.3 CAPACITANCES IN AN MOS STRUCTURE

### 2.3.3.1 GATE-VOLTAGE EQUATION

In the last subsection, charge and potential distributions in silicon were solved for in terms of the surface potential $\psi_s$ as a boundary condition. $\psi_s$ is not directly measurable, but is controlled by and can be determined from the applied gate voltage. At zero gate voltage, the bands are flat in the absence of any work function difference. ***When a voltage $V_g$ is applied to the MOS gate as shown in Fig. 2.27(a), part of it appears as a potential drop $V_{ox}$ across the oxide and the rest of it appears as a band bending $\psi_s$ in silicon.*** It is straightforward to see that

$$V_g = V_{ox} + \psi_s = \frac{-Q_s}{C_{ox}} + \psi_s, \tag{2.167}$$

where $Q_s$ is the total charge per unit area induced in the silicon, and $C_{ox}$ is the oxide capacitance per unit area given by

$$C_{ox} = \frac{\varepsilon_{ox}}{t_{ox}} \tag{2.168}$$

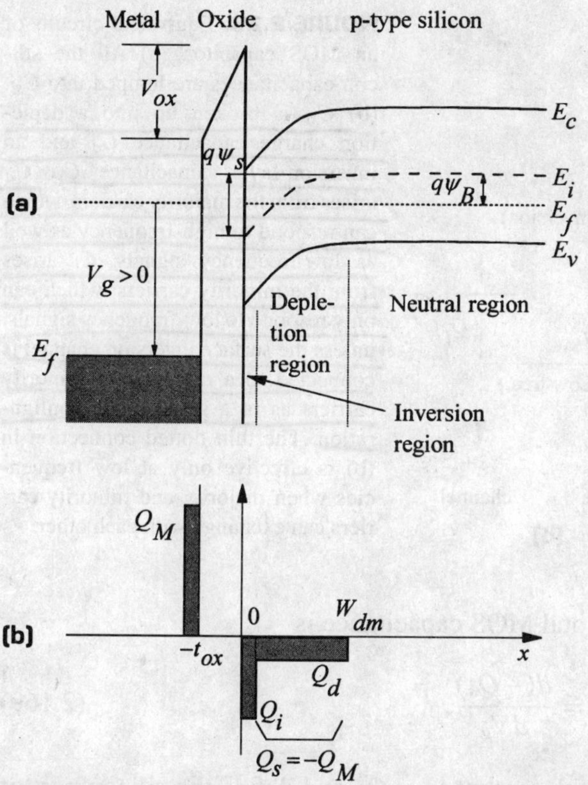


**FIGURE 2.27.** (a) Band diagram of a p-type MOS capacitor with a positive voltage applied to the gate. (b) Charge distribution under inversion condition.

for an oxide thickness $t_{ox}$. There is a negative sign in front of $Q_s$ in Eq. (2.167) because the charge on the metal gate is always equal but opposite to the charge in silicon, i.e., $Q_s$ is negative when $V_g$ is positive and vice versa.

The charge distribution in an MOS capacitor is shown schematically in Fig. 2.27(b), where the total charge $Q_s$ may include both depletion and inversion components. For simplicity of discussion, oxide and interface trapped charges are ignored here. They will be discussed in detail in Sections 2.3.6 and 2.3.7. In general, $Q_s$ is a function of $\psi_s$ given by Eq. (2.154), and plotted in Fig. 2.25. Equation (2.167) is then an implicit equation that can be solved for $\psi_s$. Under the depletion condition, $Q_s(\psi_s) = Q_d(\psi_s)$ given by Eq. (2.161), an analytical expression for $\psi_s$ can be found by solving a quadratic equation. Under extreme accumulation and inversion conditions, $-Q_s \approx C_{ox}V_g$, since both $V_g$ and $V_{ox}$ can be much larger than the silicon bandgap, $E_g/q = 1.12$ V, while $\psi_s$ is always less than $E_g$.

### 2.3.3.2 DEFINITION OF SMALL-SIGNAL CAPACITANCES

We now consider the capacitances in an MOS structure. In most cases, MOS capacitances are defined as small-signal capacitances and can easily be measured by applying a small ac voltage on top of a dc bias across the structure and sensing the out-of-phase ac current at the same frequency (the in-phase component gives the

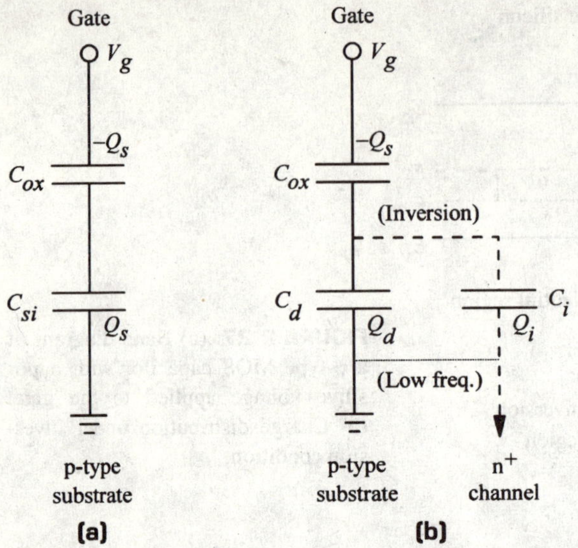


**FIGURE 2.28.** Equivalent circuits of an MOS capacitor. (a) All the silicon capacitances are lumped into $C_{si}$. (b) $C_{si}$ is broken up into a depletion charge capacitance $C_d$ and an inversion-layer capacitance $C_i$. $C_d$ arises from the majority carriers, which can respond to high-frequency as well as low-frequency signals. $C_i$ arises from the minority carriers, which can only respond to low-frequency signals, unless the surface inversion channel is connected to a reservoir of minority carriers as in a gated diode configuration. The thin dotted connection in (b) is effective only at low frequencies when majority and minority carriers can exchange with each other.

small-signal conductance). The total MOS capacitance is

$$C = \frac{d(-Q_s)}{dV_g}. \tag{2.169}$$

If we differentiate Eq. (2.167) with respect to $-Q_s$ and define the silicon part of the capacitance as

$$C_{si} = \frac{d(-Q_s)}{d\psi_s}, \tag{2.170}$$

we obtain

$$\frac{1}{C} = \frac{1}{C_{ox}} + \frac{d\psi_s}{d(-Q_s)} = \frac{1}{C_{ox}} + \frac{1}{C_{si}}. \tag{2.171}$$

In other words, the total capacitance equals the oxide capacitance and the silicon capacitance connected in series. The capacitances are defined in such a way that they are all positive quantities. An equivalent circuit is shown in Fig. 2.28(a). In reality, there is also an interface trap capacitance in parallel with $C_{si}$. It arises from charging and discharging of $Si–SiO_2$ interface traps and will be discussed in more detail in Section 2.3.7.

#### 2.3.3.3 CAPACITANCE–VOLTAGE CHARACTERISTICS: ACCUMULATION

A typical capacitance-versus-gate-voltage ($C$–$V$) curve of a p-type MOS capacitor is plotted in Fig. 2.29. In fact, there are several different curves, depending on the frequency of the applied ac signal. We start with the "low-frequency" or *quasistatic*

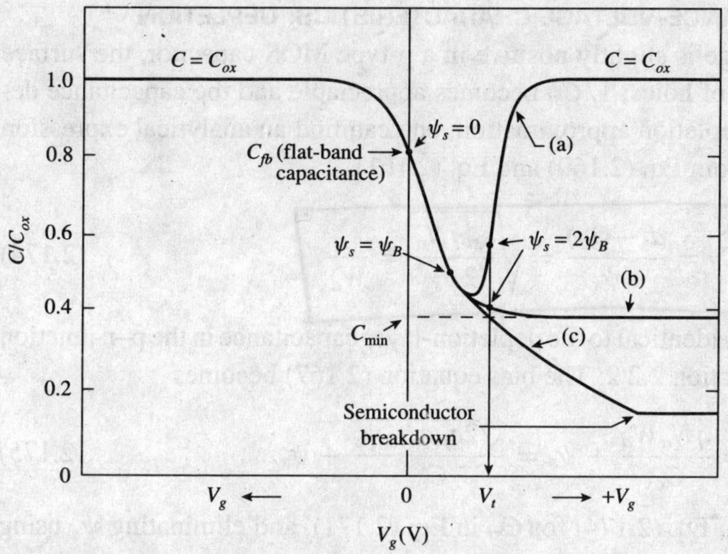


**FIGURE 2.29.** MOS capacitance–voltage curves: (a) low frequency, (b) high frequency, (c) deep depletion. (After Sze, 1981.)

$C$–$V$ curve. When the gate voltage is negative, the p-type MOS capacitor is in accumulation and $Q_s \propto \exp(-q\psi_s/2kT)$, as shown in Fig. 2.25. Therefore, $C_{si} = -dQ_s/d\psi_s = (q/2kT)Q_s = (q/2kT)C_{ox}|V_g - \psi_s|$, and the MOS capacitance is

$$\frac{1}{C} = \frac{1}{C_{ox}}\left(1 + \frac{2kT/q}{|V_g - \psi_s|}\right). \tag{2.172}$$

Since $2kT/q \approx 0.052$ V and $\psi_s$ is limited to 0.1 to 0.3 V in accumulation, the MOS capacitance rapidly approaches $C_{ox}$ when the gate voltage is 1–2 V more negative than the flat-band voltage.

#### 2.3.3.4 CAPACITANCE AT FLAT BAND

When the gate bias is zero, the MOS is near the flat-band condition; therefore, $q\psi_s/kT \ll 1$. The inversion charge term in Eq. (2.154) can be neglected and the first exponential term can be expanded into a power series. Keeping only the first three terms of the series, one obtains $Q_s = -(\varepsilon_{si}q^2N_a/kT)^{1/2}\psi_s$. From Eq. (2.171), the flat-band capacitance is given by

$$\frac{1}{C_{fb}} = \frac{1}{C_{ox}} + \sqrt{\frac{kT}{\varepsilon_{si}q^2N_a}} = \frac{1}{C_{ox}} + \frac{L_D}{\varepsilon_{si}}, \tag{2.173}$$

where $L_D$ is the Debye length defined in Eq. (2.42). $C_{fb}$ is somewhat less than $C_{ox}$.

#### 2.3.3.5 CAPACITANCE–VOLTAGE CHARACTERISTICS: DEPLETION

When the gate voltage is slightly positive in a p-type MOS capacitor, the surface starts to be depleted of holes; $1/C_{si}$ becomes appreciable and the capacitance decreases. Using the depletion approximation, one can find an analytical expression for $C$ in this case. From Eq. (2.160) and Eq. (2.161),

$$C_d = \frac{d(-Q_d)}{d\psi_s} = \sqrt{\frac{\varepsilon_{si} q N_a}{2\psi_s}} = \frac{\varepsilon_{si}}{W_d}. \tag{2.174}$$

The last expression is identical to the depletion-layer capacitance in the p–n junction case discussed in Section 2.2.2. The bias equation (2.167) becomes

$$V_g = \frac{q N_a W_d}{C_{ox}} + \psi_s = \frac{\sqrt{2\varepsilon_{si} q N_a \psi_s}}{C_{ox}} + \psi_s. \tag{2.175}$$

Substituting $C_d$ from Eq. (2.174) for $C_{si}$ in Eq. (2.171), and eliminating $\psi_s$ using Eq. (2.175), one obtains

$$C = \frac{C_{ox}}{\sqrt{1 + \left(2C_{ox}^2 V_g / \varepsilon_{si} q N_a\right)}}. \tag{2.176}$$

This equation shows how the MOS capacitance decreases with increasing $V_g$ under the depletion condition. It serves as a good approximation to the middle portions of the $C$–$V$ curves in Fig. 2.29, provided that the MOS capacitor is not biased near the flat-band or the inversion condition.

#### 2.3.3.6 LOW-FREQUENCY *C–V* CHARACTERISTICS: INVERSION

As the gate voltage increases further, however, the capacitance stops decreasing when $\psi_s = 2\psi_B$ [Eq. (2.155)] is reached and inversion occurs. ***Once the inversion layer forms, the capacitance starts to increase, since $C_{si}$ is now given by the variation of the inversion charge with respect to $\psi_s$, which is much larger than the depletion capacitance***. Assuming that the silicon charge is dominated by the inversion charge, one can carry out an approximation as in the accumulation case and show that the MOS capacitance in strong inversion is also given by Eq. (2.172). One difference is that $\psi_s$ at inversion is in the range of 0.7 to 1.0 V, significantly higher than that at accumulation. In any case, the capacitance rapidly increases back to $C_{ox}$ when the gate voltage is more than 2 to 3 V beyond the flat-band voltage, as shown in the low-frequency $C$–$V$ curve (a) in Fig. 2.29.

#### 2.3.3.7 HIGH-FREQUENCY CAPACITANCE–VOLTAGE CHARACTERISTICS

The above discussion of the low-frequency MOS capacitance assumes that the minority carrier, i.e., the inversion charge, is able to follow the applied ac signal. This is true only if the frequency of the applied signal is lower than the reciprocal of the minority-carrier response time. The minority-carrier response time can be estimated from the generation–recombination current density, $J_R = qn_i W_d/\tau$, where $\tau$ is the minority-carrier lifetime discussed in Section 2.1.4. The time it takes to generate enough minority carriers to replace something comparable to the depletion charge, $Q_d = qN_a W_d$, is on the order of $Q_d/J_R = (N_a/n_i)\tau$ (Jund and Poirier, 1966). This is typically 0.1–10 s. Therefore, ***for frequencies higher than 100 Hz or so, the inversion charge cannot respond to the applied ac signal***. Only the depletion charge (majority carriers) can respond to the signal, which means that the silicon capacitance is given by $C_d$ of Eq. (2.174) with $W_d$ equal to its maximum value, $W_{dm}$, in Eq. (2.162). The high-frequency capacitance thus approaches a constant minimum value, $C_{\min}$, at inversion given by

$$\frac{1}{C_{\min}} = \frac{1}{C_{ox}} + \sqrt{\frac{4kT \ln(N_a/n_i)}{\varepsilon_{si} q^2 N_a}}. \tag{2.177}$$

This is shown in the high-frequency $C$–$V$ curve (b) in Fig. 2.29.

Typically, $C$–$V$ curves are traced by applying a slow-varying ramp voltage to the gate with a small ac signal superimposed on it. However, if the ramp rate is fast enough that the ramping time is shorter than the minority-carrier response time, then there is insufficient time for the inversion layer to form, and the MOS capacitor is biased into deep depletion as shown by curve (c) in Fig. 2.29. In this case, the depletion width can exceed the maximum value given by Eq. (2.162), and the MOS capacitance decreases further below $C_{\min}$ until impact ionization takes place (Sze, 1981). Note that deep depletion is not a steady-state condition. If an MOS capacitor is held under such bias conditions, its capacitance will gradually increase toward $C_{\min}$ as the thermally generated minority charge builds up in the inversion layer until an equilibrium state is established. The time it takes for an MOS capacitor to recover from deep depletion and return to equilibrium is referred to as the *retention time*. It is a good indicator of the defect density in the silicon wafer and is often used to qualify processing tools in a facility.

It is possible to obtain *low-frequency-like* $C$–$V$ curves at high measurement frequencies. One way is to expose the MOS capacitor to intense illumination, which generates a large number of minority carriers in the silicon. Another commonly used technique is to form an $n^+$ region adjacent to the MOS device and connect it electrically to the p-type substrate (Grove, 1967). The $n^+$ region then acts like a reservoir of electrons which can exchange minority carriers freely with the inversion layer. In other words, the $n^+$ region is connected to the *surface channel* of the inverted MOS device. This structure is similar to that of a gated diode, to be discussed in Section 2.3.5. The small-signal capacitance between the gate and the channel ($n^+$) is given by the series combination of $C_{ox}$ and the *inversion-layer capacitance* $C_i$ as shown in Fig. 2.28(b), where $C_i$ can be expressed as

$$C_i = \frac{d(-Q_i)}{d\psi_s} = \frac{|Q_i|}{2kT/q} \tag{2.178}$$

with the help of Eq. (2.164). It is possible to measure $C_d$ and $C_i$ separately in a *split C–V* setup (Sodini *et al.*, 1982). Such a technique is used, for example, in channel mobility measurements where the inversion charge density must be determined accurately.

### 2.3.4 POLYSILICON WORK FUNCTION AND DEPLETION EFFECTS

#### 2.3.4.1 WORK FUNCTION AND FLAT-BAND VOLTAGE OF POLYSILICON GATES

In Section 2.3 so far, we have considered an ideal MOS capacitor in which there is no oxide charge and no work-function difference between the metal and silicon (Fig. 2.22). In general, however, the metal work function is not the same as the silicon work function and there is a nonzero oxide field with some band bending in silicon at zero gate voltage. This is shown, for example, for an n$^+$-polysilicon-gated p-type MOS capacitor in Fig. 2.30(a). The Fermi level of the heavily doped n$^+$ polysilicon is near the conduction-band edge, which must line up with the Fermi level of the p-type silicon at zero gate voltage. This causes the vacuum level of the bulk p-type silicon to be higher in electron energy than the vacuum level of the n$^+$ polysilicon gate by an amount equal to $E_g/2q + \psi_B$. Therefore, there is an oxide field in the direction of accelerating electrons toward the gate, and at the same

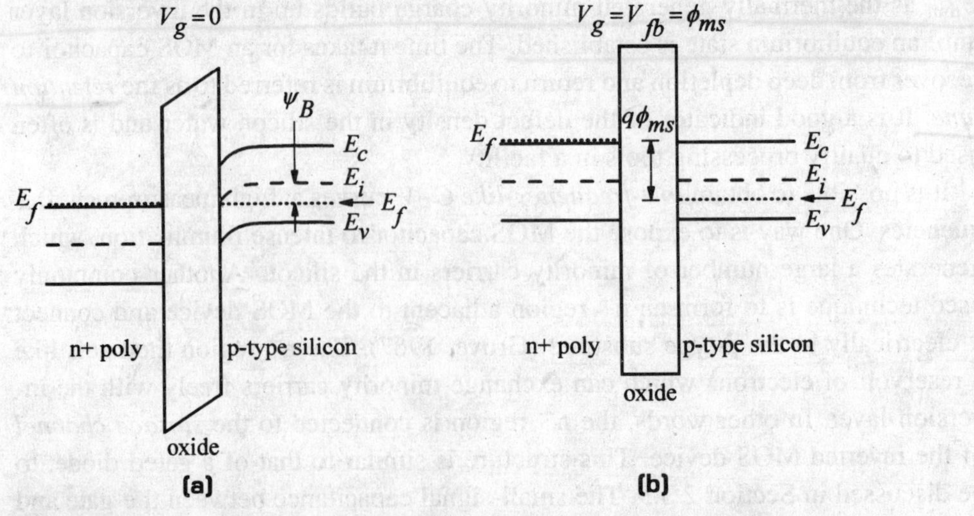


**FIGURE 2.30.** Band diagram of an n$^+$-polysilicon-gated p-type MOS capacitor biased at (a) zero gate voltage and (b) flat-band condition.

time the silicon bands bend downward (depletion) at the surface to produce a field in the same direction. To restore the flat-band condition, a negative voltage equal to the work function difference must be applied to the gate, as shown in Fig. 2.30(b). This voltage is called the *flat-band voltage*.

In general, if the work-function difference is $\phi_{ms} \equiv \phi_m - \phi_s$ and if the equivalent oxide charge per unit area at the oxide–silicon interface is $Q_{ox}$ (defined in Section 2.3.7), the flat-band voltage is given by

$$V_{fb} = \phi_{ms} - \frac{Q_{ox}}{C_{ox}}. \tag{2.179}$$

All the $C$–$V$ curves discussed in the last subsection are shifted by $V_{fb}$, and the gate voltage equation, Eq. (2.167), becomes

$$V_g = V_{fb} + \psi_s - \frac{Q_s}{C_{ox}}. \tag{2.180}$$

The rest of the discussions remain valid.

In modern VLSI technologies, $Q_{ox}/q$ at the Si–SiO$_2$ interface can be controlled to low $10^{10}$ cm$^{-2}$ (positive) for $\langle 100 \rangle$-oriented surfaces. Its contribution to the flat-band voltage is less than 50 mV for thin gate oxides used in 1-μm technology and below ($t_{ox} \leq 20$ nm). Therefore, the flat-band voltage is mainly determined by the work-function difference. Of technological importance are two cases: n$^+$ polysilicon gate on p-type silicon for n-channel MOSFET, and p$^+$ polysilicon gate on n-type silicon for p-channel MOSFET. From Eq. (2.144), the work-function difference for an n$^+$ polysilicon gate on a p-type substrate of doping concentration $N_a$ is

$$\phi_{ms} = -\frac{E_g}{2q} - \psi_B = -0.56 - \frac{kT}{q} \ln\left(\frac{N_a}{n_i}\right) \tag{2.181}$$

in volts. Similarly, the work-function difference for a p$^+$ polysilicon gate on an n-type substrate of doping concentration $N_d$ is

$$\phi_{ms} = \frac{E_g}{2q} + \psi_B = 0.56 + \frac{kT}{q} \ln\left(\frac{N_d}{n_i}\right), \tag{2.182}$$

which is symmetric to Eq. (2.181). These relations have significant implications on the scalability of MOSFET devices, as will be discussed in Chapter 4.

#### 2.3.4.2 POLYSILICON-GATE DEPLETION EFFECTS

The use of polysilicon gates is a key advance in modern CMOS technology, since it allows the source and drain regions to be self-aligned to the gate, thus eliminating parasitics from overlay errors (Kerwin *et al.*, 1969). However, if the polysilicon gate is not doped heavily enough, problems can arise from depletion of the gate itself.

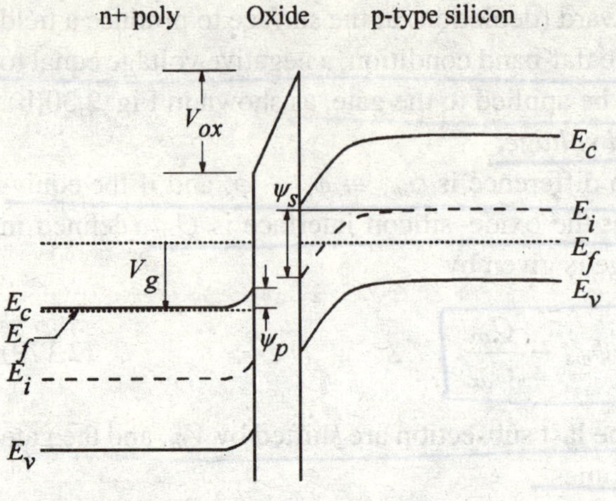


**FIGURE 2.31.** Band diagram showing polysilicon-gate depletion effects when a positive voltage is applied to the $n^+$ polysilicon gate of a p-type MOS capacitor.

This is especially a concern with the dual $n^+$–$p^+$ polysilicon-gate process in which the gates are doped by ion implantation (Wong *et al.*, 1988). Gate depletion results in an additional capacitance in series with the oxide capacitance, which in turn leads to a reduced inversion-layer charge density and degradation of the MOSFET transconductance.

The analysis of MOS capacitance in the last subsection can be extended to include the polysilicon depletion effect and quantify certain observed features in the $C$–$V$ characteristics. Consider the band diagram of an $n^+$-polysilicon-gated p-type MOS capacitor biased into inversion as shown in Fig. 2.31. Since the oxide field points in the direction of accelerating a negative charge toward the gate, the bands in the $n^+$ polysilicon bend slightly upward toward the oxide interface. This depletes the surface of electrons and forms a thin space-charge region in the polysilicon layer, which lowers the total capacitance.

#### 2.3.4.3 EFFECT OF POLYSILICON DOPING CONCENTRATION ON *C–V* CHARACTERISTICS

Typical low-frequency $C$–$V$ curves in the presence of gate depletion effects are shown in Fig. 2.32 (Rios and Arora, 1994). A distinct feature is that the capacitance at inversion does not return to the full oxide capacitance as in Fig. 2.29. Instead, the inversion capacitance exhibits a maximum value somewhat less than $C_{ox}$, depending on the effective doping concentration of the polysilicon gate. The higher the doping concentration, the less the gate depletion effect is and the closer the maximum capacitance is to the oxide capacitance.

The existence of a local maximum in the low-frequency $C$–$V$ curve can be understood semiquantitatively as follows. In Fig. 2.31, we assume $\psi_s$ to be the amount of band bending in the bulk silicon and $\psi_p$ to be that in the $n^+$ polysilicon. From charge neutrality, the total charge density $Q_p$ of the ionized donors in the depletion region of the $n^+$ polysilicon gate is equal and opposite to the combined inversion and depletion charge density $Q_s$ in the silicon substrate, i.e., $Q_p = -Q_s$.

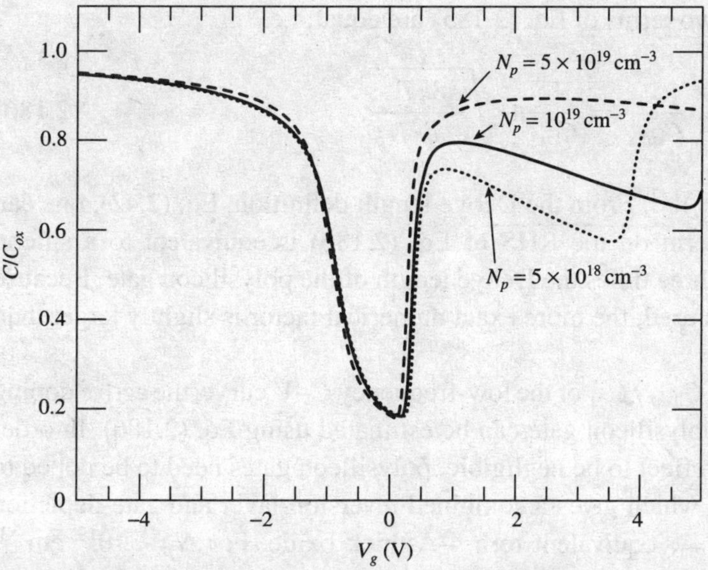


**FIGURE 2.32.** Low-frequency $C$–$V$ curves of a p-type MOS capacitor with n$^+$ polysilicon gate doped at several different concentrations. (After Rios and Arora, 1994.)

The bias equation for an applied voltage $V_g$ is obtained by adding an additional term, $\psi_p$, for the band bending in the polysilicon gate to Eq. (2.180):

$$V_g = V_{fb} + \psi_s + \psi_p - \frac{Q_s}{C_{ox}}. \tag{2.183}$$

Differentiating Eq. (2.183) with respect to $-Q_s$ and using the capacitance definitions (2.169) and (2.170), we obtain

$$\frac{1}{C} = \frac{1}{C_{ox}} + \frac{1}{C_{si}} + \frac{1}{C_p}, \tag{2.184}$$

where $C_p = -dQ_s/d\psi_p = dQ_p/d\psi_p$ is the capacitance of the polysilicon depletion region.

When the p-type substrate is biased in strong inversion, the low-frequency capacitance is given by $C_{si} = |Q_s|/(2kT/q) = Q_p/(2kT/q)$ as discussed with Eq. (2.178). From Eq. (2.174) and Eq. (2.161), the polysilicon depletion capacitance can be expressed in terms of the depletion charge density $Q_p$ as $C_p = \varepsilon_{si} q N_p / Q_p$ (depletion approximation), where $N_p$ is the doping concentration of the polysilicon gate. Substituting these expressions into Eq. (2.184) yields

$$\frac{1}{C} = \frac{1}{C_{ox}} + \frac{2kT/q}{Q_p} + \frac{Q_p}{\varepsilon_{si} q N_p}. \tag{2.185}$$

As $V_g$ becomes more positive, $C_{si}$ ($\propto Q_p$) increases but $C_p$ ($\propto 1/Q_p$) decreases. This results in a local maximum of the low-frequency capacitance at a $V_g$ or $Q_p$

value where the last two terms of Eq. (2.185) are equal; i.e.,

$$\frac{1}{C_{\max}} = \frac{1}{C_{ox}} + \sqrt{\frac{8kT}{\varepsilon_{si}q^2 N_p}}, \tag{2.186}$$

when $Q_p = (2\varepsilon_{si}kTN_p)^{1/2}$. From the Debye length definition, Eq. (2.42), one can see that the second term on the RHS of Eq. (2.186) is equivalent to a silicon capacitance of about three times the Debye length of the polysilicon gate. Because of the approximations used, the more exact numerical factor is slightly larger than that in Eq. (2.186).

From the measured $C_{\max}/C_{ox}$ of the low-frequency $C$–$V$ curve, the active doping concentration of the polysilicon gate can be estimated using Eq. (2.186). In order for the gate depletion effect to be negligible, polysilicon gates need to be doped to about $1 \times 10^{20}$ cm$^{-3}$, which gives a combined inversion-layer and gate-depletion capacitance at $C = C_{\max}$ equivalent to a 4-Å-thick oxide. For $N_p < 10^{19}$ cm$^{-3}$, another abrupt rise in the MOS capacitance may be observed at a much higher gate voltage, as shown in Fig. 2.32. This is due to the onset of inversion (to p$^+$) at the n$^+$ polysilicon surface. The above analysis is valid for p$^+$ polysilicon gates on n-type silicon as well as for n$^+$ polysilicon gates on p-type silicon. A similar approach can be carried out for n$^+$ polysilicon gates on n-type silicon and p$^+$ polysilicon gates on p-type silicon in which gate depletion occurs when the substrate is accumulated.

### 2.3.5 MOS UNDER NONEQUILIBRIUM AND GATED DIODES

An important building block of VLSI devices is the gated diode, or gate-controlled diode. Consider an MOS capacitor where there is an n$^+$ region adjacent to the gated p-type region (Grove, 1967). The n$^+$ region and the p-type region form an n$^+$–p diode. This structure is shown schematically in Fig. 2.33 and was mentioned briefly at the end of Section 2.3.3.

#### 2.3.5.1 INVERSION CONDITION OF AN MOS UNDER NONEQUILIBRIUM

As discussed in Section 2.2.1, when both the n$^+$ region and the p-type substrate are connected to the same potential (grounded), the p–n junction is in equilibrium and the Fermi level is constant across the p–n junction. If the gate voltage is large enough to invert the p-type surface, which occurs for a surface potential bending of $\psi_s(\text{inv}) = 2\psi_B$, the inverted channel is connected to the n$^+$ region and has the same potential as the n$^+$ region. In other words, ***the electron quasi-Fermi level in the channel region is the same as the Fermi level in the n$^+$ region*** as well as the Fermi level in the p-type substrate. The depletion region now extends from the p–n junction to the region under the gate between the inverted channel and the substrate, as shown in Fig. 2.33(b).

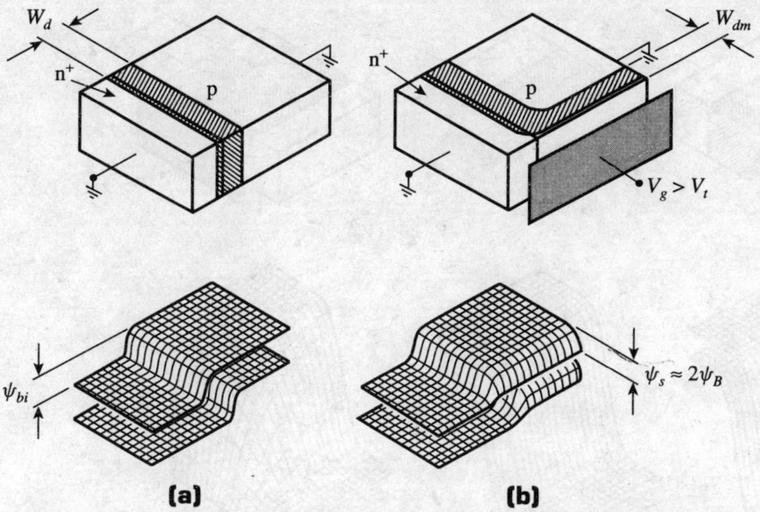


**FIGURE 2.33.** Gated diode or p-type MOS with adjacent n+ region in equilibrium (zero voltage across the p–n junction). The gate is biased at (a) flat-band and (b) inversion conditions. (After Grove, 1967.)

If, on the other hand, the p–n junction is reverse-biased at a voltage $V_R$ as shown in Fig. 2.34, the MOS is in a nonequilibrium condition in which $np \neq n_i^2$. From Eq. (2.89), the electron concentration on the p-type side of the junction is

$$n = \frac{n_i^2}{N_a} e^{-qV_R/kT}, \tag{2.187}$$

since the Fermi level of the n+ region is now $qV_R$ lower than the Fermi level in the p-type substrate. Now consider the case when a positive voltage large enough to bend the bands by $2\psi_B$ is applied to the gate, as shown in Fig. 2.34(b). This brings the conduction band at the surface $2\psi_B$ closer to the electron quasi-Fermi level. From Eq. (2.150), the electron concentration at the surface is increased by a factor of $\exp(2q\psi_B/kT) = (N_a/n_i)^2$ over that of Eq. (2.187), i.e.,

$$n = \frac{n_i^2}{N_a} e^{2q\psi_B/kT} e^{-qV_R/kT} = N_a e^{-qV_R/kT}. \tag{2.188}$$

Since this is much lower than the depletion charge density $N_a$, the surface remains depleted. Even though the positive voltage is sufficient to invert the surface in the equilibrium case, it is not enough to cause inversion in the reverse-biased case. This is because ***the reverse bias lowers the quasi-Fermi level of electrons so that even if the bands at the surface are bent as much as in the equilibrium case in Fig. 2.33(b), the conduction band is still not close enough to the quasi-Fermi level of electrons for inversion to occur.***

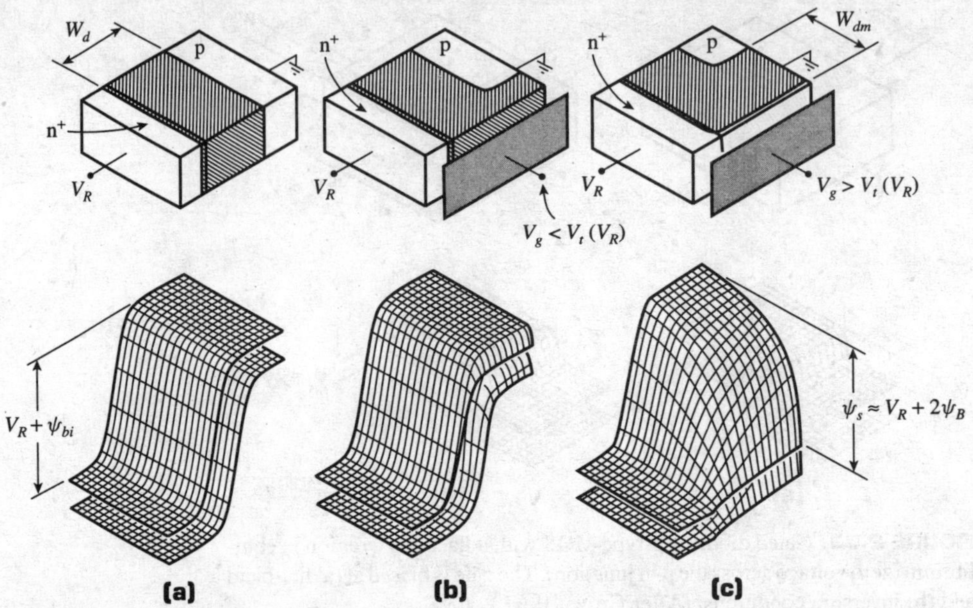


**FIGURE 2.34.** Gated diode or p-type MOS with adjacent n$^+$ region under nonequilibrium (reverse bias across the p–n junction). The gate is biased at (a) flat-band, (b) depletion, and (c) inversion conditions. (After Grove, 1967.)

To reach inversion in the nonequilibrium case, a much larger gate voltage, sufficient to bend the bands by $2\psi_B + V_R$, must be applied. This is the case shown in Fig. 2.34(c), where the electron concentration at the surface is now

$$n = \frac{n_i^2}{N_a} e^{q(2\psi_B + V_R)/kT} e^{-qV_R/kT} = N_a, \tag{2.189}$$

the same as the condition for inversion introduced in Section 2.3.2. Notice that the surface depletion layer is much wider than in the equilibrium case, just as in a reverse-biased p–n junction.

#### 2.3.5.2 BAND BENDING AND CHARGE DISTRIBUTION OF AN MOS UNDER NONEQUILIBRIUM

The above discussions are further illustrated in Fig. 2.35, where the charge distribution and band bending in a cross-section perpendicular to the gate through the neutral p-type region are shown for both the equilibrium and the nonequilibrium cases. The equilibrium case is the same as that discussed in Section 2.3.2. In the nonequilibrium case, the hole quasi-Fermi level is the same as the Fermi level in the bulk p-type silicon, but the electron quasi-Fermi level is dictated by the Fermi level in the n$^+$ region (not shown in Fig. 2.35), which is $qV_R$

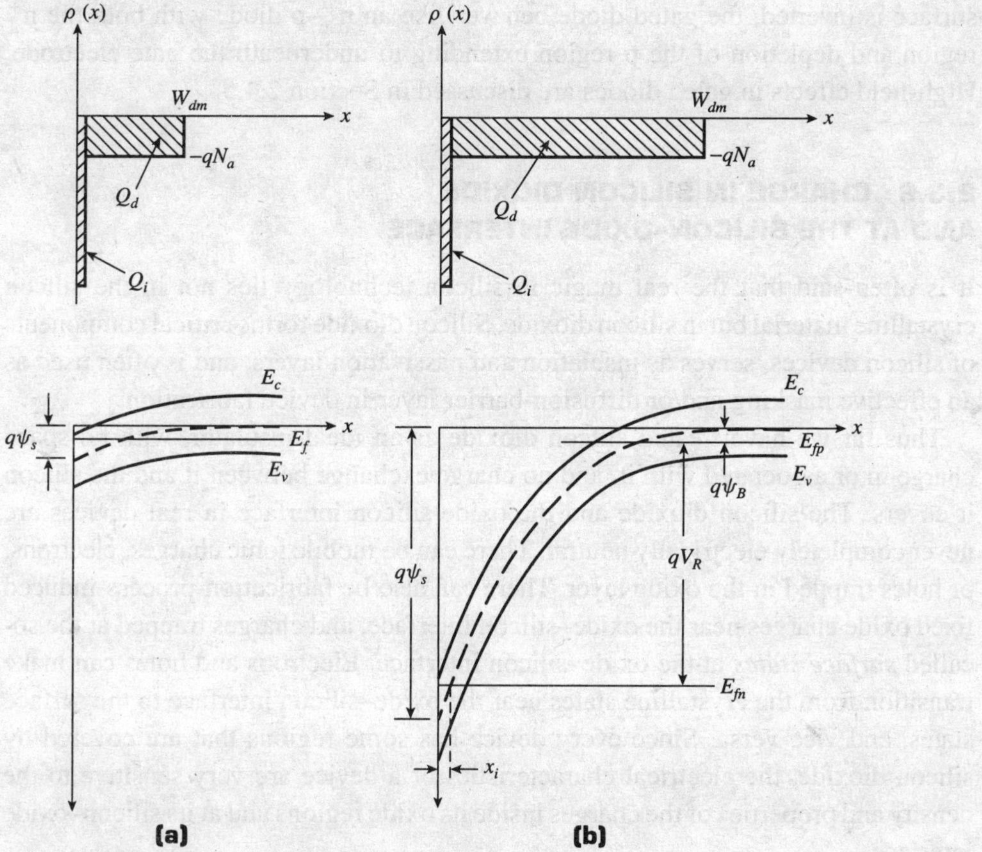


**FIGURE 2.35.** Comparison of charge distribution and energy-band variation of an inverted p-type region for (a) the equilibrium case and (b) the nonequilibrium case. (After Grove, 1967.)

lower than the p-type Fermi level. As a result, surface inversion occurs at a band bending

$$\psi_s(\text{inv}) = V_R + 2\psi_B, \tag{2.190}$$

and the maximum depletion width is a function of the reverse bias $V_R$,

$$W_{dm} = \sqrt{\frac{2\varepsilon_{si}(V_R + 2\psi_B)}{qN_a}}, \tag{2.191}$$

from Eq. (2.160).

When the surface is depleted, the gated diode behaves like an $n^+$–p diode with depletion of the p-region extending to underneath the gate electrode. When the surface is inverted, the gated diode behaves like an $n^+$–p diode with both the $n^+$ region and depletion of the p-region extending to underneath the gate electrode. High-field effects in gated diodes are discussed in Section 2.4.5.

### 2.3.6 CHARGE IN SILICON DIOXIDE AND AT THE SILICON–OXIDE INTERFACE

It is often said that the real magic in silicon technology lies not in the silicon crystalline material but in silicon dioxide. Silicon dioxide forms critical components of silicon devices, serves as insulation and passivation layers, and is often used as an effective masking and/or diffusion-barrier layer in device fabrication.

Thus far we have treated silicon dioxide as an ideal insulator, with no space charge in or associated with it, and no charge exchange between it and the silicon it covers. The silicon dioxide and the oxide-silicon interface in real devices are never completely electrically neutral. There can be mobile ionic charges, electrons, or holes trapped in the oxide layer. There can also be fabrication-process-induced fixed oxide charges near the oxide–silicon interface, and charges trapped at the so-called *surface states* at the oxide–silicon interface. Electrons and holes can make transition from the crystalline states near the oxide–silicon interface to the surface states, and vice versa. Since every device has some regions that are covered by silicon dioxide, the electrical characteristics of a device are very sensitive to the density and properties of the charges inside its oxide regions and at its silicon–oxide interface.

The nomenclature for describing the charges associated with the silicon dioxide in real devices was standardized in 1978 (Deal, 1980). The net charge per unit area is denoted by $Q$. Thus, $Q_m$ denotes the *mobile charge* per unit area, $Q_{ot}$ denotes the *oxide trapped charge* per unit area, $Q_f$ denotes the *fixed oxide charge* per unit area, and $Q_{it}$ denotes the *interface trapped charge* per unit area. The names and location of these charges are illustrated in Fig. 2.36. The properties and characteristics of these charges are discussed further below.

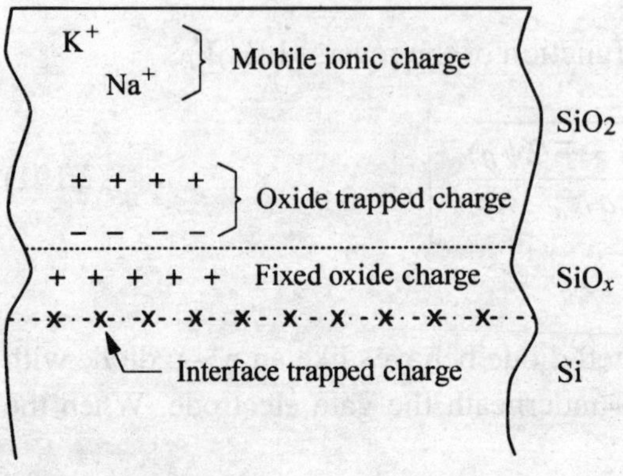


**FIGURE 2.36.** Charges and their location in thermally oxidized silicon. (After Deal, 1980.)

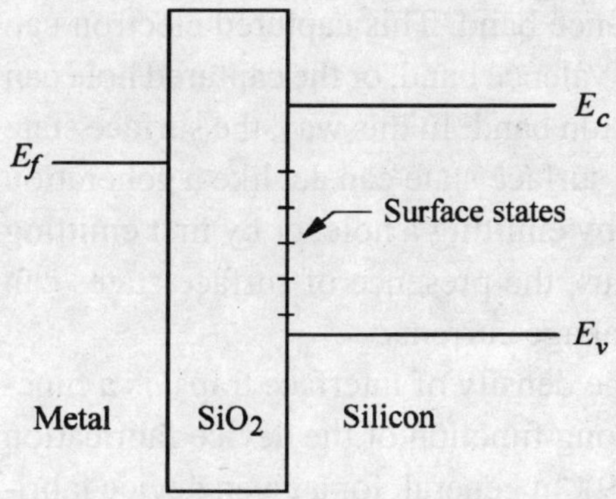


**FIGURE 2.37.** Schematic energy-band diagram of an MOS structure, illustrating the presence of surface states.

### 2.3.6.1 SURFACE STATES AND INTERFACE TRAPPED CHARGE

At the Si–$SiO_2$ interface, the lattice of bulk silicon and all the properties associated with its periodicity terminate. As a result, localized states with energy in the forbidden energy gap of silicon are introduced at or very near the Si–$SiO_2$ interface (Many *et al.*, 1965). These localized surface states are illustrated schematically in Fig. 2.37. Interface trapped charges are electrons or holes trapped in these states.

Just like impurity energy levels in bulk silicon discussed in Section 2.1.2, the probability of occupation of a surface state by an electron or by a hole is determined by the surface-state energy relative to the Fermi level. Thus, as the surface potential is changed, the energy level of a surface state, which is fixed relative to the energy-band edges at the surface, moves with it. This change relative to the Fermi level causes a change in the probability of occupation of the surface state by an electron. For instance, referring to Fig. 2.24, as the bands are bent downward, or as the surface potential is increased, more surface states move below the Fermi level and hence become occupied by electrons. This change of interface trapped charge with a change in the surface potential gives rise to an additional silicon capacitance component, which will be discussed further in Section 2.3.7.

Electrons in silicon but near an oxide–silicon interface can make transitions between the conduction-band states and the surface states. An electron in the conduction band can contribute readily to electrical conduction current, while an electron in a surface state, i.e., an interface trapped electron, does not contribute readily to electrical conduction current, except by *hopping* among the surface states or by first making a transition to the conduction band. Similarly, holes in silicon but near an oxide–silicon interface can make transitions between the valence-band states and the surface states, and trapped interface holes do not contribute readily to electrical conduction. By trapping electrons and holes, surface states can reduce the conduction current in MOSFETs. Furthermore, the trapped electrons and holes can act like charged scattering centers, located at the interface, for the mobile carriers in a surface channel, and thus lower their mobility (Sah *et al.*, 1972).

Surface states can also act like localized generation–recombination centers. Depending on the surface potential, a surface state can first capture an electron from the conduction band, or a hole from the valence band. This captured electron can subsequently recombine with a hole from the valence band, or the captured hole can recombine with an electron from the conduction band. In this way, the surface state acts like a recombination center. Similarly, a surface state can act like a generation center by first emitting an electron followed by emitting a hole, or by first emitting a hole followed by emitting an electron. Thus, the presence of surface states can lead to surface generation–recombination leakage currents.

The density of surface states, and hence the density of interface traps, is a function of silicon substrate orientation and a strong function of the device fabrication process (EMIS, 1988; Razouk and Deal, 1979). In general, for a given device fabrication process, the dependence of the interface trap density on substrate orientation is ⟨100⟩ < ⟨110⟩ < ⟨111⟩. Also, a postmetallization or "final" anneal in hydrogen, or in a hydrogen-containing ambient, at temperatures around 400°C is quite effective in minimizing the density of interface traps. Consequently, ⟨100⟩ silicon and postmetallization anneal in hydrogen are commonly used in modern VLSI device fabrication.

#### 2.3.6.2 FIXED OXIDE CHARGE

Fixed oxide charges are positive charges located in the oxide layer very close to the Si–$SiO_2$ interface. In fact, for modeling purposes, the fixed oxide charges are usually assumed to be located at the Si–$SiO_2$ interface. They are primarily due to excess silicon species introduced during oxidation and during postoxidation heat treatment (Deal *et al.*, 1967). The dependence of the density of fixed oxide charges on substrate orientation is the same as that of interface traps, namely ⟨100⟩ < ⟨110⟩ < ⟨111⟩.

The presence of fixed oxide charges at the oxide–silicon interface affects the potential in the silicon, which will be discussed in the next subsection. In addition, the fixed oxide charges act as charged scattering centers and thus reduce the mobility of the carriers in a surface inversion channel (Sah *et al.*, 1972).

#### 2.3.6.3 MOBILE IONIC CHARGE

Mobile ionic charges in $SiO_2$ are usually due to sodium or potassium contamination introduced during device fabrication. Unlike fixed oxide charges, which are not mobile, $Na^+$ and $K^+$ ions are quite mobile in $SiO_2$ and can be moved from one end of the oxide layer to the other when an electric field is applied across the oxide layer, particularly at somewhat elevated temperatures (>200°C) (Hillen and Verwey, 1986). As these positively charged ions drift close to the Si–$SiO_2$ interface, they repel holes from, and attract electrons to, the silicon surface, often causing unwanted surface electron current to flow among $n^+$ diffusion regions in a p-type substrate or well. Also, when these positively charged ions come close to the silicon surface, they can act as charged scattering centers for the carriers in the surface inversion channel, thus reducing their mobility.

In VLSI fabrication processes, ***mobile-ion contamination problems must be avoided***. This is accomplished by a combination of proper passivation, usually using phosphorosilicate glass, and "clean" fabrication technology (Hillen and Verwey, 1986).

#### 2.3.6.4 OXIDE TRAPPED CHARGE

If electron–hole pairs are generated in an oxide layer, e.g., by ionizing radiation, some of these electrons and holes can be subsequently trapped in the oxide. Also, if electrons or holes are injected into an oxide layer, by tunneling or by hot-carrier injection, some of them can be trapped in the oxide.

Electron and hole traps in $SiO_2$ can easily be introduced by bombardment with high-energy photons or particles (Bourgoin, 1989). Since bombardment by high-energy particles and photons is involved in many steps in the fabrication of modern VLSI devices (during ion implantation, plasma or reactive-ion etching, sputtering deposition, electron-beam evaporation of metal, electron-beam and x-ray lithography, etc.), electron and hole traps are often introduced in the oxide during device fabrication. Fortunately, most of these traps can be eliminated with subsequent anneals at temperatures above 550°C (Ning, 1978).

Also, depending on the oxidation condition, electron traps can be introduced during the oxide growth process itself (EMIS, 1988). For example, oxide growth in moisture-containing ambient is known to introduce electron traps (Nicollian *et al.*, 1971).

- *Capture cross section.* Traps are usually characterized by their *capture cross sections*. Electron traps with cross sections in the range of $10^{-14}$–$10^{-12}$ cm$^2$ are usually *Coulomb-attractive* traps, i.e., the trap centers are positively charged prior to electron capture (Ning *et al.*, 1975; Lax, 1960). Electron traps with cross sections in the $10^{-18}$–$10^{-14}$-cm$^2$ range are usually due to neutral traps (Lax, 1960), and those with cross sections smaller than $10^{-18}$ cm$^2$ are usually associated with *Coulomb-repulsive* traps, i.e., the trap centers are already negatively charged prior to electron capture (Balland and Barbottin, 1989). The potential wells representing these electron traps are illustrated in Fig. 2.38. Since the Coulomb-attractive and neutral centers have the largest capture cross sections, they are also the most important to include when considering the effects of

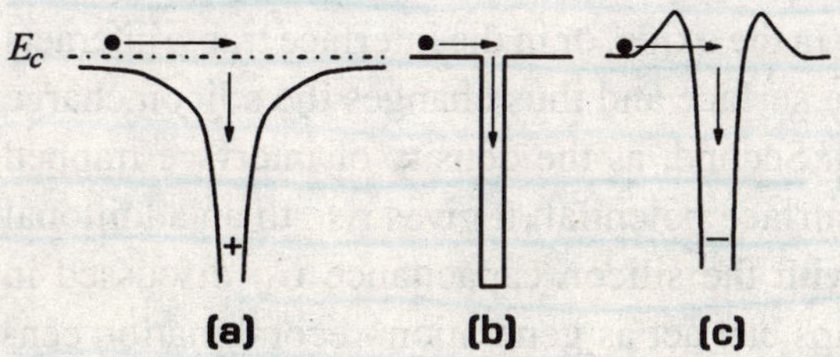


**FIGURE 2.38.** Schematics illustrating the potential wells of electron traps in silicon dioxide: (a) Coulomb-attractive trap, (b) neutral trap, and (c) Coulomb-repulsive trap.

electron traps on device characteristics. Hole traps have not been studied in as much detail as electron traps. This may be due to the fact that holes are very readily trapped when they are injected into an oxide layer (Goodman, 1966). This is consistent with the measured hole capture cross section of about $3 \times 10^{-13}$ cm$^2$, which is as large as the largest electron traps in $SiO_2$ (Ning, 1976a).

- *Temperature dependence.* Consider the capture of a mobile electron into an electron trap. The trapping process has two competing components, namely, capturing the electron into some initial high-energy state of the trap center, and reemitting that same electron from the initial captured state by thermal excitation. If an electron in an initial captured state has a higher probability of cascading down towards the ground state of the trap center than of being reemitted, the electron becomes trapped. On the other hand, if the probability of reemission by thermal excitation from the initial captured state is high enough, trapping will not occur (Lax, 1960). The capture cross section, therefore, decreases with increasing temperature, since the probability for thermal reemission increases with temperature (Lax, 1960).
- *Field dependence.* If an electric field is applied across an oxide layer, it has the effect of increasing the energy of the carriers moving in the oxide layer. As these carriers gain energy from the oxide field, the probability of their being captured in some initial trap state is lowered, since the carriers now must lose more energy in the initial capture process. At the same time, an oxide field has the effect of lowering the energy barriers for the carriers trapped in a potential well, thus increasing the probability for reemitting them from their initial captured states (Lax, 1960). As a result, the capture cross section decreases with increasing oxide field (Ning, 1976b, 1978). ***The commonly used method of injecting carriers into $SiO_2$ by tunneling at high oxide fields tends to underestimate the amount of traps*** in $SiO_2$, since the capture cross sections at such high oxide fields are much smaller than those at normal device operation.

### 2.3.7 EFFECT OF INTERFACE TRAPS AND OXIDE CHARGE ON DEVICE CHARACTERISTICS

The presence of oxide charges and interface traps has three major effects on the characteristics of devices. First, the charge in the oxide, or in the interface traps, interacts with the charge in the silicon near the surface and thus changes the silicon charge distribution and the surface potential. Second, as the density of interface trapped charge changes with changes in the surface potential, it gives rise to an additional capacitance component in parallel with the silicon capacitance $C_{si}$ discussed in Section 2.3.3. Third, the interface traps can act as generation–recombination centers, or assist in the band-to-band tunneling process, and thus contribute to the leakage current in a gated-diode structure. These effects are discussed more quantitatively below.

#### 2.3.7.1 EFFECT OF OXIDE CHARGE ON SURFACE POTENTIAL

As discussed in Section 2.3.2, the charge distribution in silicon is a function of the surface potential. Thus, the effect of oxide charge on the charge distribution in silicon can be described in terms of its effect on surface potential. In the case of an MOS structure, the effect of oxide charge is usually described in terms of the change in gate voltage, which is a readily measurable parameter, necessary to counter the effect of the oxide charge or to restore the surface potential to that of zero oxide charge.

For simplicity of illustration, let us consider an MOS structure biased at flat-band condition. Let us assume that a sheet of oxide charge $Q$ is placed at a distance $x$ from the gate electrode, and a gate voltage $\delta V_g$ has been applied to restore the MOS structure to its original, i.e., flat-band, condition. With the surface potential restored to its original value, the sheet of oxide charge has induced no change in the charge distribution in the silicon, which is a function of the surface potential, but a charge of magnitude $-Q$ on the gate electrode. This is illustrated in Fig. 2.39. Gauss's law in Eq. (2.35) implies that the electric field in the oxide between 0 and $x$ due to the sheet of oxide charge and its image charge on the gate electrode is $-Q/\varepsilon_{ox}$ (see Exercise 2.5). This is also illustrated in Fig. 2.39. The potential difference supporting this electric field is $-xQ/\varepsilon_{ox}$, which is provided by the applied gate voltage. Therefore,

$$\delta V_g = -\frac{xQ}{\varepsilon_{ox}}. \tag{2.192}$$

The gate voltage necessary to offset the effect of an arbitrary oxide charge distribution can be obtained by superposition of individual elements of the charge distribution and applying Eq. (2.192) to each element. For an oxide charge distribution of $\rho(x, \psi_s) = \rho(x) + Q_{it}(\psi_s)\delta(x - t_{ox})$, which consists of an arbitrary

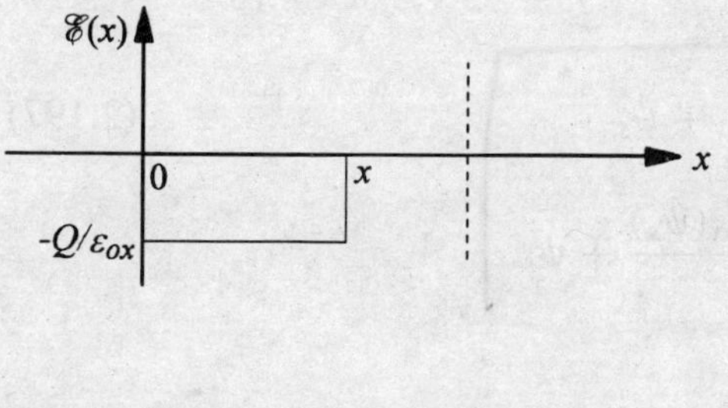


**FIGURE 2.39.** Schematic illustrating the effect of a sheet charge of areal density $Q$ within the oxide layer of an MOS capacitor biased at flat-band condition.

distribution $\rho(x)$ that is independent of the surface potential and a delta-function distribution of the interface trap charge located at $x = t_{ox}$, the gate voltage necessary to offset it is

$$\Delta V_g = \Delta V_g(\psi_s) = -\frac{1}{\varepsilon_{ox}} \int_0^{t_{ox}} x\rho(x, \psi_s)\, dx \tag{2.193}$$

$$= -\frac{1}{\varepsilon_{ox}} \left( \int_0^{t_{ox}} x\rho(x)\, dx + Q_{it}(\psi_s) t_{ox} \right).$$

According to the charge nomenclature discussed in Section 2.3.6, $\rho(x)$ includes the mobile charge, the oxide trap charge, and the fixed oxide charge. It is a common practice to define an *equivalent oxide charge per unit area*, $Q_{ox}$, by

$$Q_{ox} = Q_{ox}(\psi_s) \equiv \int_0^{t_{ox}} \frac{x}{t_{ox}} \rho(x, \psi_s)\, dx \tag{2.194}$$

$$= \int_0^{t_{ox}} \frac{x}{t_{ox}} \rho(x)\, dx + Q_{it}(\psi_s).$$

Equation (2.193) can then be rewritten in the simple form

$$\boxed{\Delta V_g(\psi_s) = -\frac{Q_{ox}(\psi_s)}{C_{ox}},} \tag{2.195}$$

where $C_{ox} = \varepsilon_{ox}/t_{ox}$ is the oxide capacitance per unit area given in Eq. (2.168). Equation (2.195) states that the effect of an arbitrary oxide charge distribution is equivalent to an oxide sheet charge $Q_{ox}(\psi_s)$ located at the oxide–silicon interface.

#### 2.3.7.2 INTERFACE-TRAP CAPACITANCE

In Section 2.3.3, the silicon part of the capacitance is defined without including any interface trapped charge. As the interface traps are filled and emptied in response to changes in the surface potential, they give rise to an *interface-trap capacitance per unit area,* $C_{it}$, defined by

$$\boxed{C_{it}(\psi_s) \equiv \frac{d|Q_{it}(\psi_s)|}{d\psi_s}.} \tag{2.196}$$

In Eq. (2.196), we have indicated explicitly that the interface-trap capacitance is a function of the surface potential.

To include the effect of $Q_{ox}$ in the operation of an MOS capacitor, Eq. (2.167) should be modified by adding to its right-hand side a $\Delta V_g$ term due to $Q_{ox}$. That is, Eq. (2.167) becomes

$$V_g = \Delta V_g(\psi_s) - \frac{Q_s}{C_{ox}} + \psi_s \tag{2.197}$$

$$= -\frac{Q_s(\psi_s) + Q_{ox}(\psi_s)}{C_{ox}} + \psi_s.$$

The total charge on the gate electrode is now $Q_s + Q_{ox}$. Equation (2.169) then becomes

$$C = -\frac{d(Q_s + Q_{ox})}{dV_g}, \tag{2.198}$$

and Eq. (2.171) becomes

$$\frac{1}{C} = \frac{1}{C_{ox}} + \frac{1}{C_{si} + C_{it}}. \tag{2.199}$$

That is, the interface-trap capacitance is in parallel with the silicon capacitance.

As discussed in the previous subsection, the probability of a surface state being filled with an electron is governed by its energy level relative to the Fermi level. ***Only those interface traps that can be filled and emptied at a rate faster than the capacitance-measurement signal can contribute to $C_{it}$***. Traps too slow to follow the capacitance-measurement signal will not contribute to $C_{it}$. Therefore the observed $C_{it}$ is not only a function of the interface-trap density but also a function of the MOS capacitor gate voltage, which controls the surface potential and hence the probability of occupation of the traps, and a function of the frequency at which the capacitance measurement is made. In order for a device to have predictable and reproducible capacitance characteristics, it is important to employ fabrication processes that minimize interface-trap density.

#### 2.3.7.3 SURFACE GENERATION–RECOMBINATION CENTERS

As discussed in the previous subsection, interface states can serve as generation–recombination centers. In the case of a gated-diode structure, the surface generation–recombination current adds to the diode leakage current. The magnitude of the surface leakage current depends on whether or not the surface states are *exposed*, i.e., whether or not the silicon surface is depleted (Grove and Fitzgerald, 1966). If the surface is inverted, the surface states are all filled with minority carriers and do not function efficiently as generation centers. Similarly, if the surface is in accumulation, the surface states are all filled with majority carriers and do not function efficiently as generation centers either. ***Only when the silicon surface is depleted will the surface states function efficiently as generation centers***. Thus, surface leakage current can be suppressed by biasing the gate to keep the silicon surface either in inversion or in accumulation.

As recombination centers, surface states can degrade the minority-carrier lifetime of devices. Consequently, devices where long minority-carrier lifetimes are required are usually designed to confine the minority carriers in them away from the silicon surface. In addition, the device fabrication processes are usually optimized to minimize the density of surface states.

#### 2.3.7.4 SURFACE-STATE- OR TRAP-ASSISTED BAND-TO-BAND TUNNELING

As will be discussed in Section 2.4.2, band-to-band tunneling occurs when the electric field across a p–n junction is sufficiently large. For a gated diode, or for a p–n diode with silicon surface components, the presence of surface states or interface traps in the high-field region can enhance the band-to-band tunneling current very significantly. Thus, gate-induced drain leakage currents in MOSFETs, which will be discussed in Section 2.4.5, and emitter–base diode tunneling currents in bipolar transistors, which will be discussed in Section 6.3.4, depend strongly on the density of interface states at the oxide–silicon interface of these devices.

## 2.4 HIGH-FIELD EFFECTS

In the presence of an electric field, carriers gain energy from the field as they drift along. These carriers in turn lose energy by emitting phonons. As the field increases, the average energy of the carriers increases. At sufficiently high fields, a number of physical phenomena which have important implications on the design and operation of VLSI devices can occur. In the case of high fields in silicon, these phenomena include impact ionization, or generation of electron–hole pairs; junction breakdown; band-to-band tunneling; and injection of hot carriers from locations near the silicon–oxide interface into the silicon dioxide region. In the case of high fields in silicon dioxide, the important phenomena include tunneling through the oxide layer and dielectric breakdown. The basic physics of these phenomena as they relate to VLSI devices is discussed in this section.

### 2.4.1 IMPACT IONIZATION AND AVALANCHE BREAKDOWN

Consider the depletion region of a p–n diode. At sufficiently high fields, an electron in the conduction band can gain enough energy to "lift" an electron from the valence band into the conduction band, thus generating one free electron in the conduction band and one free hole in the valence band. This process is known as *impact ionization*. Similarly, a hole in the valence band can gain enough energy to cause impact ionization. If the field is high enough, these secondary electrons and holes can themselves cause further impact ionization, thus beginning a process of carrier multiplication in the high-field region. The p–n diode breaks down when the multiplication process runs away or becomes an avalanche. The equations relating the rate of impact ionization to the condition for avalanche breakdown are derived below.

Consider a reverse-biased p–n diode with an electric field in its depletion region high enough to cause impact ionization. Let $x = 0$ and $x = W$ be the locations of

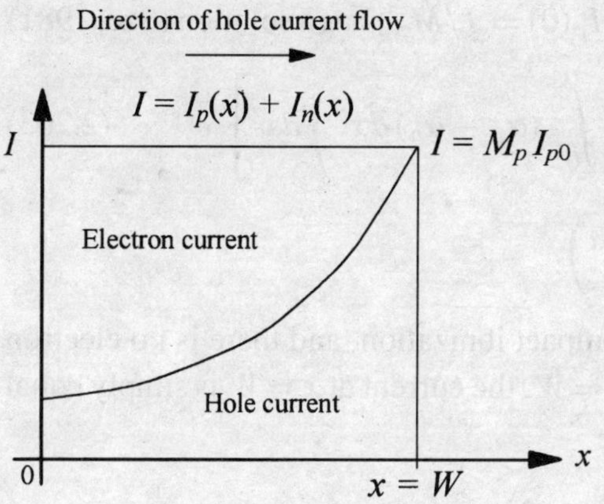


**FIGURE 2.40.** Schematic illustration of the steady-state currents caused by hole-initiated impact ionization within the depletion region of a p–n diode.

the two boundaries of the depletion region. Suppose there is a hole current $I_{p0}$ entering the depletion region at $x = 0$, as illustrated in Fig. 2.40. This hole current will generate electron–hole pairs. The secondary electrons and holes in turn cause further impact ionization as they traverse the depletion region. Thus, the hole current will increase with distance, reaching a value of $M_p I_{p0}$ at $x = W$, where $M_p$ is the multiplication factor for holes. At steady state, the total current $I$ is constant and independent of distance, i.e., $I = M_p I_{p0}$. Within the depletion region, the total current is the sum of the hole and electron currents (Moll, 1964), i.e.,

$$I = I_p + I_n. \tag{2.200}$$

These current components are illustrated in Fig. 2.40. The field is such that holes move towards the right ($x = W$) and electrons move towards the left ($x = 0$).

Consider a differential distance between $x$ and $x + dx$. There are $I_p(x)/q$ holes and $I_n(x)/q$ electrons crossing this differential distance per unit time. In crossing this differential distance, the holes cause $\alpha_p(x)I_p(x)\,dx/q$ electron–hole pairs to be generated, where $\alpha_p$ is the *hole-initiated rate of electron–hole pair generation per unit distance*. Similarly, the number of electron–hole pairs generated by the electrons is $\alpha_n(x)I_n(x)\,dx/q$, where $\alpha_n$ is the *electron-initiated rate of electron–hole pair generation per unit distance*. Thus the increase in the hole current as the electrons and holes cross the differential distance $dx$ is

$$dI_p = \alpha_p I_p\,dx + \alpha_n I_n\,dx. \tag{2.201}$$

Equations (2.200) and (2.201) give

$$\frac{dI_p}{dx} = (\alpha_p - \alpha_n)I_p + \alpha_n I, \tag{2.202}$$

which, subject to the boundary condition $I_p(0) = I/M_p$, has a solution (Sze, 1981)

$$I_p(x) = I\left[\frac{1}{M_p} + \int_0^x \alpha_n \exp\left(-\int_0^{x'} (\alpha_p - \alpha_n)\,dx''\right) dx'\right] \times \exp\left(\int_0^x (\alpha_p - \alpha_n)\,dx'\right). \tag{2.203}$$

Since we are considering hole-initiated impact ionization, and there is no electron current entering the depletion region at $x = W$, the current at $x = W$ is simply equal to $I$. Therefore, Eq. (2.203) gives

$$\frac{1}{M_p} = \exp\left(-\int_0^W (\alpha_p - \alpha_n)\,dx\right) - \int_0^W \alpha_n \exp\left(-\int_0^x (\alpha_p - \alpha_n)\,dx'\right) dx. \tag{2.204}$$

Similarly, for impact ionization initiated by electrons, the electron multiplication factor $M_n$ is given by

$$\frac{1}{M_n} = \exp\left(-\int_0^W (\alpha_n - \alpha_p)\,dx\right) - \int_0^W \alpha_p \exp\left(-\int_x^W (\alpha_n - \alpha_p)\,dx'\right) dx. \tag{2.205}$$

Avalanche breakdown occurs when carrier multiplication by impact ionization runs away, i.e., when the multiplication factors become infinite. It is shown in Appendix 5 that ***the condition for avalanche breakdown is the same whether the breakdown process is initiated by electrons or by holes***. That is, when a p–n junction breaks down, it does not matter if the avalanche breakdown process is initiated by an electron or by a hole.

In theory, Eqs. (2.204) and (2.205) can be used to calculate the multiplication factors, and hence the breakdown voltage. For the special case of $\alpha_n = \alpha_p =$ constant, avalanche breakdown occurs when the product $\alpha_n W = \alpha_p W$ approaches unity. In practice, however, the ionization rates, as well as the junction doping profiles, are simply not known accurately enough for calculation of breakdown voltages to be made with sufficient accuracy for VLSI device design purposes. Breakdown voltages in modern VLSI devices are usually determined experimentally.

#### 2.4.1.1 EMPIRICAL IMPACT-IONIZATION RATES

The measured ionization rates are often expressed in the ***empirical*** form of

$$\alpha = A \exp(-b/\mathscr{E}), \tag{2.206}$$

**TABLE 2.2** Impact-Ionization Rates in Silicon

| Data | Field Range (V/cm) | $\alpha_n$ (cm$^{-1}$) | | $\alpha_p$ (cm$^{-1}$) | |
|---|---|---|---|---|---|
| | | $A_n$ (cm$^{-1}$) | $b_n$ (V/cm) | $A_p$ (cm$^{-1}$) | $b_p$ (V/cm) |
| van Overstraeten | $1.75 \times 10^5 < \mathscr{E} < 4.0 \times 10^5$ | $7.03 \times 10^5$ | $1.231 \times 10^6$ | $1.582 \times 10^6$ | $2.036 \times 10^6$ |
| and de Man | $4.0 \times 10^5 < \mathscr{E} < 6.0 \times 10^5$ | $7.03 \times 10^5$ | $1.231 \times 10^6$ | $6.71 \times 10^5$ | $1.693 \times 10^6$ |
| Grant | $2.0 \times 10^5 < \mathscr{E} < 2.4 \times 10^5$ | $2.6 \times 10^6$ | $1.43 \times 10^6$ | $2.0 \times 10^6$ | $1.97 \times 10^6$ |
| | $2.4 \times 10^5 < \mathscr{E} < 5.3 \times 10^5$ | $6.2 \times 10^5$ | $1.08 \times 10^6$ | $2.0 \times 10^6$ | $1.97 \times 10^6$ |
| | $5.3 \times 10^5 < \mathscr{E}$ | $5.0 \times 10^5$ | $0.99 \times 10^6$ | $5.6 \times 10^5$ | $1.32 \times 10^6$ |

where $A$ and $b$ are constants, and $\mathscr{E}$ is the electric field (Chynoweth, 1957). There is quite a bit of spread in the measured impact-ionization rates reported in the literature. However, the most recent measurements give similar results (van Overstraeten and de Man, 1970; Grant, 1973). These results are shown in Table 2.2 and plotted in Fig. 2.41.

Two points are clear from Fig. 2.41. First, $\alpha_n$ is much larger than $\alpha_p$, particularly at low electric fields. This is due to the effective mass of holes being much larger than that of electrons. Second, the impact-ionization rates increase very rapidly with electric field. For the depletion region of a p–n diode where the electric field is not constant, it is the small region surrounding the maximum-field point that contributes most to the impact-ionization currents. Thus, to minimize impact ionization in a p–n diode, the maximum electric field should be minimized. As mentioned in Section 2.2.2, doping-profile grading, or using lightly doped regions or i-layers, can effectively reduce the peak electric field in a p–n junction.

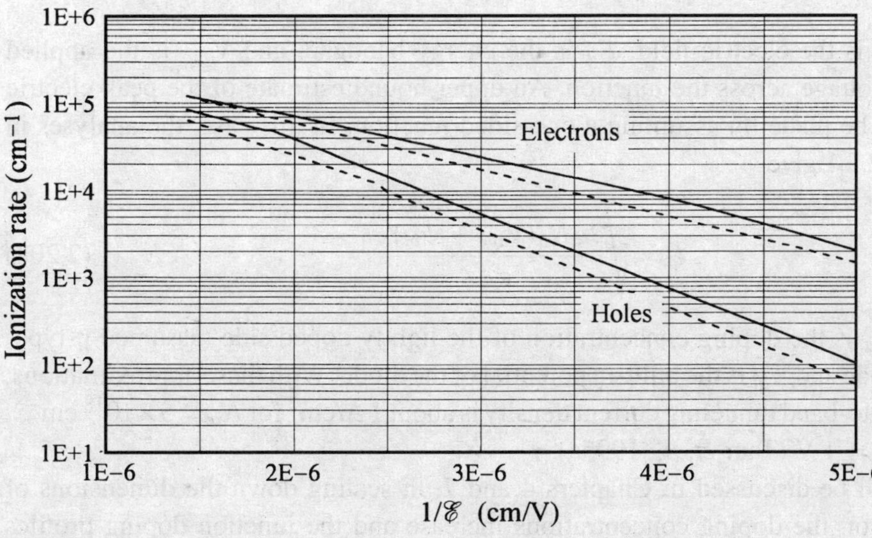


**FIGURE 2.41.** Impact-ionization rates in silicon. The solid curves are data of Grant (1973), and the dash curves are data of van Overstraeten and de Man (1970).

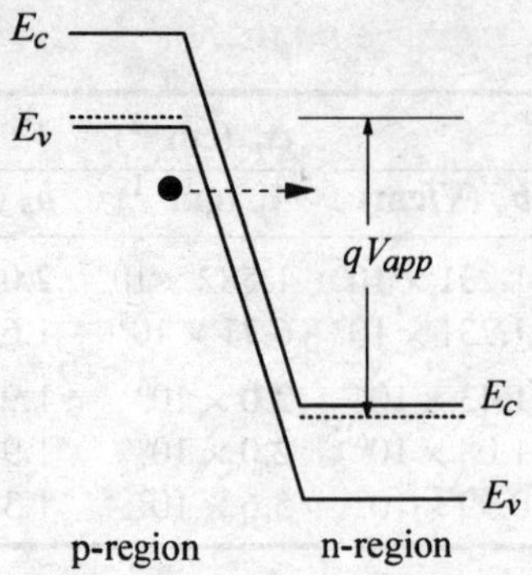


**FIGURE 2.42.** Schematic illustrating band-to-band tunneling in a p–n junction.

Impact-ionization rates decrease as temperature increases (Grant, 1973). This is due to the increased lattice scattering at higher temperatures. The data in Table 2.2 and Fig. 2.41 are for room temperature.

## 2.4.2 BAND-TO-BAND TUNNELING

When the electric field across a reverse-biased p–n junction approaches $10^6$ V/cm, significant current flow can occur due to tunneling of electrons from the valence band of the p-region into the conduction band of the n-region. This phenomenon is illustrated schematically in Fig. 2.42. In silicon this tunneling process usually involves the emission or absorption of phonons (Kane, 1961; Chynoweth *et al.*, 1960), and the tunneling current density is given by (Fair and Wivell, 1976)

$$J_{b-b} = \frac{\sqrt{2m^*}q^3 \mathscr{E} V_{app}}{4\pi^3 \hbar^2 E_g^{1/2}} \exp\left( -\frac{4\sqrt{2m^*}E_g^{3/2}}{3q\mathscr{E}\hbar} \right), \tag{2.207}$$

where $\mathscr{E}$ is the electric field, $E_g$ is the energy bandgap, and $V_{app}$ is the applied reverse voltage across the junction. An upper-bound estimate of the peak electric field can be made by assuming a one-sided junction. In this case, the analyses in Section 2.2.1 give

$$\mathscr{E} = \sqrt{\frac{2qN_a(V_{app} + \psi_{bi})}{\varepsilon_{si}}}, \tag{2.208}$$

where $N_a$ is the doping concentration of the lightly doped side (assumed p-type) of the diode and $\psi_{bi}$ is the built-in potential of the diode. With these approximations, the band-to-band tunneling current density is about 1 A/cm$^2$ for $N_a = 5 \times 10^{18}$ cm$^{-3}$, and $V_{app} = 1$ V (Taur *et al.*, 1995a).

As will be discussed in Chapters 4 and 7, in scaling down the dimensions of a transistor, the doping concentrations increase and the junction doping profiles become more abrupt, and hence the band-to-band tunneling effect increases. Once the leakage current due to band-to-band tunneling is appreciable, it increases very rapidly with electric field. For modern VLSI devices, band-to-band tunneling is becoming one of the most important leakage-current components, particularly for applications such as DRAM where leakage currents must be kept extremely low.

### 2.4.3 TUNNELING INTO AND THROUGH SILICON DIOXIDE

Consider an MOS capacitor as discussed in Section 2.3. For simplicity, the gate electrode is assumed to be heavily doped n-type polysilicon. When biased at the flat-band condition, the energy-band diagram is as shown in Fig. 2.43(a), where $\phi_{ox}$ denotes the Si–$SiO_2$ interface energy barrier for electrons, which, as indicated in Fig. 2.22, is about 3.1 eV. When a large positive bias is applied to the gate electrode, electrons in the strongly inverted surface can tunnel into or through the oxide layer and hence give rise to a gate current. Similarly, if a large negative voltage is applied to the gate electrode, electrons from the $n^+$ polysilicon can tunnel into or through the oxide layer and again give rise to a gate current.

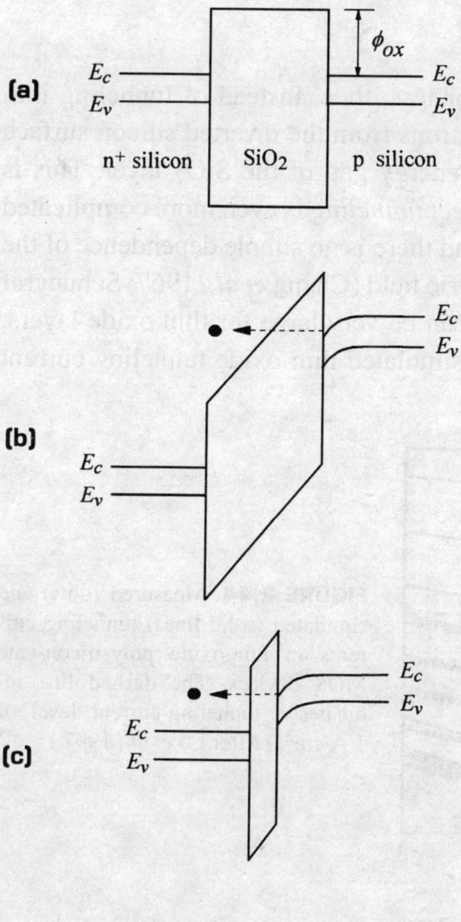


**FIGURE 2.43.** Tunneling effects in an MOS capacitor structure: (a) energy-band diagram of an n-type polysilicon-gate MOS structure at flat band; (b) Fowler–Nordheim tunneling; (c) direct tunneling.

#### 2.4.3.1 FOWLER–NORDHEIM TUNNELING

*Fowler–Nordheim tunneling* occurs when electrons tunnel into the conduction band of the oxide layer. Figure 2.43(b) illustrates Fowler–Nordheim tunneling of electrons from the silicon surface inversion layer. The complete theory of Fowler–Nordheim tunneling is rather complicated (Good and Müller, 1956). For the simple case where the effects of finite temperature and image-force barrier lowering (which will be discussed later) are ignored the tunneling current density is given by (Lenzlinger and Snow, 1969)

$$J_{FN} = \frac{q^3 \mathscr{E}_{ox}{}^2}{16\pi^2 \hbar \phi_{ox}} \exp\left(-\frac{4\sqrt{2m^*}\phi_{ox}^{3/2}}{3\,\hbar\, q\, \mathscr{E}_{ox}}\right), \tag{2.209}$$

where $\mathscr{E}_{ox}$ is the electric field in the oxide. Equation (2.209) shows that Fowler–Nordheim tunneling current is characterized by a straight line in a plot of $\log(J/\mathscr{E}_{ox}^2)$ versus $1/\mathscr{E}_{ox}$. At an oxide field of 8 MV/cm, the measured Fowler–Nordheim tunneling current density is about $5 \times 10^{-7}$ A/cm$^2$ (Lenzlinger and Snow, 1969), which is very small. Thus, for normal device operation, Fowler–Nordheim tunneling current is negligible.

#### 2.4.3.2 DIRECT TUNNELING

If the oxide layer is very thin, say 4 nm or less, then, instead of tunneling into the conduction band of the $SiO_2$ layer, electrons from the inverted silicon surface can tunnel directly through the forbidden energy gap of the $SiO_2$ layer. This is illustrated in Fig. 2.43(c). The theory of *direct tunneling* is even more complicated than that of Fowler–Nordheim tunneling, and there is no simple dependence of the tunneling current density on voltage or electric field (Chang *et al.*, 1967; Schuegraf *et al.*, 1992). The direct-tunneling current can be very large for thin oxide layers. Figure 2.44 is a plot of the measured and simulated thin-oxide tunneling current

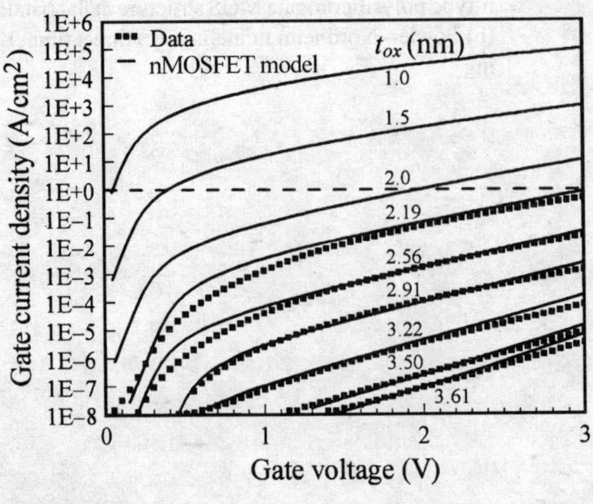


**FIGURE 2.44.** Measured (dots) and simulated (solid lines) tunneling currents in thin-oxide polysilicon-gate MOS devices. The dashed line indicates a tunneling-current level of 1 A/cm$^2$. (After Lo *et al.*, 1997.)

versus voltage in polysilicon-gate MOSFETs (Lo *et al.*, 1997). For the gate-voltage range shown in Fig. 2.44, the current is primarily a direct-tunneling current. Direct-tunneling current is important in MOSFETs of very small dimensions, where the gate oxide layers can be as thin as 2–3 nm (Taur *et al.*, 1995a).

#### 2.4.3.3 IMAGE-FORCE-INDUCED BARRIER LOWERING

The energy barrier $\phi_{ox}$ shown in Fig. 2.43(a) is the difference in energy between the conduction band of $SiO_2$ and the conduction band of Si. However, as electrons are emitted from Si into $SiO_2$, the actual energy barrier to emission is smaller than $\phi_{ox}$ by an amount equal to

$$\Delta\phi = \sqrt{\frac{q^3 \mathscr{E}_{ox}}{4\pi\,\varepsilon_{ox}}}. \tag{2.210}$$

This is the so-called *Schottky effect* or *image-force-induced barrier-lowering effect* (Sze, 1981). The actual energy barrier to emission is $\phi_{ox} - \Delta\phi$.

For $\mathscr{E}_{ox} = 1 \times 10^6$ V/cm, $\Delta\phi = 0.19$ eV. Thus, for practical oxide fields of $10^6$ V/cm or larger, image-force barrier lowering is not negligible compared to the interface energy barrier of 3.1 eV. Image-force barrier lowering is included in the more accurate theories of Fowler–Nordheim tunneling (Lenzlinger and Snow, 1969) and direct tunneling (Chang *et al.*, 1967). Interested readers are referred to these references for details.

### 2.4.4 INJECTION OF HOT CARRIERS FROM SILICON INTO SILICON DIOXIDE

If a region of high electric field is located near the Si–$SiO_2$ interface, and if the electric field is high enough, some of the electrons or holes can gain sufficient energy from the electric field to surmount the interface barrier and enter the $SiO_2$ layer. In general, injection from Si into $SiO_2$ is much more likely for hot electrons than for hot holes because (a) electrons can gain energy from the electric field much more readily than holes due to their smaller effective mass, and (b) the Si–$SiO_2$ interface energy barrier is larger for holes ($\approx$4.6 eV) than for electrons ($\approx$3.1 eV), as indicated in Fig. 2.22.

The process of hot-electron or hot-hole injection from silicon into silicon dioxide is much too complex to model quantitatively. Thus far, quantitative agreement has been shown only for the special case of hot electrons traveling perpendicularly towards the Si–$SiO_2$ interface, and only with Monte Carlo models that take into account the correct band structures, all the relevant scattering processes, and nonlocal transport properties (Fischetti *et al.*, 1995). This one-dimensional injection process is illustrated in Fig. 2.45. Fortunately, a very simple but empirical model of *lucky*

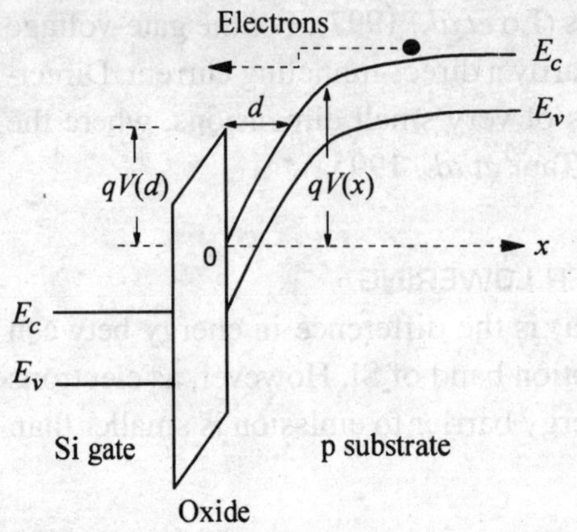


**FIGURE 2.45.** Schematic illustrating hot electrons traveling perpendicularly towards the Si–$SiO_2$ interface and being injected into the $SiO_2$ layer.

*electrons* proposed by Shockley (1961) describes the measured data surprisingly well.

In this lucky-electron model (Ning *et al.*, 1977), the probability that a hot electron at a distance $d$ from the Si–$SiO_2$ interface will be emitted into the $SiO_2$ layer is expressed as

$$P(d) = A\exp(-d/\lambda), \tag{2.211}$$

where $\lambda$ is an effective mean free path for energy loss by hot electrons in silicon, and $A$ is a constant for fitting to the experimental data. The relation among the parameter $d$, the effective hot-electron emission energy barrier, and the electron potential energy is illustrated in Fig. 2.45. The parameter $d$ can be obtained as follows. Referring to Fig. 2.45, $qV(x)$ is the potential energy of an electron at $x$. An electron at $x = d$ has just enough potential energy to overcome the effective energy barrier for emission if it can travel from $x = d$ to the interface at $x = 0$ without undergoing any energy-losing collision. That is, $qV(d)$ is equal to the effective energy barrier for emission.

It was determined empirically (Ning *et al.*, 1977) that the effective energy barrier for emission can be written as

$$qV(d) = \phi_{ox} - \Delta\phi - \alpha\mathscr{E}_{ox}^{2/3}, \tag{2.212}$$

where the first two terms are the Si–$SiO_2$ interface energy barrier and the image-force barrier lowering discussed in the previous subsections, and the third term is introduced to allow for the fact that hot electrons without enough energy to surmount the image-force-lowered energy barrier can still tunnel into the $SiO_2$ layer. It was found that setting $\alpha = 1 \times 10^{-5}$ e-$(\text{cm}^2\text{-V})^{1/3}$ and $A = 2.9$ fits a wide range of measured emission probabilities.

The temperature dependence of the hot-electron injection process is contained in the temperature dependence of the effective mean free path (Crowell and

Sze, 1966)

$$\lambda(T) = \lambda_0 \tanh(E_R/2kT), \tag{2.213}$$

where $E_R = 63$ meV is the optical-phonon energy, and $\lambda_0$ is the low-temperature limit of $\lambda$. It was found empirically that $\lambda_0 = 10.8$ nm (Ning *et al.*, 1977).

### 2.4.5 HIGH-FIELD EFFECTS IN GATED DIODES

Thus far, the effects of high fields have been considered for p–n diodes and MOS capacitors separately. In a gated diode structure, both the location of the peak-field region and the magnitude of the peak field vary with gate voltage. Let us consider a gated $n^+$–p diode. As discussed in Section 2.3.5, when the gate is biased to invert the silicon surface, the inverted surface region has about the same potential as the $n^+$ region, and the gated diode behaves like a large-area $n^+$–p diode. If the p-region is uniformly doped, the depletion-layer width is about the same below the $n^+$ silicon region as below the surface inversion region; hence the electric field is rather uniformly distributed and is about the same as in a simple p–n diode. This is illustrated schematically in Fig. 2.46(a).

When the gate is biased to cause an accumulation layer to form at the silicon surface, the silicon surface under the gate has about the same potential as the p-type substrate. Due to the presence of the accumulated holes at the surface, the surface behaves like a p-region more heavily doped than the substrate, thus causing the depletion layer at the surface to become narrower than elsewhere. This is illustrated schematically in Fig. 2.46(b). The narrowing of the depletion layer at or near the intersection of the p–n junction and the Si–$SiO_2$ interface causes *field crowding*, or an increase of the local electric field, and greatly increases the high-field effects in and near that location.

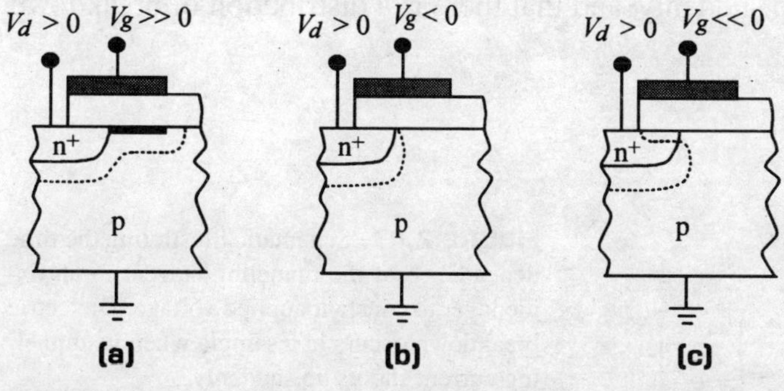


**FIGURE 2.46.** Schematics illustrating a gated $n^+$–p diode when the surface is (a) inverted and (b) accumulated, and (c) when the surface of the $n^+$ region is depleted or inverted. The dashed lines indicate the boundary of the depletion region.

When the negative gate bias is large enough, the $n^+$ region under the gate can become depleted, and even inverted. This is illustrated in Fig. 2.46(c). In this case, the gate and the $n^+$ region behave like an MOS capacitor with a heavily doped n-type "substrate." There is more field crowding, and the peak field increases.

As the electric field in and around the gated p–n junction is increased by the gate voltage, all the high-field effects, such as avalanche multiplication and band-to-band tunneling, can increase very dramatically. Thus the leakage current of a reverse-biased gated diode can increase dramatically when the gate voltage begins to cause field crowding in and around the junction region. When this occurs in the drain junction of a MOSFET, the increased junction leakage current is called *gate-induced drain leakage*, or GIDL (Chan *et al.*, 1987a; Noble *et al.*, 1989). For deep-submicron CMOS devices, ***GIDL is a very important leakage-current component that must be minimized.***

It should be noted that for the gated $n^+$–p diode considered here, when the gate is biased to accumulate the silicon surface, the oxide field favors the injection of hot holes from the silicon substrate into the silicon dioxide layer (Verwey, 1972). Similarly, for a gated $p^+$–n diode, the oxide field favors the injection of hot electrons when the gate is biased to accumulate the silicon surface. Thus, injection of majority carriers, instead of minority carriers, from the silicon substrate into the silicon dioxide layer takes place when significant gate-voltage-induced avalanche multiplication occurs in a gated diode.

### 2.4.6 DIELECTRIC BREAKDOWN

As discussed in Section 2.4.3, significant electron tunneling can take place when a large electric field is applied across an oxide layer. Figure 2.47 illustrates schematically the typical time dependence of the tunneling current when a large but constant voltage is applied across an oxide layer. It shows that there is a gradual decrease of the current with time until the oxide suffers a *dielectric breakdown*, at which point the current shoots up suddenly, and that there is a distribution of breakdown

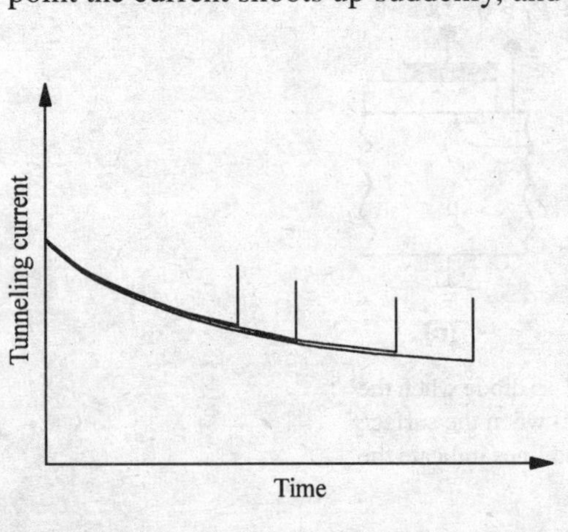


**FIGURE 2.47.** Schematic illustrating the time dependence of the tunneling current in an oxide layer at constant applied voltage. Dielectric breakdown occurs in a sample when its tunneling current shoots up suddenly.

voltage for the various samples (Harari, 1978). In general, an oxide layer ceases to be a good electrical insulator after it suffers dielectric breakdown.

Two types of dielectric breakdown characteristics are often observed in silicon devices. In one type, the breakdown occurs rather abruptly, either on the current–time plot or on a current–voltage plot. In another type, the breakdown occurs rather gradually, or *softly*, on similar plots. In the case of soft breakdown, the current appears to keep on increasing, often with the maximum current limited by the compliance limit of the current meter. As a result, in practice, an oxide layer is often said to have broken down once its current has exceeded some arbitrary but conveniently measurable limit. Oxide films thicker than about 100 nm often break down abruptly, while those thinner than about 10 nm often break down softly.

The quality of an oxide film is often measured in terms of the electric field at which dielectric breakdown occurs. "Good-quality" thick (>100 nm) $SiO_2$ films typically break down at fields greater than about 10 MV/cm, while "good-quality" thin (<10 nm) $SiO_2$ films usually show larger breakdown fields, often in excess of 15 MV/cm.

Dielectric breakdown in a device must be avoided. In bipolar transistors, because there are normally no thin oxide components, the electric fields across the oxide layers are usually so small that dielectric breakdown is not a concern. In CMOS devices, the maximum oxide field varies widely, depending on the application. For devices used in logic and memory circuits, the maximum electric field is typically in the 3–5-MV/cm range in normal operation, and can reach as high as 5–7 MV/cm in special operations (such as during a device burn-in process). For devices used in electrically programmable nonvolatile memories, where normal operation involves tunneling through a thin dielectric layer, the maximum electric field across the thin dielectric layer could be in excess of 10 MV/cm. Therefore, dielectric breakdown is a real concern in CMOS devices.

As a result of its importance, dielectric breakdown in thin oxide films has been, and still is, a subject of extensive research. The physical mechanisms involved in, and leading to, the dielectric breakdown process are very complex. They involve impact ionization in the oxide layer, injection of holes from the anode (the electrode which acts as electron sink), creation of electron and hole traps in the oxide, electron and hole trapping, creation of surface states at the oxide–silicon interface, and the interaction of many or all of these processes (DiStefano and Shatzkes, 1974; Harari, 1978; Chen *et al.*, 1986; DiMaria *et al.*, 1993). In this subsection, we cover only those aspects of the dielectric-breakdown process that are more directly relevant to the design and operation of VLSI devices. The reader is referred to the vast literature on the subject for more details.

#### 2.4.6.1 TUNNELING INTO AN ELECTRON TRAP

As discussed in Section 2.3.6, there are electron and hole traps in silicon dioxide. As electrons tunnel into an oxide layer, some of the electrons can get trapped.

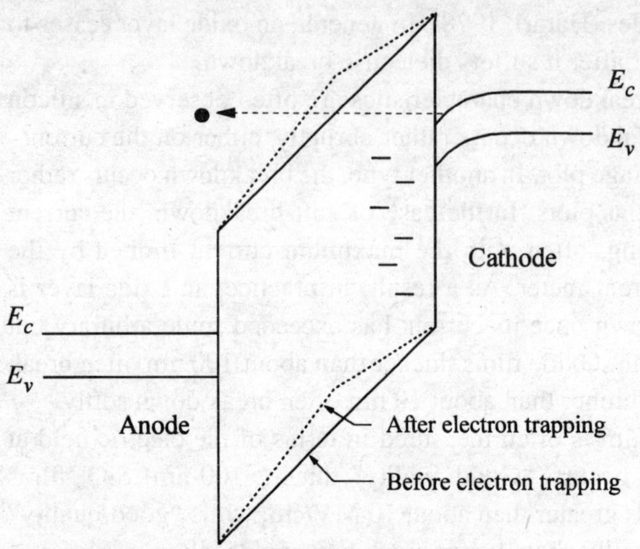


**FIGURE 2.48.** Schematic illustrating the trapping of tunneling electrons. As electrons are trapped, the oxide field near the cathode (electron source) is decreased, while the oxide field near the anode (electron sink) is increased.

The trapped electrons modify the oxide field so that the field near the cathode (the electrode that acts as an electron source) is decreased, while the field near the anode (the electrode that acts as an electron sink) is increased. This is illustrated in Fig. 2.48. According to Eq. (2.209) for Fowler–Nordheim tunneling, as the oxide field near the cathode is decreased, the tunneling current decreases. This explains the time dependence of the constant-voltage tunneling current in Fig. 2.47. Also, electron trapping during the tunneling process often leads to a hysteresis in the measured current–voltage plot.

### 2.4.6.2 HOLE GENERATION, INJECTION, AND TRAPPING DURING ELECTRON TUNNELING

As an electron travels in the conduction band of an oxide layer, it gains energy from the oxide field. If the voltage drop across the oxide layer is larger than the bandgap energy of silicon dioxide, which, as indicated in Fig. 2.22, is about 9 eV, the electron can gain sufficient energy to cause impact ionization in the oxide. The holes generated by impact ionization can be trapped in the oxide.

Holes can be injected indirectly into the oxide layer during electron tunneling as well. A tunneling electron arriving at the anode could cause impact ionization in the anode near the oxide–anode interface. Depending on the energy of the tunneling electron, the hole thus generated could be from deep down in the valence band of the anode, and thus could be "hot." As discussed in Section 2.4.4,

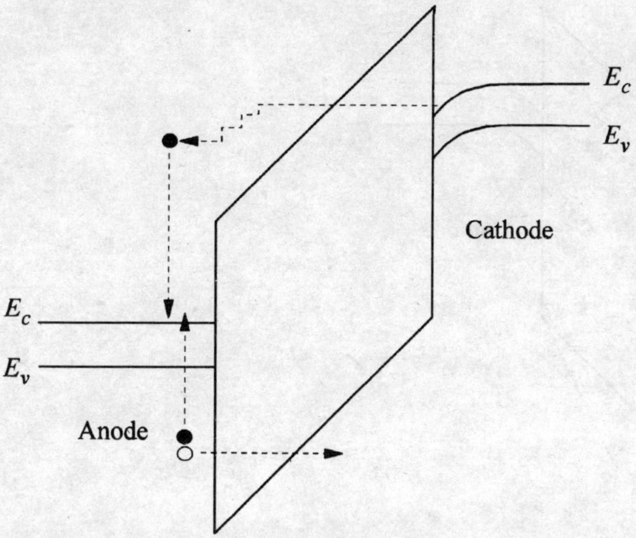


**FIGURE 2.49.** Schematic illustrating the generation of an electron–hole pair in the anode by a tunneling electron. The hole thus generated can then be injected (by tunneling in this example) into the oxide layer.

a hot hole in the silicon near the silicon–oxide interface can have a high probability of being injected into the oxide layer. This process is illustrated in Fig. 2.49. The injected hole can be trapped in the oxide layer as it travels towards the cathode.

The trapped holes in the oxide layer cause an increase in the oxide field near the cathode, and a decrease in the oxide field near the anode. This is illustrated in Fig. 2.50. Again, as can be seen from Eq. (2.209) for Fowler–Nordheim tunneling, a small increase in the oxide field near the cathode can cause a large increase in the tunneling current. Thus, hole trapping in the oxide near the cathode provides positive feedback to the electron tunneling process. Dielectric breakdown occurs when this positive feedback leads to a runaway of the electron tunneling current at some local *weak spots* of the oxide (DiStefano and Shatzkes, 1974).

It should be pointed out that the increase in tunneling current due to hole trapping is usually not readily observable, except near breakdown, because it is masked by the decrease in tunneling current due to electron trapping. The rise of the tunneling current near breakdown is so rapid that it appears as a current spike in current-versus-time plots, as illustrated in Fig. 2.47.

Since weak spots are not intrinsic to silicon dioxide, different oxide samples can have different numbers of weak spots and/or weak spots of different weakness. Therefore, for the same applied voltage and oxide thickness, different oxide samples can break down at different times. This explains the observed spread in breakdown events in Fig. 2.47.

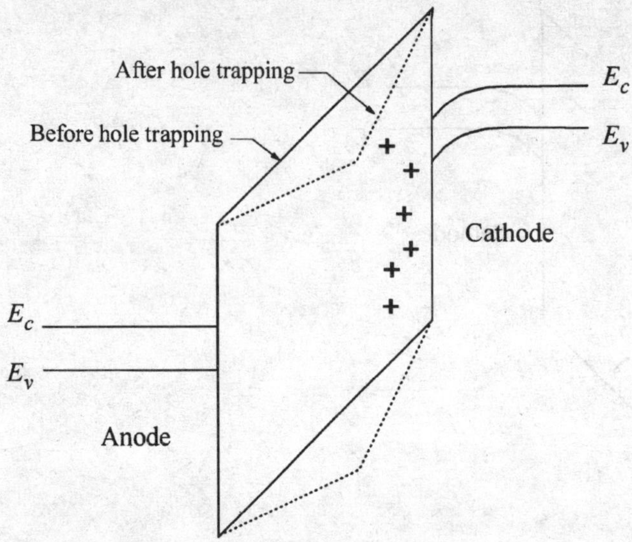


**FIGURE 2.50.** Schematic showing the trapping of holes in the oxide layer. The trapped holes enhance the electric field near the cathode, and decrease the electric field near the anode.

#### 2.4.6.3 TRAP AND INTERFACE-STATE GENERATION

The positive feedback described above is only one critical part of the process leading to catastrophic breakdown of an oxide layer. Prior to catastrophic breakdown, the tunneling electrons, directly or indirectly, also cause several other physical processes to take place, with the net effect of "softening" the oxide before final breakdown. These include trap generation, which enhances the trapping of electrons and holes, and interface-state generation, which is a sign that some of the electron bonds in the oxide, at least those near the oxide–silicon interface, are "weakened." The physics of trap generation and interface-state generation is very complex (Harari, 1978; DiMaria *et al.*, 1993; Hsu and Ning, 1991). It has been, and still is, a subject of extensive research. The reader is referred to the vast literature on the subject.

#### 2.4.6.4 TIME TO BREAKDOWN AND CHARGE TO BREAKDOWN

The breakdown characteristics of an oxide film are often described in terms of its *time to breakdown*, which measures the time needed for the film to break down, or its *charge to breakdown*, which measures the integrated total tunneling charge just before breakdown. Although both time to breakdown and charge to breakdown are used, most recent publications describe dielectric breakdown of silicon dioxide only in terms of charge to breakdown. It appears that it is easier to develop physical models relating charge to breakdown to the other physical processes, such as hole current, trapping, trap generation, and interface-state generation (Schuegraf and Hu,

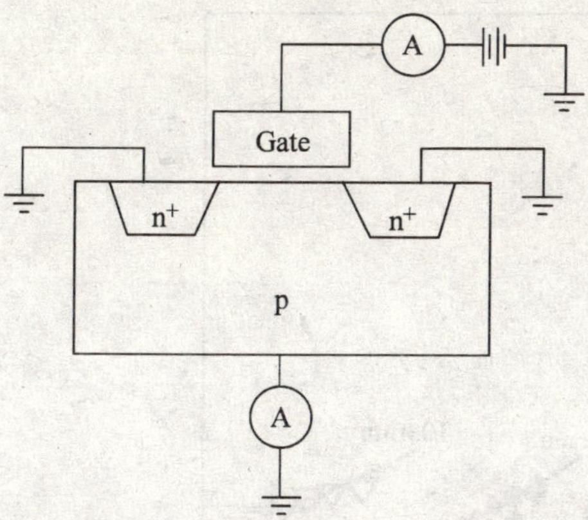


**FIGURE 2.51.** Schematic illustrating the bias configuration of an n-channel MOSFET for measuring the charge to breakdown and its hole-charge component.

1994; DiMaria and Stathis, 1997) than to develop physical models relating time to breakdown to these processes. Therefore, we will not discuss time to breakdown any further here.

As discussed above, a tunneling electron current can generate a hole current. Thus, the charge to breakdown, $Q_{BD}$, is the sum of the charges due to electrons and holes. If an MOS capacitor structure is used to measure $Q_{BD}$, then, due to the two-terminal nature of the device, only the total charge can be measured. However, if an n-channel MOSFET or an $n^+$–p gated-diode structure is used to measure $Q_{BD}$, then both the total charge and the hole-charge component can be determined. For the case of an n-channel MOSFET, the bias configuration for such measurements is illustrated in Fig. 2.51. Integration of the gate current gives the total charge, and integration of the substrate current gives the charge due to the holes.

For a given oxide film, the charge to breakdown is often plotted as a function of oxide voltage. Figure 2.52 is a typical plot for oxide thickness in the 2.5–10-nm range (Schuegraf and Hu, 1994). It shows that, for a given oxide thickness, $Q_{BD}$ decreases with increasing oxide voltage. It has also been shown that $Q_{BD}$ is about the same for n-channel and p-channel MOSFETs (DiMaria and Stathis, 1997). Many papers are being published attempting to explain these and other dielectric-breakdown-related data (Schuegraf and Hu, 1994; DiMaria *et al.*, 1993; Chen *et al.*, 1986; DiMaria and Stathis, 1997). The reader is referred to the literature for details.

#### 2.4.6.5 DIELECTRIC BREAKDOWN IN MOSFETs

It should be pointed out that most published charge-to-breakdown plots, such as that shown in Fig. 2.52, are for typical but good-quality oxides. In practice, the collected charge-to-breakdown data are much more scattered and occasionally show values much less than those shown in Fig. 2.52. When that happens, it is usually due to some extrinsic defects in the oxide and not due to the intrinsic quality of the oxide being bad.

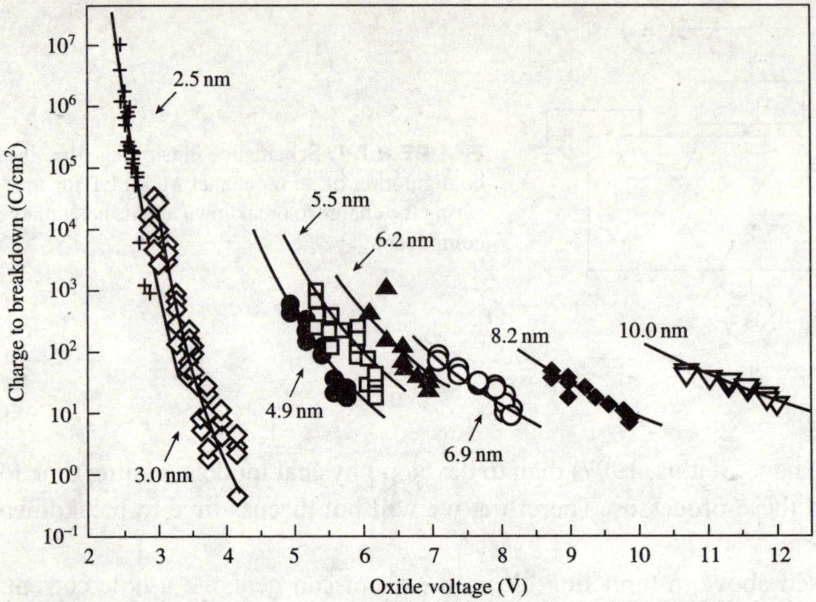


**FIGURE 2.52.** Typical plot of charge to breakdown versus oxide voltage for several oxide thickness values. (After Schuegraf and Hu, 1994.)

The design of practical MOSFETs used in VLSI products has been guided by some scaling principles, to be discussed in Chapter 4. In scaling, both the power-supply voltage and the physical dimensions of a MOSFET are reduced in some coordinated manner. Thus, for 5-, 3.3-, 2.5-, and 1.5-V MOSFET devices, the corresponding oxide thicknesses are about 20, 10, 7, and 3.5 nm, respectively (Taur *et al.*, 1995). For reliable device operation, the product of the oxide tunneling current and the device lifetime should be less than the corresponding $Q_{BD}$-value. It can be readily inferred from the magnitude of the tunneling current illustrated in Fig. 2.44, and from the charge-to-breakdown values indicated in Fig. 2.52, that, for these voltages and oxide thicknesses, dielectric breakdown is not a factor limiting the power-supply voltage at all, provided that the fabrication process produces oxides of quality comparable to that shown in Fig. 2.52.

## EXERCISES

**2.1** Show that the values of the Fermi–Dirac distribution function, Eq. (2.1), at a pair of energies symmetric about the Fermi energy $E_f$, are complementary, i.e., show that $f(E_f - \Delta E) + f(E_f + \Delta E) = 1$, independent of temperature.

**2.2** For a given donor level $E_d$ and concentration $N_d$ of an n-type silicon, solve the Fermi energy $E_f$ from the charge neutrality condition, Eq. (2.10) (neglecting the hole term). Show that $E_c - E_f$ approaches the complete ionization value, Eq. (2.11), under the condition of shallow donor level with low to moderate concentration. What happens if the condition is not satisfied?

**2.3** Use the density of states $N(E)$ derived in Appendix 3 to evaluate the average kinetic energy of electrons in the conduction band:

$$\langle \text{K.E.} \rangle = \frac{\int_{E_c}^{\infty} (E - E_c) N(E) f(E)\, dE}{\int_{E_c}^{\infty} N(E) f(E)\, dE}.$$

**(a)** For a nondegenerate semiconductor in which $f(E)$ can be approximated by the Maxwell–Boltzmann distribution, Eq. (2.2), show that $\langle \text{K.E.} \rangle = \frac{3}{2}kT$.

**(b)** For a degenerate semiconductor at 0 K, show that $\langle \text{K.E.} \rangle = \frac{3}{5}(E_f - E_c)$.

**2.4** The 3-D Gauss's law is obtained after a volume integration of the 3-D Poisson's equation and takes the form

$$\oiint_S \mathscr{E} \cdot d\mathbf{S} = \frac{Q}{\varepsilon_{si}},$$

where the left-hand side is an integral of the normal electric field over a closed surface $S$, and $Q$ is the net charge enclosed within $S$. Use it to derive the electric field at a distance $r$ from a point charge $Q$ (Coulomb's law). What is the electric potential in this case?

**2.5** **(a)** Use Gauss's law to show that the electric field at a point above a uniformly charged sheet of charge density $Q_s$ per unit area is $Q_s/2\varepsilon$, where $\varepsilon$ is the permittivity of the medium.

**(b)** For two oppositely charged parallel plates with surface charge densities $Q_s$ and $-Q_s$, show that the electric field is uniform and equals $Q_s/\varepsilon$ in the region between the two plates and is zero in the regions outside the two plates.

**2.6** The total depletion charge and inversion charge densities of a p-type MOS capacitor can be expressed as

$$Q_d = -qN_a \int_0^{W_d} \left(1 - e^{-q\psi/kT}\right) dx = -qN_a \int_0^{\psi_s} \frac{1 - e^{-q\psi/kT}}{\mathscr{E}}\, d\psi,$$

and

$$Q_i = -q\frac{n_i^2}{N_a} \int_0^{W_d} \left(e^{q\psi/kT} - 1\right) dx = -q\frac{n_i^2}{N_a} \int_0^{\psi_s} \frac{e^{q\psi/kT} - 1}{\mathscr{E}}\, d\psi,$$

using Eqs. (2.149) and (2.150). Here $\mathscr{E} = -d\psi/dx$ is given by Eq. (2.153).

**(a)** Write down the expressions for the small-signal depletion capacitance, $C_d = -dQ_d/d\psi_s$, and the small-signal inversion capacitance (low frequency), $C_i = -dQ_i/d\psi_s$, in silicon as represented in the equivalent circuit in Fig. 2.28.

**(b)** Show that $C_d + C_i = C_{si}$, where $C_{si} = -dQ_s/d\psi_s$ is evaluated using Eq. (2.154).

**(c)** Show that $C_d \approx C_i$ at the condition of strong inversion, $\psi_s = 2\psi_B$. (This allows one to use a split $C$–$V$ measurement to determine the gate voltage where $\psi_s = 2\psi_B$.)

**(d)** From the behavior of $C_d$ beyond strong inversion, explain the "screening" of depletion charge (incremental) by the inversion layer.

**2.7** Near the surface of an MOS capacitor biased well into strong inversion, only the $\exp(q\psi/kT)$ term in the square-root expression of Eq. (2.163) needs to be kept (classical model). Solve $\psi(x)$ under the boundary condition $\psi(0) = \psi_s$. Express the inversion electron concentration $n(x)$ in terms of the surface concentration $n(0)$ given by Eq. (2.165).

**2.8** Solve the gate voltage equation (2.167) for $\psi_s(V_g)$ under the depletion condition in which $Q_s(\psi_s) = Q_d(\psi_s)$ given by Eq. (2.161). Show that for incremental changes, $\Delta\psi_s = \Delta V_g/(1 + C_d/C_{ox})$, where $C_d$ is the depletion charge capacitance given by Eq. (2.174).

**2.9** When the gate voltage greatly exceeds the threshold for strong inversion, a first-order solution of $\psi_s(V_g)$ can be obtained from the coupled equations (2.180) and (2.154), by keeping only the inversion charge term. Show that

$$\psi_s \approx 2\psi_B + \frac{2kT}{q} \ln\left(\frac{C_{ox}(V_g - V_{fb} - 2\psi_B)}{\sqrt{2\varepsilon_{si}kTN_a}}\right)$$

under these circumstances. Estimate how much higher $\psi_s$ can be over $2\psi_B$ by substituting some typical values in the logarithmic expression.

**2.10** In the split $C$–$V$ measurement in Fig. 2.28(b), show that the $n^+$ channel part of the small-signal gate capacitance is

$$\frac{dQ_i}{dV_g} = \frac{C_{ox}C_i}{C_{ox} + C_i + C_d}.$$

Sketch the functional behavior of $dQ_i/dV_g$ versus $V_g$, and from it describe the behavior of $Q_i$ versus $V_g$.

**2.11** The multiplication factors for holes and for electrons are given by Eqs. (2.204) and (2.205), respectively. For the special case of constant $\alpha_p$ and $\alpha_n$, show that $M_p \to \infty$ occurs when the depletion-layer width approaches the value of $W = \ln(\alpha_n/\alpha_p)/(\alpha_n - \alpha_p)$. Also show that the condition for $M_n \to \infty$ gives the same result for $W$.

**2.12** Prove the following mathematical identities:

$$\int_0^W f(x) \exp\left(-\int_0^x f(x')\,dx'\right) dx = 1 - \exp\left(-\int_0^W f(x')\,dx'\right)$$

and

$$\int_0^W f(x)\exp\left(-\int_x^W f(x')dx'\right)dx = 1 - \exp\left(-\int_0^W f(x')\,dx'\right).$$

These identities are used in Appendix 5 to show that the condition for hole-initiated avalanche breakdown, namely $1/M_p \to 0$, is the same as that for electron-initiated avalanche breakdown, namely $1/M_n \to 0$.

**2.13** The depletion-layer capacitance $C_d$ of a uniformly doped abrupt p–n diode and its dependence on doping concentration and applied voltage are given in Eqs. (2.65), (2.66), and (2.68). Sketch $1/C_d^2$ as a function of the applied reverse-bias voltage $V_{app}$. Show how this plot can be used to determine $N_a$ and $N_d$.

**2.14** The depletion-layer capacitance of a one-sided p–n diode is often used to determine the doping profile of the lightly doped side. Consider an n$^+$–p diode, with a nonuniform p-side doping concentration of $N_a(x)$. If $Q_d(V)$ is the depletion-layer charge per unit area at bias voltage $V$, the capacitance per unit area at bias voltage $V$ is $C = dQ_d/dV$. In terms of the depletion-layer width $W$, we have $C(V) = \varepsilon_{si}/W$, where $W$ is a function of $V$. (For simplicity, we have dropped the subscripts in $C$, $W$, and $V$ here.) Show that the doping concentration at the depletion-layer edge is given by

$$N_a(W) = \frac{2}{q\varepsilon_{si}\, d(1/C^2)/dV}.$$

**2.15** The charge distribution of a p–i–n diode is shown schematically in Fig 2.14. The i-layer thickness is $d$. The depletion-layer capacitance is given by Eq. (2.81), namely $C_d = \varepsilon_{si}/W_d$, where $W_d = x_n + x_p$ is the total depletion-layer width. Derive this result from $C_d = dQ_d/dV$.

**2.16** Consider a p–n diode. Assume the junction is located at $x = 0$, with the n-region to the left (i.e., $x < 0$) and the p-region to the right (i.e., $x > 0$) of the junction. The distribution of the excess electrons is given by Eq. (2.101), and the electron current density entering the p-region is given by Eq. (2.102). Derive the equation for the distribution of the excess holes in the n-region and the equation for the hole current density entering the n-region.

**2.17** The minimum leakage current of a reverse-biased diode is determined by its saturation current components. The saturation currents depend on the dopant concentrations of the diode, as well as on the widths of the quasineutral p- and n-regions. They also depend on whether or not the heavy-doping effect is included. This exercise is designed to show the magnitude of these effects.

**(a)** Consider an n$^+$–p diode, with an emitter doping concentration of $10^{20}$ cm$^{-3}$ and a base doping concentration of $10^{17}$ cm$^{-3}$. Assume both the emitter and the base to be wide compared with their corresponding minority-carrier diffusion lengths. Ignore the heavy-doping effect and calculate the electron and hole saturation current densities [see Eq. (2.110)].

**(b)** In most modern MOSFET and bipolar devices, the $n^+$–p diodes have the $n^+$-region width small compared with its hole diffusion length. If we assume the quasineutral $n^+$ region to have a width of 0.1 μm, again ignoring the heavy-doping effect, estimate the hole saturation current density [see Eq. (2.114)].

**(c)** It is discussed in Section 6.1.2 and shown in Fig 6.3 that the effect of heavy doping should be included once the doping concentration is larger than $10^{17}$ cm$^{-3}$ for p-type silicon, and large than $10^{18}$ cm$^{-3}$ for n-type silicon. Heavy-doping effect is usually included simply by replacing the intrinsic-carrier concentration $n_i$ by an effective intrinsic-carrier concentration $n_{ie}$, where $n_i$ and $n_{ie}$ are related by

$$n_{ie}^2 = n_i^2 \exp(\Delta E_g / kT).$$

The empirical parameter $\Delta E_g$ is called the apparent bandgap narrowing due to the heavy-doping effect, and its values are plotted in Fig. 6.3. Repeat (b) including the effects of heavy doping.

**2.18** Consider an $n^+$–p diode, with the $n^+$ emitter side being wide compared with its hole diffusion length and the p base side being narrow compared with its electron diffusion length. The diffusion capacitance due to electron storage in the base is $C_{Dn}$, and that due to hole storage in the emitter is $C_{Dp}$. Assume the emitter to have a doping concentration of $10^{20}$ cm$^{-3}$ and the base to have a width of 100 nm and a doping concentration of $10^{17}$ cm$^{-3}$.

**(a)** If the heavy-doping effect is ignored, the capacitance ratio is [see Eq. (2.143)]

$$\frac{C_{Dn}}{C_{Dp}} = \frac{N_E}{N_B}\frac{W_B}{2L_{pE}} \qquad \text{(heavy-doping effect ignored).}$$

Evaluate this ratio for the $n^+$–p diode.

**(b)** When the heavy-doping effect cannot be ignored, it is usually included simply by replacing the intrinsic-carrier concentration $n_i$ by an effective intrinsic-carrier concentration $n_{ie}$ [see part (c) of Exercise 2.17]. Show that when the heavy-doping effect is included, the capacitance ratio becomes

$$\frac{C_{Dn}}{C_{Dp}} = \left(\frac{n_{ieB}^2}{n_{ieE}^2}\right)\left(\frac{N_E}{N_B}\right)\frac{W_B}{2L_{pE}} \qquad \text{(heavy-doping effect included),}$$

where the subscript $B$ denotes quantities in the base and the subscript $E$ denotes quantities in the emitter. Evaluate this ratio for the $n^+$–p diode.

(This exercise demonstrates that heavy-doping effects cannot be ignored in any quantitative modeling of the switching speed of a diode.)

**2.19** As electrons are injected from silicon into silicon dioxide, some of these electrons become trapped in the oxide. Let $N_T$ be the electron trap density, $n_T$ be the density of trapped electrons, and $j_G/q$ be the injected electron particle current density. The rate equation governing $n_T(t)$ is

$$\frac{dn_T}{dt} = \frac{j_G}{q}\sigma(N_T - n_T),$$

where $\sigma$ is the capture cross section of the traps. If the initial condition for $n_T$ is $n_T(t = 0) = 0$, show that the time dependence of the trapped electron density is given by

$$n_T(t) = N_T\{1 - \exp[-\sigma N_{inj}(t)]\},$$

where

$$N_{inj}(t) \equiv \int_0^t \frac{j_G(t')}{q}\,dt'$$

is the number of injected electrons per unit area. Assume $N_T = 5 \times 10^{12}\ \text{cm}^{-3}$ and $\sigma = 1 \times 10^{-13}\ \text{cm}^2$, sketch a log–log plot of $n_T$ as a function of $N_{inj}$. (The capture cross section is often measured by fitting to such a plot.)

# 3 MOSFET DEVICES

The metal–oxide–semiconductor field-effect transistor (MOSFET) is the building block of VLSI circuits in microprocessors and dynamic memories. Because the current in a MOSFET is transported predominantly by carriers of one polarity only (e.g., electrons in an n-channel device), the MOSFET is usually referred to as a unipolar or majority-carrier device. Throughout this chapter, n-channel MOSFETs are used as an example to illustrate device operation and derive drain-current equations. The results can easily be extended to p-channel MOSFETs by exchanging the dopant types and reversing the voltage polarities.

The basic structure of a MOSFET is shown in Fig. 3.1. It is a four-terminal device with the terminals designated as *gate* (subscript $g$), *source* (subscript $s$), *drain* (subscript $d$), and *substrate* or *body* (subscript $b$). An n-channel MOSFET, or nMOSFET, consists of a p-type silicon substrate into which two $n^+$ regions, the source and the drain, are formed (e.g., by ion implantation). The gate electrode is usually made of metal or heavily doped polysilicon and is separated from the substrate by a thin silicon dioxide film, the *gate oxide*. The gate oxide is usually formed by thermal oxidation of silicon. In VLSI circuits, a MOSFET is surrounded by a thick oxide called the *field oxide* to isolate it from the adjacent devices. The surface region under the gate oxide between the source and drain is called the *channel* region and is critical for current conduction in a MOSFET. The basic operation of a MOSFET device can be easily understood from the MOS capacitor discussed in Section 2.3. When there is no voltage applied to the gate or when the gate voltage is zero, the p-type silicon surface is either in accumulation or in depletion and there is no current flow between the source and drain. The MOSFET device acts like two back-to-back p–n junction diodes with only low-level leakage currents present. When a sufficiently large positive voltage is applied to the gate, the silicon surface is inverted to n-type, which forms a conducting channel between the $n^+$ source and drain. If there is a voltage difference between them, an electron current will flow from the source to the drain. A MOSFET device therefore operates like a switch ideally suited for digital circuits. ***Since the gate electrode is electrically insulated from the substrate, there is effectively no dc gate current, and the channel is capacitively***

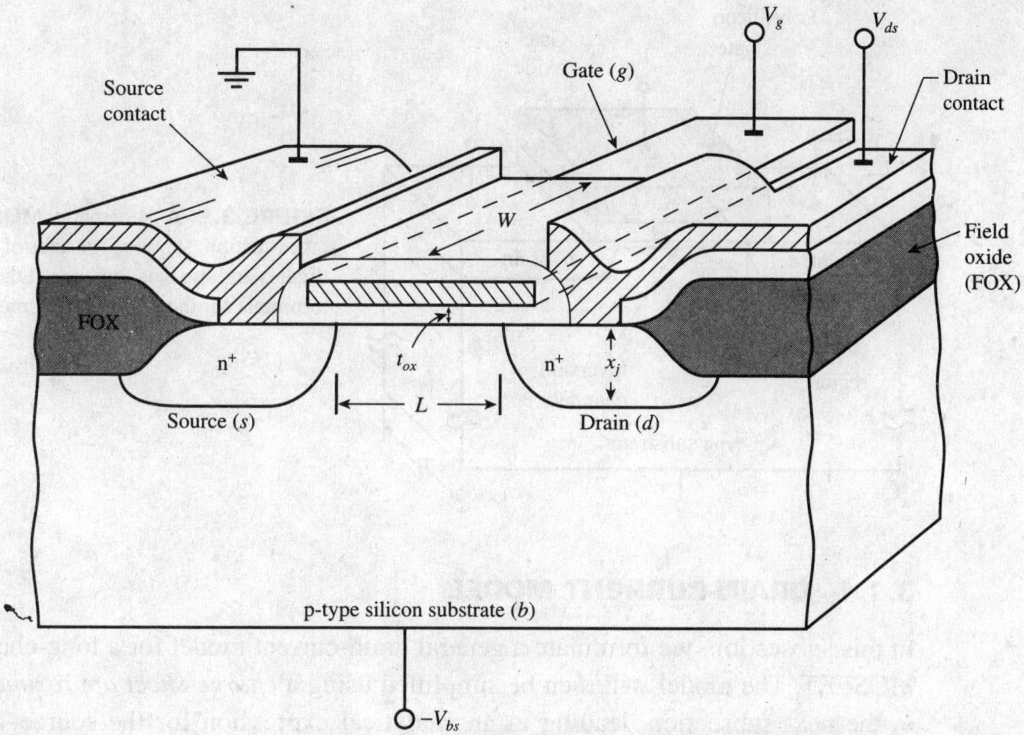


**FIGURE 3.1.** Three-dimensional view of basic MOSFET device structure. (After Arora, 1993.)

***coupled to the gate via the electric field in the oxide*** (hence the name *field-effect transistor*).

## 3.1 LONG-CHANNEL MOSFETs

This section describes the basic characteristics of a long-channel MOSFET, which will serve as the foundation for understanding the more important but more complex short-channel MOSFETs in Section 3.2. First, a general MOSFET current model based on the *gradual channel approximation* (GCA) is formulated in Section 3.1.1. The GCA is valid for most regions of MOSFET operation except beyond the pinch-off or saturation point. A *charge-sheet approximation* is then introduced in Section 3.1.2 to obtain analytical expressions for the source–drain current in the linear and saturation regions. Current characteristics in the subthreshold region are discussed in Section 3.1.3. Section 3.1.4 addresses the threshold-voltage dependence on substrate bias and temperature. Section 3.1.5 presents an empirical model for electron and hole mobilities in a MOSFET channel. Lastly, intrinsic MOSFET capacitances and inversion-layer capacitance effects (neglected in the charge sheet approximation) are covered in Section 3.1.6.

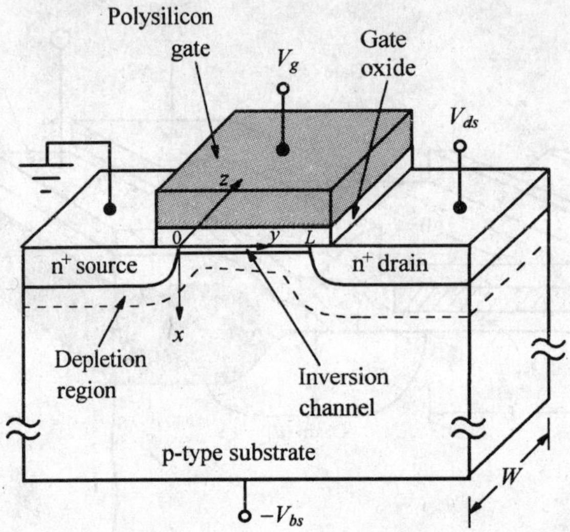


**FIGURE 3.2.** A schematic MOSFET cross section, showing the axes of coordinates and the bias voltages at the four terminals for the drain-current model.

### 3.1.1 DRAIN-CURRENT MODEL

In this subsection, we formulate a general drain-current model for a long-channel MOSFET. The model will then be simplified using a *charge-sheet approximation* in the next subsection, leading to an analytical expression for the source–drain current. Figure 3.2 shows the schematic cross section of an n-channel MOSFET in which the source is the $n^+$ region on the left, and the drain is the $n^+$ region on the right. A thin oxide film separates the gate from the channel region between the source and drain. We choose an $x$–$y$ coordinate system consistent with Section 2.3 on MOS capacitors, namely, the $x$-axis is perpendicular to the gate electrode and is pointing into the p-type substrate with $x = 0$ at the silicon surface. The $y$-axis is parallel to the channel or the current flow direction, with $y = 0$ at the source and $y = L$ at the drain. $L$ is called the *channel length* and is a key parameter in a MOSFET device. The MOSFET is assumed to be uniform along the $z$-axis over a distance called the *channel width*, $W$, determined by the boundaries of the thick field oxide.

Conventionally, the source voltage is defined as the ground potential. The drain voltage is $V_{ds}$, the gate voltage is $V_g$, and the p-type substrate is biased at $-V_{bs}$. Initially, we assume $V_{bs} = 0$, i.e., the substrate contact is grounded to the source potential. Later on, we will discuss the effect of substrate bias on MOSFET characteristics. The p-type substrate is assumed to be uniformly doped with an acceptor concentration $N_a$.

As defined in Section 2.3, $\psi(x, y)$ is the band bending, or intrinsic potential, at $(x, y)$ with respect to the bulk intrinsic potential. We further assume that $V(y)$ is the electron quasi-Fermi potential at a point $y$ along the channel with respect to the Fermi potential of the $n^+$ source. Since there is no bias between the source and the substrate, the Fermi potential of the source is the same as that of the bulk substrate. Therefore, ***in terms of the channel-to-substrate diode, V(y) plays the same role as the reverse bias $V_R$ in Section 2.3.5 on MOS capacitors under nonequilibrium***. As was discussed in Section 2.2.3, the quasi-Fermi potential stays essentially constant across the depletion region, i.e., $V(y)$ does not change with $x$ in the direction perpendicular to the surface. At the drain end of the channel, $V(y = L) = V_{ds}$.

#### 3.1.1.1 INVERSION CHARGE DENSITY AS A FUNCTION OF QUASI-FERMI POTENTIAL

From Eq. (2.150) and Eq. (2.187), the electron concentration at any point $(x, y)$ is given by

$$n(x, y) = \frac{n_i^2}{N_a} e^{q(\psi - V)/kT}. \tag{3.1}$$

Following the same approach as in Section 2.3.2, one obtains an expression for the electric field similar to that of Eq. (2.153):

$$\mathscr{E}^2(x, y) = \left(\frac{d\psi}{dx}\right)^2 = \frac{2kTN_a}{\varepsilon_{si}}\left[\left(e^{-q\psi/kT} + \frac{q\psi}{kT} - 1\right) + \frac{n_i^2}{N_a^2}\left(e^{-qV/kT}(e^{q\psi/kT} - 1) - \frac{q\psi}{kT}\right)\right]. \tag{3.2}$$

The condition for surface inversion, Eq. (2.190), becomes

$$\psi(0, y) = V(y) + 2\psi_B, \tag{3.3}$$

which is a function of $y$. From Eq. (2.191), the maximum depletion layer width is

$$W_{dm}(y) = \sqrt{\frac{2\varepsilon_{si}[V(y) + 2\psi_B]}{qN_a}}, \tag{3.4}$$

which is also a function of $y$.

#### 3.1.1.2 GRADUAL-CHANNEL APPROXIMATION

One of the key assumptions in any 1-D MOSFET model is the ***gradual channel approximation (GCA), which assumes that the variation of the electric field in the y-direction (along the channel) is much less than the corresponding variation in the x-direction (perpendicular to the channel)*** (Pao and Sah, 1966). This allows us to reduce Poisson's equation to the 1-D form ($x$-component only) as in Eq. (2.147). The GCA is valid for most of the channel regions except beyond the *pinch-off* point, which will be discussed later. One further assumes that both the hole current and the generation and recombination current are negligible, so that the current continuity equation can be applied to the electron current in the $y$-direction. In other words, the total drain-to-source current $I_{ds}$ is the same at any point along the channel. From Eq. (2.45), the electron current density at a point $(x, y)$ is

$$J_n(x, y) = -q\mu_n n(x, y)\frac{dV(y)}{dy}, \tag{3.5}$$

where $n(x, y)$ is the electron density, and $\mu_n$ is the electron mobility in the channel. The carrier mobility in the channel is generally much lower than the mobility in the bulk, due to additional surface scattering mechanisms, as will be addressed in Section 3.1.5. ***With V(y) defined as the quasi-Fermi potential, i.e., playing the role of $\phi_n$ in Eq. (2.45), Eq. (3.5) includes both the drift and diffusion currents***. The total current at a point $y$ along the channel is obtained by multiplying Eq. (3.5) with the channel width $W$ and integrating over the depth of the inversion layer. The integration is carried out from $x = 0$ to $x_i$, the bottom of the inversion layer where $\psi = \psi_B$:

$$I_{ds}(y) = qW\int_0^{x_i} \mu_n n(x, y)\frac{dV}{dy}\,dx. \tag{3.6}$$

There is a sign change, as we define $I_{ds} > 0$ to be the drain-to-source current in the $-y$ direction. Since $V$ is a function of $y$ only, $dV/dy$ can be taken outside the integral. We also assume that $\mu_n$ can be taken outside the integral by defining an *effective mobility*, $\mu_{eff}$, at some average gate and drain fields. What remains in the integral is the electron concentration, $n(x, y)$. Its integration over the inversion layer gives the inversion charge per unit gate area, $Q_i$:

$$Q_i(y) = -q\int_0^{x_i} n(x, y)\,dx. \tag{3.7}$$

Equation (3.6) then becomes

$$I_{ds}(y) = -\mu_{eff}W\frac{dV}{dy}Q_i(y) = -\mu_{eff}W\frac{dV}{dy}Q_i(V). \tag{3.8}$$

In the last step, $Q_i$ is expressed as a function of $V$; $V$ is interchangeable with $y$, since $V$ is a function of $y$ only. Multiplying both sides of Eq. (3.8) by $dy$ and integrating from 0 to $L$ (source to drain) yield

$$\int_0^L I_{ds}\,dy = \mu_{eff}W\int_0^{V_{ds}} [-Q_i(V)]\,dV. \tag{3.9}$$

Current continuity requires that $I_{ds}$ be a constant, independent of $y$. Therefore, the drain-to-source current is

$$I_{ds} = \mu_{eff} \frac{W}{L} \int_0^{V_{ds}} [-Q_i(V)]\, dV. \tag{3.10}$$

#### 3.1.1.3 PAO AND SAH'S DOUBLE INTEGRAL

An alternative form of $Q_i(V)$ can be derived if $n(x, y)$ is expressed as a function of $(\psi, V)$ using Eq. (3.1), i.e.,

$$n(x, y) = n(\psi, V) = \frac{n_i^2}{N_a} e^{q(\psi - V)/kT}, \tag{3.11}$$

and substituted into Eq. (3.7):

$$\begin{aligned} Q_i(V) &= -q \int_{\psi_s}^{\psi_B} n(\psi, V) \frac{dx}{d\psi}\, d\psi \\ &= -q \int_{\psi_B}^{\psi_s} \frac{(n_i^2/N_a) e^{q(\psi - V)/kT}}{\mathscr{E}(\psi, V)}\, d\psi. \end{aligned} \tag{3.12}$$

Here, $\psi_s$ is the surface potential at $x = 0$ and $\mathscr{E}(\psi, V) = -d\psi/dx$ is given by the square root of Eq. (3.2). Substituting Eq. (3.12) into Eq. (3.10) yields

$$I_{ds} = q \mu_{eff} \frac{W}{L} \int_0^{V_{ds}} \left( \int_{\psi_B}^{\psi_s} \frac{(n_i^2/N_a) e^{q(\psi - V)/kT}}{\mathscr{E}(\psi, V)}\, d\psi \right) dV. \tag{3.13}$$

This is referred to as *Pao and Sah's double integral* (Pao and Sah, 1966). The boundary value $\psi_s$ is determined by two coupled equations: Eq. (2.180) and $Q_s = -\varepsilon_{si}\mathscr{E}_s(\psi_s)$ or Gauss's law, where $\mathscr{E}_s(\psi_s)$ is obtained by letting $\psi = \psi_s$ in Eq. (3.2). In inversion, only two of the terms in Eq. (3.2) are significant and need to be kept. The merged equation is then

$$\begin{aligned} V_g &= V_{fb} + \psi_s - \frac{Q_s}{C_{ox}} \\ &= V_{fb} + \psi_s + \frac{\sqrt{2\varepsilon_{si} k T N_a}}{C_{ox}} \left[ \frac{q\psi_s}{kT} + \frac{n_i^2}{N_a^2} e^{q(\psi_s - V)/kT} \right]^{1/2}, \end{aligned} \tag{3.14}$$

which is an implicit equation for $\psi_s(V)$. Equations (3.14) and (3.13) can only be solved numerically.

### 3.1.2 MOSFET *I–V* CHARACTERISTICS

In this subsection, we derive the basic expressions for long-channel current in the *linear* and *saturation* regions.

#### 3.1.2.1 CHARGE-SHEET APPROXIMATION

In order to derive an analytical solution for the drain current, we simplify the general model using the charge-sheet approximation (Brews, 1978) in which the inversion-layer thickness is treated as zero. ***It assumes that all the inversion charges are located at the silicon surface like a sheet of charge and that there is no potential drop or band bending across the inversion layer***. Furthermore, the depletion approximation is applied to the bulk depletion region. After the onset of inversion, the surface potential is pinned at $\psi_s = 2\psi_B + V(y)$, as indicated by Eq. (3.3). From Eq. (3.4), the bulk depletion charge density is

$$Q_d = -qN_a W_{dm} = -\sqrt{2\varepsilon_{si} q N_a (2\psi_B + V)}. \tag{3.15}$$

The total charge density in the silicon is given by Eq. (2.180),

$$Q_s = -C_{ox}(V_g - V_{fb} - \psi_s) = -C_{ox}(V_g - V_{fb} - 2\psi_B - V). \tag{3.16}$$

The inversion charge density is then the difference of the above two equations,

$$\begin{aligned} Q_i &= Q_s - Q_d \\ &= -C_{ox}(V_g - V_{fb} - 2\psi_B - V) + \sqrt{2\varepsilon_{si} q N_a (2\psi_B + V)}. \end{aligned} \tag{3.17}$$

Substituting Eq. (3.17) into Eq. (3.10) and carrying out the integration, we obtain the drain current as a function of the gate and drain voltages:

$$\begin{aligned} I_{ds} = \mu_{eff} C_{ox} \frac{W}{L} \Bigg[ & \left( V_g - V_{fb} - 2\psi_B - \frac{V_{ds}}{2} \right) V_{ds} \\ & - \frac{2\sqrt{2\varepsilon_{si} q N_a}}{3C_{ox}} \left[ (2\psi_B + V_{ds})^{3/2} - (2\psi_B)^{3/2} \right] \Bigg]. \end{aligned} \tag{3.18}$$

Equation (3.18) represents the basic $I$–$V$ characteristics of a MOSFET device based on the charge-sheet model. It indicates that, for a given $V_g$, the drain current $I_{ds}$ first increases linearly with the drain voltage $V_{ds}$ (called the *linear* or *triode* region), then gradually levels off to a saturated value (*saturation* region). These two distinct regions are further examined below.

#### 3.1.2.2 CHARACTERISTICS IN THE LINEAR (TRIODE) REGION

When $V_{ds}$ is small, one can expand Eq. (3.18) into a power series in $V_{ds}$ and keep only the lowest-order (first-order) terms:

$$\begin{aligned} I_{ds} &= \mu_{eff} C_{ox} \frac{W}{L} \left( V_g - V_{fb} - 2\psi_B - \frac{\sqrt{4\varepsilon_{si} q N_a \psi_B}}{C_{ox}} \right) V_{ds} \\ &= \mu_{eff} C_{ox} \frac{W}{L} (V_g - V_t) V_{ds}, \end{aligned} \tag{3.19}$$

where $V_t$ is the *threshold voltage* given by

$$V_t = V_{fb} + 2\psi_B + \frac{\sqrt{4\varepsilon_{si} q N_a \psi_B}}{C_{ox}}. \tag{3.20}$$

Comparing this equation with Eq. (2.175) and Eq. (2.180), one can see that ***$V_t$ is simply the gate voltage when the surface potential or band bending reaches $2\psi_B$ and the silicon charge (the square root) is equal to the bulk depletion charge for that potential.*** As a reminder, $2\psi_B = (2kT/q)\ln(N_a/n_i)$, which is typically 0.6–0.9 V. When $V_g$ is below $V_t$, there is very little current flow and the MOSFET is said to be in the *subthreshold* region, to be discussed in Section 3.1.3. Equation (3.19) indicates that, ***in the linear region, the MOSFET simply acts like a resistor with a sheet resistivity, $\rho_{sh} = 1/[\mu_{eff} C_{ox}(V_g - V_t)]$, modulated by the gate voltage***. The threshold voltage $V_t$ can be determined by plotting $I_{ds}$ versus $V_g$ at low drain voltages, as shown in Fig. 3.3. The extrapolated intercept of the linear portion of the $I_{ds}(V_g)$ curve with the $V_g$-axis gives the approximate value of $V_t$. In reality, such a *linearly extrapolated threshold voltage* ($V_{on}$) is slightly higher than the "$2\psi_B$" $V_t$ due to inversion-layer capacitance and other effects, as will be addressed in Section 3.1.6. Notice that the $I_{ds}(V_g)$ curve is not linear near the threshold voltage. This is because the charge-sheet approximation, on which Eq. (3.19) is based, is no longer valid in that regime. Low-drain $I_{ds}(V_g)$ curves are also used to extract the *effective channel length* of a MOSFET, as will be discussed in Chapter 4.

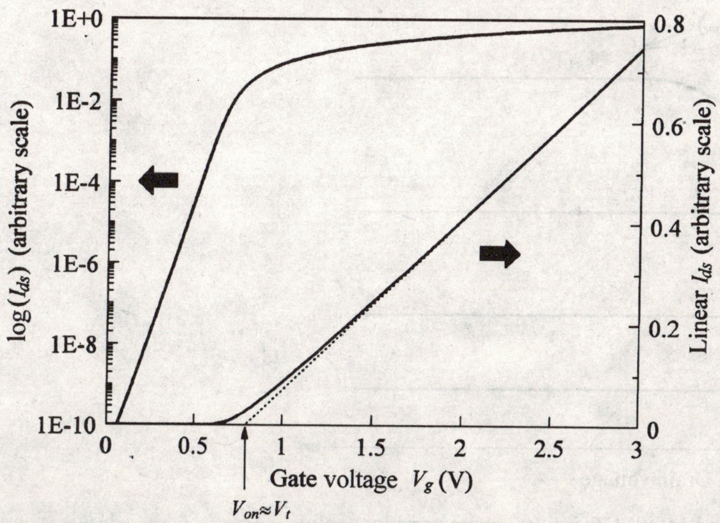


**FIGURE 3.3.** Typical MOSFET $I_{ds}$–$V_g$ characteristics at low drain bias voltages. The same current is plotted on both linear and logarithmic scales. The dotted line illustrates the determination of the linearly extrapolated threshold voltage, $V_{on}$.

### 3.1.2.3 CHARACTERISTICS IN THE SATURATION REGION

For larger values of $V_{ds}$, the second-order terms in the power series expansion of Eq. (3.18) are also important and must be kept. A good approximation to the drain current is then

$$I_{ds} = \mu_{eff} C_{ox} \frac{W}{L}\left((V_g - V_t)V_{ds} - \frac{m}{2}V_{ds}^2\right), \tag{3.21}$$

where

$$m = 1 + \frac{\sqrt{\varepsilon_{si} q N_a / 4\psi_B}}{C_{ox}} = 1 + \frac{C_{dm}}{C_{ox}} = 1 + \frac{3t_{ox}}{W_{dm}} \tag{3.22}$$

is the *body-effect coefficient*. Here $m$ typically lies between 1.1 and 1.4 and is related to the *body effect* to be discussed in Section 3.1.4. Equation (3.22) shows several alternative expressions for $m$. The one in terms of the capacitance ratio follows from Eq. (2.174), where $C_{dm}$ is the bulk depletion capacitance at $\psi_s = 2\psi_B$. Alternatively, $m$ can be expressed in terms of a thickness ratio, since $C_{dm} = \varepsilon_{si}/W_{dm}$, $C_{ox} = \varepsilon_{ox}/t_{ox}$, and $\varepsilon_{si}/\varepsilon_{ox} \approx 3$. The threshold voltage, given by Eq. (3.20), can be expressed in terms of $m$ as $V_t = V_{fb} + (2m - 1)2\psi_B$. ***As $V_{ds}$ increases, $I_{ds}$ follows a parabolic curve, as shown in Fig. 3.4, until a maximum or saturation value is reached.*** This occurs when $V_{ds} = V_{dsat} = (V_g - V_t)/m$, at which

$$I_{ds} = I_{dsat} = \mu_{eff} C_{ox} \frac{W}{L}\frac{(V_g - V_t)^2}{2m}. \tag{3.23}$$

Equation (3.23) reduces to the well-known expression for the MOSFET

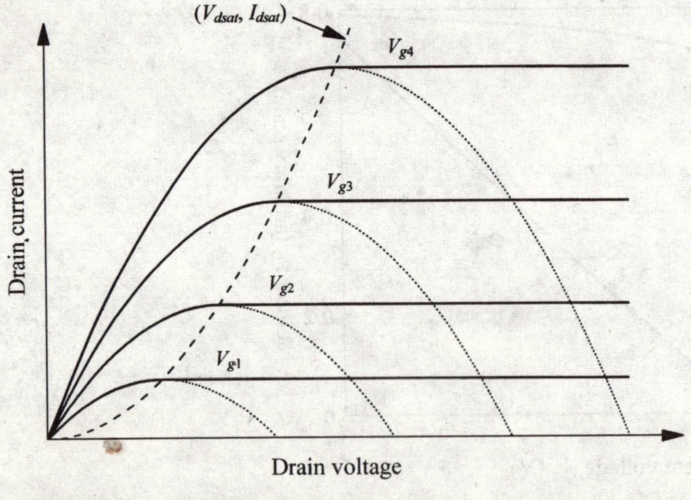


**FIGURE 3.4.** Long-channel MOSFET $I_{ds}$–$V_{ds}$ characteristics (solid curves) for several different values of $V_g$. The dashed curve shows the trajectory of drain voltage beyond which the current saturates. The dotted curves help to illustrate the parabolic behavior of the characteristics before saturation.

saturation current when the bulk depletion charge is neglected (valid for low substrate doping) so $m = 1$. The dashed curve in Fig. 3.4 shows the trajectory of $V_{dsat}$ through the various $I_{ds}$–$V_{ds}$ curves for different $V_g$. Equation (3.21), or Eq. (3.18), is valid only for $V_{ds} \le V_{dsat}$. ***Beyond $V_{dsat}$, $I_{ds}$ stays constant at $I_{dsat}$, independent of $V_{ds}$.***

#### 3.1.2.4 THE ONSET OF PINCH-OFF AND CURRENT SATURATION

The saturation of drain current can be understood from the inversion charge density, Eq. (3.17). For $V \le 2\psi_B$, one can expand the square-root term of Eq. (3.17) into a power series in $V$ and keep only the two lowest terms,

$$Q_i(V) = -C_{ox}(V_g - V_t - mV). \tag{3.24}$$

$Q_i(V)$ is plotted in Fig. 3.5. Equation (3.10) states that the drain current is proportional to the area under the $-Q_i(V)$ curve between $V = 0$ and $V_{ds}$. When $V_{ds}$ is small (linear region), the inversion charge density at the drain end of the channel is only slightly lower than that at the source end. As the drain voltage increases (for a fixed gate voltage), the current increases, but the inversion charge density at the drain decreases until finally it goes to zero when $V_{ds} = V_{dsat} = (V_g - V_t)/m$. At this point, $I_{ds}$ reaches its maximum value. In other words, ***the surface channel vanishes at the drain end of the channel when saturation occurs***. This is called *pinch-off* and is illustrated in Fig. 3.6. When $V_{ds}$ increases beyond saturation, the pinch-off point moves toward the source, but the drain current remains essentially the same. This is because for $V_{ds} > V_{dsat}$, the voltage at the pinch-off point remains

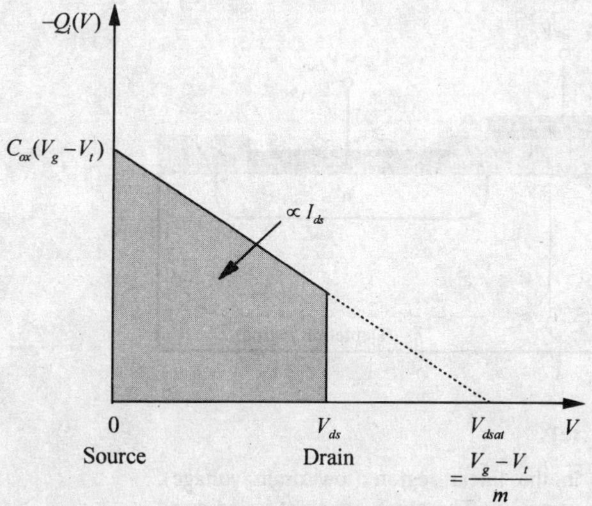


**FIGURE 3.5.** Inversion charge density as a function of the quasi-Fermi potential of a point in the channel. Before saturation, the drain current is proportional to the shaded area integrated from zero to the drain voltage.

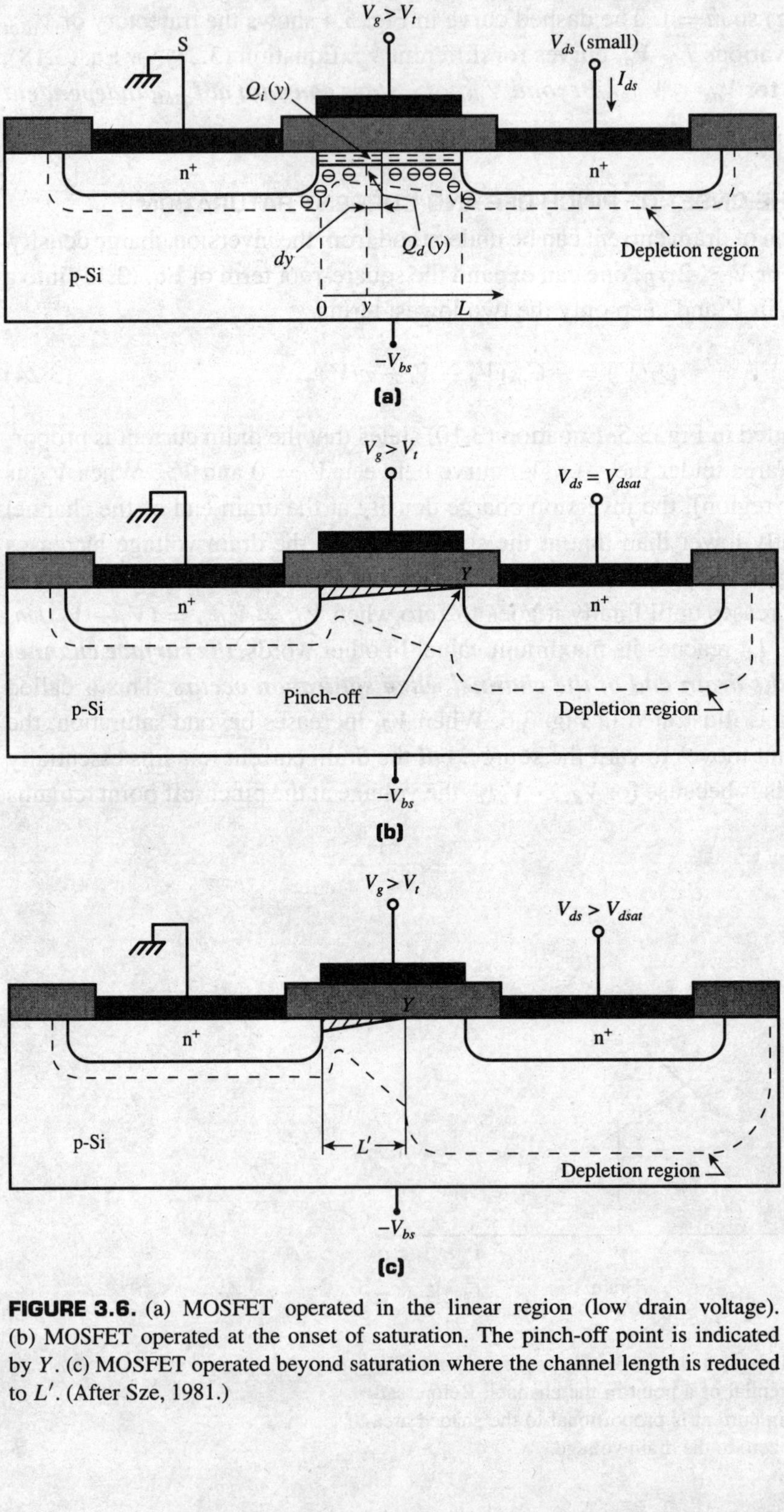


**FIGURE 3.6.** (a) MOSFET operated in the linear region (low drain voltage). (b) MOSFET operated at the onset of saturation. The pinch-off point is indicated by $Y$. (c) MOSFET operated beyond saturation where the channel length is reduced to $L'$. (After Sze, 1981.)

at $V_{dsat}$ and the current, given by

$$\int_0^{L'} I_{ds}\,dy = \mu_{eff} W \int_0^{V_{dsat}} [-Q_i(V)]\,dV, \tag{3.25}$$

stays the same apart from a slight decrease in $L$ (to $L'$), as shown in Fig. 3.6. This phenomenon is called *channel length modulation* and will be discussed in association with short-channel MOSFETs in Section 3.2.

Further insight into the MOSFET behavior at pinch-off can be gained by examining the function $V(y)$. Integrating from 0 to $y$ after multiplying both sides of Eq. (3.8) by $dy$ yields

$$\begin{aligned} I_{ds} y &= \mu_{eff} W \int_0^{V} [-Q_i(V)]\,dV \qquad (3.26) \\ &= \mu_{eff} C_{ox} W \left( (V_g - V_t)V - \frac{m}{2} V^2 \right). \end{aligned}$$

Substituting $I_{ds}$ from Eq. (3.21) into Eq. (3.26), one can solve for $V(y)$:

$$V(y) = \frac{V_g - V_t}{m} - \sqrt{\left(\frac{V_g - V_t}{m}\right)^2 - 2\frac{y}{L}\left(\frac{V_g - V_t}{m}\right) V_{ds} + \frac{y}{L} V_{ds}^2}. \tag{3.27}$$

Both $V(y)$ and $-Q_i/mC_{ox} = (V_g - V_t)/m - V(y)$ are plotted in Fig. 3.7 for several values of $V_{ds}$. At low $V_{ds}$, $V(y)$ varies almost linearly between the source and drain. As $V_{ds}$ increases, the inversion charge density at the drain decreases due to the

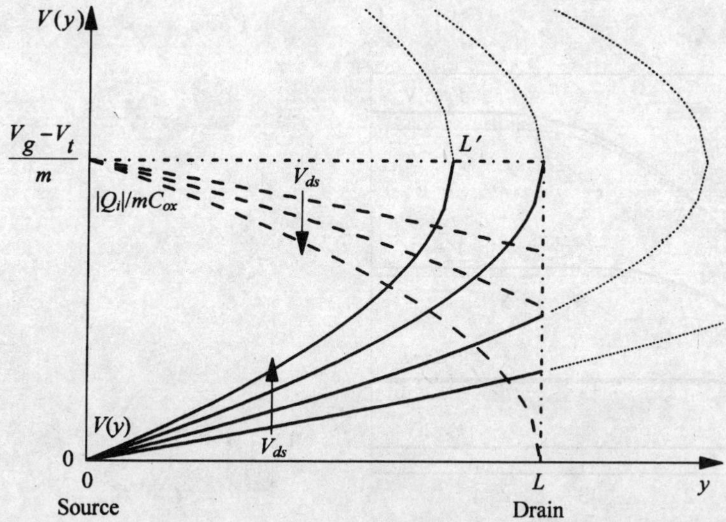


**FIGURE 3.7.** Quasi-Fermi potential versus distance between the source and the drain for several $V_{ds}$-values from the linear region to beyond saturation. The dashed curves show the corresponding variation of inversion charge density along the channel. The dotted curves help visualize the parabolic behavior of the characteristics.

lowering of the electron quasi-Fermi level. This is accompanied by a corresponding increase of $dV/dy$ to maintain current continuity. When $V_{ds}$ reaches $V_{dsat} = (V_g - V_t)/m$, we have $Q_i(y = L) = 0$ and $V(y)$ exhibits a singularity at the drain, where $dV/dy = \infty$. This implies that ***the electric field in the y-direction changes more rapidly than the field in the x-direction and the gradual channel approximation breaks down***. In other words, beyond the pinch-off point, carriers are no longer confined to the surface channel, and a 2-D Poisson's equation must be solved for carrier injection from the pinch-off point into the drain depletion region (El-Mansy and Boothroyd, 1977).

Strictly speaking, if $V_{ds} > 2\psi_B$, neither Eq. (3.17) nor Eq. (3.18) can be expanded into a power series in $V_{ds}$. A more general form of the saturation voltage is obtained by letting $Q_i = 0$ in Eq. (3.17) and solving for $V = V_{dsat}$ [equivalent to solving $dI_{ds}/dV_{ds} = 0$ by differentiating Eq. (3.18)]:

$$V_{dsat} = V_g - V_{fb} - 2\psi_B + \frac{\varepsilon_{si} q N_a}{C_{ox}^2} - \sqrt{\frac{2\varepsilon_{si} q N_a}{C_{ox}^2}\left(V_g - V_{fb} + \frac{\varepsilon_{si} q N_a}{2C_{ox}^2}\right)}. \tag{3.28}$$

The corresponding saturation current can be found by substituting Eq. (3.28) for $V_{ds}$ in Eq. (3.18). The mathematics is rather tedious (Brews, 1981). A few selected curves are plotted in Fig. 3.8 and compared with those calculated from Eq. (3.21). It turns out that Eq. (3.21) serves as a good approximation to the drain current over a much wider range of voltages than expected. Even for a drain voltage several times greater than $2\psi_B$, the current is only slightly (≈5%) underestimated.

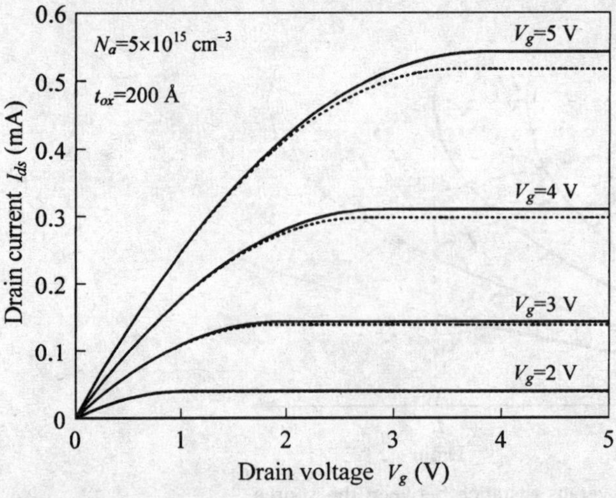


**FIGURE 3.8.** $I_{ds}$–$V_{ds}$ curves calculated from the full equation (3.18) (solid curves), compared with the parabolic approximation (3.21) (dotted curves).

### 3.1.2.5 pMOSFET *I–V* CHARACTERISTICS

So far we have used an n-channel device as an example to discuss MOSFET operation and $I$–$V$ characteristics. A p-channel MOSFET operates similarly, except that it is fabricated inside an n-well with implanted $p^+$ source and drain regions (cf. Fig. 3.2), and that the polarities of all the voltages and currents are reversed. For example, $I_{ds}$–$V_{ds}$ characteristics for a pMOSFET (cf. Fig. 3.8) have negative gate and drain voltages with respect to the source terminal for a hole current to flow from the source to the drain.

Since the source of a pMOSFET is at the highest potential compared with the other terminals, it is usually connected to the power supply $V_{dd}$ in a CMOS circuit so that all the voltages are positive (or zero). In that case, the device conducts if the gate voltage is lower than $V_{dd} - V_t$, where $V_t$ (>0) is the magnitude of the threshold voltage of the pMOSFET. The ohmic contact to the n-well is also connected to $V_{dd}$, in contrast to an nMOSFET, where the p-type substrate is usually tied to the ground potential. This leaves the n-well-to-p-substrate junction reverse biased as required. More about nMOSFET and pMOSFET bias conditions in a CMOS circuit configuration will be given in Section 5.1.

## 3.1.3 SUBTHRESHOLD CHARACTERISTICS

Depending on the gate and source–drain voltages, a MOSFET device can be biased in one of the three regions shown in Fig. 3.9. Linear and saturation region characteristics have been described in the previous subsection. In this subsection, we discuss the characteristics of a MOSFET device in the subthreshold region where $V_g < V_t$. In Fig. 3.3, the drain current on a linear scale appears to approach zero immediately below the threshold voltage. On a logarithmic scale, however, the descending drain current remains at nonnegligible levels for several tenths of a volt below $V_t$. This is because the inversion charge density does not drop to zero abruptly. Rather,

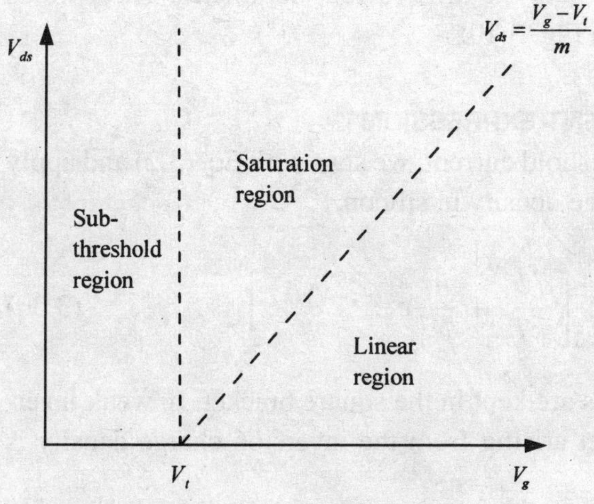


**FIGURE 3.9.** Three regions of MOSFET operation in the $V_{ds}$–$V_g$ plane.

it follows an exponential dependence on $\psi_s$ or $V_g$, as is evident from Eq. (3.11). Subthreshold behavior is of particular importance in low-voltage, low-power applications, such as in digital logic and memory circuits, because it describes how a MOSFET device switches off. The subthreshold region immediately below $V_t$, in which $\psi_B \leq \psi_s \leq 2\psi_B$, is also called the *weak inversion* region.

#### 3.1.3.1 DRIFT AND DIFFUSION COMPONENTS OF DRAIN CURRENT

***Unlike the strong inversion region, in which the drift current dominates, subthreshold conduction is dominated by the diffusion current.*** Both current components are included in Pao and Sah's double integral, Eq. (3.13). In general, current continuity only applies to the total current, not to its individual components. In other words, the fractional ratio between the drift and the diffusion components may vary from one point of the channel to another. At low drain bias voltages, however, it is possible to separate the drift and diffusion components using the implicit $\psi_s(V)$ relation, Eq. (3.14). When $qV/kT \ll 1$, only the first-order terms of $V$ need to be kept. In Eq. (3.8), $Q_i(V)$ can be replaced by its zeroth-order value, $Q_i(V=0)$; hence $V$ must vary linearly from the source to the drain, as required by current continuity. Since the total current is proportional to $dV/dy$ and the drift current is proportional to the electric field or $d\psi_s/dy$, the drift fraction of the current is given by the change of surface potential (band bending) with respect to the quasi-Fermi potential, i.e., $d\psi_s/dV$. This can be evaluated from Eq. (3.14) in the limit of $V \to 0$:

$$\frac{d\psi_s}{dV} = \frac{(n_i^2/N_a^2)e^{q\psi_s/kT}}{1 + (n_i^2/N_a^2)e^{q\psi_s/kT} + (C_{ox}^2/\varepsilon_{si}qN_a)(|Q_s|/C_{ox})}, \tag{3.29}$$

where $|Q_s|/C_{ox}$ is the voltage drop across the oxide given by the last term of Eq. (3.14). It is clear that in weak inversion where $\psi_B < \psi_s < 2\psi_B$, the numerator is much less than unity and the diffusion component dominates. Conversely, beyond strong inversion, $d\psi_s/dV \approx 1$ and the drift current dominates. These kinds of behavior are further illustrated in Fig. 3.10.

#### 3.1.3.2 SUBTHRESHOLD CURRENT EXPRESSION

To find an expression for the subthreshold current, we start with Eq. (3.2) and apply Gauss's law to obtain the total charge density in silicon,

$$-Q_s = \varepsilon_{si}\mathscr{E}_s = \sqrt{2\varepsilon_{si}kTN_a}\left[\frac{q\psi_s}{kT} + \frac{n_i^2}{N_a^2}e^{q(\psi_s - V)/kT}\right]^{1/2}, \tag{3.30}$$

where only the two significant terms are kept in the square bracket. In weak inversion, the second term in the bracket arising from the inversion charge density is

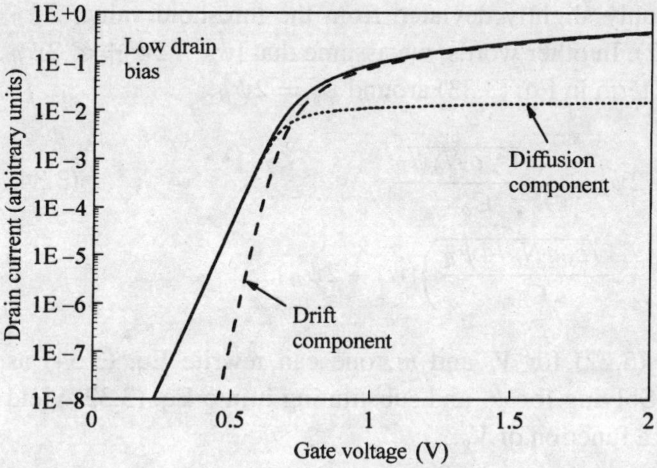


**FIGURE 3.10.** Drift and diffusion components of current in an $I_{ds}$–$V_g$ plot. Their sum is the total current represented by the solid curve.

much less than the first term from the depletion charge density. Equation (3.30) can then be expanded into a power series: the zeroth-order term is the depletion charge density $-Q_d$, and the first-order term gives the inversion charge density,

$$-Q_i = \sqrt{\frac{\varepsilon_{si} q N_a}{2\psi_s}} \left(\frac{kT}{q}\right)\left(\frac{n_i}{N_a}\right)^2 e^{q(\psi_s - V)/kT}. \tag{3.31}$$

The surface potential $\psi_s$ is related to the gate voltage through Eq. (3.14). Since the inversion charge density is small, $\psi_s$ can be considered as a function of $V_g$ only, independent of $V$. This also means that the electric field along the channel direction is small; hence the drift current is negligible.

Substituting $Q_i$ into Eq. (3.10) and carrying out the integration, we obtain the drain current in the subthreshold region:

$$I_{ds} = \mu_{eff} \frac{W}{L} \sqrt{\frac{\varepsilon_{si} q N_a}{2\psi_s}} \left(\frac{kT}{q}\right)^2 \left(\frac{n_i}{N_a}\right)^2 e^{q\psi_s/kT}(1 - e^{-qV_{ds}/kT}). \tag{3.32}$$

$\psi_s$ can be expressed in terms of $V_g$ using Eq. (3.14), where only the depletion charge term needs to be kept:

$$V_g = V_{fb} + \psi_s + \frac{\sqrt{2\varepsilon_{si} q N_a \psi_s}}{C_{ox}}. \tag{3.33}$$

It is straightforward to solve a quadratic equation for $\psi_s$. To further simplify the result, we consider $\psi_s$ as only slightly deviated from the threshold value, $2\psi_B$ (Swanson and Meindl, 1972). In other words, we assume that $|\psi_s - 2\psi_B| \ll 2\psi_B$ and expand the square-root term in Eq. (3.33) around $\psi_s = 2\psi_B$:

$$V_g = V_{fb} + 2\psi_B + \frac{\sqrt{4\varepsilon_{si} q N_a \psi_B}}{C_{ox}} + \left(1 + \frac{\sqrt{\varepsilon_{si} q N_a / 4\psi_B}}{C_{ox}}\right)(\psi_s - 2\psi_B). \tag{3.34}$$

Using Eq. (3.20) and Eq. (3.22) for $V_t$ and $m$, one can rewrite Eq. (3.34) as $V_g = V_t + m(\psi_s - 2\psi_B)$. Solving for $\psi_s$ and substituting it into Eq. (3.32) yield the subthreshold current as a function of $V_g$:

$$I_{ds} = \mu_{eff} \frac{W}{L} \sqrt{\frac{\varepsilon_{si} q N_a}{4\psi_B}} \left(\frac{kT}{q}\right)^2 e^{q(V_g - V_t)/mkT} (1 - e^{-qV_{ds}/kT}), \tag{3.35}$$

or

$$I_{ds} = \mu_{eff} C_{ox} \frac{W}{L} (m - 1) \left(\frac{kT}{q}\right)^2 e^{q(V_g - V_t)/mkT} (1 - e^{-qV_{ds}/kT}). \tag{3.36}$$

#### 3.1.3.3 SUBTHRESHOLD SLOPE

The subthreshold current is independent of the drain voltage once $V_{ds}$ is larger than a few $kT/q$, as would be expected for diffusion-dominated current transport. The dependence on gate voltage, on the other hand, is exponential with an *inverse subthreshold slope* (Fig. 3.10),

$$S = \left(\frac{d(\log_{10} I_{ds})}{dV_g}\right)^{-1} = 2.3\frac{mkT}{q} = 2.3\frac{kT}{q}\left(1 + \frac{C_{dm}}{C_{ox}}\right), \tag{3.37}$$

of typically 70–100 mV/decade. Here $m = 1 + (C_{dm}/C_{ox})$ from Eq. (3.22). If the $Si–SiO_2$ interface trap density is high, the subthreshold slope may be more graded than that given by Eq. (3.37), since the capacitance associated with the interface trap is in parallel with the depletion-layer capacitance $C_{dm}$. It should be noted that for $\psi_s$ substantially below $2\psi_B$, e.g., when $V_g$ is a few tenths of a volt below $V_t$, Eq. (3.37) tends to underestimate the inverse subthreshold slope by 5–10%. As a result, the subthreshold current can be 2 to 4 times higher than that given by Eq. (3.36). For VLSI circuits, a steep subthreshold slope is desirable for the ease of switching the transistor current off. However, ***except for a slight dependence on bulk doping concentration through $C_{dm}$, the subthreshold slope is rather insensitive to device***

***parameters***. It is only a function of temperature. This has significant implications on device scaling as will be discussed in Chapter 4.

### 3.1.4 SUBSTRATE BIAS AND TEMPERATURE DEPENDENCE OF THRESHOLD VOLTAGE

The threshold voltage is one of the key parameters of a MOSFET device. In this subsection, we examine the dependence of threshold voltage on substrate bias and temperature.

#### 3.1.4.1 SUBSTRATE SENSITIVITY (BODY EFFECT)

The drain-current equation in Section 3.1.2 was derived assuming zero substrate bias ($V_{bs}$). If $V_{bs} \neq 0$, one can modify the previously discussed MOSFET equations by considering that applying $-V_{bs}$ to the substrate is equivalent to raising all other voltages (namely, gate, source, and drain voltages) by $+V_{bs}$ while keeping the substrate grounded. This is shown in Fig. 3.11.

Using the charge-sheet model as before, Eq. (3.17) becomes

$$Q_i = -C_{ox}(V_g + V_{bs} - V_{fb} - 2\psi_B - V) + \sqrt{2\varepsilon_{si} q N_a (2\psi_B + V)}, \qquad (3.38)$$

where $V$ is the reverse bias voltage between a point in the channel and the substrate.

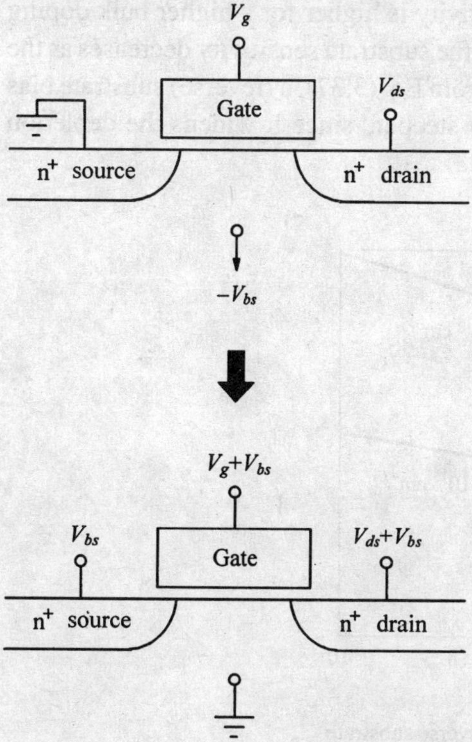


**FIGURE 3.11.** Equivalent circuits used to evaluate the effect of substrate bias on MOSFET $I$–$V$ characteristics.

The current is obtained by integrating $Q_i$ from $V_{bs}$ (source) to $V_{bs} + V_{ds}$ (drain):

$$I_{ds} = \mu_{eff} C_{ox} \frac{W}{L} \left[ \left( V_g - V_{fb} - 2\psi_B - \frac{V_{ds}}{2} \right) V_{ds} - \frac{2\sqrt{2\varepsilon_{si} q N_a}}{3C_{ox}} \left[ (2\psi_B + V_{bs} + V_{ds})^{3/2} - (2\psi_B + V_{bs})^{3/2} \right] \right]. \tag{3.39}$$

At low drain voltages (linear region), the current is still given by Eq. (3.19), except that the threshold voltage is now

$$V_t = V_{fb} + 2\psi_B + \frac{\sqrt{2\varepsilon_{si} q N_a (2\psi_B + V_{bs})}}{C_{ox}}. \tag{3.40}$$

It can be seen from Eq. (3.40) that ***the effect of (reverse) substrate bias is to widen the bulk depletion region and raise the threshold voltage***. Figure 3.12 plots $V_t$ as a function of $V_{bs}$. The slope of the curve,

$$\frac{dV_t}{dV_{bs}} = \frac{\sqrt{\varepsilon_{si} q N_a / 2(2\psi_B + V_{bs})}}{C_{ox}}, \tag{3.41}$$

is referred to as the *substrate sensitivity*. At $V_{bs} = 0$, the slope equals $C_{dm}/C_{ox}$, or $m - 1$ [Eq. (3.22)]. The substrate sensitivity is higher for a higher bulk doping concentration. It is clear from Fig. 3.12 that the substrate sensitivity decreases as the substrate (reverse) bias voltage increases. From Eq. (3.37), a (reverse) substrate bias also makes the subthreshold slope slightly steeper, since it widens the depletion region and lowers $C_{dm}$.

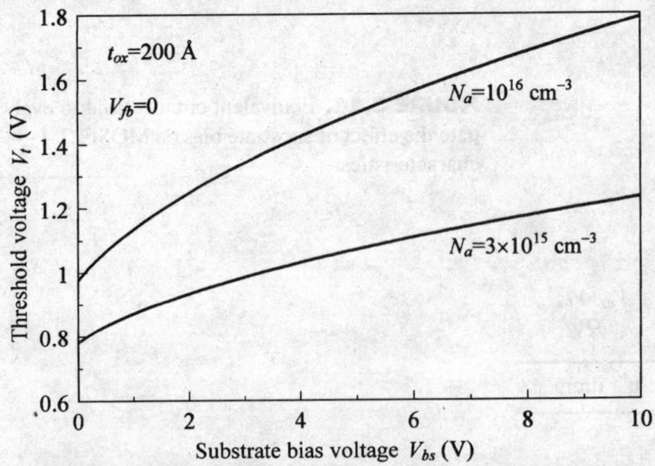


**FIGURE 3.12.** Threshold-voltage variation with reverse substrate bias for two uniform substrate doping concentrations.

### 3.1.4.2 TEMPERATURE DEPENDENCE OF THRESHOLD VOLTAGE

Next, we examine the temperature dependence of the threshold voltage. The flat-band voltage of an nMOSFET with n$^+$ polysilicon gate is $V_{fb} = -E_g/2q - \psi_B$ [Eq. (2.181)], assuming there is no oxide charge. Substituting it into Eq. (3.20) yields the threshold voltage,

$$V_t = -\frac{E_g}{2q} + \psi_B + \frac{\sqrt{4\varepsilon_{si} q N_a \psi_B}}{C_{ox}}, \tag{3.42}$$

at zero substrate bias. The temperature dependence of $V_t$ is related to the temperature dependence of $E_g$ and $\psi_B$:

$$\begin{aligned}\frac{dV_t}{dT} &= -\frac{1}{2q}\frac{dE_g}{dT} + \left(1 + \frac{\sqrt{\varepsilon_{si} q N_a/\psi_B}}{C_{ox}}\right)\frac{d\psi_B}{dT} \\ &= -\frac{1}{2q}\frac{dE_g}{dT} + (2m-1)\frac{d\psi_B}{dT}.\end{aligned} \tag{3.43}$$

$d\psi_B/dT$ stems from the temperature dependence of the intrinsic carrier concentration, which can be evaluated using Eq. (2.37) and Eq. (2.7):

$$\begin{aligned}\frac{d\psi_B}{dT} &= \frac{d}{dT}\left[\frac{kT}{q}\ln\left(\frac{N_a}{\sqrt{N_c N_v}\, e^{-E_g/2kT}}\right)\right] \\ &= -\frac{k}{q}\ln\left(\frac{\sqrt{N_c N_v}}{N_a}\right) - \frac{kT}{q\sqrt{N_c N_v}}\frac{d\sqrt{N_c N_v}}{dT} + \frac{1}{2q}\frac{dE_g}{dT}.\end{aligned} \tag{3.44}$$

Since both $N_c$ and $N_v$ are proportional to $T^{3/2}$, we have $d(N_c N_v)^{1/2}/dT = \frac{3}{2}(N_c N_v)^{1/2}/T$. Substituting Eq. (3.44) into Eq. (3.43) yields

$$\frac{dV_t}{dT} = -(2m-1)\frac{k}{q}\left[\ln\left(\frac{\sqrt{N_c N_v}}{N_a}\right) + \frac{3}{2}\right] + \frac{m-1}{q}\frac{dE_g}{dT}. \tag{3.45}$$

From Section 2.1.1 and Table 2.1, $dE_g/dT \approx -2.7 \times 10^{-4}$ eV/K and $(N_c N_v)^{1/2} \approx 2.4 \times 10^{19}$ cm$^{-3}$. For $N_a \sim 10^{16}$ cm$^{-3}$ and $m \approx 1.1$, ***$dV_t/dT$ is typically −1 mV/K.*** Note that the temperature coefficient decreases slightly as $N_a$ increases: for $N_a \sim 10^{18}$ cm$^{-3}$ and $m \approx 1.3$, $dV_t/dT$ is about −0.7 mV/K. These numbers imply that, at an elevated temperature of, for example, 100°C, the threshold voltage is 55–75 mV lower than at room temperature. Since digital VLSI circuits often operate at elevated temperatures due to heat generation, this effect, plus the degradation of subthreshold slope with temperature, causes the leakage current at $V_g = 0$ to increase considerably over its room-temperature value. Typically, the off-state leakage current of a MOSFET at 100°C is 30–50 times larger than the leakage current at 25°C. These are important design considerations, to be addressed in detail in Chapter 4.

### 3.1.5 MOSFET CHANNEL MOBILITY

The carrier mobility in a MOSFET channel is significantly lower than that in bulk silicon, due to additional scattering mechanisms. Lattice or phonon scattering is aggravated by the presence of crystalline discontinuity at the surface boundary, and surface roughness scattering severely degrades mobility at high normal fields. Channel mobility is also affected by processing conditions that alter the Si–$SiO_2$ interface properties (e.g., oxide charge and interface traps, as discussed in Section 2.3.6).

#### 3.1.5.1 EFFECTIVE MOBILITY AND EFFECTIVE NORMAL FIELD

In Section 3.1.1, the channel mobility was treated as a constant by defining an effective mobility as

$$\mu_{eff} = \frac{\int_0^{x_i} \mu_n n(x)\, dx}{\int_0^{x_i} n(x)\, dx}, \tag{3.46}$$

which is essentially an average value weighted by the carrier concentration in the inversion layer. Empirically, it has been found that ***when $\mu_{eff}$ is plotted against an effective normal field $\mathscr{E}_{eff}$, there exists a universal relationship independent of the substrate bias, doping concentration, and gate oxide thickness*** (Sabnis and Clemens, 1979). The effective normal field is defined as the average electric field perpendicular to the Si–$SiO_2$ interface experienced by the carriers in the channel. Using Gauss's law, one can express $\mathscr{E}_{eff}$ in terms of the depletion and inversion charge densities:

$$\mathscr{E}_{eff} = \frac{1}{\varepsilon_{si}}\left(|Q_d| + \frac{1}{2}|Q_i|\right), \tag{3.47}$$

where $|Q_d| + \frac{1}{2}|Q_i|$ is the total silicon charge inside a Gaussian surface through the middle of the inversion layer. Using Eq. (2.161) and Eq. (3.20), the depletion charge can be expressed as

$$|Q_d| = \sqrt{4\varepsilon_{si} q N_a \psi_B} = C_{ox}(V_t - V_{fb} - 2\psi_B). \tag{3.48}$$

Substituting this expression and $|Q_i| \approx C_{ox}(V_g - V_t)$ into Eq. (3.47) yields

$$\mathscr{E}_{eff} = \frac{V_t - V_{fb} - 2\psi_B}{3t_{ox}} + \frac{V_g - V_t}{6t_{ox}}, \tag{3.49}$$

where $C_{ox} = \varepsilon_{ox}/t_{ox}$ and $\varepsilon_{si} \approx 3\varepsilon_{ox}$ were used. Equation (3.49) can further be simplified if the gate electrode is n$^+$ polysilicon (for nMOSFETs) such that $V_{fb} = -E_g/2q - \psi_B$. For submicron CMOS technologies, $\psi_B = 0.30$–$0.42$ V. Therefore, the effective normal field can be expressed in terms of explicit device parameters as

$$\mathscr{E}_{eff} = \frac{V_t + 0.2}{3t_{ox}} + \frac{V_g - V_t}{6t_{ox}}. \tag{3.50}$$

Equation (3.50) is valid for low drain voltages. At high drain voltages, $Q_i$ decreases toward the drain end of the channel. To estimate the average effective field in that case, the second term in Eq. (3.50) should be reduced accordingly.

#### 3.1.5.2 ELECTRON MOBILITY DATA

A typical set of data on mobility versus effective normal field for nMOSFETs is shown in Fig. 3.13 (Takagi *et al.*, 1988). At room temperature, the mobility follows a $\mathscr{E}_{eff}^{-1/3}$ dependence below $5 \times 10^5$ V/cm. A simple, approximate expression for this case is (Baccarani and Wordeman, 1983)

$$\mu_{eff} \approx 32500 \times \mathscr{E}_{eff}^{-1/3}. \tag{3.51}$$

Beyond $\mathscr{E}_{eff} = 5 \times 10^5$ V/cm, $\mu_{eff}$ decreases much more rapidly with increasing $\mathscr{E}_{eff}$ because of increased surface roughness scattering as carriers are distributed closer to the surface under high normal fields. For each doping concentration, there exists an effective field below which the mobility falls off the universal curve. This is

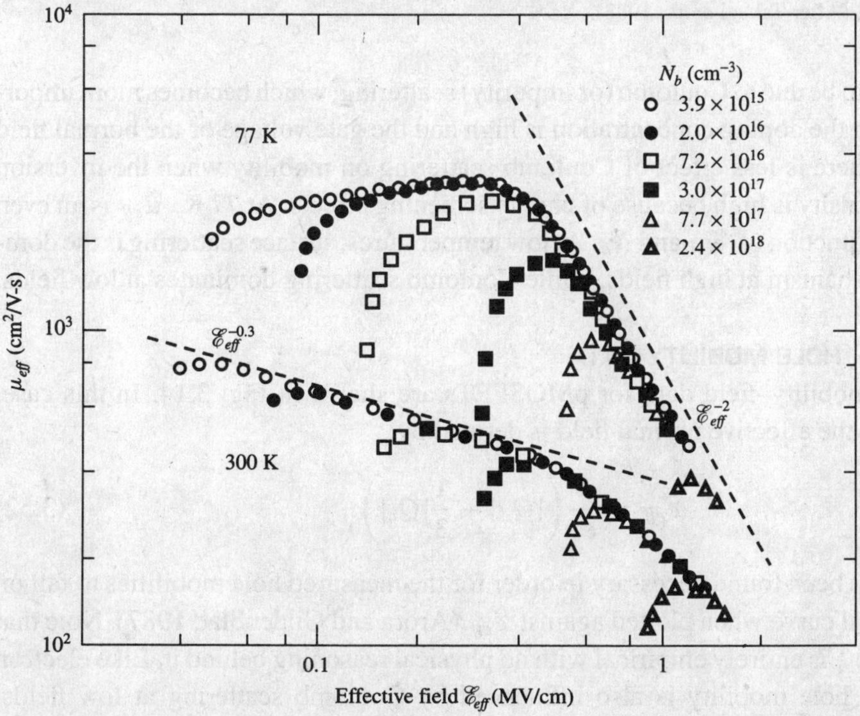


**FIGURE 3.13.** Measured electron mobility at 300 and 77 K versus effective normal field for several substrate doping concentrations. (After Takagi *et al.*, 1988).

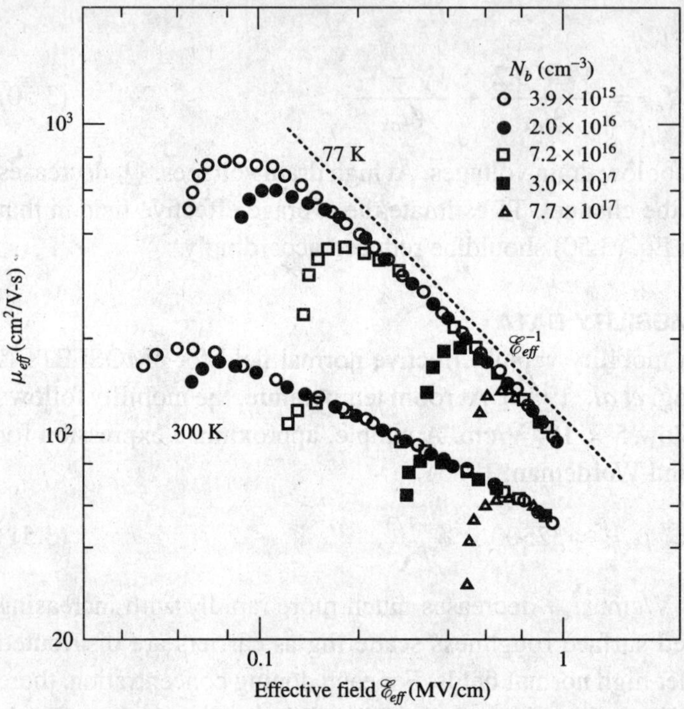


**FIGURE 3.14.** Measured hole mobility at 300 K and 77 K versus effective normal field (with a factor $\frac{1}{3}$) for several substrate doping concentrations. (After Takagi *et al.*, 1988).

believed to be due to Coulomb (or impurity) scattering, which becomes more important when the doping concentration is high and the gate voltage or the normal field is low. There is less effect of Coulomb scattering on mobility when the inversion charge density is high because of charge screening effects. At 77 K, $\mu_{eff}$ is an even stronger function of $\mathscr{E}_{eff}$ and $N_a$. At low temperatures, surface scattering is the dominant mechanism at high fields, while Coulomb scattering dominates at low fields.

### 3.1.5.3 HOLE MOBILITY DATA

Similar mobility–field data for pMOSFETs are shown in Fig. 3.14. In this case, however, the effective normal field is defined by

$$\mathscr{E}_{eff} = \frac{1}{\varepsilon_{si}}\left(|Q_d| + \frac{1}{3}|Q_i|\right), \tag{3.52}$$

which has been found necessary in order for the measured hole mobilities to fall on a universal curve when plotted against $\mathscr{E}_{eff}$ (Arora and Gildenblat, 1987). Note that the factor $\frac{1}{3}$ is entirely empirical with no physical reasoning behind it. Like electron mobility, hole mobility is also influenced by Coulomb scattering at low fields, depending on the doping concentration. The field dependence is also stronger at 77 K, but not quite as strong as in the electron case. It should be noted that the hole mobility data were taken from surface-channel pMOSFETs with $p^+$ polysilicon gate. Buried-channel pMOSFETs with $n^+$ polysilicon gate have a higher mobility (by about 30%) for the same threshold voltage. This is because the normal field in a buried-channel device, given by Eq. (3.49) with $V_{fb} = +E_g/2q - \psi_B$, is much lower. However, buried-channel MOSFETs cannot be scaled to as short a channel length as surface-channel MOSFETs, as will be discussed in Chapter 4.

At higher temperatures, the MOSFET channel mobility decreases because of increased phonon scattering. The temperature dependence is similar to that of bulk mobility discussed in Section 2.1.3, i.e., $\mu_{eff} \propto T^{-3/2}$.

### 3.1.6 MOSFET CAPACITANCES AND INVERSION-LAYER CAPACITANCE EFFECT

In this subsection, we discuss the intrinsic capacitances of a MOSFET device in different regions of operation and the effect of finite inversion-layer capacitance, which has been neglected in the charge-sheet model, on linear $I_{ds}$–$V_g$ characteristics.

#### 3.1.6.1 INTRINSIC MOSFET CAPACITANCES

The capacitance of a MOSFET device plays a key role in the switching delay of a logic gate, since for a given current the capacitance determines how fast the gate can be charged (or discharged) to a certain potential which turns on (or off) the source-to-drain current. MOSFET capacitances can be divided into two main categories: intrinsic capacitances and parasitic capacitances. This sub-subsection focuses on the intrinsic MOSFET capacitances arising from the inversion and depletion charges in the channel region. Parasitic capacitances are discussed in Section 5.2. As in the earlier drain-current discussions, gate capacitances are also considered separately in the three regions of MOSFET operation: subthreshold region, linear region, and saturation region, as shown in Fig. 3.9.

- *Subthreshold region.* In the subthreshold region, the inversion charge is negligible. Only the depletion charge needs to be supplied when the gate potential is changed. Therefore, the intrinsic gate-to-source–drain capacitance is essentially zero (the extrinsic gate-to-source–drain overlap capacitance is discussed in Section 5.2.2), while the gate-to-body capacitance is given by the serial combination of $C_{ox}$ and $C_d$ (Fig. 2.28), i.e.,

$$C_g = WL\left(\frac{1}{C_{ox}} + \frac{1}{C_d}\right)^{-1} \approx WLC_d, \tag{3.53}$$

where $C_d$ is the depletion capacitance per unit area given by Eq. (2.174). For high drain biases, the surface potential and therefore the depletion width at the drain end of the channel become larger, according to Eq. (3.4). The average

$C_d$ to be used in Eq. (3.53) should then be slightly lower than that evaluated at the source end.

- *Linear region.* Once the surface channel forms, there is no more capacitive coupling between the gate and the body due to screening by the inversion charge. All the gate capacitances are to the channel, i.e., to the source and drain terminals. Within the framework of the charge-sheet model, Eq. (3.24), the inversion charge density $Q_i$ at low drain biases varies linearly from $-C_{ox}(V_g - V_t)$ at the source end to $-C_{ox}(V_g - V_t - mV_{ds})$ at the drain end. The total inversion charge under the gate is then $-WLC_{ox}(V_g - V_t - mV_{ds}/2)$, and the gate to channel capacitance is simply given by the oxide capacitance,

$$C_g = WLC_{ox}. \tag{3.54}$$

- *Saturation region.* When $V_{ds}$ is appreciable, the inversion charge density $Q_i(y)$ varies parabolically along the channel, as shown in Fig. 3.7. At the pinch-off condition, $V_{ds} = V_{dsat} = (V_g - V_t)/m$, and $Q_i = 0$ at the drain. In this case,

$$Q_i(y) = -C_{ox}(V_g - V_t)\sqrt{1 - \frac{y}{L}}, \tag{3.55}$$

from Eqs. (3.24) and (3.27). The total inversion charge obtained by integrating Eq. (3.55) in both the channel length ($y$) and the channel width directions is then $-\frac{2}{3}WLC_{ox}(V_g - V_t)$, and the gate-to-channel capacitance in the saturation region is

$$C_g = \frac{2}{3}WLC_{ox}. \tag{3.56}$$

### 3.1.6.2 INVERSION-LAYER CAPACITANCE

In Section 3.1.2 and in the discussions above, MOSFET $I$–$V$ relations and capacitances were derived based on the charge-sheet approximation that all the inversion charge is located at the silicon surface and the surface potential is pinned at $\psi_s(\text{inv}) = 2\psi_B$ once the inversion layer forms. In reality, the inversion layer has a finite thickness (Fig. 2.26), and the surface potential still increases slightly with $V_g$ even beyond $2\psi_B$ (Fig. 2.25). In other words, there is a finite inversion-layer capacitance, $C_i = -dQ_i/d\psi_s$, in series with the oxide capacitance. As a result, the inversion charge density is less than that given by Eq. (3.17). The error is illustrated in the $Q_i$–$V_g$ curves in Fig. 3.15. The dashed line represents $|Q_i| = C_{ox}(V_g - V_t)$ from the charge-sheet model. The solid curve is a more exact solution calculated numerically from Eq. (3.12) and Eq. (3.14). The discrepancy extends to high gate voltages but is more serious for low-voltage operation. An approximate expression for the inversion charge density taking this effect into account can be derived by

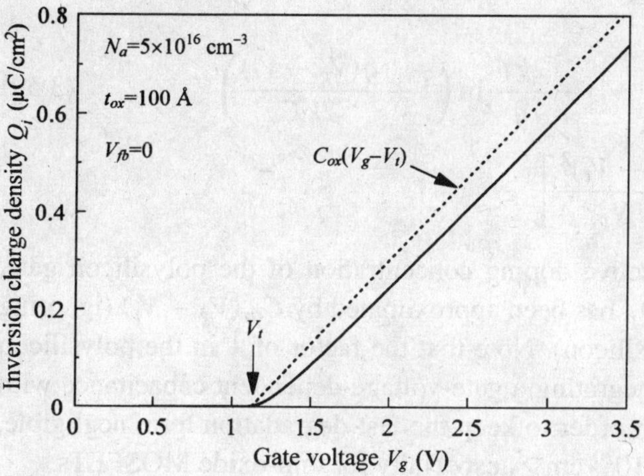


**FIGURE 3.15.** $Q_i$–$V_g$ curve (solid line) calculated from Pao and Sah's model for zero drain voltage, compared with that of the charge-sheet approximation (dotted line). $V_t$ indicates the $2\psi_B$ threshold.

considering the small-signal capacitances in Fig. 2.28(b) (Wordeman, 1986),

$$\frac{d(-Q_i)}{dV_g} = \frac{C_{ox}C_i}{C_{ox}+C_i+C_d} \approx C_{ox}\left(1 - \frac{1}{1+C_i/C_{ox}}\right). \tag{3.57}$$

Here $C_d \approx 0$ after the onset of strong inversion because of screening by the inversion charge. From Eq. (2.178), $C_i \approx |Q_i|/(2kT/q)$. Since $|Q_i| \approx C_{ox}(V_g - V_t)$, one can write $C_i/C_{ox} = (V_g - V_t)/(2kT/q)$. Substituting it into Eq. (3.57) and integrating with respect to $V_g$, one obtains

$$-Q_i = C_{ox}\left[(V_g - V_t) - \frac{2kT}{q}\ln\left(1 + \frac{q(V_g - V_t)}{2kT}\right)\right], \tag{3.58}$$

which agrees well with the numerically calculated curve in Fig. 3.15.

#### 3.1.6.3 EFFECT OF POLYSILICON-GATE DEPLETION ON INVERSION CHARGE

Depletion of polysilicon gates, discussed in Section 2.3.4, can also have an effect on the $Q_i$–$V_g$ curve if the gate is not doped highly enough. To first order, the depletion region in polysilicon acts like a large capacitor in series with the oxide capacitor, which further degrades inversion charge density for a given applied gate voltage. In contrast to the inversion-layer capacitance effect, however, the gate depletion effect becomes more severe at high gate voltages. Assuming an n$^+$ polysilicon gate on nMOSFET (and vice versa for pMOSFET) and following a similar approach to that in Eq. (2.185), one can add an additional term to Eq. (3.58) for polysilicon depletion effects:

$$-Q_i = C_{ox}\left[(V_g - V_t) - \frac{2kT}{q}\ln\left(1 + \frac{q(V_g - V_t)}{2kT}\right) - \frac{C_{ox}^2(V_g - V_t)^2}{2\varepsilon_{si} q N_p}\right]. \tag{3.59}$$

Here $N_p$ is the electrically active doping concentration of the polysilicon gate, and the gate charge density $Q_p$ has been approximated by $C_{ox}(V_g - V_t)$ (ignoring the depletion charge in bulk silicon). Note that the factor of $\frac{1}{2}$ in the polysilicon depletion term arises from integrating a gate-voltage-dependent capacitance with respect to the gate voltage. In order to keep the last degradation term negligible, $N_p$ should be in the range of $10^{20}$ cm$^{-3}$, especially for thin-oxide MOSFETs.

### 3.1.6.4 LINEAR $I_{ds}$–$V_g$ CHARACTERISTICS

Given $Q_i(V_g)$ and $\mu_{eff}(V_g)$ [Eqs. (3.51) and (3.50)], the low-drain-bias (linear) $I_{ds}$–$V_g$ curve is simply

$$I_{ds}(V_g) = \mu_{eff}(V_g)\frac{W}{L}Q_i(V_g)V_{ds}. \tag{3.60}$$

An example is shown in Fig. 3.16, where $I_{ds}$ is calculated assuming no polysilicon depletion with $Q_i(V_g)$ from Fig. 3.15 (solid curve). Both the drain current and the *linear transconductance*, defined by $g_m \equiv dI_{ds}/dV_g$, are degraded significantly at high gate voltages because of the decrease of mobility with increasing normal field. There is a point of maximum slope or linear transconductance about 0.5 V above the threshold voltage. It is conventional to define the linearly extrapolated threshold

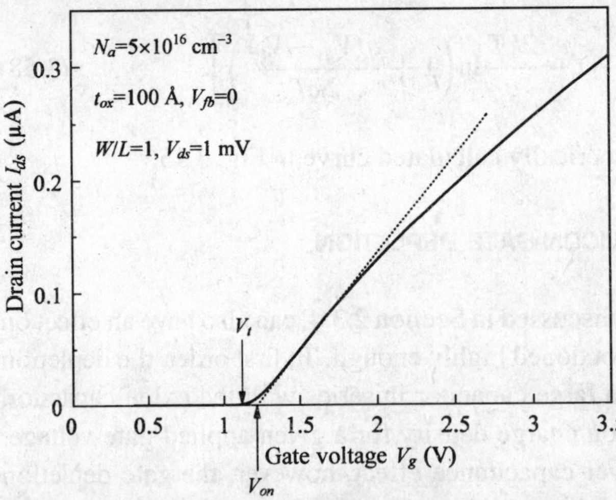


**FIGURE 3.16.** Calculated low-drain $I_{ds}$–$V_g$ curve with inversion-layer capacitance and mobility degradation effects. The dotted line shows the linearly extrapolated threshold voltage $V_{on}$.

voltage, $V_{on}$, by the intercept of a tangent through this point. For a second-order correction in $V_{ds}$ based on Eq. (3.21), $V_{on}$ is obtained by subtracting $mV_{ds}/2$ from the intercept. Because of the combined inversion-layer capacitance and mobility degradation effects, the linearly extrapolated threshold voltage, $V_{on}$, is typically (2–4)$kT/q$ higher than the threshold voltage $V_t$ at $\psi_s(\text{inv}) = 2\psi_B$. One should be careful not to mix up $V_{on}$ with $V_t$, which is used in Eq. (3.36) for estimating subthreshold currents. At $V_g = V_{on}$, the extrapolated subthreshold current (along the same subthreshold slope in a semilog plot) is about 10× of that of Eq. (3.36) for $V_g = V_t$. This current is rather insensitive to temperature but does depend on the technology generation.

A commonly used expression for the low-drain, linear $I_{ds}$–$V_g$ characteristics takes the form

$$I_{ds}(V_g) = \mu'_{eff}(V_g)C_{ox}\frac{W}{L}\left(V_g - V_{on} - \frac{m}{2}V_{ds}\right)V_{ds}. \tag{3.61}$$

Note that the inversion-layer capacitance effect is lumped into $\mu'_{eff}(V_g)$, which in general is different from $\mu_{eff}(V_g)$ in Eq. (3.60) except at high gate voltages.

## 3.2 SHORT-CHANNEL MOSFETs

It is clear from Section 3.1 that for a given supply voltage, the MOSFET current increases with decreasing channel length. The intrinsic capacitance of a short-channel MOSFET is also lower, which makes it easier to switch. However, for a given process, the channel length cannot be arbitrarily reduced even if allowed by lithography. Short-channel MOSFETs differ in many important aspects from long-channel devices discussed in Section 3.1. This section covers the basic features of short-channel devices that are important for device design consideration. These features are: (a) short-channel effect, (b) velocity saturation, (c) channel length modulation, (d) source–drain series resistance, and (e) MOSFET breakdown.

### 3.2.1 SHORT-CHANNEL EFFECT

The *short-channel effect* (SCE) is the decrease of the MOSFET threshold voltage as the channel length is reduced. An example is shown in Fig. 3.17 (Taur *et al.*, 1985). The short-channel effect is especially pronounced when the drain is biased at a voltage equal to that of the power supply (high drain bias). In a CMOS VLSI technology, channel length varies statistically from chip to chip, wafer to wafer, and lot to lot due to process tolerances. The short-channel effect is therefore an important consideration in device design; one must ensure that the threshold voltage does not become too low for the minimum-channel-length device on the chip.

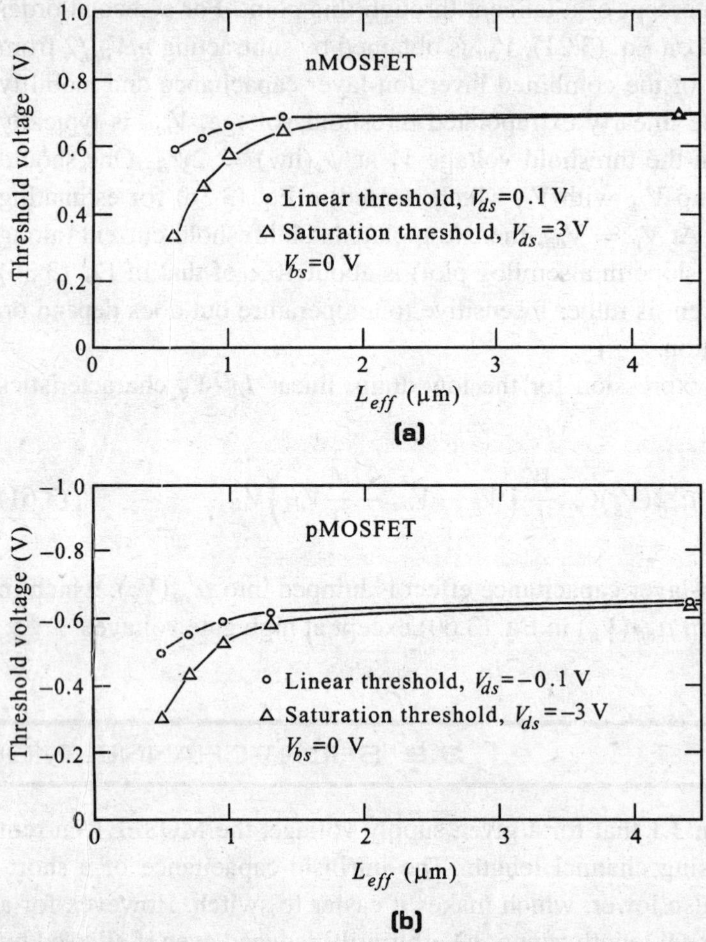


**FIGURE 3.17.** Short-channel threshold roll-off: Measured low- and high-drain threshold voltages of n- and p-MOSFETs versus channel length. (After Taur *et al.*, 1985.)

#### 3.2.1.1 2-D POTENTIAL CONTOURS AND THE CHARGE-SHARING MODEL

The key difference between a short-channel and a long-channel MOSFET is that the field pattern in the depletion region of a short-channel MOSFET is two-dimensional, as shown in Fig. 3.18. The constant-potential contours in a long-channel device in Fig. 3.18(a) are largely parallel to the oxide–silicon interface or along the channel length direction ($y$-axis), so that the electric field is one-dimensional (along the vertical direction or $x$-axis) over the most part of the device. The constant-potential contours in a short-channel device in Fig. 3.18(b), however, are more curvilinear, and the resulting electric field pattern is of a two-dimensional nature. In other words, both the $x$- and $y$-components of the electric field are appreciable in a short-channel MOSFET. It is also important to note that, for a given gate voltage, there is more band bending (higher $\psi$) at the silicon–oxide interface in a short-channel device than in a long-channel device. Specifically, the maximum surface potential

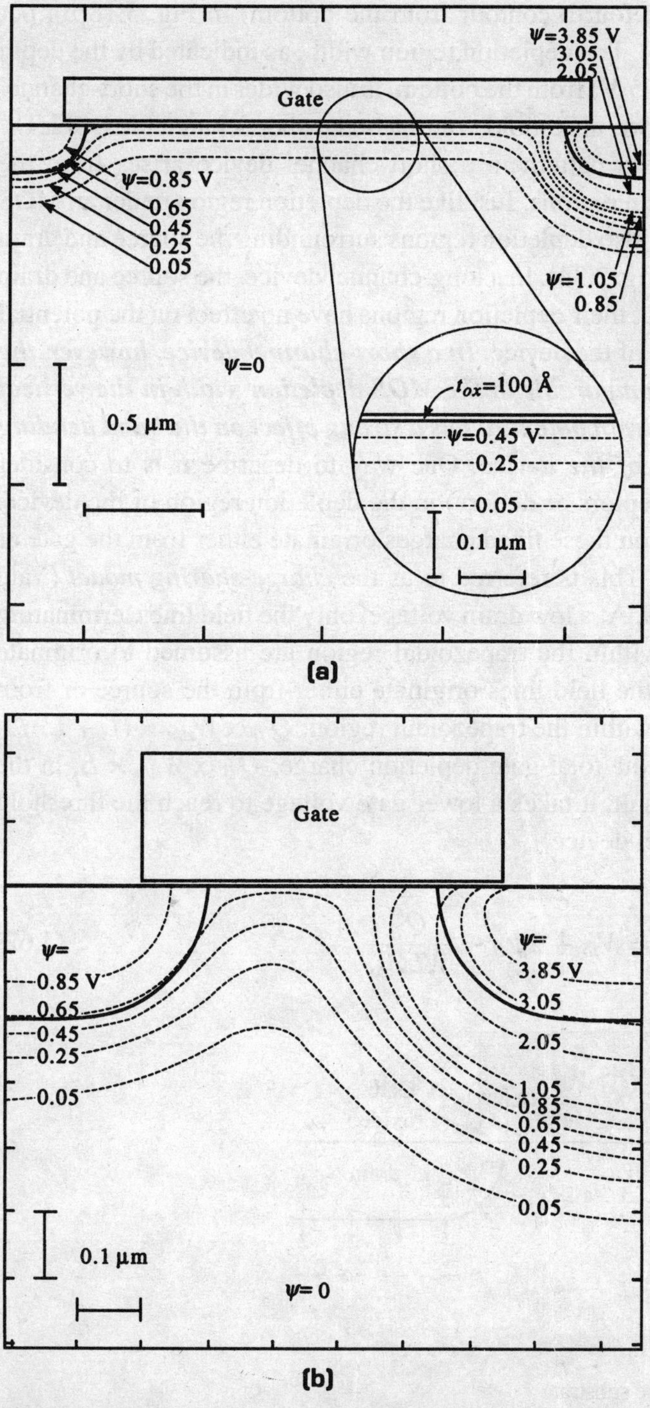


**FIGURE 3.18.** Simulated constant potential contours of (a) a long-channel and (b) a short-channel nMOSFET. The contours are labeled by the band bending with respect to the neutral p-type region. The solid lines indicate the location of the source and drain junctions (metallurgical). The drain is biased at 3.0 V. Both devices are biased at the same gate voltage slightly below the threshold.

is slightly over 0.65 V (the fourth contour from the bottom) in Fig. 3.18(b), but below 0.65 V in Fig. 3.18(a). The depletion region width, as indicated by the depth of the first contour ($\psi = 0.05$ V) from the bottom, is also wider in the short-channel case. These all point to a lower threshold voltage in the short-channel MOSFET.

The two-dimensional field pattern in a short-channel device arises from the proximity of source and drain regions. Just like the depletion region under an MOS gate (Section 2.3), there are also depletion regions surrounding the source and drain junctions (Section 2.2 and Fig. 2.34). In a long-channel device, the source and drain are far enough separated that their depletion regions have no effect on the potential or field pattern in most part of the device. ***In a short-channel device, however, the source–drain distance is comparable to the MOS depletion width in the vertical direction, and the source–drain potential has a strong effect on the band bending over a significant portion of the device.*** One way to describe it is to consider the net charge (ionized acceptors or donors) in the depletion region of the device. The field lines terminating on these fixed charges originate either from the gate or from the source and drain. This is referred to as the *charge-sharing model* (Yau, 1974), as shown in Fig. 3.19. At a low drain voltage, only the field lines terminating on the depletion charges within the trapezoidal region are assumed to originate from the gate. The rest of the field lines originate either from the source or from the drain. The total charge within the trapezoidal region, $Q'_B \propto W_{dm} \times (L + L')/2$, is proportionally less than the total gate depletion charge, $Q_B \propto W_{dm} \times L$, in the long-channel case. As a result, it takes a lower gate voltage to reach the threshold condition of a short-channel device,

$$V_t = V_{fb} + 2\psi_B + \frac{Q'_B}{WLC_{ox}}, \tag{3.62}$$

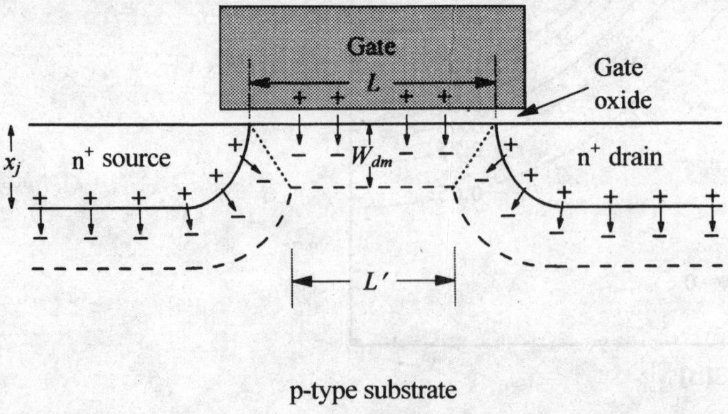


**FIGURE 3.19.** Schematic diagram of the charge-sharing model. The dashed lines indicate the boundary of the gate and source–drain depletion regions. The arrows represent electric field lines that originate from a positive charge and terminate on a negative charge. The dotted lines partition the depletion charge and form the two sides of the trapezoid discussed in the text. (After Yau, 1974.)

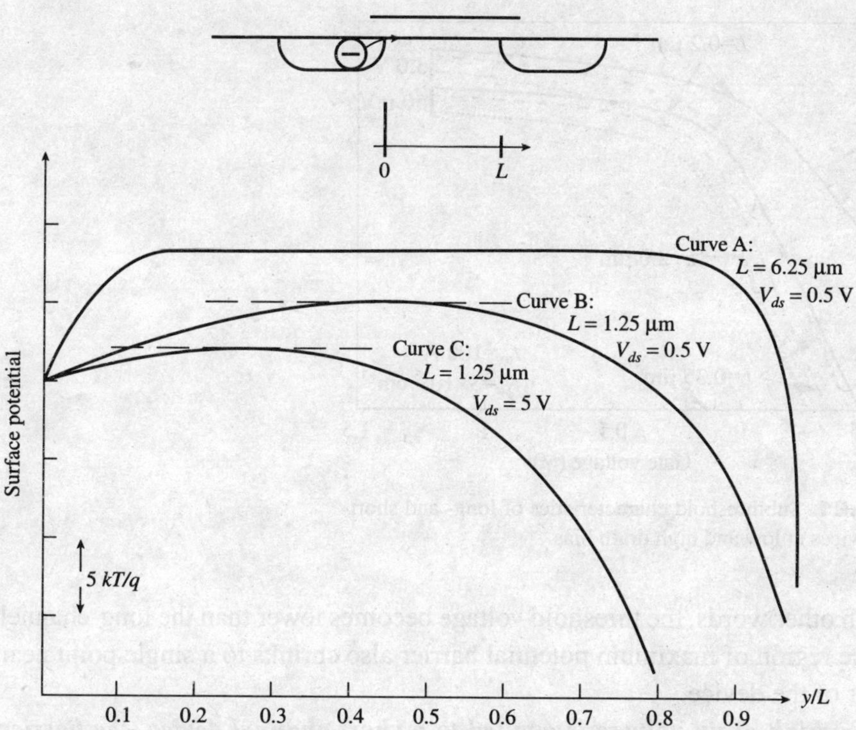


**FIGURE 3.20.** Surface potential versus lateral distance (normalized to the channel length $L$) from the source to the drain for (a) a long-channel MOSFET, (b) a short-channel MOSFET at low drain bias, and (c) a short-channel MOSFET at high drain bias. The gate voltage is the same for all three cases. (After Troutman, 1979.)

based on Eq. (3.20). Even though a simple, analytical expression for threshold voltage can be obtained from the charge-sharing model, the division of depletion charge between the gate and the source and drain is somewhat arbitrary. Furthermore, there is no simple way of dealing with high-drain-bias conditions, especially in the subthreshold region.

#### 3.2.1.2 DRAIN-INDUCED BARRIER LOWERING

The physics of the short-channel effect can be understood from a different angle by considering the potential barrier (to electrons for an n-channel MOSFET) at the surface between the source and drain, as shown in Fig. 3.20 (Troutman, 1979). Under off conditions, this potential barrier (p-type region) prevents electron current from flowing to the drain. The surface potential is mainly controlled by the gate voltage. When the gate voltage is below the threshold voltage, there are only a limited number of electrons injected from the source over the barrier and collected by the drain (subthreshold current). In the long-channel case, the potential barrier is flat over most part of the device. Source and drain fields only affect the very ends of the channel. As the channel length is shortened, however, the source and drain fields penetrate deeply into the middle of the channel, which lowers the potential barrier between the source and drain. This causes a substantial increase of the subthreshold

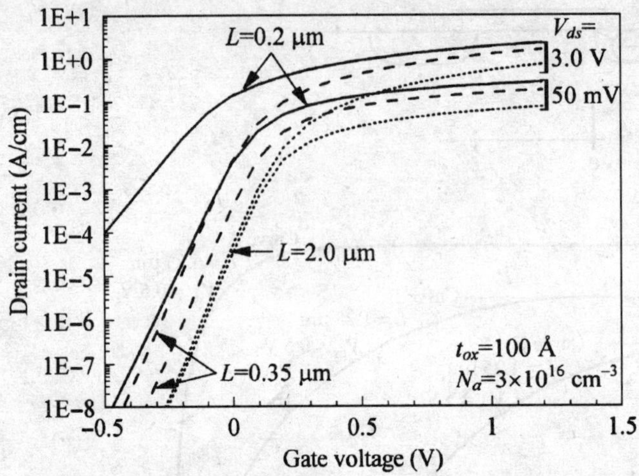


**FIGURE 3.21.** Subthreshold characteristics of long- and short-channel devices at low and high drain bias.

current. In other words, the threshold voltage becomes lower than the long-channel value. The region of maximum potential barrier also shrinks to a single point near the center of the device.

***When a high drain voltage is applied to a short-channel device, the barrier height is lowered even more, resulting in further decrease of the threshold voltage.*** The point of maximum barrier also shifts toward the source end as shown in Fig. 3.20. This effect is referred to as *drain-induced barrier lowering* (DIBL). It explains the experimentally observed increase of subthreshold current with drain voltage in a short-channel MOSFET. Figure 3.21 shows the subthreshold characteristics of long- and short-channel devices at different drain bias voltages. For long-channel devices, the subthreshold current is independent of drain voltage ($\geq 2kT/q$), as expected from Eq. (3.36). For short-channel devices, however, there is a parallel shift of the curve to a lower threshold voltage for high drain bias conditions. At even shorter channel lengths, the subthreshold slope starts to degrade as the surface potential is more controlled by the drain than by the gate. Eventually, the device reaches the *punch-through* condition when the gate totally loses control of the channel and high drain current persists independent of gate voltage.

#### 3.2.1.3 2-D POISSON'S EQUATION AND LATERAL FIELD PENETRATION

Further insight into the role of the lateral electric field, $\mathscr{E}_y = -\partial\psi_i/\partial y$, in a short-channel MOSFET can be gained by examining the two-dimensional Poisson's equation,

$$\frac{\partial^2\psi_i}{\partial x^2} + \frac{\partial^2\psi_i}{\partial y^2} = -\frac{\rho}{\varepsilon_{si}}. \tag{3.63}$$

In the depletion region of an nMOSFET, mobile carrier densities are negligible.

Only ionized acceptors need to be considered. For a uniformly doped background concentration $N_a$, Poisson's equation can be written in terms of the electric fields as

$$\frac{\partial \mathscr{E}_x}{\partial x} + \frac{\partial \mathscr{E}_y}{\partial y} = \frac{\rho}{\varepsilon_{si}} = -\frac{qN_a}{\varepsilon_{si}}, \tag{3.64}$$

where $\mathscr{E}_x = -\partial\psi_i/\partial x$ is the electric field in the vertical direction. The depletion charge density, $\rho = -qN_a$, from ionized acceptors can be considered as being split into two parts: the first part, $\varepsilon_{si}\partial\mathscr{E}_x/\partial x$, is controlled by the gate field in the vertical direction; the second part, $\varepsilon_{si}\partial\mathscr{E}_y/\partial y$, is controlled by the source–drain field in the lateral direction (Nguyen and Plummer, 1981). In a long-channel device, the lateral field is negligible over most of the channel, and almost all of the depletion charge is controlled by the gate field. In a short-channel device, the lateral field becomes appreciable. Figure 3.22(a) shows an example of the magnitude of the lateral field

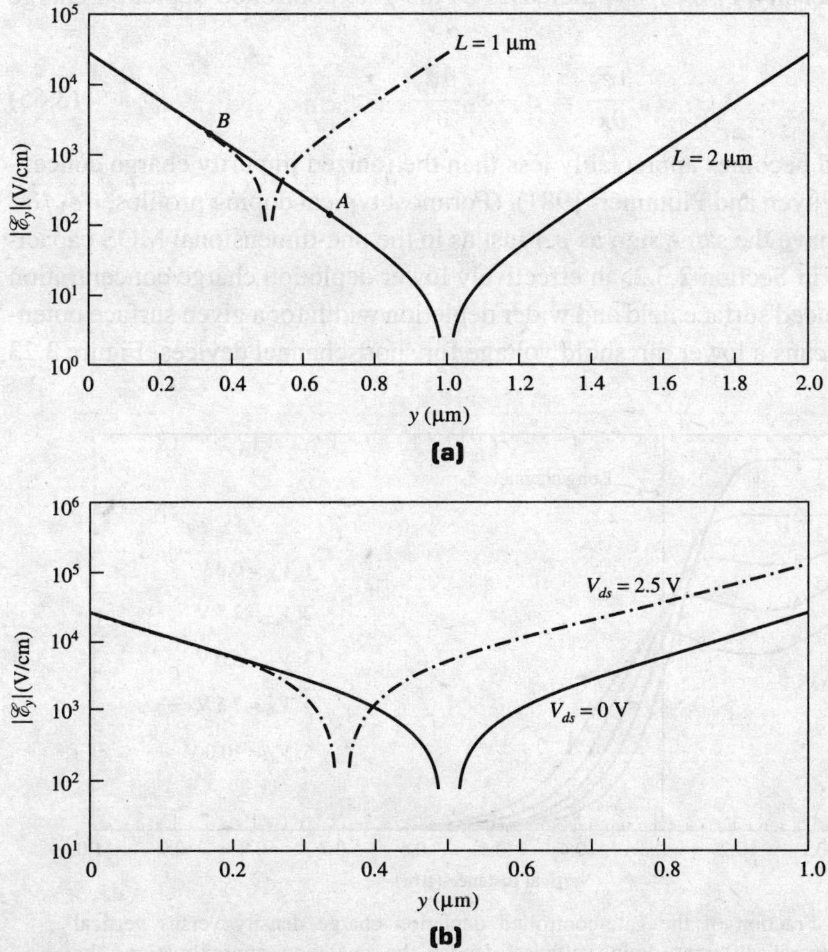


**FIGURE 3.22.** Simulated lateral field as a function of lateral distance along a horizontal cut through the gate-depletion layer for (a) long- and short-channel devices and (b) low and high drain bias voltages. (After Nguyen, 1984.)

along the channel length direction as obtained from a 2-D numerical simulation. The lateral field is highest at the source and drain junctions, decreasing exponentially toward the middle of the channel. At low drain voltages, the source and drain fields cancel each other exactly at the center of the device. ***When the channel length becomes shorter, the characteristic length of the exponential decay remains unchanged, while the magnitude of the lateral field near the middle of the device increases significantly. This depicts the penetration of source and drain fields into the channel region of a short-channel MOSFET***. Application of a high drain voltage [Fig. 3.22(b)] does not change the source field but does increase the drain field. This shifts the zero-field point toward the source, thus making it asymmetric, and at the same time raises the lateral field intensity even further. The zero-field point corresponds to the point of the least band bending in Fig. 3.18, as well as the point of maximum potential barrier in Fig. 3.20.

With the increase of the lateral field strength, the source–drain controlled depletion charge density, $\varepsilon_{si}\,\partial\mathscr{E}_y/\partial y$, increases as the gate-controlled depletion charge density,

$$\varepsilon_{si}\frac{\partial\mathscr{E}_x}{\partial x} = \rho - \varepsilon_{si}\frac{\partial\mathscr{E}_y}{\partial y}, \tag{3.65}$$

decreases and becomes appreciably less than the ionized impurity charge concentration, $\rho$ (Nguyen and Plummer, 1981). (For most typical doping profiles, $\partial\mathscr{E}_x/\partial x$ and $\partial\mathscr{E}_y/\partial y$ have the same sign as $\rho$.) Just as in the one-dimensional MOS capacitor discussed in Section 2.3.2, an effectively lower depletion charge concentration results in reduced surface field and wider depletion width for a given surface potential, which means a lower threshold voltage for short-channel devices. Figure 3.23

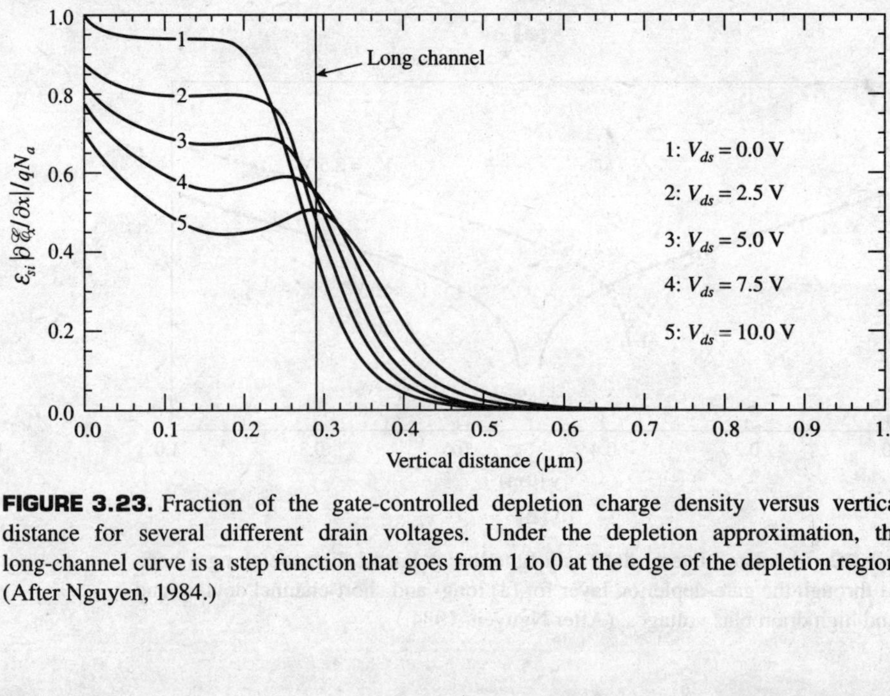


**FIGURE 3.23.** Fraction of the gate-controlled depletion charge density versus vertical distance for several different drain voltages. Under the depletion approximation, the long-channel curve is a step function that goes from 1 to 0 at the edge of the depletion region. (After Nguyen, 1984.)

shows the fraction of depletion charge controlled by the gate field versus the vertical distance from the surface. As the drain voltage increases, the effective gate-controlled charge density decreases significantly below the long-channel value. Even though the depletion region becomes slightly wider, the integrated charge density (area under the curve), and therefore the threshold voltage, decreases.

#### 3.2.1.4 AN ANALYTICAL EXPRESSION FOR SHORT-CHANNEL THRESHOLD VOLTAGE

With a few approximations, an analytical solution to the two-dimensional Poisson's equation can be obtained using a simplified short-channel MOSFET geometry in Appendix 6 (Nguyen, 1984). The region of interest is a rectangular box of length equal to the channel length $L$ defined as the distance between the source and the drain (Fig. 3.19). In the vertical direction, the box consists of an oxide region of thickness $t_{ox}$ and a silicon region of depth given by the depletion-layer width $W_d$ [Eq. (2.160)]. To eliminate the discontinuity of $\partial\psi_i/\partial x$ across the silicon–oxide boundary, the oxide is replaced by an equivalent region of the same dielectric constant as silicon, but with a thickness equal to $(\varepsilon_{si}/\varepsilon_{ox})t_{ox} = 3t_{ox}$. The entire rectangular region can then be treated as a homogeneous material of height $W_d + 3t_{ox}$ and dielectric constant $\varepsilon_{si}$. This is a good approximation when the oxide is thin compared with the depletion depth $W_d$, as is the case with most practical CMOS technologies.

The boundary conditions of the electrostatic potential at the source and drain boundaries are $\psi_{bi}$ and $\psi_{bi} + V_{ds}$, respectively, with the potential in the neutral p-type region defined as zero. Here $\psi_{bi}$ is the built-in potential of the source- or drain-to-substrate junction, and $V_{ds}$ is the drain voltage. For an abrupt $n^+$–p junction, $\psi_{bi} = E_g/2q + \psi_B$, where $\psi_B$ is given by Eq. (2.37). Typically, $\psi_{bi} \approx$ 0.8–0.9 V.

Under subthreshold conditions, current conduction is dominated by diffusion and is mainly controlled by the point of highest barrier for electrons along the channel, as shown in Fig. 3.18(b) and Fig. 3.20. The threshold voltage of a short-channel device is defined as the gate voltage at which the minimum electrostatic potential (maximum barrier for electrons) at the surface equals $2\psi_B$. It is shown in Appendix 6 that this occurs at a gate voltage lower than the long-channel threshold voltage by an amount

$$\Delta V_t = \frac{24t_{ox}}{W_{dm}}\sqrt{\psi_{bi}(\psi_{bi} + V_{ds})}e^{-\pi L/2(W_{dm}+3t_{ox})}. \tag{3.66}$$

Here $W_{dm}$ is the minimum depletion width at the threshold condition in a short-channel device, as indicated in Fig. 3.19. In order to distinguish it from $W_{dm}$, the depletion width at the threshold condition in a long-channel device will be designated as $W_{dm}^0$, as in Appendix 6. If $L$ is not too short, the body-effect coefficient $m$

can be approximated by

$$m \equiv 1 + \frac{\varepsilon_{si}/W_{dm}^0}{C_{ox}} \approx 1 + \frac{3t_{ox}}{W_{dm}}, \tag{3.67}$$

and Eq. (3.66) can be expressed as

$$\Delta V_t = 8(m-1)\sqrt{\psi_{bi}(\psi_{bi} + V_{ds})}e^{-\pi L/2mW_{dm}}. \tag{3.68}$$

Typically, $m \approx 1.1$–$1.4$. These analytical short-channel threshold roll-off expressions are good approximations if the source and drain junctions are deeper than the maximum gate depletion region, i.e., if $x_j \geq W_{dm}$.

***Because of the exponential facor, the threshold voltage roll-off with channel length is very sensitive to the gate depletion width $W_{dm}$.*** For very short channel lengths, the minimum depletion width is larger than the long-channel value, as can be seen from Figs. 3.18 and 3.23. If the short-channel effect is not too severe, however, $W_{dm}$ can be approximated to the first order by its long-channel value,

$$W_{dm}^0 = \sqrt{\frac{4\varepsilon_{si}kT\ln(N_a/n_i)}{q^2N_a}} \tag{3.69}$$

from Eq. (2.162). $W_{dm}^0$ is plotted in Fig. 3.24 versus $N_a$. ***To avoid excessive short-channel effects, the substrate (or well) doping concentration in a CMOS device design should be chosen such that the minimum channel length, $L_{min}$, is about 2–3 times $W_{dm}$.*** This is similar to the well-known criterion that the minimum channel length should be greater than the sum of the source and the drain depletion widths, i.e., $L_{\min} \geq W_S + W_D$, where $W_S$ and $W_D$ are given by Eqs. (A6.7) and (A6.8). Another observation of Eq. (3.66) is that a reverse substrate bias, $-V_{bs}$, aggravates short-channel effects, since it changes $\psi_{bi}$ to a higher value, $\psi_{bi} + V_{bs}$. By the same

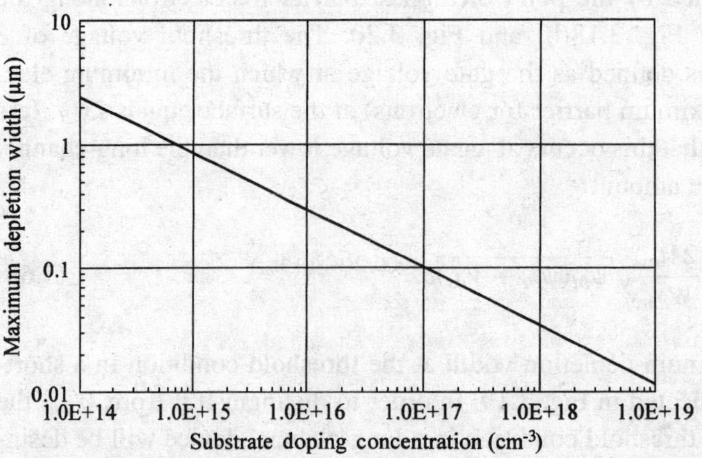


**FIGURE 3.24.** Depletion region width at $2\psi_B$ threshold condition versus doping concentration for uniformly doped substrates.

token, the substrate sensitivity [Eq. (3.41)] of a short-channel MOSFET is less than that of a long-channel MOSFET. This can be understood from the charge-sharing model (Fig. 3.19) that as $W_{dm}$ increases due to a reverse substrate bias, the gate-controlled depletion charge within the trapezoid region increases in proportion to $W_{dm}$ in a long-channel device, but less than in proportion to $W_{dm}$ in a short-channel device. More on the short-channel substrate sensitivity can be found in Appendix 6.

It should be noted that Eq. (3.66) is, of course, just an approximation. In general, short-channel device design should be carried out with a two-dimensional device simulator for more accurate results. Further details on channel profile and threshold design are discussed in Section 4.2.

### 3.2.2 VELOCITY SATURATION

As discussed in Section 3.1.2, when the drain voltage increases in a long-channel MOSFET, the drain current first increases, then becomes saturated at a voltage equal to $V_{dsat} = (V_g - V_t)/m$ with the onset of pinch-off at the drain. ***In a short-channel device, the saturation of drain current may occur at a much lower voltage due to velocity saturation***. This causes the saturation current $I_{dsat}$ to deviate from the $1/L$ dependence depicted in Eq. (3.23) for long-channel devices. Velocity–field relationships in bulk silicon are plotted in Fig. 2.9. Saturation velocities of electrons and holes in a MOSFET channel are slightly lower than their bulk values. $v_{sat} \approx 7\text{–}8 \times 10^6$ cm/s for electrons and $v_{sat} \approx 6\text{–}7 \times 10^6$ cm/s for holes have been reported in the literature (Coen and Muller, 1980; Taur *et al.*, 1993a). Figure 3.25

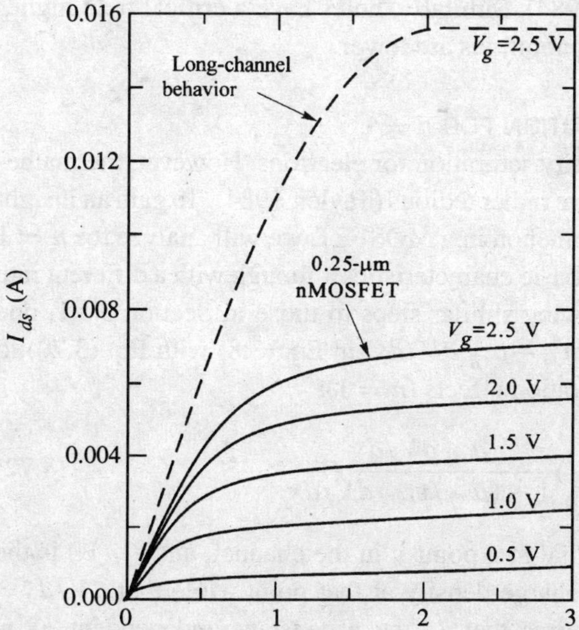


**FIGURE 3.25.** Experimental $I$–$V$ curves of a 0.25-µm nMOSFET (solid lines). The device width is 9.5 µm. The dashed curve shows the long-channel-like drain current expected for this channel length if there were no velocity saturation. (After Taur *et al.*, 1993a.)

shows the experimentally measured $I_{ds}$–$V_{ds}$ curves of a 0.25-μm nMOSFET. The dashed curve represents the long-channel-like current given by Eq. (3.23) for $V_g =$ 2.5 V. Due to velocity saturation, the drain current saturates at a drain voltage much lower than $(V_g - V_t)/m$, thus severely limiting the saturation current of a short-channel device.

#### 3.2.2.1 VELOCITY–FIELD RELATIONSHIP

Experimental measurements show that the velocity–field relationship for electrons and holes takes the empirical form (Caughey and Thomas, 1967)

$$v = \frac{\mu_{eff}\mathscr{E}}{[1 + (\mathscr{E}/\mathscr{E}_c)^n]^{1/n}}, \tag{3.70}$$

where $n = 2$ for electrons and $n = 1$ for holes. $n\,(\geq 1)$ is a measure of how rapidly the carriers approach saturation. The parameter $\mathscr{E}_c$ is called the *critical field*. When the field strength is comparable to or greater than $\mathscr{E}_c$, velocity saturation becomes important. At low fields, $v = \mu_{eff}\mathscr{E}$, which is simply Ohm's law. As $\mathscr{E} \rightarrow \infty$, $v = v_{sat} = \mu_{eff}\mathscr{E}_c$. Therefore,

$$\mathscr{E}_c = \frac{v_{sat}}{\mu_{eff}}. \tag{3.71}$$

It was discussed in Section 3.1.5 that the effective mobility $\mu_{eff}$ is a function of the vertical (or normal) field $\mathscr{E}_{eff}$. Since $v_{sat}$ is a constant independent of $\mathscr{E}_{eff}$, the critical field $\mathscr{E}_c$ is a function of $\mathscr{E}_{eff}$ as well. More specifically, ***for a higher vertical field, the effective mobility is lower, but the critical field for velocity saturation becomes higher*** (Sodini *et al.*, 1984). Similarly, holes have a critical field higher than that of electrons, since hole mobilities are lower.

#### 3.2.2.2 AN ANALYTICAL SOLUTION FOR $n = 1$

It is more important to treat velocity saturation for electrons. However, the mathematics in solving the $n = 2$ case are rather tedious (Taylor, 1984). To gain an insight into the velocity saturation phenomenon in a MOSFET, we will analyze the $n = 1$ case instead, which has the same basic characteristics although with a different rate of approaching saturation. Following similar steps to those in Section 3.1.1, one replaces the low-field drift velocity, $-\mu_{eff}\,dV/dy$, in Eq. (3.8) with Eq. (3.70) to allow for high-field velocity saturation effects ($n = 1$):

$$I_{ds} = -WQ_i(V)\frac{\mu_{eff}\,dV/dy}{1 + (\mu_{eff}/v_{sat})\,dV/dy}. \tag{3.72}$$

Here $V$ is the quasi-Fermi potential at a point $y$ in the channel, and $Q_i(V)$ is the integrated (vertically) inversion charge density at that point. Note that $dV/dy = -\mathscr{E} > 0$. Current continuity requires that $I_{ds}$ be a constant, independent of $y$.

Rearranging Eq. (3.72), one obtains

$$I_{ds} = -\left(\mu_{eff} W Q_i(V) + \frac{\mu_{eff} I_{ds}}{v_{sat}}\right)\frac{dV}{dy}. \tag{3.73}$$

Multiplying by $dy$ on both sides and integrating from $y = 0$ to $L$ and from $V = 0$ to $V_{ds}$, one solves for $I_{ds}$:

$$I_{ds} = \frac{-\mu_{eff}(W/L)\int_0^{V_{ds}} Q_i(V)\,dV}{1 + (\mu_{eff} V_{ds}/v_{sat} L)}. \tag{3.74}$$

The numerator is simply the long-channel current, Eq. (3.10), without velocity saturation. It is clear that if the "average" field along the channel, $V_{ds}/L$, is much less than the critical field $\mathscr{E}_c = v_{sat}/\mu_{eff}$, the drain current is hardly affected by velocity saturation. When $V_{ds}/L$ becomes comparable to or greater than $\mathscr{E}_c$, however, the drain current is significantly reduced. If one uses the approximate expression (3.24) in the charge-sheet model for $Q_i(V)$,

$$Q_i(V) = -C_{ox}(V_g - V_t - mV), \tag{3.75}$$

the integration in Eq. (3.74) can be carried out to yield

$$I_{ds} = \frac{\mu_{eff} C_{ox}(W/L)\left[(V_g - V_t)V_{ds} - (m/2)V_{ds}^2\right]}{1 + (\mu_{eff} V_{ds}/v_{sat} L)}. \tag{3.76}$$

#### 3.2.2.3 SATURATION DRAIN VOLTAGE AND CURRENT

For a given $V_g$, $I_{ds}$ increases with $V_{ds}$ until a maximum current is reached. Beyond this point, the drain current is saturated. The saturation voltage, $V_{dsat}$, can be found by solving $dI_{ds}/dV_{ds} = 0$:

$$V_{dsat} = \frac{2(V_g - V_t)/m}{1 + \sqrt{1 + 2\mu_{eff}(V_g - V_t)/(m v_{sat} L)}}. \tag{3.77}$$

This expression is always less than the long-channel saturation voltage, $(V_g - V_t)/m$. Substituting Eq. (3.77) into Eq. (3.76), one finds the saturation current,

$$I_{dsat} = C_{ox} W v_{sat}(V_g - V_t)\frac{\sqrt{1 + 2\mu_{eff}(V_g - V_t)/(m v_{sat} L)} - 1}{\sqrt{1 + 2\mu_{eff}(V_g - V_t)/(m v_{sat} L)} + 1}. \tag{3.78}$$

Example curves of $I_{dsat}$ versus $V_g - V_t$ are plotted in Fig. 3.26 for several different channel lengths. In the long-channel case, the solid curve calculated from Eq. (3.78) is not too different from the dashed curve representing the drain current without velocity saturation. In fact, it can be shown that Eq. (3.78) reduces to the long-channel saturation current [Eq. (3.23)],

$$I_{dsat} = \mu_{eff} C_{ox} \frac{W}{L}\frac{(V_g - V_t)^2}{2m}, \tag{3.79}$$

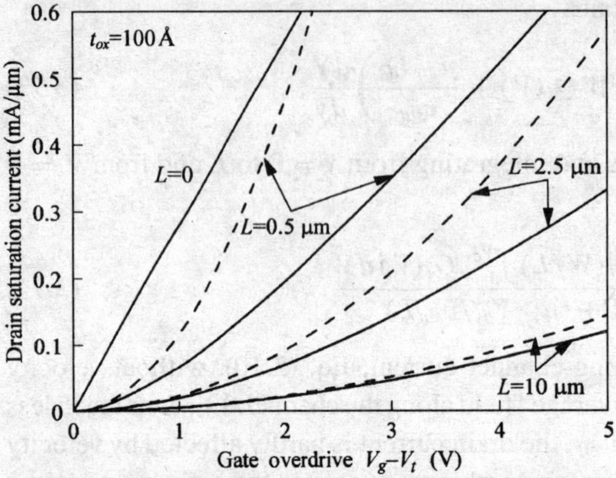


**FIGURE 3.26.** Saturation current calculated from Eq. (3.78) versus $V_g - V_t$ for several different channel lengths (solid curves). The dashed curves are the corresponding "long-channel-like" saturation currents calculated from Eq. (3.79), i.e., by letting $v_{sat} \to \infty$ in Eq. (3.78). The $L = 0$ line represents the limiting case imposed by velocity saturation, Eq. (3.80).

when $V_g - V_t \ll m v_{sat} L / 2\mu_{eff}$. As the channel length becomes shorter, the velocity-saturated current (solid curves) is significantly less than that of Eq. (3.79) (dashed curves) over an increasing range of gate voltage. In the limit of $L \to 0$, Eq. (3.78) becomes the velocity-saturation-limited current,

$$I_{dsat} = C_{ox} W v_{sat} (V_g - V_t), \tag{3.80}$$

as indicated by the straight line labeled $L = 0$ in Fig. 3.26. ***Note that Eq. (3.80) is independent of channel length L and varies linearly with $V_g - V_t$ instead of quadratically as in the long-channel case***. This is consistent with observations of the experimental curves in Fig. 3.25. For very short channel lengths, the saturation voltage, Eq. (3.77), can be approximated by

$$V_{dsat} = \sqrt{2 v_{sat} L (V_g - V_t) / m \mu_{eff}}, \tag{3.81}$$

which decreases with channel length.

#### 3.2.2.4 PINCH-OFF POINT AT VELOCITY SATURATION

It is instructive to examine the charge and field behavior at the drain end of the channel when $V_{ds} = V_{dsat}$. From Eq. (3.75),

$$Q_i(y = L) = -C_{ox}(V_g - V_t - m V_{dsat}). \tag{3.82}$$

Substituting $V_{dsat}$ from Eq. (3.77), one finds

$$Q_i(y = L) = -C_{ox}(V_g - V_t)\frac{\sqrt{1 + 2\mu_{eff}(V_g - V_t)/(mv_{sat}L)} - 1}{\sqrt{1 + 2\mu_{eff}(V_g - V_t)/(mv_{sat}L)} + 1}. \tag{3.83}$$

Comparison with Eq. (3.78) yields $I_{dsat} = -W v_{sat} Q_i(y = L)$, i.e., the carrier drift velocity at the drain end of the channel is equal to the saturation velocity. From Eq. (3.72), this means that the lateral field along the channel, $dV/dy$, approaches infinity at the drain. Just as in the long-channel pinch-off situation discussed in Subsection 3.1.2, such a singularity leads to the breakdown of the gradual-channel approximation which assumed that the lateral field changes slowly in comparison with the vertical field. In other words, ***beyond the saturation point, carriers which are traveling at saturation velocity are no longer confined to the surface channel.*** Their transport must then be described by a 2-D Poisson's equation to be elaborated in Section 3.2.3. A key difference between pinch-off in long-channel devices and velocity saturation in short-channel devices is that in the latter case, the inversion charge density at the drain, Eq. (3.83), does not vanish.

#### 3.2.2.5 VELOCITY OVERSHOOT

All the MOSFET current formulations discussed thus far, including the mobility definition and velocity saturation, are under the realm of the drift–diffusion approximation, which treats carrier transport in some average fashion always in thermal equilibrium with the silicon lattice. ***The drift–diffusion model breaks down in ultrashort-channel devices where high field or rapid spatial variation of potential is present.*** In such cases, the scattering events are no longer localized, and some fraction of the carriers may acquire much higher than thermal energy over a portion of the device, for example, near the drain. These carriers are not in thermal equilibrium with the silicon lattice and are generally referred to as *hot carriers*. Under these circumstances, it is possible for the carrier velocity to exceed the saturation velocity. This phenomenon is called *velocity overshoot*.

A more rigorous treatment of the carrier transport under spatially nonuniform high-field conditions has been carried out by a Monte Carlo solution of the Boltzmann transport equation for the electron distribution function (Laux and Fischetti, 1988). Figure 3.27 shows the calculated saturation transconductance of nMOSFETs versus channel length, together with experimental results (Sai-Halasz *et al.*, 1988). Local velocity overshoot near the drain starts to occur below 0.2-μm channel length. At channel lengths near 0.05 μm, velocity overshoot takes place over a substantial portion of the device such that the terminal saturation transconductance exceeds the velocity saturation limited value,

$$g_{msat} \equiv \frac{dI_{dsat}}{dV_g} = C_{ox} W v_{sat}, \tag{3.84}$$

from Eq. (3.80).

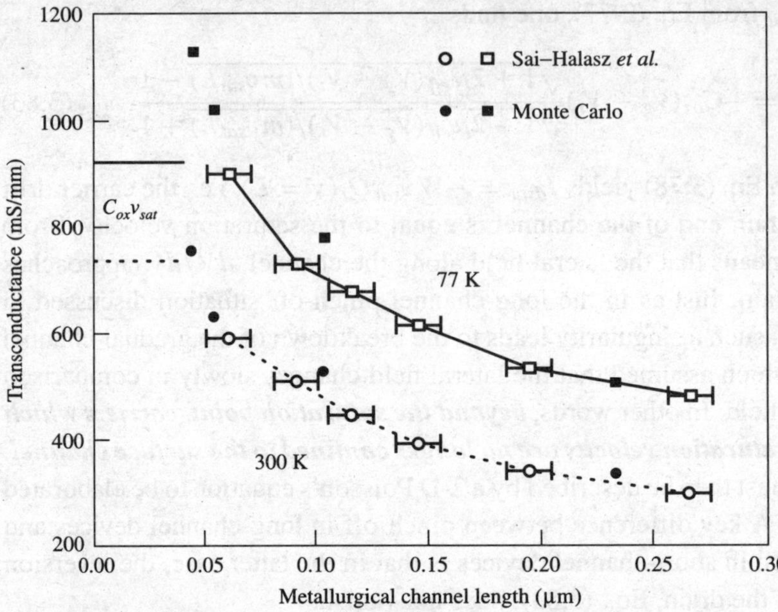


**FIGURE 3.27.** Measured (open symbols) and calculated (solid symbols) saturation transconductance versus channel length. The gate oxide is 45 Å thick. Absolute upper bounds for the transconductance in the absence of velocity overshoot at 300 and 77 K are indicated by the lines labeled $C_{ox}v_{sat}$. (After Laux and Fischetti, 1988.)

It should be noted that while the carrier velocity can reach rather high values in the high-field region near the drain, MOSFET currents are mainly controlled by the average carrier velocity near the source end of the channel, where the inversion charge density $Q_i$ is $C_{ox}(V_g - V_t)$, independent of drain voltage. Carrier velocity near the source, in turn, is determined by both the thermal velocity as the carriers are injected from the source and the field and scattering rates (mobility) in the channel region near the source (Lundstrom, 1997). Velocity overshoot near the drain helps raise MOSFET currents only to the extent that it increases the field near the source. Once the device approaches the *ballistic* limit, the current depends only on the injection velocity from the source, independent of the field and scattering parameters. In other words, there is an upper limit on the MOSFET current set by thermal injection from the source, and velocity overshoot in the channel does not extend this limit. Such a limiting current takes the same form as Eq. (3.80), except that the parameter $v_{sat}$ should be interpreted as the source thermal velocity $v_T$ (Lundstrom, 1997), which for electrons can be 1.5–2 times $v_{sat}$ in a heavily doped degenerate $n^+$ source.

### 3.2.3 CHANNEL LENGTH MODULATION

In this subsection, we discuss the characteristics of short-channel MOSFETs biased beyond saturation. In a long-channel device, the drain current stays constant

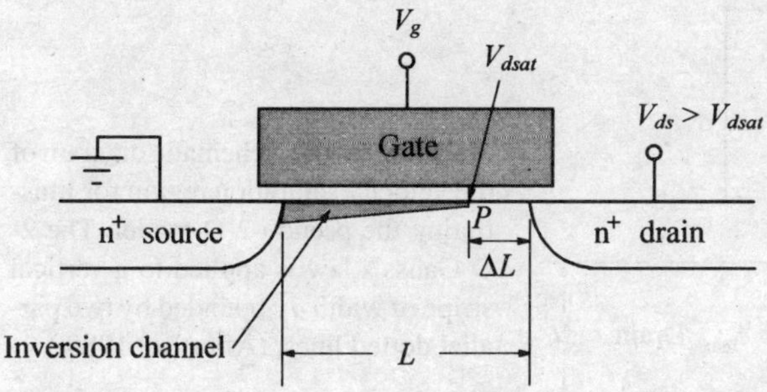


**FIGURE 3.28.** Schematic diagram showing channel length modulation when a MOSFET is biased beyond saturation. The surface channel collapses at point $P$, where carriers reach saturation velocity.

when the drain voltage exceeds $V_{dsat}$, as shown in Fig. 3.8. The output conductance, $dI_{ds}/dV_{ds}$, is zero in the saturation region. In contrast, the drain current of a short-channel MOSFET can still increase slightly beyond the pinch-off or the velocity saturation point with a nonzero output conductance, as is evident from the experimental curves in Fig. 3.25. This arises because of two factors: the short-channel effect and channel length modulation. The short-channel effect was discussed in Section 3.2.1; when the drain voltage increases beyond saturation in a short-channel device, the threshold voltage decreases, and therefore the drain current increases. In this subsection, we focus on channel length modulation.

#### 3.2.3.1 DRAIN CURRENT BEYOND THE SATURATION POINT

According to the one-dimensional model in the preceding subsection, the electric field along the channel approaches infinity at the saturation point. In practice, the field remains finite. However, its magnitude becomes comparable to the vertical field, so that the gradual-channel approximation breaks down and carriers are no longer confined to the surface channel. ***As the drain voltage increases beyond the saturation voltage $V_{dsat}$, the saturation point where the surface channel collapses begins to move slightly toward the source***, as shown in Fig. 3.28. The voltage at the saturation point remains constant at $V_{dsat}$, independent of $V_{ds}$. The voltage difference $V_{ds} - V_{dsat}$ is dropped across the region between the saturation point and the drain. Carriers injected from the surface channel into this region travel at saturation velocity until collected by the drain junction. The distance between the saturation point and the drain, $\Delta L$, is referred to as the amount of channel length modulation by the drain voltage. Since the one-dimensional model is still valid between the source and the saturation point where the voltage remains at $V_{dsat}$, the device acts as if its channel length were shortened by $\Delta L$. The drain current is then obtained simply by replacing $L$ with $L - \Delta L$ in Eq. (3.78). In the long-channel limit, this increases the drain current by a factor of $(1 - \Delta L/L)^{-1}$, i.e.,

$$I_{ds} = \frac{I_{dsat}}{1 - (\Delta L/L)}. \tag{3.85}$$

Since $\Delta L$ increases with increasing drain voltage, the drain current continues to increase in the saturation region.

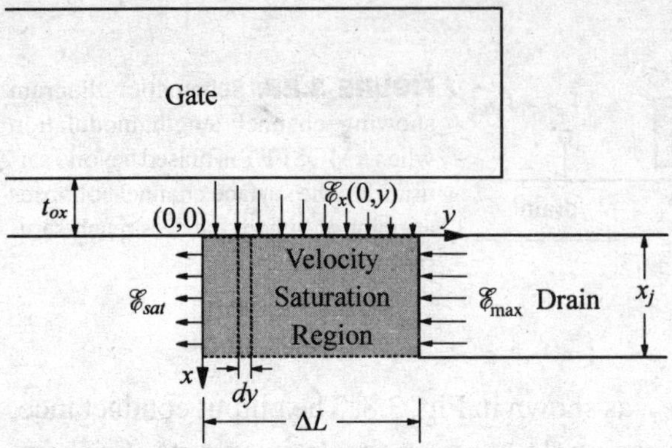


**FIGURE 3.29.** Schematic diagram of the velocity saturation region for illustrating the pseudo-2-D model. The 2-D Gauss's law is applied to a vertical stripe of width $dy$ bounded by two parallel dotted lines. (After Ko, 1989.)

#### 3.2.3.2 A PSEUDO-2-D MODEL FOR THE VELOCITY SATURATION REGION

To find out $\Delta L$ as a function of $V_{ds} - V_{dsat}$, we adopt a pseudo-2-D model (Ko *et al.*, 1981) that yields simple results yet captures the essential physics taking place beyond the saturation point. Figure 3.29 shows a schematic cross section of the region of interest. The velocity saturation region is bounded by $y = 0$ (saturation point) to $y = \Delta L$ (drain), and $x = 0$ (surface) to $x = x_j$ (drain junction depth). The mobile carriers are assumed to spread to a depth equal to the drain junction depth $x_j$. It is also assumed that the heavily doped drain junction is infinitely abrupt with a square corner.

Along the surface, the quasi-Fermi level $V(y)$ increases from $V_{dsat}$ at $y=0$ to $V_{ds}$ at $y=\Delta L$. This results in a reduction of the potential drop $V_{ox}$ across the oxide, since the total band offset,

$$V_g - V_{fb} = V_{ox}(y) + \psi_s(y) = V_{ox}(y) + 2\psi_B + V(y), \tag{3.86}$$

is constant for a fixed gate voltage. Here the surface potential $\psi_s$ is assumed to be pinned at $2\psi_B + V$ as given by Eq. (3.3) for strong inversion. This is valid as long as $V(y) \le (V_g - V_t)/m$, the long-channel pinch-off voltage. It then follows that the vertical field at the silicon surface,

$$\mathscr{E}_x(0, y) = \frac{\varepsilon_{ox}}{\varepsilon_{si}} \mathscr{E}_{ox}(y) = \frac{\varepsilon_{ox}}{\varepsilon_{si}} \frac{V_{ox}(y)}{t_{ox}}, \tag{3.87}$$

also decreases toward the drain, as depicted in Fig. 3.29. The silicon–oxide boundary condition, Eq. (2.146), was applied here with $\mathscr{E}_{ox}$ being the oxide field. At $y = 0$, all the silicon charges are still controlled by the gate, so that the one-dimensional Gauss's law is applicable:

$$\mathscr{E}_x(0, 0) = \frac{qN_a x_j + Q_i(y = 0)}{\varepsilon_{si}}, \tag{3.88}$$

where $Q_i$ $(> 0)$ is the mobile (electron) charge density per unit area. It is assumed here that the junction depth $x_j$ is comparable to the depletion width $W_{dm}$. Since carriers are already traveling at saturation velocity such that $I_{ds} = WQ_i v_{sat}$, the mobile charge density,

$$Q_i(y) = q\int_0^{x_j} n(x, y)\,dx, \tag{3.89}$$

has to remain constant, i.e., independent of $y$, toward the drain in order to maintain current continuity. Therefore, ***as the vertical field $\mathscr{E}_x(0, y)$ and the gate-controlled charge decrease toward the drain, some of the mobile charge spreads deep and becomes controlled by the drain***. The physics is similar to that of the 2-D fields discussed in Section 3.2.1. The difference is that fixed depletion charges are involved in the short-channel effect, while mobile charges are involved in the saturation region. As a result of the drain gradually taking control of the mobile charge, the electric field, $\mathscr{E}_y$, originating from the drain increases toward the drain.

Assuming that $\mathscr{E}_y$ is uniform in the $x$-direction and neglecting the vertical field at the bottom boundary ($x = x_j$), one can apply the two-dimensional Gauss's law to a thin slice of width $dy$ and length $x_j$ located at $y$ (Fig. 3.29):

$$\mathscr{E}_x(0, y)\,dy - \mathscr{E}_y(y + dy)x_j + \mathscr{E}_y(y)x_j = \frac{qN_a x_j\,dy + Q_i\,dy}{\varepsilon_{si}} \tag{3.90}$$

Expanding $\mathscr{E}_y(y + dy)$ into $\mathscr{E}_y(y) + (d\mathscr{E}_y/dy)\,dy$ and making use of Eq. (3.88), we obtain

$$-x_j\frac{d\mathscr{E}_y}{dy} = \mathscr{E}_x(0, 0) - \mathscr{E}_x(0, y). \tag{3.91}$$

From Eqs. (3.87) and (3.86), the vertical field difference can be expressed as

$$\begin{aligned}\mathscr{E}_x(0, 0) - \mathscr{E}_x(0, y) &= \frac{\varepsilon_{ox}}{\varepsilon_{si}t_{ox}}[V_{ox}(0) - V_{ox}(y)] \\ &= \frac{\varepsilon_{ox}}{\varepsilon_{si}t_{ox}}[V(y) - V(0)].\end{aligned} \tag{3.92}$$

Since $V(0) = V_{dsat}$ and $\mathscr{E}_y = -dV/dy$, substituting Eq. (3.92) into Eq. (3.91) yields

$$\frac{d^2V}{dy^2} = \frac{\varepsilon_{ox}}{\varepsilon_{si}t_{ox}x_j}[V(y) - V_{dsat}], \tag{3.93}$$

or

$$\frac{d^2V}{dy^2} = \frac{V(y) - V_{dsat}}{l^2}, \tag{3.94}$$

where the characteristic length $l$ is given by

$$l = \sqrt{\frac{\varepsilon_{si}}{\varepsilon_{ox}}t_{ox}x_j} \approx \sqrt{3t_{ox}x_j}. \tag{3.95}$$

Equation (3.94) is a linear, second-order differential equation which can be solved with the boundary conditions $V(0) = V_{dsat}$ and $\mathscr{E}_y(0) = -[dV/dy]_{y=0} = -\mathscr{E}_{sat}$:

$$V(y) = V_{dsat} + l\mathscr{E}_{sat} \sinh\left(\frac{y}{l}\right). \tag{3.96}$$

Mathematically, there is no unambiguous definition for $\mathscr{E}_{sat}$, the lateral field at the saturation point, since carriers do not reach saturation velocity until $\mathscr{E}_y = \infty$. In practice, ***carriers traveling close to the saturation velocity start moving away from the surface when the lateral field becomes appreciable compared to the vertical field***. A good choice for $\mathscr{E}_{sat}$ is a field strength on the order of or several times the critical field $\mathscr{E}_c$ defined by Eq. (3.71). For example, $\mathscr{E}_{sat} = 2\mathscr{E}_c = 2v_{sat}/\mu_{eff}$, which is on the order of $5 \times 10^4$ V/cm for electrons, has been used in the literature (Ko, 1982). This is a reasonable value, since the vertical field in a MOSFET device typically lies in the range of $10^5$–$10^6$ V/cm.

#### 3.2.3.3 PEAK FIELD AT THE DRAIN

Once $V(y)$ is known, $\Delta L$ can be found by solving $V(y = \Delta L) = V_{ds}$:

$$\Delta L = l \ln\left[\frac{V_{ds} - V_{dsat}}{l\mathscr{E}_{sat}} + \sqrt{\left(\frac{V_{ds} - V_{dsat}}{l\mathscr{E}_{sat}}\right)^2 + 1}\,\right]. \tag{3.97}$$

It is then straightforward to substitute $\Delta L$ into Eq. (3.85) or, more accurately, replace $L$ with $L - \Delta L$ in Eq. (3.78), to obtain the source–drain current beyond saturation. From Eq. (3.96), the electric field along the channel is given by

$$\mathscr{E}_y(y) = -\frac{dV}{dy} = -\mathscr{E}_{sat} \cosh\left(\frac{y}{l}\right), \tag{3.98}$$

which increases exponentially toward the drain. An example is shown in Fig. 3.30. The peak field is reached at the drain, where

$$\mathscr{E}_{\max} \equiv \mathscr{E}_y(y = \Delta L) = -\sqrt{\left(\frac{V_{ds} - V_{dsat}}{l}\right)^2 + \mathscr{E}_{sat}^2}. \tag{3.99}$$

This field can be as high as mid-$10^5$ to $10^6$ V/cm and is responsible for a variety of hot-carrier effects such as impact ionization, substrate current, and oxide degradation.

### 3.2.4 SOURCE–DRAIN SERIES RESISTANCE

In the discussion of MOSFET current thus far, it was assumed that the source and drain regions were perfectly conducting. In reality, as the current flows from the

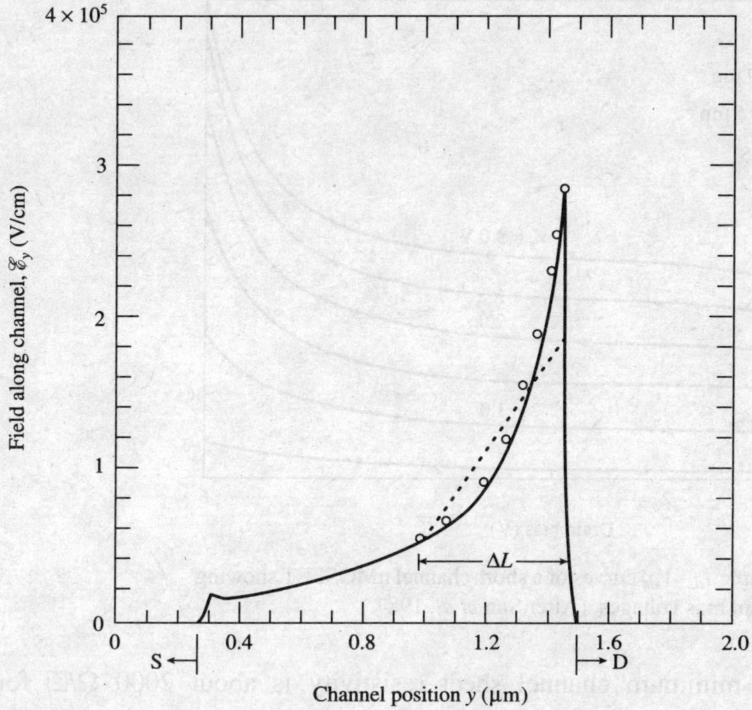


**FIGURE 3.30.** Calculated channel field versus distance between the source and drain. The velocity saturation region extends from a point where $\mathscr{E}_y \approx 5 \times 10^4$ V/cm to the drain. (After Ko *et al.*, 1981.)

channel to the terminal contact, there is a small voltage drop in the source and drain regions due to the finite silicon resistivity and metal contact resistance. In a long-channel device, the source–drain parasitic resistance is negligible compared with the channel resistance. In a short-channel device, however, the source–drain series resistance can be an appreciable fraction of the channel resistance and can therefore cause significant current degradation.

The most severe current degradation by series resistance occurs in the linear region (low $V_{ds}$) when the gate voltage is high. This is because the MOSFET channel resistance,

$$R_{ch} \equiv \frac{V_{ds}}{I_{ds}} = \frac{L}{\mu'_{eff} C_{ox} W (V_g - V_{on} - m V_{ds}/2)} \tag{3.100}$$

from Eq. (3.61), is the lowest under such bias conditions. It is instructive to estimate the sheet resistivity of a MOSFET channel,

$$\rho_{ch} \equiv R_{ch}\frac{W}{L} = \frac{1}{\mu'_{eff}(\varepsilon_{ox}/t_{ox})(V_g - V_{on} - m V_{ds}/2)} \approx \frac{1}{\mu_{eff}\varepsilon_{ox}\mathscr{E}_{ox}}. \tag{3.101}$$

Since the maximum oxide field $\mathscr{E}_{ox}$ is typically 2–5 MV/cm for most VLSI

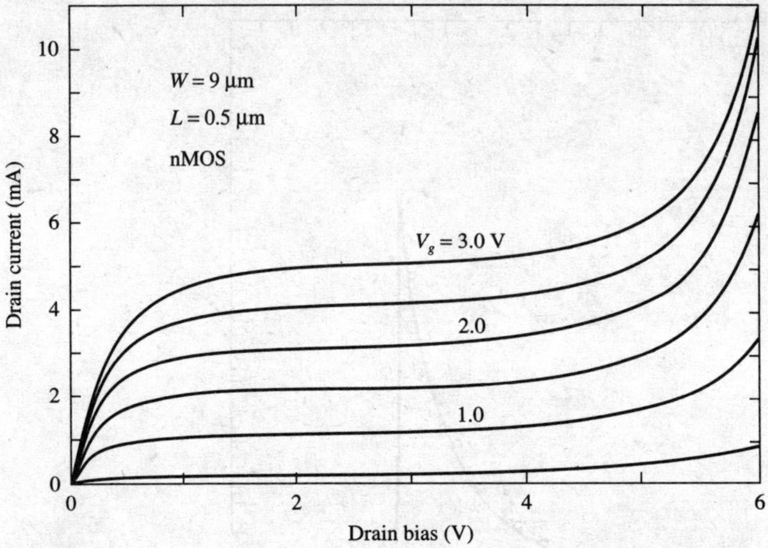


**FIGURE 3.31.** Example $I_{ds}$–$V_{ds}$ curves of a short-channel nMOSFET showing breakdown at high drain bias voltages. (After Sun *et al.* 1987.)

technologies, the minimum channel sheet resistivity is about 2000 Ω/□ for nMOSFETs and 7000 Ω/□ for pMOSFETs.

The MOSFET current in the saturation region is least affected by the resistance degradation of source–drain voltage, since $I_{ds}$ is essentially independent of $V_{ds}$ in saturation. The saturation current is only affected through gate-voltage degradation by the voltage drop between the source contact and the source end of the channel (Fig. 4.20).

The effect of series resistance on linear $I_{ds}$–$V_g$ curves used in channel-length extraction will be addressed in Section 4.3. Various contributions to the source–drain series resistance and their effect on circuit performance will be discussed in detail in Chapter 5.

### 3.2.5 MOSFET BREAKDOWN

Breakdown occurs in a short-channel MOSFET when the drain voltage exceeds a certain value, as shown in Fig. 3.31. It was discussed in Section 3.2.3 that the peak electric field given by Eq. (3.99) in the saturation region can attain large values at high drain voltages. When the field exceeds mid-$10^5$ V/cm, impact ionization (Section 2.4.1) takes place at the drain, leading to an abrupt increase of drain current. The breakdown voltage of nMOSFETs is usually lower than that of pMOSFETs because electrons have a higher rate of impact ionization (Fig. 2.37) and because $n^+$ source and drain junctions are more abrupt than $p^+$ junctions. There is also a weak dependence of the breakdown voltage on channel length; shorter devices have a lower breakdown voltage.

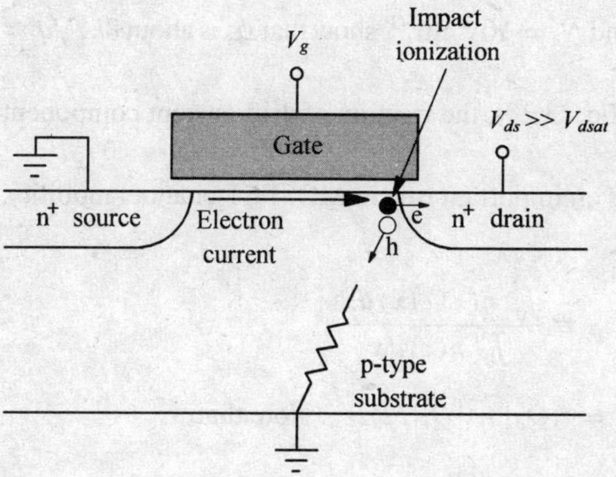


**FIGURE 3.32.** Schematic diagram showing impact ionization at the drain.

The breakdown process in an nMOSFET is shown schematically in Fig. 3.32. Electrons gain energy from the field as they move down the channel. Before they lose energy through collisions, they possess high kinetic energy and are capable of generating secondary electrons and holes by impact ionization. The generated electrons are attracted to the drain, adding to the drain current, while the holes are collected by the substrate contact, resulting in a substrate current. The substrate current in turn can produce a voltage (*IR*) drop from the spreading resistance in the bulk, which tends to forward-bias the source junction. This lowers the threshold voltage of the MOSFET and triggers a positive feedback effect, which further enhances the channel current. Substrate current is usually a good indicator of hot carriers generated by low-level impact ionization before runaway breakdown occurs.

Breakdown often results in permanent damage to the MOSFET as large amounts of hot carriers are injected into the oxide in the gate-to-drain overlap region. MOSFET breakdown is particularly a problem for VLSI technology during the elevated-voltage burn-in process. It can be relieved to some extent by using a lightly doped drain (LDD) structure (Ogura *et al.*, 1982), which introduces additional series resistance and reduces the peak field in a MOSFET. However, drain current and therefore device performance are traded off as a result. Ultimately, the devices should operate at a power-supply voltage far enough below the breakdown condition. This is one of the key CMOS design considerations in Chapter 4.

## EXERCISES

**3.1** It is commonly assumed that the surface potential $\psi_s$ is pinned at $2\psi_B$ once the inversion layer is formed. In fact, $\psi_s$ still rises slightly as the gate voltage and inversion charge density increase. Use Eq. (3.14) for $V = 0$ to show that a second-order correction term takes the form

$$\psi_s = 2\psi_B + \frac{2kT}{q} \ln\left(\frac{C_{ox}(V_g - V_{fb} - 2\psi_B)}{\sqrt{2\varepsilon_{si} kT N_a}}\right).$$

For $V_g = 5$ V, $t_{ox} = 200$ , and $N_a = 10^{16}$ cm$^{-3}$, show that $\psi_s$ is about $8kT/q \approx$ 0.2 V over $2\psi_B$.

**3.2** Fill in the steps that lead to Eq. (3.29), the fraction of drift current component in the limit of $V \to 0$.

**3.3** The effective field $\mathscr{E}_{eff}$ plays an important role in MOSFET channel mobility. Show that the definition

$$\mathscr{E}_{eff} \equiv \frac{\int_0^{x_i} n(x)\,\mathscr{E}(x)\,dx}{\int_0^{x_i} n(x)\,dx},$$

leads to Eq. (3.47), i.e., $\mathscr{E}_{eff} = (|Q_d| + |Q_i|/2)/\varepsilon_{si}$. Note that

$$|Q_i| = q\int_0^{x_i} n(x)\,dx$$

and

$$\mathscr{E}(x) = \frac{1}{\varepsilon_{si}}\left(|Q_d| + q\int_x^{x_i} n(x')\,dx'\right)$$

from Gauss's law. The inversion-layer depth $x_i$ is assumed to be much smaller than the bulk depletion width.

**3.4** An alternative threshold definition is based on the rate of change of inversion charge density with gate voltage. Equation (3.57) from Fig. 2.28(b) states that $d|Q_i|/dV_g$ is given by the serial combination of $C_{ox}$ and $C_i \equiv d|Q_i|/d\psi_s$. Below threshold, $C_i \ll C_{ox}$, so that $d|Q_i|/dV_g \approx C_i$ and $Q_i$ increases exponentially with $V_g$. Above threshold, $C_i \gg C_{ox}$, so that $d|Q_i|/dV_g \approx C_{ox}$ and $Q_i$ increases linearly with $V_g$. The change of behavior occurs at an *inversion charge threshold voltage*, $V_t^{inv}$, where $C_i = C_{ox}$. Show that at $V_g = V_t^{inv}$ one has $d|Q_i|/dV_g = C_{ox}/2$ and $Q_i \approx (2kT/q)C_{ox}$. Note that such an inversion charge threshold is independent of depletion charge and is slightly higher than the conventional $2\psi_B$ threshold.

**3.5** From Eq. (3.59) (neglecting the second term from inversion charge capacitance), show that the fractional loss of inversion charge due to the polysilicon depletion effect is $\Delta Q_i/Q_i \approx C_{ox}/2C_p$, where $C_p$ is the small-signal polysilicon-depletion capacitance defined in Eq. (2.184). Explain why there is a factor-of-two difference between the loss of charge and the loss of capacitance.

**3.6** *Charge-sharing model* (Yau, 1974): In Fig. 3.19, assume that both the source and drain depletion depths are equal to the gate depletion width $W_{dm}$ (low-drain-bias condition) and that the junction curvatures under the gate edges are cylindrical. Show that

$$\frac{L + L'}{2L} = 1 - \frac{x_j}{L}\left(\sqrt{1 + \frac{2W_{dm}}{x_j}} - 1\right).$$

In the linear region, the threshold voltage is largely determined by the total integrated depletion charge under the gate, instead of by the highest barrier in the channel as in the subthreshold regime. Show that the short-channel threshold roll-off is given by

$$\Delta V_t(\mathrm{SCE}) = \frac{qN_a W_{dm}}{C_{ox}} \left( \sqrt{1 + \frac{2W_{dm}}{x_j}} - 1 \right) \frac{x_j}{L}.$$

Note the $1/L$ dependence, in contrast with the exponential dependence in Eq. (3.66) for $\Delta V_t(\mathrm{SCE})$ under subthreshold conditions.

**3.7** The small-signal transconductance in the saturation region is defined as $g_{msat} \equiv dI_{dsat}/dV_g$. Derive an expression for $g_{msat}$ using Eq. (3.78) based on the $n = 1$ velocity saturation model. Show that $g_{msat}$ approaches the saturation-velocity-limited value, Eq. (3.84), when $L \to 0$. What becomes of the expression for $g_{msat}$ in the long-channel limit when $v_{sat} \to \infty$?

**3.8** From Eq. (3.78) based on the $n = 1$ velocity saturation model, what is the carrier velocity at the source end of the channel? What are the limiting values when $L \to 0$ and when $v_{sat} \to \infty$?

**3.9** Following a similar approach as in the text for the $n = 1$ velocity saturation model, derive an integral equation for the $n = 2$ velocity saturation model from which $I_{ds}$ can be solved. It is very tedious to carry out the integration analytically (Taylor, 1984). Interested readers may attempt performing it numerically on a computer.

**3.10** Assuming the $n = 1$ velocity saturation model, show that the total integrated inversion charge under the gate is

$$Q_i(\mathrm{total}) = -WLC_{ox}(V_g - V_t) \frac{\sqrt{1 + 2\mu_{eff}(V_g - V_t)/(mv_{sat}L)} + \frac{1}{3}}{\sqrt{1 + 2\mu_{eff}(V_g - V_t)/(mv_{sat}L)} + 1}$$

in the saturation region. Evaluate the intrinsic gate-to-channel capacitance, and show that it approaches Eq. (3.56) in the long-channel limit.

# 4 CMOS DEVICE DESIGN

This chapter examines the key device design issues in a modern CMOS VLSI technology. It begins with an extensive review of the concept of MOSFET scaling. Two important CMOS device design parameters, threshold voltage and channel length, are then discussed in detail.

## 4.1 MOSFET SCALING

CMOS technology evolution in the past twenty years has followed the path of device scaling for achieving density, speed, and power improvements. MOSFET scaling was propelled by the rapid advancement of lithographic techniques for delineating fine lines of 1 μm width and below. In Section 3.2.1, we discussed that reducing the source-to-drain spacing, i.e., the channel length of a MOSFET, led to short-channel effects. For digital applications, the most undesirable short-channel effect is a reduction in the gate threshold voltage at which the device turns on, especially at high drain voltages. Full realization of the benefits of the new high-resolution lithographic techniques therefore requires the development of new device designs, technologies, and structures which can be optimized to keep short-channel effects under control at very small dimensions. Another necessary technological advancement for device scaling is in ion implantation, which not only allows the formation of very shallow source and drain regions but also is capable of accurately introducing a sharply profiled, low concentration of doping atoms for optimum channel profile design.

### 4.1.1 CONSTANT-FIELD SCALING

In constant-field scaling (Dennard *et al.*, 1974), it was proposed that one can keep short-channel effects under control by scaling down the vertical dimensions (gate insulator thickness, junction depth, etc.) along with the horizontal dimensions, while also proportionally decreasing the applied voltages and increasing the substrate doping concentration (decreasing the depletion width). This is shown schematically in Fig. 4.1. ***The principle of constant-field scaling lies in scaling the device voltages***

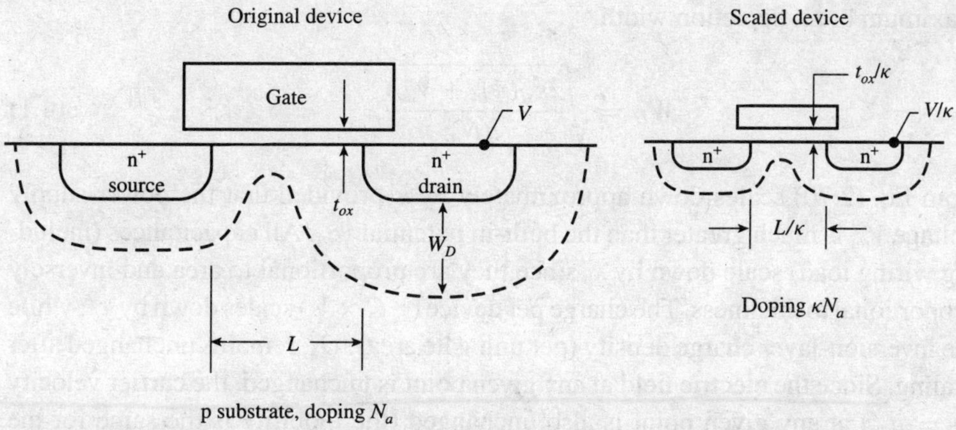


**FIGURE 4.1.** Principles of MOSFET constant-electric-field scaling. (After Dennard, 1986.)

***and the device dimensions (both horizontal and vertical) by the same factor, $\kappa$ (> 1), so that the electric field remains unchanged.*** This assures that the reliability of the scaled device is not worse than that of the original device.

#### 4.1.1.1 RULES FOR CONSTANT-FIELD SCALING

Table 4.1 shows the scaling rules for various device parameters and circuit performance factors. The doping concentration must be increased by the scaling factor $\kappa$ in order to keep Poisson's equation (3.64) invariant with respect to scaling. The

**TABLE 4.1** Scaling of MOSFET Device and Circuit Parameters

| | **MOSFET Device and Circuit Parameters** | **Multiplicative Factor ($\kappa$ > 1)** |
|---|---|---|
| Scaling assumptions | Device dimensions ($t_{ox}, L, W, x_j$) | $1/\kappa$ |
| | Doping concentration ($N_a, N_d$) | $\kappa$ |
| | Voltage ($V$) | $1/\kappa$ |
| Derived scaling behavior of device parameters | Electric field ($\mathscr{E}$) | 1 |
| | Carrier velocity ($v$) | 1 |
| | Depletion-layer width ($W_d$) | $1/\kappa$ |
| | Capacitance ($C = \varepsilon A/t$) | $1/\kappa$ |
| | Inversion-layer charge density ($Q_i$) | 1 |
| | Current, drift ($I$) | $1/\kappa$ |
| | Channel resistance ($R_{ch}$) | 1 |
| Derived scaling behavior of circuit parameters | Circuit delay time ($\tau \sim CV/I$) | $1/\kappa$ |
| | Power dissipation per circuit ($P \sim VI$) | $1/\kappa^2$ |
| | Power–delay product per circuit ($P\tau$) | $1/\kappa^3$ |
| | Circuit density ($\propto 1/A$) | $\kappa^2$ |
| | Power density ($P/A$) | 1 |

maximum drain depletion width,

$$W_D = \sqrt{\frac{2\varepsilon_{si}(\psi_{bi} + V_{dd})}{qN_a}}, \tag{4.1}$$

from Eq. (2.70) scales down approximately by $\kappa$ provided that the power-supply voltage $V_{dd}$ is much greater than the built-in potential $\psi_{bi}$. All capacitances (including wiring load) scale down by $\kappa$, since they are proportional to area and inversely proportional to thickness. The charge per device ($\sim C \times V$) scales down by $\kappa^2$, while the inversion-layer charge density (per unit gate area), $Q_i$, remains unchanged after scaling. Since the electric field at any given point is unchanged, the carrier velocity ($v = \mu\mathscr{E}$) at any given point is also unchanged (the mobility is the same for the same vertical field). Therefore, any velocity saturation effects will be similar in the original and the scaled devices.

The drift current per MOSFET width, obtained by integrating the first term of the electron current density equation (2.43) over the inversion layer thickness, is

$$\frac{I_{drift}}{W} = Q_i v = Q_i \mu \mathscr{E}, \tag{4.2}$$

and is unchanged with respect to scaling. This means that the drift current scales down by $\kappa$, consistent with the behavior of both the linear and the saturation MOSFET currents in Eq. (3.19) and Eq. (3.23). A key implicit assumption is that the threshold voltage also scales down by $\kappa$. Note that the velocity saturated current, Eq. (3.78), also scales the same way, since both $v_{sat}$ and $\mu_{eff}$ are constants, independent of scaling. However, the diffusion current per unit MOSFET width, obtained by integrating the second term of the current density equation (2.43) and given by

$$\frac{I_{diff}}{W} = D_n \frac{dQ_i}{dx} = \mu_n \frac{kT}{q} \frac{dQ_i}{dx}, \tag{4.3}$$

scales up by $\kappa$, since $dQ_i/dx$ is inversely proportional to the channel length. Therefore, the diffusion current does not scale down the same way as the drift current. This has significant implications in the nonscaling of MOSFET subthreshold currents, as will be discussed in Section 4.1.3.

#### 4.1.1.2 EFFECT OF SCALING ON CIRCUIT PARAMETERS

With both the voltage and the current scaled down by the same factor, it follows that the active channel resistance [e.g., Eq. (3.100)] of the scaled-down device remains unchanged. It is further assumed that parasitic resistance is either negligible or unchanged in scaling. The circuit delay, which is proportional to $RC$ or $CV/I$, then scales down by $\kappa$. ***This is the most important conclusion of constant-field scaling: once the device dimensions and the power-supply voltage are scaled*** *down, the circuit speeds up by the same factor. Moreover, power dissipation per circuit, which is proportional to VI, is reduced by* $\kappa^2$. Since the circuit density has increased by $\kappa^2$, the power density, i.e., the active power per chip area, remains unchanged in the scaled-down device. This has important technological implications in that, in contrast to bipolar devices (Chapters 6, 7, and 8), packaging of the scaled CMOS devices does not require more elaborate heat-sinking. The power–delay product of the scaled CMOS circuit shows a dramatic improvement by a factor of $\kappa^3$ (Table 4.1).

#### 4.1.1.3 THRESHOLD VOLTAGE

It was assumed earlier that the threshold voltage should be decreased by the scaling factor, $\kappa$, in proportion to the power-supply voltage. This is examined using the threshold equation (3.40) for a uniformly doped substrate:

$$V_t = V_{fb} + 2\psi_B + \frac{\sqrt{2\varepsilon_{si} q N_a (2\psi_B + V_{bs})}}{C_{ox}}, \qquad (4.4)$$

where $V_{bs}$ is the substrate bias voltage. In silicon technology, the material-related parameters (energy gap, work function, etc.) do not change with scaling; hence, in general, $V_t$ does not scale. However, in a conventional process, $n^+$-polysilicon gates are used for n-channel MOSFETs, and $V_{fb} = -E_g/2q - \psi_B$ from Eq. (2.181). It turns out that the first two terms on the RHS of Eq. (4.4) add up to approximately −0.15 V, which can be neglected. One can then argue (Dennard *et al.*, 1974) that by adjusting $V_{bs}$ so that $2\psi_B + V_{bs}$ scales down by $\kappa$, the last term of Eq. (4.4), and therefore $V_t$, will also scale down by $\kappa$. However, at the present level of technology development, $V_{bs}$ has been reduced to zero for most logic applications, though a reverse-biased body–source junction is still used for some dynamic memory array devices. Further reduction of the $2\psi_B + V_{bs}$ term with scaling would require a forward bias on the substrate. This is not commonly used in VLSI technologies, although it has been attempted in experimental devices (Sai-Halasz *et al.*, 1990). In practice, nonuniform doping profiles have been employed to tailor the threshold voltage of scaled devices, as will be discussed in Section 4.2.

p-channel MOSFETs with $p^+$-polysilicon gates scale similarly to their counterparts. However, in buried-channel devices, e.g., when an $n^+$-polysilicon gate is used for p-channel MOSFETs, the sum of the first two terms in Eq. (4.4) is nearly 1 V and therefore cannot be neglected. For this reason, it is difficult to scale buried-channel devices to low threshold voltages. More about threshold voltage design can be found in Section 4.2.

### 4.1.2 GENERALIZED SCALING

Even though constant-field scaling provides a basic guideline to the design of scaled MOSFETs, the requirement of reducing the voltage by the same factor as the

**TABLE 4.2** CMOS VLSI Technology Generations

| Feature Size (μm) | Power-Supply Voltage (V) | Gate Oxide Thickness (Å) | Oxide Field (MV/cm) |
|---|---|---|---|
| 2 | 5 | 350 | 1.4 |
| 1.2 | 5 | 250 | 2.0 |
| 0.8 | 5 | 180 | 2.8 |
| 0.5 | 3.3 | 120 | 2.8 |
| 0.35 | 3.3 | 100 | 3.3 |
| 0.25 | 2.5 | 70 | 3.6 |

device physical dimension is too restrictive. Because of subthreshold nonscaling and reluctance to depart from the standardized voltage levels of the previous generation, the power-supply voltage was seldom scaled in proportion to channel length. Table 4.2 lists the supply voltage and device parameters of several generations of CMOS VLSI technology. It is clear that the oxide field has been increasing over the generations rather than staying constant. For device design purposes, therefore, it is necessary to develop a more general set of guidelines that allows the electric field to increase. In such a generalized scaling (Baccarani *et al.*, 1984), it is desired that both the vertical and the lateral electric fields change by the same multiplication factor so that the shape of the electric field pattern is preserved. This assures that 2-D effects, such as short-channel effects, do not become worse when scaling to a smaller dimension. Higher fields, however, do cause reliability concerns as mentioned in Section 2.4.

#### 4.1.2.1 RULES FOR GENERALIZED SCALING

If we assume that the electric field intensity changes by a factor of $\alpha$, i.e., $\mathscr{E} \to \alpha\mathscr{E}$, while the device physical dimensions (both lateral and vertical) scale down by $\kappa$ ($>1$) in generalized scaling, the potential or voltage will change by a factor equal to the ratio $\alpha/\kappa$. If $\alpha = 1$, it reduces back to constant-field scaling. To keep Poisson's equation invariant under the transformation, $(x, y) \to (x, y)/\kappa$ and $\psi \to \psi/(\kappa/\alpha)$ within the depletion region,

$$\frac{\partial^2(\alpha\psi/\kappa)}{\partial(x/\kappa)^2} + \frac{\partial^2(\alpha\psi/\kappa)}{\partial(y/\kappa)^2} = \frac{qN_a'}{\varepsilon_{si}}, \tag{4.5}$$

$N_a'$ should be scaled to $(\alpha\kappa)N_a$. In other words, the doping concentration must be scaled up by an extra factor of $\alpha$ to control the depletion-region depth and thus avoid increased short-channel effects due to the higher electric field. Table 4.3 shows the generalized scaling rules of other device and circuit parameters.

Since the electric field intensity is usually increased in generalized scaling, the carrier velocity tends to increase as well. How much the velocity increases depends on how velocity-saturated the original device is. In the long-channel limit, carrier

**TABLE 4.3** Generalized MOSFET Scaling

| | MOSFET Device and Circuit Parameters | Multiplicative Factor ($\kappa > 1$) | |
|---|---|---|---|
| Scaling assumptions | Device dimensions ($t_{ox}$, $L$, $W$, $x_j$) | $1/\kappa$ | |
| | Doping concentration ($N_a$, $N_d$) | $\alpha\kappa$ | |
| | Voltage ($V$) | $\alpha/\kappa$ | |
| Derived scaling behavior of device parameters | Electric field ($\mathscr{E}$) | $\alpha$ | |
| | Depletion-layer width ($W_d$) | $1/\kappa$ | |
| | Capacitance ($C = \varepsilon A/t$) | $1/\kappa$ | |
| | Inversion-layer charge density ($Q_i$) | $\alpha$ | |
| | | Long Ch. | Vel. Sat. |
| | Carrier velocity ($v$) | $\alpha$ | 1 |
| | Current, drift ($I$) | $\alpha^2/\kappa$ | $\alpha/\kappa$ |
| Derived scaling behavior of circuit parameters | Circuit delay time ($\tau \sim CV/I$) | $1/\alpha\kappa$ | $1/\kappa$ |
| | Power dissipation per circuit ($P \sim VI$) | $\alpha^3/\kappa^2$ | $\alpha^2/\kappa^2$ |
| | Power–delay product per circuit ($P\tau$) | $\alpha^2/\kappa^3$ | |
| | Circuit density ($\propto 1/A$) | $\kappa^2$ | |
| | Power density ($P/A$) | $\alpha^3$ | $\alpha^2$ |

velocities are far from saturation and will increase by the same factor, $\alpha$, as the electric field. The drift current, which is proportional to $WQ_iv$, will then change by a factor of $\alpha^2/\kappa$. This is consistent with the scaling behavior of long-channel currents, Eq. (3.19) and Eq. (3.23). On the other hand, if the original device is fully velocity-saturated, the carrier velocity cannot increase any more, in spite of the higher field in the scaled device. The current in this case will change only by a factor of $\alpha/\kappa$, consistent with the velocity-saturated current, Eq. (3.80). The circuit delay scales down by a factor between $\kappa$ and $\alpha\kappa$, depending on the degree of velocity saturation. The most serious issue with generalized scaling is the increase of the power density by a factor of $\alpha^2$ to $\alpha^3$. This puts a great burden on VLSI packaging technology to dissipate the extra heat generated on the chip. The power–delay product is also a factor of $\alpha^2$ higher than for constant-field scaling.

#### 4.1.2.2 CONSTANT-VOLTAGE SCALING

Even though Poisson's equation within the depletion region is invariant under generalized scaling, the same is not true in the inversion layer when mobile charges are present. This is because mobile charge densities are exponential functions of potential which do not scale linearly with either physical dimensions or voltage. Furthermore, even in the depletion region, not all the boundary conditions scale consistently under generalized scaling. ***This is due to the fact that the band bending at the source junction is given by the built-in potential (Appendix 6), which***

***does not scale with voltage. Strictly speaking, the shape of the field pattern is preserved only if $\alpha = \kappa$, i.e., constant-voltage scaling.*** Under constant-voltage scaling, the electric field scales up by $\kappa$ and the doping concentration $N_a$ scales up by $\kappa^2$. The maximum gate depletion width (long-channel) [Eq. (3.69)],

$$W_{dm}^0 = \sqrt{\frac{4\varepsilon_{si}kT\ln(N_a/n_i)}{q^2N_a}}, \tag{4.6}$$

then scales down by $\kappa$. Here $\ln(N_a/n_i)$ is a weak function of $N_a$ and can be treated as a constant. This allows the short-channel $V_t$ roll-off [Eq. (3.66)],

$$\Delta V_t = \frac{24t_{ox}}{W_{dm}}\sqrt{\psi_{bi}(\psi_{bi}+V_{ds})}e^{-\pi L/2(W_{dm}+3t_{ox})}, \tag{4.7}$$

to remain unchanged, as both $t_{ox}$ and $W_{dm}$ are scaled down by the same factor as the channel length $L$. Both the power-supply voltage and the threshold voltage [Eq. (4.4)],

$$V_t = V_{fb} + 2\psi_B + \frac{\sqrt{2\varepsilon_{si}qN_a(2\psi_B+V_{bs})}}{C_{ox}}, \tag{4.8}$$

also remain unchanged. From Eq. (2.166), the inversion-layer charge per unit area is related to the electron concentration at the silicon surface, $n(0)$, by

$$Q_i = \sqrt{2\varepsilon_{si}kTn(0)}. \tag{4.9}$$

Since $Q_i$ scales up by $\kappa$ in constant-voltage scaling, $n(0)$ scales up by $\kappa^2$. Therefore, the mobile charge density scales the same way as the fixed charge density $N_a$. The inversion-layer thickness, being proportional to $Q_i/qn(0)$, scales down by $\kappa$ just like other linear dimensions. The Debye length, $L_D = (\varepsilon_{si}kT/q^2N_a)^{1/2}$, also scales down by $\kappa$ under constant-voltage scaling.

Although constant-voltage scaling leaves the solution of Poisson's equation for the electrostatic potential unchanged except for a constant multiplicative factor in the electric field, it cannot be practiced without limit, since the power density increases by a factor of $\kappa^2$ to $\kappa^3$. Higher fields also cause hot-electron and oxide reliability problems. In reality, CMOS technology evolution has followed mixed steps of constant-voltage and constant-field scaling, as is evident in Table 4.2.

### 4.1.3 NONSCALING EFFECTS

#### 4.1.3.1 PRIMARY NONSCALING FACTORS

From the above discussions, it is clear that although constant-field scaling provides a basic framework for shrinking CMOS devices to gain higher density and speed without degrading reliability and power, there are several factors that scale neither with the physical dimensions nor with the operating voltage. ***The primary reason for the nonscaling effects is that neither the thermal voltage kT/q nor the silicon bandgap $E_g$ changes with scaling***. The former leads to subthreshold nonscaling; i.e., the threshold voltage cannot be scaled down like other parameters. The latter leads to nonscalability of the built-in potential, depletion-layer width, and short-channel effect.

From Eq. (3.36), the *off current* of a MOSFET is given by

$$I_{ds}(V_g = 0, V_{ds} = V_{dd}) = \mu_{eff} C_{ox} \frac{W}{L}(m-1)\left(\frac{kT}{q}\right)^2 e^{-qV_t/mkT} \tag{4.10}$$

Because of the exponential dependence, the threshold voltage cannot be scaled down significantly without causing a substantial increase in the off current. In fact, even if the threshold voltage is held unchanged, the off current per device still increases by a factor of $\kappa$ (from the $C_{ox}$ factor) when the physical dimensions are scaled down by $\kappa$. This imposes a serious limitation on how low the threshold voltage can be, especially in dynamic circuits and random-access memories. The threshold voltage limitation in turn sets a lower limit on the power-supply voltage $V_{dd}$, since the circuit delay increases rapidly with the ratio $V_t/V_{dd}$ when the latter exceeds about 0.3, as will be discussed in Chapter 5.

Another nonscaling factor related to $kT/q$ is the inversion-layer thickness, which is unchanged in constant-field scaling. Since the inversion-layer capacitance arising from the finite thickness is in series with the oxide capacitance, the total gate capacitance per unit area of the scaled device increases by a factor less than $\kappa$ (Baccarani and Wordeman, 1983). This degrades the inversion charge density and therefore the current, especially at low gate voltages, as can be seen from Eq. (3.58).

Because both the junction built-in potential [Eq. (2.69)] and the maximum surface potential [Eq. (2.155)] are in the range of 0.6–1.0 V and do not change significantly with device scaling, the depletion-region widths, Eq. (4.1) and Eq. (4.6), do not scale quite as much as other linear dimensions. This results in worse short-channel effects in the scaled MOSFET, as is evident from Eq. (4.7). To compensate for these effects, the doping concentration must increase more than that suggested by constant-field scaling or generalized scaling.

#### 4.1.3.2 SECONDARY NONSCALING FACTORS

Because of subthreshold nonscaling, the voltage level cannot be scaled down as much as the linear dimensions, and the electric field has increased as a result. This triggers several secondary nonscaling effects. First, in our discussions so far, it was implicitly assumed that carrier mobilities are constant, independent of scaling. However, as discussed in Section 3.1.5, the mobility decreases with increasing electric field:

$$\mu_{eff} \approx 32500\mathscr{E}_{eff}^{-1/3}, \tag{4.11}$$

in units of $cm^2/V$-s for $\mathscr{E}_{eff} \le 5 \times 10^5$ V/cm. Beyond $\mathscr{E}_{eff} = 5 \times 10^5$ V/cm, the mobility decreases even faster due to surface roughness scattering (Fig. 3.13). Since it is inevitable that the electric field increases with scaling, carrier mobilities are degraded in scaled MOSFETs. As a result, both the current and the delay improve less than the factors listed in Table 4.3 for generalized scaling. Furthermore, higher fields tend to push device operation more into the velocity-saturated regime. This means that the current gain and the delay improvement are closer to the velocity-saturated column of Table 4.3, and there is little to gain by operating at an even higher field or voltage.

***The most serious problems associated with the higher field intensity are reliability and power***. The power density increases by a factor of $\alpha^2$ to $\alpha^3$ as discussed before. Reliability problems arise from higher oxide fields, higher channel fields, and higher current densities. Even under the fully velocity-saturated condition, the current density increases by $\alpha\kappa$. This aggravates the problem of electromigration in aluminum lines, which is already becoming worse under constant-field scaling (Dennard *et al.*, 1974). Higher fields also drive gate oxides closer to the breakdown condition, making it difficult to maintain oxide integrity. In fact, in order to curb the growing oxide field, the gate oxide thickness has been reduced less than the lateral device dimensions, e.g., the channel length, as is evident in Table 4.2. This means that the channel doping concentration must be increased more than called for in Table 4.3 to keep short-channel effects [Eq. (4.7)] under control. In other words, the maximum gate depletion width $W_{dm}$ must be reduced more than the oxide thickness $t_{ox}$. This triggers another set of nonscaling effects, including the subthreshold slope $\propto m = 1 + (3t_{ox}/W_{dm})$, and the substrate sensitivity $dV_t/dV_{bs} = m - 1$ [Eq. (3.41)]. These will be discussed in detail in Section 4.2.3.

#### 4.1.3.3 OTHER NONSCALING FACTORS

In practice, there is yet another set of nonscaling factors encountered in CMOS technology evolution. One kind of nonscaling effects is related to the gate and source–drain doping levels. If not properly scaled up, they may lead to gate depletion and source–drain series resistance problems. From Eq. (2.185), polysilicon gate depletion contributes a capacitance $C_p = \varepsilon_{si} q N_p / Q_p$ in series with the oxide capacitance $C_{ox}$. As $C_{ox}$ increases by a factor of $\kappa$ while $Q_p$ remains unchanged in constant-field scaling, $N_p$ must scale up by $\kappa$ also to keep $C_p$ in step with $C_{ox}$. In generalized scaling, $N_p$ must scale up even more (by $\alpha\kappa$). In reality, this is seldom done, for process reasons. The total gate capacitance then scales up by less than $C_{ox}$, leading to degradation of the inversion charge density and transconductance. Similarly, it is difficult to scale up the source–drain doping level and make the profile more abrupt while scaling down the junction depth. In practice, the source–drain series resistance has not been reduced in proportion to channel resistance, Eq. (3.100). This causes loss of current drive as the parasitic component becomes a more significant fraction of the total resistance in the scaled device.

Another class of nonscaling factors arise from process tolerances. The full benefit of scaling cannot be realized unless all process tolerances are reduced by the same factor as the device parameters. These include channel length tolerance, oxide thickness tolerance, threshold voltage tolerance, etc. ***It is a key requirement and challenge in VLSI technology development to keep the tolerance to a constant percentage of the device parameter as the dimension is scaled down***. This could be a major factor in manufacturing costs as one tries to control a few hundred angstroms of channel length or a few atomic layers of gate oxide.

## 4.2 THRESHOLD VOLTAGE

This section focuses on a key design parameter in CMOS technology: threshold voltage. Although the threshold voltage was introduced in Chapter 3, the discussions there were restricted to the case of uniform doping. In this section, threshold voltages under nonuniform doping conditions are discussed, leading to the design of MOSFET channel profile.

### 4.2.1 THRESHOLD-VOLTAGE REQUIREMENT

#### 4.2.1.1 VARIOUS DEFINITIONS OF THRESHOLD VOLTAGE

First, we examine the various definitions of threshold voltage and the threshold-voltage requirement from a technology point of view. There are quite a number of different ways to define the threshold voltage of a MOSFET device. In Chapter 3 we followed the most commonly used definition [$\psi_s(\text{inv}) = 2\psi_B$] of $V_t$. The advantage of this definition lies in its popularity and ease of incorporation into analytical solutions. However, it is not directly measurable from experimental $I$–$V$ characteristics (it can be determined from a split $C$–$V$ measurement; see Exercise 2.6). In Section 3.1.6, we introduced the linearly extrapolated threshold voltage, $V_{on}$, determined by the intercept of a tangent through the maximum-slope (linear transconductance) point of the low-drain $I_{ds}$–$V_g$ curve. This is easily measured experimentally, but is about $3kT/q$ higher than the $2\psi_B$ threshold voltage, due to inversion-layer capacitance effects illustrated in Fig. 3.16.

Another commonly employed definition of threshold voltage is based on the subthreshold current, Eq. (3.36). For $V_{ds} > 2kT/q$ and $V_g < V_t$,

$$I_{ds}(V_g) = \mu_{eff} C_{ox} \frac{W}{L}(m-1)\left(\frac{kT}{q}\right)^2 e^{q(V_g - V_t)/mkT}. \tag{4.12}$$

For a given constant current level $I_0$ (say, 50 nA/□), one can define a threshold voltage $V_t^{sub}$ such that $I_{ds}(V_g = V_t^{sub}) = I_0(W/L)$. The advantages of such a threshold-voltage definition are twofold. First, it is easy to extract from hardware data and is therefore suitable for automated measurement of a large number of devices. Second, the device off current, $I_{off} = I_{ds}(V_g = 0)$, can be directly calculated from $I_0$, $V_t^{sub}$, and the subthreshold slope. However, there is a serious problem when the definition $I_{ds}(V_g = V_t^{sub}) = I_0(W/L)$ is used on a short-channel device whose exact channel length is not known. Even if the channel length is extracted from the currents in the linear region (Section 4.3.2), it is not necessarily the same channel length needed in Eq. (4.12) for the subthreshold region (Nguyen, 1984). This is because of the different current conduction mechanisms involved (diffusion versus drift). In subsequent discussions, we will adhere to the $2\psi_B$ definition of $V_t$. In general, $V_t$ depends on temperature (temperature coefficient), substrate bias (body-effect coefficient), channel length, and drain voltage (short-channel effect, or SCE).

#### 4.2.1.2 OFF-CURRENT REQUIREMENT AND MINIMUM THRESHOLD VOLTAGE

It is evident from Chapter 3 that the lower the threshold voltage, the higher the current drive, hence the faster the switching speed. From a CMOS performance point of view, it is desirable to have a threshold voltage as low as possible. This, of course, is counterbalanced by the off-current requirement that the MOSFET be turned off properly at $V_g = 0$. To meet the maximum-off-current requirement, one needs to consider the worst case when the threshold voltage is the lowest. Many tolerance factors must be taken into consideration in the worst-case design of a VLSI technology, for example, process tolerances (film thickness, implant dose, etc.), dimension tolerances (lithography, etching, etc.), operating temperature range, and bias conditions. Since the threshold voltage increases with substrate bias and decreases with temperature as discussed in Section 3.1.4, the worst-case condition is at zero substrate bias and maximum operating temperature, $T_{\max}$. Depending on the application, $T_{\max}$ is typically 100°C or so, driven by both environmental factors and heat dissipation of the VLSI chip in operation. From Eq. (4.12), the minimum threshold voltage for a maximum off current $I_{off}$ at $T = T_{\max}$ is

$$V_{t\,\min} = \frac{mkT_{\max}}{q} \ln\left(\frac{I_{ds}(V_g = V_t)}{I_{off}}\right), \tag{4.13}$$

where

$$I_{ds}(V_g = V_t) = \mu_{eff} C_{ox} \frac{W}{L}(m-1)\left(\frac{kT_{\max}}{q}\right)^2 \tag{4.14}$$

is the drain current at threshold voltage. Note that $I_{ds}(V_g = V_t)$ is rather insensitive to temperature, since $\mu_{eff} \propto T_{\max}^{-3/2}$. However, it does depend on technology. For example, $I_{ds}(V_g = V_t)/W$ varies from $10^{-8}$ A/μm for 1-μm CMOS technology to $10^{-6}$ A/μm for 0.1-μm CMOS technology. (Note that these numbers are for nMOSFETs; pMOSFET currents are about 3 times lower. Also note that the extrapolated subthreshold currents at the linearly extrapolated threshold voltage $V_{on}$ are about 10 times higher than these numbers, as discussed at the end of Section 3.1.6.). Taking the median value of $10^{-7}$ A/μm for $I_{ds}(V_g = V_t)/W$ and assuming a worst-case off-current requirement of $I_{off}/W = 10^{-8}$ A/μm, one obtains $V_{t\,\min} \approx 0.1$ V for $T_{\max} = 100°$C and $m = 1.3$. For a given $I_{off}/W$, $V_{t\,\min}$ increases as the channel length is scaled down. However, this is opposite to the downward scaling trend of the power-supply voltage. As a result, the device designer often encounters a tradeoff between the performance and the off-current requirement in scaled CMOS technologies (Mii *et al.*, 1994).

The above figures are acceptable for CMOS logic technologies. In a dynamic memory technology, however, the off-current requirement is much more stringent for the access transistor in the cell: on the order of $I_{off}/W = 10^{-12}$ A/μm or so (Dennard, 1984). This means $V_{t\,\min} \approx 0.5$ V for the DRAM access device. It should be understood that Eq. (4.12) is an analytical expression based on some simplifying approximations. It is used here for the purpose of illustration. More exact values of off current for a particular design should be obtained from numerical simulations.

#### 4.2.1.3 THRESHOLD-VOLTAGE TOLERANCES

Equation (4.13) gives the minimum threshold voltage at the highest operating temperature for the worst-case process conditions. To obtain the nominal design threshold voltage at room temperature, one needs to add $\Delta V_t$ due to the temperature difference as well as the sum of all $V_t$ tolerances to $V_{t\,\min}$. In other words,

$$V_t(\text{nominal, } 23°\text{C}) = V_{t\,\min} + \Delta V_t(\text{temp.}) + \sqrt{[\Delta V_t(\text{SCE})]^2 + [\Delta V_t(\text{process})]^2}, \tag{4.15}$$

where $\Delta V_t$(SCE) and $\Delta V_t$(process) are the $3\sigma$ tolerances of $V_t$ reduction from short-channel effects and from process variations, respectively. They are added in an root-mean-square (RMS) fashion, since there are no correlations between them. $\Delta V_t$(temp.) is about 55–75 mV, since $dV_t/dT \approx -0.7$ to $-1$ mV/°C, insensitive to technology (Section 3.1.4).

An approximate expression for $\Delta V_t$(SCE) is given by Eq. (3.68) for the worst-case high drain bias:

$$\Delta V_t(\text{SCE}) = 8(m-1)\sqrt{\psi_{bi}(\psi_{bi} + V_{ds})}\, e^{-\pi L_{\min}/2mW_{dm}}, \tag{4.16}$$

where $m \approx 1 + (3t_{ox}/W_{dm})$, and $L_{\min}$ is the $3\sigma$ worst-case channel length of the manufacturing process. The preexponential factor in Eq. (4.16) does not scale much with technology, since $\psi_{bi}$ is largely determined by the silicon bandgap and does not vary significantly with either device dimension or supply voltage. From 1-μm CMOS technology with $\psi_{bi} \approx 0.7$ V and $V_{ds} = 5$ V to 0.1-μm CMOS technology

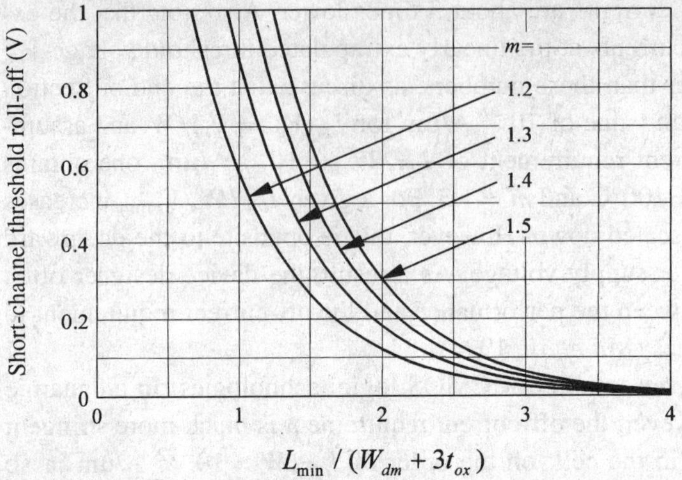


**FIGURE 4.2.** Short-channel threshold voltage roll-off, $\Delta V_t$(SCE), versus $L_{min}/mW_{dm}$ or $L_{min}/(W_{dm} + 3t_{ox})$ for several typical values of $m$, based on an approximate analytical expression.

with $\psi_{bi} \approx 0.94$ V and $V_{ds} = 1.5$ V, the square-root factor in Eq. (4.16) changes only slightly, from 2.0 to 1.5 V. Taking an average value of 1.75 V, one can plot $\Delta V_t$(SCE) as a function of $L_{min}/mW_{dm}$, or equivalently, as a function of $L_{min}/(W_{dm} + 3t_{ox})$, as in Fig. 4.2, for several possible values of $m$. Because of the exponential factor in Eq. (4.16), $\Delta V_t$(SCE) is very sensitive to $L_{min}/mW_{dm}$. ***A good choice of*** $\boldsymbol{L_{min}/mW_{dm}}$ ***is 2***, which gives $\Delta V_t(\text{SCE}) \approx 0.2$ V for a median value of $m = 1.3$. Lower values of $L_{min}/mW_{dm}$ result in too severe a short-channel effect or large $\Delta V_t$(SCE). Higher values of $L_{min}/mW_{dm}$ improve the short-channel effect but also raise the junction capacitance (smaller $W_{dm}$ for a given $L_{min}$) or increase the oxide field (smaller $t_{ox}$ for a given $V_{dd}$). The last quantity in Eq. (4.15), $\Delta V_t$(process) due to $t_{ox}$ and $N_a$ variations, can be estimated from the threshold equation (3.20). In most cases, $3\sigma$ variations of $t_{ox}$ and $N_a$ can be controlled to within 5–10% of their nominal values. Therefore, $\Delta V_t$(process) should be less than 10% of $V_t$, or less than about 50 mV, and $[\Delta V_t(\text{process})]^2$ can be neglected in comparison with $[\Delta V_t(\text{SCE})]^2$ in Eq. (4.15). The threshold voltage requirements in above examples are then $V_t$(nominal, 23°C ) $\approx 0.4$ V for logic technologies and $\approx 0.8$ V for DRAM technologies.

Another consideration that may further limit how low the threshold voltage can be is the *burn-in* procedure. Burn-in is required in most VLSI technologies to remove early failures and ensure product reliability. It is usually carried out at elevated temperatures and overvoltages to accelerate the degradation process. Both of these conditions further lower the threshold voltage and aggravate the leakage currents. Ideally, the burn-in procedure should be designed so that it does not require a compromise on the device performance.

### 4.2.2 NONUNIFORM DOPING

When the MOSFET channel is uniformly doped, the maximum gate depletion width (long-channel),

$$W_{dm}^0 = \sqrt{\frac{4\varepsilon_{si}\psi_B}{qN_a}}, \tag{4.17}$$

and the threshold voltage,

$$V_t = V_{fb} + 2\psi_B + \frac{\sqrt{4\varepsilon_{si}qN_a\psi_B}}{C_{ox}}, \tag{4.18}$$

are coupled through the parameter $N_a$, and therefore cannot be varied independently (for a given $V_{fb}$ and $t_{ox}$). It was discussed in the previous subsection that in order to control short-channel effects, $W_{dm}$ or $W_{dm}^0$ should be on the order of $L_{\min}/(2m)$. The doping concentration that satisfies this requirement may not give the desired threshold voltage that satisfies the off-current requirement. Nonuniform channel doping gives the device designer an additional degree of freedom to tailor the profile for meeting both requirements. Such an optimization is made possible by the ion implantation technology.

#### 4.2.2.1 INTEGRAL SOLUTION TO POISSON'S EQUATION

In this subsection, the surface potential, electric field, and threshold voltage for the case of nonuniform channel doping are solved for under the depletion approximation. Mathematically, a general expression can be derived as follows. For a nonuniform p-type doping profile $N(x)$, the electric field is obtained by integrating Poisson's equation once (neglecting mobile carriers in the depletion region):

$$\mathscr{E}(x) = \frac{q}{\varepsilon_{si}}\int_x^{W_d} N(x)\,dx, \tag{4.19}$$

where $W_d$ is the depletion-layer width. Integrating again gives the surface potential,

$$\psi_s = \frac{q}{\varepsilon_{si}}\int_0^{W_d}\int_x^{W_d} N(x')\,dx'\,dx. \tag{4.20}$$

Using integration by parts, one can show that Eq. (4.20) is equivalent to (Brews, 1979)

$$\psi_s = \frac{q}{\varepsilon_{si}}\int_0^{W_d} xN(x)\,dx. \tag{4.21}$$

The maximum depletion-layer width (long-channel) $W_{dm}^0$ is determined by the condition $\psi_s = 2\psi_B$ when $W_d = W_{dm}^0$. ***The threshold voltage of a nonuniformly doped MOSFET is then determined by both the integral (depletion charge density) and the first moment of N(x) within (0, $W_{dm}^0$).***

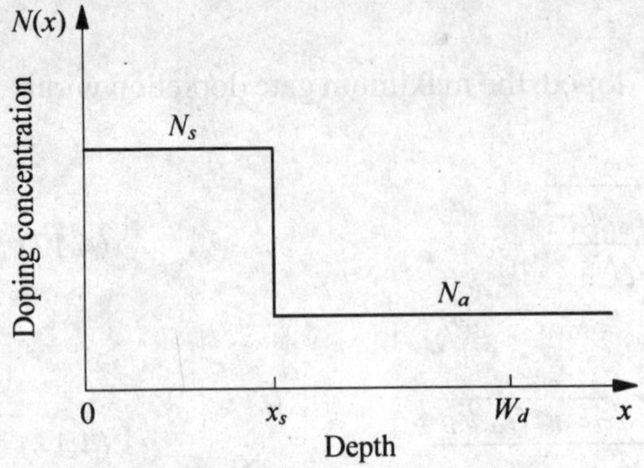


**FIGURE 4.3.** A schematic diagram showing the high–low step doping profile. $x = 0$ denotes the silicon–oxide interface.

#### 4.2.2.2 A HIGH–LOW STEP PROFILE

Consider the idealized step doping profile shown in Fig. 4.3 (Rideout *et al.*, 1975). It can be formed by making one or more low-dose, shallow implants into a uniformly doped substrate of concentration $N_a$. After drive-in, the implanted profile is approximated by a region of constant doping $N_s$ that extends from the surface to a depth $x_s$. If the entire depletion region at the threshold condition is contained within $x_s$, the MOSFET can be considered as uniformly doped with a concentration $N_s$. The case of particular interest analyzed here is when the depletion width $W_d$ exceeds $x_s$, so that part of the depletion region has a charge density $N_s$ and part of it $N_a$. The integration in Eq. (4.21) can be easily carried out for this profile to yield the surface potential, or the band bending at the surface,

$$\psi_s = \frac{qN_s}{2\varepsilon_{si}}x_s^2 + \frac{qN_a}{2\varepsilon_{si}}\left(W_d^2 - x_s^2\right). \tag{4.22}$$

Inversely, the depletion layer width $W_d$ can be found as a function of the surface potential $\psi_s$:

$$W_d = \sqrt{\frac{2\varepsilon_{si}}{qN_a}\left(\psi_s - \frac{q(N_s - N_a)x_s^2}{2\varepsilon_{si}}\right)}. \tag{4.23}$$

This is less than the depletion width in the uniformly doped ($N_a$) case for the same surface potential. The electric field at the surface is obtained by evaluating the integral in Eq. (4.19) with $x = 0$:

$$\mathscr{E}_s = \frac{qN_s x_s}{\varepsilon_{si}} + \frac{qN_a(W_d - x_s)}{\varepsilon_{si}} \tag{4.24}$$

From Gauss's law, the total depleted charge per unit area in silicon is given by

$$Q_s = -\varepsilon_{si}\mathscr{E}_s = -qN_s x_s - qN_a(W_d - x_s), \tag{4.25}$$

as would be expected from Fig. 4.3. ***The effect of the nonuniform surface doping***

***is then to increase the depletion charge within $0 \le x \le x_s$ by $(N_s - N_a)x_s$ and, at the same time, reduce the depletion layer width as indicated by Eq. (4.23).***

For an applied gate voltage $V_g$, the quantities $\psi_s$ and $Q_s$ are related by Eq. (3.14):

$$V_g = V_{fb} + \psi_s - \frac{Q_s}{C_{ox}} = V_{fb} + \psi_s + \frac{qN_s x_s + qN_a(W_d - x_s)}{C_{ox}}. \tag{4.26}$$

Substituting Eq. (4.23) for $W_d$ yields

$$V_g = V_{fb} + \psi_s + \frac{1}{C_{ox}}\sqrt{2\varepsilon_{si} q N_a \left(\psi_s - \frac{q(N_s - N_a)x_s^2}{2\varepsilon_{si}}\right)} + \frac{q(N_s - N_a)x_s}{C_{ox}}. \tag{4.27}$$

By definition, the threshold voltage is the gate voltage at which $\psi_s = 2\psi_B$, i.e.,

$$V_t = V_{fb} + 2\psi_B + \frac{1}{C_{ox}}\sqrt{2\varepsilon_{si} q N_a \left(2\psi_B - \frac{q(N_s - N_a)x_s^2}{2\varepsilon_{si}}\right)} + \frac{q(N_s - N_a)x_s}{C_{ox}}. \tag{4.28}$$

The maximum depletion width (long-channel) at threshold is given by Eq. (4.23) with $\psi_s = 2\psi_B$:

$$W_{dm}^0 = \sqrt{\frac{2\varepsilon_{si}}{qN_a}\left(2\psi_B - \frac{q(N_s - N_a)x_s^2}{2\varepsilon_{si}}\right)}. \tag{4.29}$$

There is some ambiguity as to whether $2\psi_B$ is defined in terms of $N_s$ or $N_a$. We adopt the convention that $2\psi_B$ is defined in terms of the p-type concentration at the depletion-layer edge, i.e., $2\psi_B = (2kT/q)\ln(N_a/n_i)$. In fact, it makes very little difference which concentration we use, since $2\psi_B$ is a rather weak function of the doping concentration anyway. Further refinement of the threshold condition would require a numerical simulation of the specific profile.

In Section 3.1.3, we showed that the subthreshold slope is given by $2.3mkT/q$, where $m = dV_g/d\psi_s$ at $\psi_s = 2\psi_B$. Here $m$ is also referred to as the body-effect coefficient, $1 + (C_{dm}/C_{ox})$, defined in Eq. (3.22). In the nonuniformly doped case, $m$ can be evaluated from Eq. (4.27):

$$m \equiv \frac{dV_g}{d\psi_s}(\psi_s = 2\psi_B) \tag{4.30}$$

$$= 1 + \frac{\sqrt{2\varepsilon_{si} q N_a}}{2C_{ox}}\left(2\psi_B - \frac{q(N_s - N_a)x_s^2}{2\varepsilon_{si}}\right)^{-1/2}.$$

It can be expressed in terms of $W_{dm}^0$ using Eq. (4.29):

$$m = 1 + \frac{\varepsilon_{si}/W_{dm}^0}{C_{ox}} = 1 + \frac{C_{dm}}{C_{ox}} = 1 + \frac{3t_{ox}}{W_{dm}^0}. \tag{4.31}$$

These expressions are consistent with Eq. (3.67) for a uniformly doped channel. Similarly, the threshold voltage in the presence of a substrate bias $-V_{bs}$ is given by Eq. (4.28) with the $2\psi_B$ term in the square root replaced by $2\psi_B + V_{bs}$. Using Eq. (4.29), one can show that the substrate sensitivity is

$$\frac{dV_t}{dV_{bs}} = \frac{\varepsilon_{si}/W_{dm}^0}{C_{ox}} = \frac{C_{dm}}{C_{ox}} = m - 1. \tag{4.32}$$

Therefore, ***all the previous expressions for the depletion capacitance, subthreshold slope, and body-effect coefficient in terms of $W_{dm}^0$ for the uniformly doped case remain valid for the nonuniformly doped case***. The only difference is that the maximum depletion layer width $W_{dm}^0$ in the nonuniformly doped case is given by Eq. (4.29) instead of Eq. (4.17).

#### 4.2.2.3 GRAPHICAL INTERPRETATION

The above solutions of potential, field, and threshold voltage for a nonuniformly doped channel are best illustrated graphically by plotting the electric field versus depth as shown in Figs. 4.4 and 4.5. We start with the uniformly doped case in Fig. 4.4(a), where $\mathscr{E}(x)$ is a straight line with a negative slope whose magnitude is proportional to the substrate doping concentration $N_a$. The $x$-intercept gives the depletion-layer width where $\mathscr{E} = 0$. The $y$-intercept gives the surface electric field $\mathscr{E}_s$, which from Gauss's law is proportional to the total depletion charge per unit area. Since $\mathscr{E} = -d\psi/dx$, the triangular area under $\mathscr{E}(x)$ equals the surface potential, or the band bending $\psi_s$. As the gate voltage increases, both $W_d$ and $\mathscr{E}_s$ increase, and so does $\psi_s$ until it reaches $2\psi_B$. At this point, surface inversion occurs and the depletion-layer width has reached its maximum value. The threshold voltage is largely determined by the $y$-intercept or the surface field $\mathscr{E}_s$ when $\psi_s = 2\psi_B$, since

$$V_t = V_{fb} + 2\psi_B + \frac{\varepsilon_{si}\mathscr{E}_s}{C_{ox}}, \tag{4.33}$$

and the first two terms on the RHS nearly cancel each other for n$^+$-polysilicon gates on nMOSFETs and vice versa, as discussed in Section 4.1.1. For a lower $N_a$, the magnitude of the $\mathscr{E}(x)$ slope decreases, and therefore $V_t$ decreases while $W_{dm}^0$ increases as depicted in Fig. 4.4(b). The two triangular areas under the different $\mathscr{E}(x)$ lines are approximately the same at the threshold condition, since $2\psi_B$ is a rather weak function of $N_a$ and can be considered as a constant for practical purposes.

Figure 4.5(a) shows an $\mathscr{E}(x)$ plot for the nonuniformly doped case of a high–low step profile discussed above. With a higher doping $N_s$ in the surface

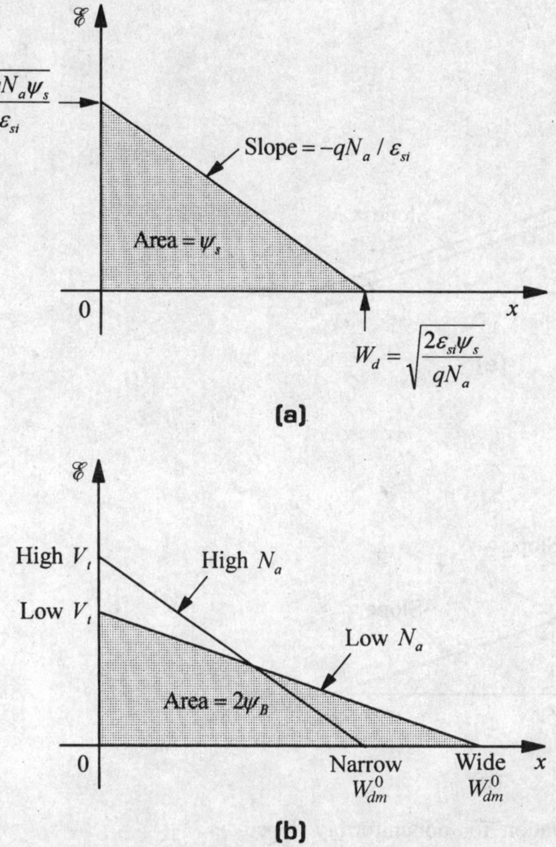


**FIGURE 4.4.** Graphical interpretation of relationships among doping concentration, depletion width, surface potential, and threshold voltage. (a) Uniformly doped case. The shaded area equals the surface potential $\psi_s$. (b) Two uniformly doped cases at $\psi_s = 2\psi_B$, one with a lower doping than the other. Both triangular areas equal $2\psi_B$ at threshold. $V_t$ is directly related to the $y$-intercept.

region $0 \le x \le x_s$, the function $\mathscr{E}(x)$, instead of being a straight line, changes slope at $x = x_s$. At threshold, the shaded area under the two-sloped line $\mathscr{E}(x)$ equals $2\psi_B$ just as in the uniformly doped case. In other words, the shaded area in Fig. 4.5(a) equals the triangular area under the straight line $\mathscr{E}(x)$, but the $x$- and $y$-intercepts change quite differently from the uniformly doped cases depicted in Fig. 4.4(b). How much $V_t$ shifts in response to a given change of $W_{dm}^0$ depends on where $x_s$ is. If $x_s$ is very close to the surface and $N_s$ is high, it is possible to increase $V_t$ with very little or almost no change in $W_{dm}^0$. ***This additional degree of freedom allows $V_t$ and $W_{dm}^0$ to decouple from the uniformly doped relations (4.17) and (4.18).*** Mathematically, the threshold shift and the change of depletion width for the nonuniformly doped case are given by Eq. (4.28) and Eq. (4.29).

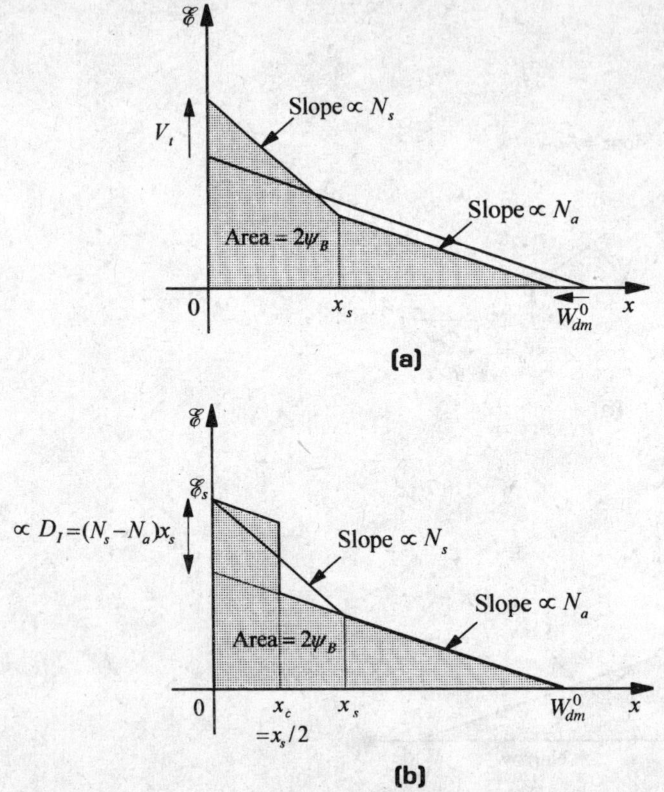


**FIGURE 4.5.** Graphical interpretation for nonuniformly doped cases. (a) High–low step profile compared with a uniformly doped profile ($N_a$). (b) A delta-function profile equivalent to the high–low profile in (a).

#### 4.2.2.4 GENERALIZATION TO A GAUSSIAN PROFILE

Using the graphical representation in Fig. 4.5(b), one can show that the nonuniform step doping profile discussed above is equivalent to the delta-function profile shown in Fig. 4.6 with an equivalent dose of

$$D_I = (N_s - N_a)x_s \tag{4.34}$$

centered at $x_c = x_s/2$. This is because both the area under $\mathscr{E}(x)$ and the $y$-intercept (i.e., $\mathscr{E}_s$ or $V_t$) are identical between the two cases. The same result follows from Eq. (4.19) and Eq. (4.21).

Similar arguments apply to a general Gaussian (or other symmetric) profile with a dopant distribution,

$$N(x) = \frac{D_I}{\sqrt{2\pi}\,\sigma} \exp\left(-\frac{(x - x_c)^2}{2\sigma^2}\right), \tag{4.35}$$

where $\sigma$ is the implant straggle. The effect of such an implanted profile on threshold voltage and depletion-layer width is equivalent to that of the step doping profile

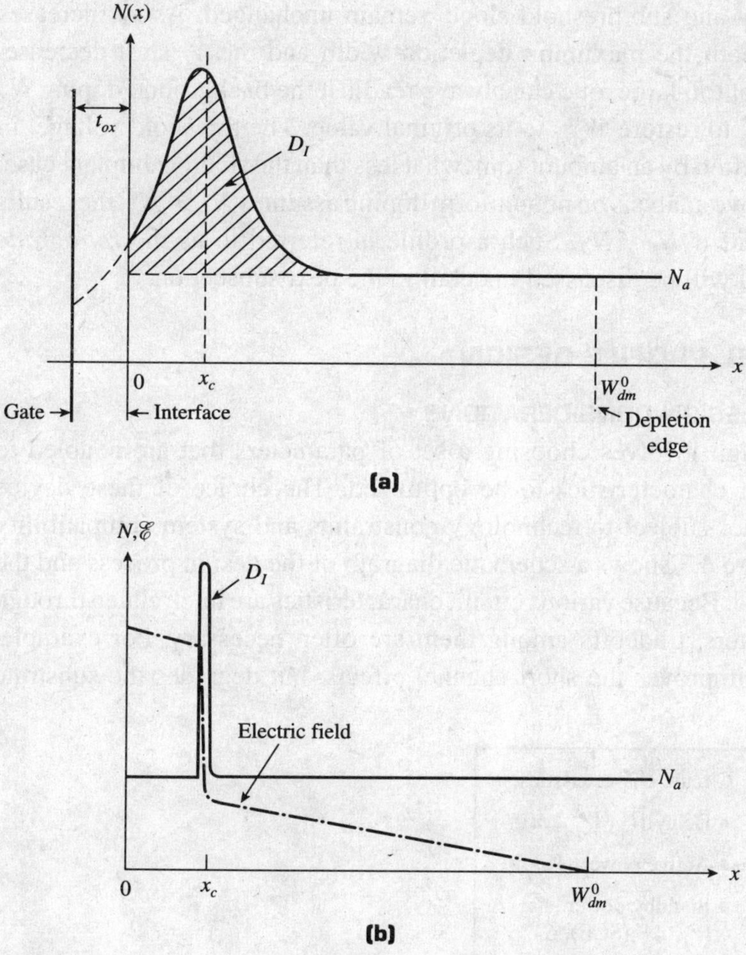


**FIGURE 4.6.** Schematic diagrams showing (a) an implanted Gaussian profile and (b) a delta-function profile equivalent to (a). (After Brews, 1979.)

discussed above, independent of $\sigma$. Substituting Eq. (4.34) and $x_c = x_s/2$ into the threshold voltage equation (4.28) yields

$$V_t = V_{fb} + 2\psi_B + \frac{1}{C_{ox}}\sqrt{2\varepsilon_{si} q N_a \left(2\psi_B - \frac{qD_I x_c}{\varepsilon_{si}}\right)} + \frac{qD_I}{C_{ox}}. \tag{4.36}$$

Similarly, the maximum depletion width, Eq. (4.29), becomes

$$W_{dm}^0 = \sqrt{\frac{2\varepsilon_{si}}{qN_a}\left(2\psi_B - \frac{qD_I x_c}{\varepsilon_{si}}\right)}. \tag{4.37}$$

For a given implanted dose $D_I$, the resulting threshold voltage shift depends on the location of the implant, $x_c$. ***For shallow surface implants, $x_c = 0$, there is no change in the depletion width. The $V_t$ shift is simply given by $qD_I/C_{ox}$, as with a sheet of charge at the silicon–oxide interface***. All other device parameters, e.g., substrate sensitivity and subthreshold slope, remain unchanged. As $x_c$ increases for a given dose, both the maximum depletion width and the $V_t$ shift decrease. However, if $x_c$ is not too large, one can always readjust the background doping $N_a$ to a lower value $N_a'$ to restore $W_{dm}^0$ to its original value. The threshold voltage, in the meantime, is shifted by an amount somewhat less than the shallow implant case.

Although the above analysis on nonuniform doping assumes $N_s > N_a$, the results remain equally valid if $N_s < N_a$. Such a profile is referred to as the *retrograde channel doping* and will be discussed in detail in the next subsection.

### 4.2.3 CHANNEL PROFILE DESIGN

#### 4.2.3.1 CMOS DESIGN CONSIDERATIONS

CMOS device design involves choosing a set of parameters that are coupled to a variety of circuit characteristics to be optimized. The choice of these device parameters is further subject to technology constraints and system compatibility requirements. Figure 4.7 shows a schematic diagram of the design process and the parameters involved. Because various circuit characteristics are interrelated through the device parameters, tradeoffs among them are often necessary. For example, reduction of $W_{dm}$ improves the short-channel effects, but degrades the substrate

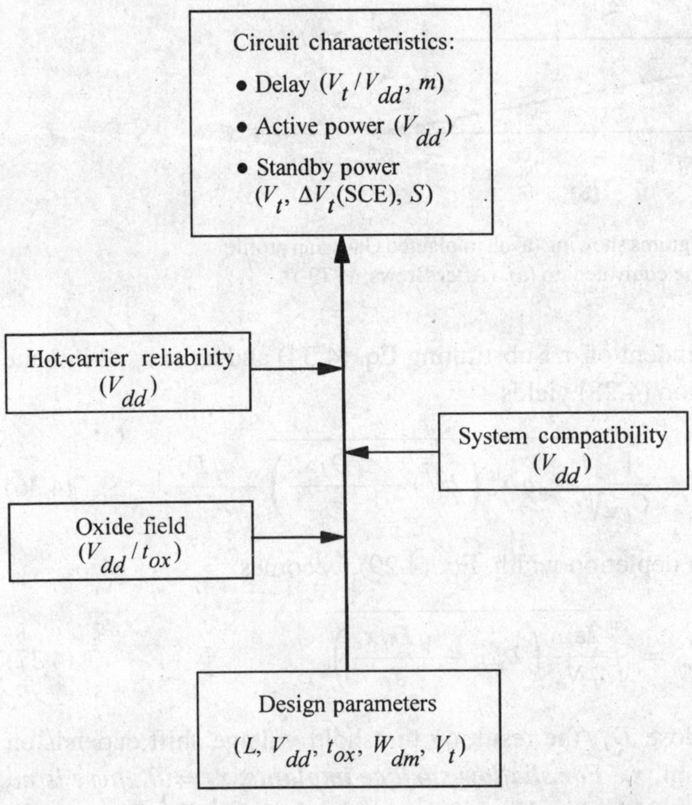


**FIGURE 4.7.** A CMOS design flowchart showing device parameters, technology constraints, and circuit objectives.

sensitivity; thinner $t_{ox}$ increases the current drive, but causes reliability concerns, etc. There is no unique way of designing CMOS devices for a given technology generation. Nevertheless, we attempt here to give a general guideline of how these device parameters should be chosen.

It was discussed in Section 4.2.1 that in order to keep short-channel effects under control, a good choice of the maximum gate depletion width $W_{dm}$ is such that $L_{\min}/mW_{dm} \approx 2$, where $L_{\min}$ is the minimum channel length. Since $L_{\min}$ is some fraction shorter than the nominal channel length $L$ of the technology, this requirement can be stated as $L/mW_{dm} > 2$, or equivalently, as $W_{dm} + 3t_{ox} < L/2$. Here the body-effect coefficient is $m = 1 + (3t_{ox}/W_{dm}^0) \approx 1 + (3t_{ox}/W_{dm})$ from Eq. (4.31), for either a uniformly doped or a nonuniformly doped channel. Since both the inverse subthreshold slope, $2.3mkT/q$, and the substrate sensitivity, $dV_t/dV_{bs} = m - 1$, degrade with higher $m$, $m$ should be kept close to 1. A larger $m$ also results in a lower saturation current in the long-channel limit [Eq. (3.23)]. Typically, one requires $m < 1.5$, or $3t_{ox}/W_{dm} < \frac{1}{2}$. These design considerations are illustrated in Fig. 4.8. A lower limit on $t_{ox}$ imposed by technology constraints is $V_{dd}/\mathcal{E}_{ox}^{\max}$, where $\mathcal{E}_{ox}^{\max}$ is the maximum oxide field. ***For a given L and $V_{dd}$, the allowable parameter space in the $t_{ox}$–$W_{dm}$ design plane is a triangular area bounded by requirements on the SCE, oxide field, and subthreshold slope (also substrate sensitivity).***

#### 4.2.3.2 TRENDS OF POWER-SUPPLY VOLTAGE AND THRESHOLD VOLTAGE

Channel profile design is largely dictated by threshold-voltage requirements. The lower limit of threshold voltage is given by off-current specifications outlined in Section 4.2.1: $V_t \geq 0.4$ V. The upper limit of threshold voltage is imposed by circuit delay or performance considerations. It will be shown in Chapter 5 that CMOS delay degrades rapidly once $V_t$ exceeds 25% of $V_{dd}$. Therefore, one should keep $V_t \leq V_{dd}/4$ if possible (Mii *et al.*, 1994). Figure 4.9 shows the trends in

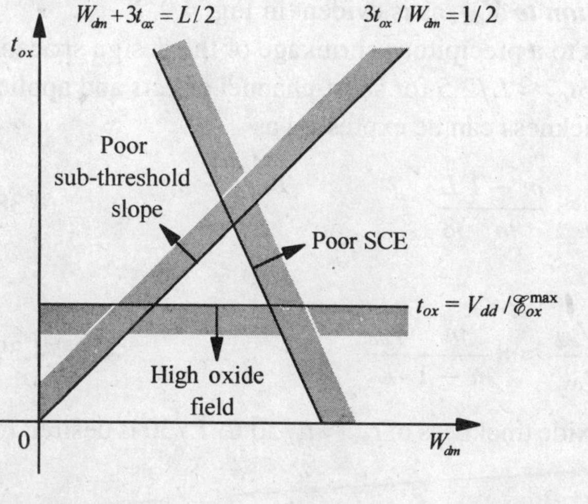


**FIGURE 4.8.** The $t_{ox}$–$W_{dm}$ design plane. Some tradeoff among the various factors can be made within the parameter space bounded by SCE, body-effect, and oxide-field considerations.

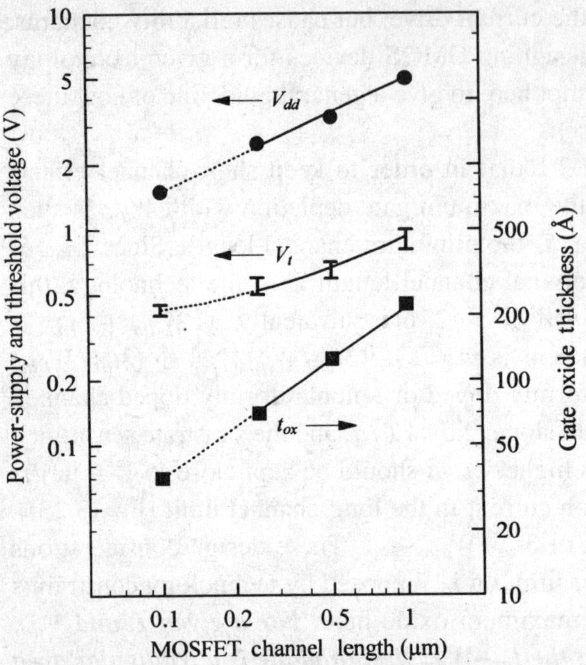


**FIGURE 4.9.** Trends of power-supply voltage, threshold voltage, and gate oxide thickness versus channel length for CMOS technologies from 1 to 0.1 µm. (After Taur *et al.*, 1995a.)

power-supply voltage, threshold voltage, and oxide thickness for CMOS logic technologies from 1.0- to 0.1-µm channel length (Taur *et al.*, 1995a). When $V_{dd}$ is high, there is plenty of design room to choose a threshold voltage that satisfies both requirements: $0.4\text{ V} \le V_t \le V_{dd}/4$. For example, $V_{dd} = 5$ V and $V_t = 0.8$–1.0 V for 1-µm CMOS technology; and $V_{dd} = 3.3$ V and $V_t = 0.6$–0.7 V for 0.5-µm CMOS technology. When $V_{dd}$ is reduced toward shorter channel lengths, it becomes increasingly difficult to satisfy both the performance and the off-current requirements. One often faces a tradeoff of leakage current versus circuit speed. This stems from subthreshold nonscalability. For this reason and for compatibility with the standardized power-supply voltage of earlier-generation systems, ***the general trend is that $V_{dd}$ has not been scaled down in proportion to L, and $V_t$ has not been scaled down in proportion to $V_{dd}$***, as is evident in Fig. 4.9.

A higher value of $V_{dd}/L$ leads to a precipitous shrinkage of the design space in Fig. 4.8. If one assumes $W_{dm} + 3t_{ox} \approx L/2.5$ for short-channel effects and applies $3t_{ox}/W_{dm} \approx m - 1$, the oxide thickness can be expressed as

$$t_{ox} \approx \frac{m-1}{m}\frac{L}{8}, \tag{4.38}$$

and the oxide field as

$$\mathscr{E}_{ox} \equiv \frac{V_{dd}}{t_{ox}} \approx 8\frac{m}{m-1}\frac{V_{dd}}{L}. \tag{4.39}$$

Equation (4.38) implies that an oxide thickness of $t_{ox} \approx L/50$ to $L/30$ is desired for control of short-channel effects. From Eq. (4.39), the increase of $V_{dd}/L$ inevitably leads to higher oxide fields. This trend is clearly seen in Table 4.2. Some drain engineering, such as a lightly doped drain (LDD) structure (Ogura *et al.*, 1982) is also necessary to relieve hot-electron reliability problems at higher voltages. However, yield and reliability considerations constrain the maximum oxide field to about 5 MV/cm, since oxide breakdown occurs at slightly beyond 10 MV/cm as mentioned in Section 2.4. When that limit is reached, the power-supply voltage must be reduced for thinner oxides and shorter channel lengths. (Another reason for voltage reduction comes from the active-power consideration to be addressed in Chapter 5.) When $V_{dd}$ becomes less than about 2 V, a tradeoff between off current and device delay is necessary. For example, $V_{dd} = 1.5$ V and $V_t = 0.4$ V for 0.1-μm CMOS devices (Taur *et al.*, 1993c). More details on this tradeoff will be given in Chapter 5.

#### 4.2.3.3 EFFECT OF GATE WORK FUNCTION

The gate work function has a major effect on channel profile design, since, through the $V_{fb}$ term, it has a strong influence on the MOSFET threshold voltage:

$$V_t = V_{fb} + 2\psi_B + \frac{-Q_d}{C_{ox}}. \tag{4.40}$$

When n$^+$-polysilicon gates are used for n-channel MOSFETs, $V_{fb} = -E_g/2q - \psi_B$. This results in a low threshold voltage. For example, consider a 1-μm nMOSFET with $t_{ox} = 250$ Å. A p-type doping of $N_a = 10^{16}$ cm$^{-3}$, which gives $\psi_B = 0.35$ V and a maximum depletion width [Eq. (4.17)] of $W_{dm}^0 = 0.3$ μm, is sufficient to control the short-channel effect. For a uniformly doped channel, $-Q_d = qN_aW_{dm}^0$. Since $V_{fb} = -0.91$ V for the n$^+$-polysilicon gate, the threshold voltage calculated from Eq. (4.40) is $V_t = 0.14$ V. This is not high enough to satisfy the off-current requirement discussed in Section 4.2.1. ***A nonuniform high–low channel doping as described in the last subsection can be used to increase $|Q_d|$ and therefore the threshold voltage without significantly altering the gate depletion width.*** This is usually carried out with a shallow, p-type ($^{11}$B or $^{11}$B $^{19}$F$_2$) implant for nMOSFETs. Since $W_{dm}^0$ remains essentially unchanged, neither the substrate sensitivity, $dV_t/dV_{bs} = m - 1 = 3t_{ox}/W_{dm}^0$, nor the subthreshold slope, $2.3mkT/q$, is degraded.

Although high–low channel doping allows a higher threshold voltage without degrading the substrate sensitivity, the surface field at threshold, $\mathscr{E}_s = |Q_d|/\varepsilon_{si}$, becomes higher due to the increased depletion charge. This results in degradation of channel mobility, as discussed in Section 3.1.5. Ideally, the threshold voltage can be adjusted by choosing a proper gate work function without increased fields. For example, if a midgap-work-function gate is used in the 1-μm case above, then $V_{fb} = -\psi_B$ and $V_t = 0.7$ V without additional doping in the channel. This will result in the same electric field and channel mobility as the uniform,

$10^{16}$-cm$^{-3}$-doped case. A midgap-work-function gate is also symmetrical for nMOSFETs and pMOSFETs. In reality, however, no midgap-work-function gate material has been used in VLSI production, although that has been attempted in research laboratories (Davari *et al.*, 1987). Technology issues such as compatibility of gate material with thin gate oxides are the main obstacles.

#### 4.2.3.4 BURIED-CHANNEL MOSFETs

Further positive shifts in $V_t$ are possible if a p$^+$-polysilicon gate is used for nMOSFETs, which gives $V_{fb} = +E_g/2q - \psi_B$. For the above example, $V_{fb} = +0.21$ V and $V_t = 1.26$ V for a uniform, $10^{16}$-cm$^{-3}$-doped channel. Now the threshold voltage is too high! To reduce $V_t$, the channel must be counterdoped to lower $|Q_d|$. This means a shallow n-type implant for nMOSFETs, and an n–p junction is formed near the surface. At zero gate voltage, the n-type region is depleted of electrons by the gate field so there is no conduction between the source and drain. The surface field at threshold is lower than in the uniform, $10^{16}$-cm$^{-3}$-doped case, since $|Q_d|$ is lower. This improves channel mobility. The n-type counterdoping can be increased to the point that $Q_d$ becomes positive and the third term on the RHS of Eq. (4.40) becomes negative. When this happens, the surface field at threshold is negative and the MOSFET is called a *buried-channel device*, as inversion first takes place at a point of maximum potential below the surface.

In reality, buried-channel nMOSFETs have not been utilized in VLSI manufacturing, since p$^+$-polysilicon gates have a higher resistance and are more difficult to process because of possible boron penetration problems (Sun *et al.*, 1989). However, their counterparts, n$^+$-polysilicon gated pMOSFETs, have been employed in VLSI manufacturing for CMOS technologies of 0.5-μm channel length and above. In those technologies, n$^+$-polysilicon gates are used for both n- and pMOSFETs, and boron or $BF_2$ channel implants are made for both types of devices (Taur *et al.*, 1985). Figure 4.10 shows the band diagram of a buried-channel pMOSFET. It can be seen that at the threshold, the surface field is negative (pushing holes away from the surface), and the channel for holes is formed at a potential minimum slightly below the surface. As the gate voltage increases beyond threshold, the field changes sign and the channel moves to the surface, but the effective field is still lower than that of a conventional surface-channel device. ***Although a buried-channel device offers higher mobilities, its short-channel effect is inherently worse than that of a surface-channel device*** (Nguyen and Plummer, 1981). This is because the counterdoping (especially boron) at the surface tends to diffuse deeper into the silicon during subsequent thermal cycles in the process. As the channel length and power-supply voltage are scaled down, a lower threshold voltage is required. It becomes increasingly more difficult to build a buried-channel device, since higher counterdoping in the channel invariably leads to wider gate depletion widths and poorer short-channel effects. ***For CMOS logic technologies of 0.25-μm channel length***

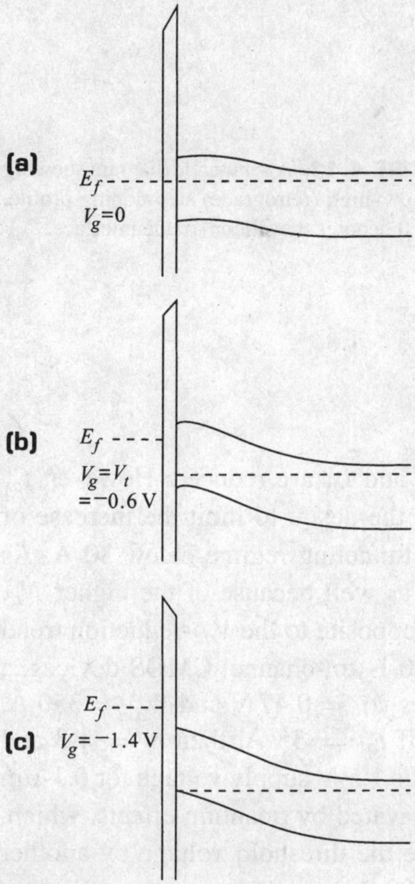


**FIGURE 4.10.** Band diagram of a buried-channel pMOSFET with n$^+$-polysilicon gate. A shallow p-type layer is implanted at the surface to lower the magnitude of threshold voltage. The gate voltage is (a) below threshold, (b) at threshold, and (c) above threshold. (After Taur *et al.*, 1985.)

***and below, dual polysilicon gates (n$^+$ polysilicon for nMOSFET and p$^+$ polysilicon for pMOSFET) are used, so that both types of devices are surface-channel devices*** (Wong *et al.*, 1988).

#### 4.2.3.5 RETROGRADE (LOW–HIGH) CHANNEL PROFILE

When the channel length is scaled to 0.15 μm and below, a much higher doping concentration is needed in the channel to reduce $W_{dm}$ and control short-channel effects. If a uniform profile were used, the depletion-charge term would increase disproportionately and the threshold voltage would become too high even with dual polysilicon gates. This can be seen by writing the threshold voltage equation (4.18) for the uniformly doped case as

$$V_t = V_{fb} + 2\psi_B + 2(m-1)\,2\psi_B, \tag{4.41}$$

using Eq. (4.17) and Eq. (4.31). For an n$^+$-polysilicon-gated nMOSFET, $V_{fb} = -E_g/2q - \psi_B$, and therefore

$$V_t = -0.56\ \mathrm{V} + (4m-3)\psi_B. \tag{4.42}$$

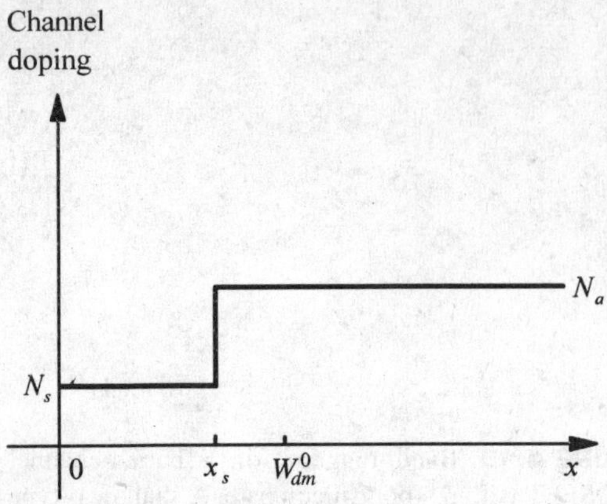


**FIGURE 4.11.** A schematic diagram showing the low–high (retrograde) step doping profile. $x = 0$ denotes the silicon–oxide interface.

As the channel length scales down, both $W_{dm}^0$ and $t_{ox}$ are reduced. However, $t_{ox}$ tends not to scale as much as $W_{dm}^0$ because of the desire to limit the increase of oxide fields and to delay entry into the direct tunneling regime below 30 Å. As a result, $m$ becomes higher. $\psi_B$ has increased as well because of the higher $N_a$. Both of these factors tend to raise $V_t$, which is opposite to the $V_{dd}$-reduction trend required for shorter devices. For example, for 0.1-μm-channel CMOS devices, a channel doping of $N_a = 10^{18}$ cm$^{-3}$, which gives $\psi_B = 0.47$ V and $W_{dm}^0 = 350$ Å, is needed to control the short-channel effect. If $t_{ox} = 35$ Å, then $m = 1.3$ and $V_t = 0.47$ V, which is too high with respect to the 1.5-V supply voltage for 0.1-μm devices (Fig. 4.9). The problem is further aggravated by quantum effects, which, as will be discussed in Section 4.2.4, can raise the threshold voltage by another 0.1–0.2 V at such high fields (van Dort *et al.*, 1994).

***To reduce the threshold voltage without significantly increasing the gate depletion width, a retrograde channel profile, i.e., a low–high doping profile as shown schematically in Fig. 4.11, is required*** (Sun *et al.*, 1987; Shahidi *et al.*, 1989). Such a profile is formed using higher-energy implants that peak below the surface. It is assumed that the maximum gate depletion width extends into the higher-doped region. All the equations in Section 4.2.2 remain valid for $N_s < N_a$. For simplicity, we assume an ideal retrograde channel profile for which $N_s = 0$. Equation (4.28) then becomes

$$V_t = V_{fb} + 2\psi_B + \frac{qN_a}{C_{ox}}\sqrt{\frac{4\varepsilon_{si}\psi_B}{qN_a} + x_s^2} - \frac{qN_a x_s}{C_{ox}}. \tag{4.43}$$

Similarly, Eq. (4.29) gives the maximum depletion width (long-channel),

$$W_{dm}^0 = \sqrt{\frac{4\varepsilon_{si}\psi_B}{qN_a} + x_s^2}. \tag{4.44}$$

The net effect of low–high doping is that the threshold voltage is reduced, but

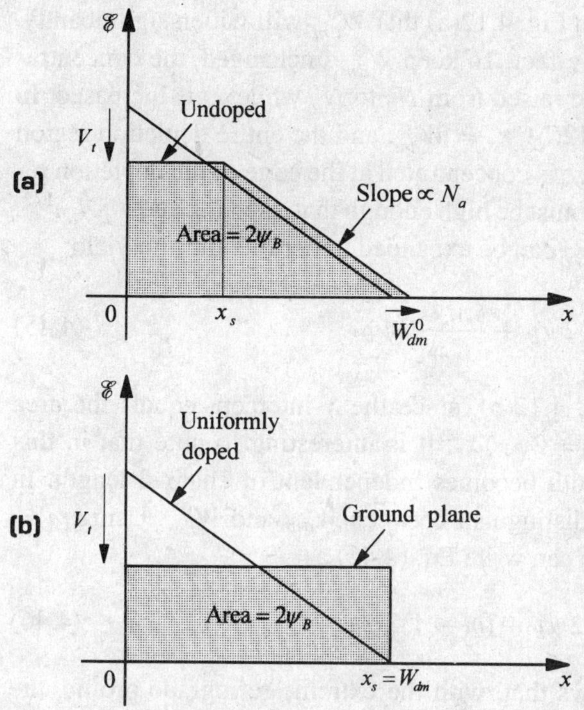


**FIGURE 4.12.** Graphical interpretation of retrograde doping profiles. (a) A low–high step profile compared with a uniformly doped profile ($N_a$). (b) An extreme retrograde profile that degenerates into a ground-plane MOSFET; the band bending in this case is given by the rectangular area, which equals $2\psi_B$ at threshold.

the depletion width has increased, just opposite to that of high-low doping. Note that Eq. (4.44) has the same form as Eq. (2.76) for a p–i–n diode discussed in Section 2.2.2. All other expressions, such as those for the subthreshold slope and the substrate sensitivity, in Section 4.2.2 apply with $W_{dm}^0$ replaced by Eq. (4.44).

A graphical representation of the retrograde channel profile is shown in Fig. 4.12(a). As described in Section 4.2.2.3, when the electric field is plotted against depth at threshold condition, the $x$-intercept is the maximum depletion width while the $y$-intercept is proportional to the depletion charge (third) term of $V_t$ in Eq. (4.40). The area under the $\mathscr{E}(x)$ curve equals $2\psi_B$. With a retrograde doping profile, it is possible to reduce the $y$-intercept, and hence $V_t$, with only a slight increase in the depletion width while keeping the area under the curve unchanged. Note that $\mathscr{E}(x)$ is flat within the undoped region, $0 < x < x_s$, where there is no depletion charge.

#### 4.2.3.6 EXTREME RETROGRADE PROFILE AND GROUND-PLANE MOSFET

Two limiting cases are worth discussing. If $x_s \ll (4\varepsilon_{si}\psi_B/qN_a)^{1/2}$, then $W_{dm}^0$ remains essentially unchanged from the uniformly doped value [Eq. (4.44)], while $V_t$ is lowered by a net amount equal to $qN_a x_s/C_{ox}$ [Eq. (4.43)]. To reduce $V_t$ even further, $x_s$ must increase, assuming there is no counterdoping of the channel. If $N_a$

stays the same, it can be seen from Fig. 4.12(a) that $W_{dm}^0$ will widen significantly, which degrades the short-channel effect. To keep $W_{dm}^0$ unchanged, the concentration (i.e., slope) beyond $x_s$ must be raised from $N_a$ to $N_a'$ while $x_s$ is increased. In the limiting case shown in Fig. 4.12(b), $x_s = W_{dm}^0$, and the entire depletion region is undoped. All the depletion charge is concentrated at the edge of the depletion region. In order for this to occur, $N_a'$ must be high enough that $x_s \gg (4\varepsilon_{si}\psi_B/qN_a')^{1/2}$. With $N_a$ replaced by $N_a'$, Eq. (4.43) can be expanded under this limit to yield

$$V_t = V_{fb} + 2\psi_B + \frac{\varepsilon_{si}/x_s}{C_{ox}} 2\psi_B. \tag{4.45}$$

This result is expected from Fig. 4.12(b), since the $y$-intercept equals the area divided by the $x$-intercept, or $\mathscr{E}_s = 2\psi_B/x_s$. It is interesting to note that in this case, the maximum depletion width becomes independent of channel length. In other words, there is no need to distinguish between $W_{dm}$ and $W_{dm}^0$. Using $m = 1 + 3t_{ox}/W_{dm} = 1 + 3t_{ox}/x_s$, one can write Eq. (4.45) as

$$V_t = V_{fb} + 2\psi_B + (m-1)2\psi_B. \tag{4.46}$$

Comparison with Eq. (4.41) shows that, with the extreme retrograde profile, the depletion-charge term of $V_t$ is reduced to half of the uniformly doped value. This is also clear from Fig. 4.12(b). Substituting $V_{fb} = -E_g/2q - \psi_B$ in Eq. (4.46) yields

$$V_t = -0.56\text{ V} + (2m-1)\psi_B. \tag{4.47}$$

For the above 0.1-μm MOSFET example, $\psi_B = 0.47$ V and $m = 1.3$, which gives $V_t = 0.19$ V. This value is low enough for $V_{dd} = 1.5$ V, even with the quantum correction of $V_t$ (Section 4.2.4) taken into account. If there is a substrate bias $-V_{bs}$ present, the factor $2\psi_B$ in the last term of Eq. (4.46) is replaced by $2\psi_B + V_{bs}$, i.e.,

$$V_t = V_{fb} + 2m\psi_B + (m-1)V_{bs}. \tag{4.48}$$

Further reduction of $V_t$ can be accomplished by either counterdoping the channel or forward-biasing the substrate. A forward substrate bias also helps improve short-channel effects, as it effectively reduces the built-in potential, $\psi_{bi}$ in Eq. (3.66), between the source–drain and the p-type substrate. However, forward substrate bias also causes source junction leakage, increases the drain-to-substrate capacitance, and degrades the subthreshold slope and body effect.

Since $\psi_B$ is a weak function of $N_a'$, the above results are independent of the exact value of $N_a'$ as long as it is high enough to satisfy $x_s \gg (4\varepsilon_{si}\psi_B/qN_a')^{1/2}$. All the essential device characteristics, such as SCE ($W_{dm}$), subthreshold slope ($m$), and threshold voltage, are determined by the depth of the undoped layer, $x_s$. ***The limiting case of retrograde channel profile therefore degenerates into a ground-plane MOSFET*** (Yan *et al.*, 1991). The band diagram and charge distribution of

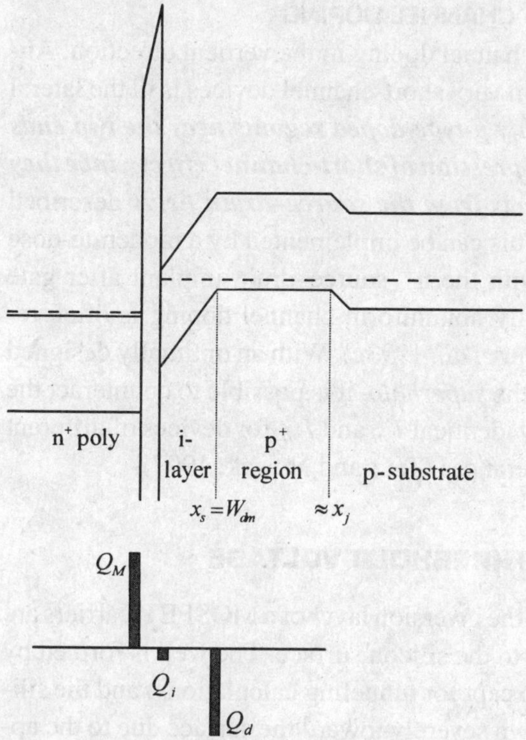


**FIGURE 4.13.** Band diagram and charge distribution of an extreme retrograde-doped or ground-plane nMOSFET at threshold condition.

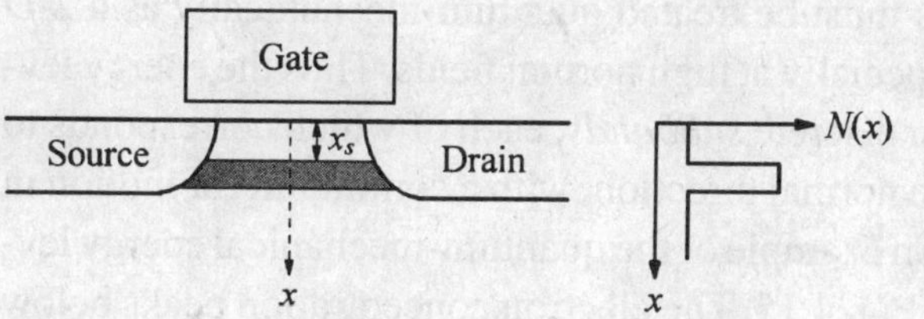


**FIGURE 4.14.** Schematic cross section of a low–high–low, or pulse-shaped, or delta-doped MOSFET. The doping concentration along the dashed line is depicted in the profile to the right. The highly doped region corresponds to the shaded area in the cross section.

such a device at threshold condition are shown schematically in Fig. 4.13. Note that the field is constant (no potential curvature) in the undoped region between the surface and $x_s$. There is an abrupt change of field at $x = x_s$, where a delta function of depletion charge (area $= 2\varepsilon_{si}\psi_B/x_s$) is located. Beyond $x_s$, the bands are essentially flat. It is desirable not to extend the $p^+$ region under the source and drain junctions, since that increases the parasitic capacitance. The ideal channel doping profile is then that of a low–high–low type shown in Fig. 4.14, in which the narrow $p^+$region is used only to confine the gate depletion width. Such a profile is also referred to as *pulse-shaped doping* or *delta doping* in the literature. The integrated dose of the $p^+$ region must be at least $2\varepsilon_{si}\psi_B/qx_s$ to provide the gate depletion charge needed. It is advisable to use somewhat higher than the minimum dose to supply additional depletion charge for the source–drain fields in short-channel devices. However, too high a $p^+$ dose or concentration may result in band-to-band tunneling leakage between the source or drain and the substrate, as mentioned in Section 2.4.2.

#### 4.2.3.7 LATERALLY NONUNIFORM CHANNEL DOPING

So far we have discussed nonuniform channel doping in the vertical direction. Another type of nonuniform doping used in very short-channel devices is in the lateral direction. ***For nMOSFETs, more highly p-type-doped regions near the two ends of the channel are beneficial to the suppression of short-channel effect, since they help compensate charge-sharing effects from the source–drain fields*** described in Section 3.2.1 (Ogura *et al.*, 1982). This can be implemented by a moderate-dose p-type implant carried out together with the $n^+$ source–drain implant after gate patterning. Such a self-aligned, laterally nonuniform channel doping is often referred to as *halo* or *pocket* implants (Taur *et al.*, 1993c). With an optimally designed 2-D nonuniform doping profile called the *superhalo*, it is possible to counteract the short-channel effect and achieve nearly identical $I_{on}$ and $I_{off}$ for devices of different channel lengths within the process tolerances (Taur and Nowak, 1997).

### 4.2.4 QUANTUM EFFECT ON THRESHOLD VOLTAGE

It was discussed in Section 2.3.2 that in the inversion layer of a MOSFET, carriers are confined in a potential well very close to the silicon surface. The well is formed by the oxide barrier (essentially infinite except for tunneling calculations) and the silicon conduction band, which bends down severely toward the surface due to the applied gate field. Because of the confinement of motion in the direction normal to the surface, inversion-layer electrons must be treated quantum-mechanically as a 2-D gas (Stern and Howard, 1967), especially at high normal fields. Thus the energy levels of the electrons are grouped in discrete *subbands*, each of which corresponds to a quantized level for motion in the normal direction, with a continuum for motion in the plane parallel to the surface. An example of the quantum-mechanical energy levels and band bending is shown in Fig. 4.15. The electron concentration peaks below the silicon–oxide interface and goes to nearly zero at the interface, as dictated by the boundary condition of the electron wave function. This is in contrast to the classical model in which the electron concentration peaks at the surface, as shown in Fig. 4.16. Quantum-mechanical behavior of inversion-layer electrons affects MOSFET operation in two ways. ***First, at high fields, threshold voltage becomes higher, since more band bending is required to populate the lowest subband, which is some energy above the bottom of the conduction band. Second, once the inversion layer forms below the surface, it takes a higher gate-voltage overdrive to produce a given level of inversion charge density***. In other words, the effective gate oxide thickness is slightly larger than the physical thickness. This reduces the transconductance and the current drive of a MOSFET.

#### 4.2.4.1 TRIANGULAR POTENTIAL APPROXIMATION FOR THE SUBTHRESHOLD REGION

A full solution of the silicon inversion layer involves numerically solving coupled Poisson's and Schrödinger's equations self-consistently (Stern and Howard, 1967).

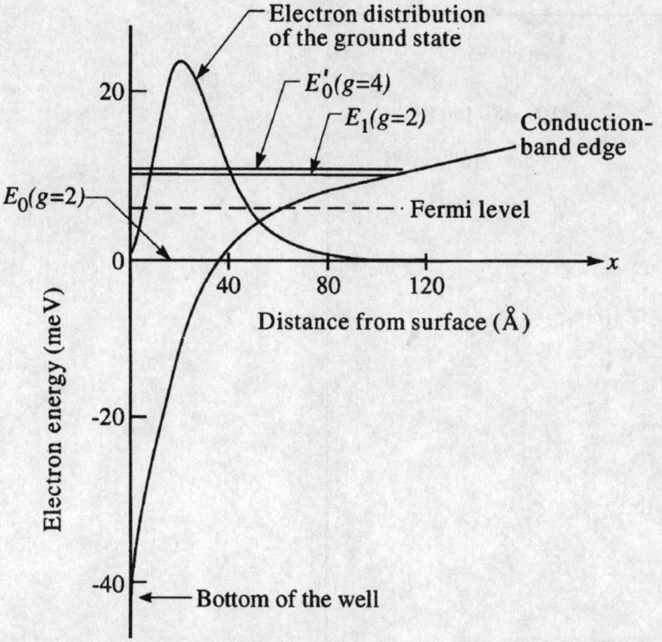


**FIGURE 4.15.** An example of quantum-mechanically calculated band bending and energy levels of inversion-layer electrons near the surface of an MOS device. The ground state is about 40 meV above the bottom of the conduction band at the surface. The dashed line indicates the Fermi level for $10^{12}$ electrons/cm$^2$ in the inversion layer. (After Stern and Howard, 1967.)

Under subthreshold conditions when the inversion charge density is low, band bending is solely determined by the depletion charge. It is then possible to decouple the two equations and obtain some insight into the quantum-mechanical (QM) effect on the threshold voltage. Since the inversion electrons are located in a narrow region close to the surface where the electric field is nearly constant ($\mathscr{E}_s$), it is a good approximation to consider the potential well as composed of an infinite oxide barrier for $x < 0$, and a triangular potential $V(x) = q\mathscr{E}_s x$ due to the depletion charge for $x > 0$. The Schrödinger equation is solved with the boundary conditions that the electron wave function goes to zero at $x = 0$ and at infinity. The solutions are Airy functions with eigenvalues $E_j$ given by (Stern, 1972)

$$E_j = \left[\frac{3hq\mathscr{E}_s}{4\sqrt{2m_x}}\left(j + \frac{3}{4}\right)\right]^{2/3}, \qquad j = 0, 1, 2, \ldots, \tag{4.49}$$

where $h = 6.63 \times 10^{-34}$ J-s is Planck's constant, and $m_x$ is the effective mass of electrons perpendicular to the surface. Note that MKS units are used throughout this subsection (i.e., length must be in meters, not centimeters). The average distance from the surface for electrons in the $j$th subband is given by

$$x_j = \frac{2E_j}{3q\mathscr{E}_s}. \tag{4.50}$$

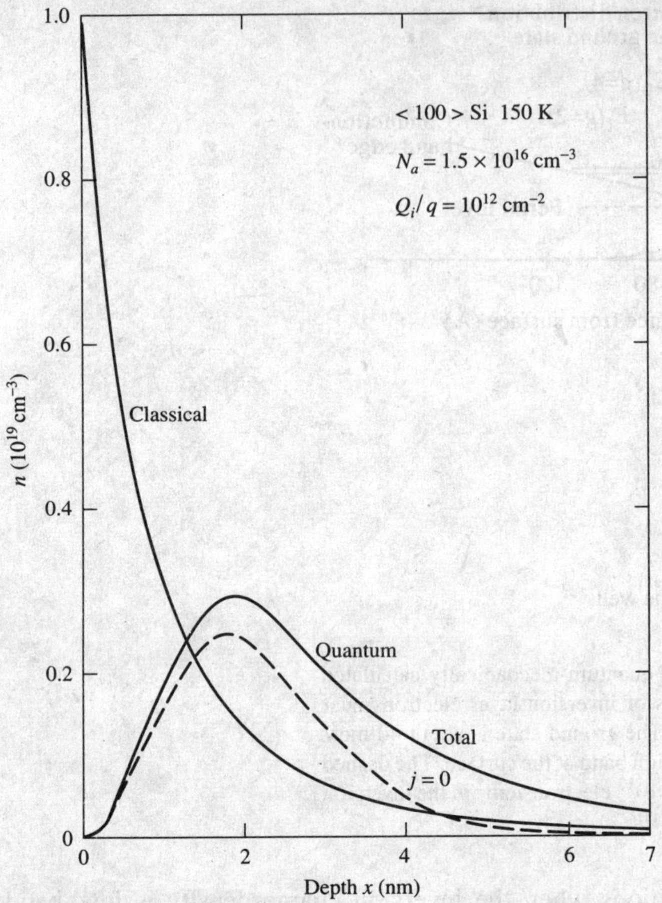


**FIGURE 4.16.** Classical and quantum-mechanical electron density versus depth for a ⟨100⟩ silicon inversion layer. The dashed curve shows the electron density distribution for the lowest subband. (After Stern, 1974.)

For silicon in the ⟨100⟩ direction, there are two groups of subbands, or *valleys*. The lower valley has a twofold degeneracy ($g = 2$) with $m_x = 0.92m_0$, where $m_0 = 9.1 \times 10^{-31}$ kg is the free-electron mass. These energy levels are designated as $E_0, E_1, E_2 \ldots$. The higher valley has a fourfold degeneracy ($g' = 4$) with $m'_x = 0.19m_0$. The energy levels are designated as $E'_0, E'_1, E'_2, \ldots$. Note that

$$E'_j = \left[\frac{3hq\mathscr{E}_s}{4\sqrt{2m'_x}}\left(j + \frac{3}{4}\right)\right]^{2/3}, \qquad j = 0, 1, 2, \ldots. \tag{4.51}$$

At room temperature, several subbands in both valleys are occupied near threshold, with a majority of the electrons in the lowest subband of energy $E_0$ above the bottom of the conduction band. From Appendix 7, the total inversion charge per unit area is expressed as (Stern and Howard, 1967)

$$Q_i^{QM} = \frac{4\pi qkT}{h^2}\left(gm_d \sum_j \ln\left(1 + e^{(E_f - E_c' - E_j)/kT}\right) + g'm_d' \sum_j \ln\left(1 + e^{(E_f - E_c' - E_j')/kT}\right)\right), \tag{4.52}$$

where $m_d = 0.19m_0$ and $m_d' = 0.42m_0$ are the density-of-states effective masses of the two valleys, and $E_f - E_c'$ is the difference between the Fermi level and the bottom of the conduction band at the surface. It is shown in Appendix 7 that in the subthreshold region, Eq. (4.52) can be simplified to

$$Q_i^{QM} = \frac{4\pi qkTn_i^2}{h^2 N_c N_a}\left(2m_d \sum_j e^{-E_j/kT} + 4m_d' \sum_j e^{-E_j'/kT}\right) e^{q\psi_s/kT}, \tag{4.53}$$

where $N_c$ is the effective density of states in the conduction band.

#### 4.2.4.2 THRESHOLD-VOLTAGE SHIFT DUE TO QUANTUM EFFECT

When $\mathscr{E}_s < 10^4$–$10^5$ V/cm at room temperature, both the lowest energy level $E_0$ and the spacings between the subbands are comparable to or less than $kT$. A large number of subbands are occupied, and $Q_i^{QM}$ is essentially the same as the classical inversion charge density per unit area given by Eq. (3.31) for the subthreshold region,

$$Q_i = \frac{kTn_i^2}{\mathscr{E}_s N_a} e^{q\psi_s/kT}. \tag{4.54}$$

(The expression has been generalized to cover nonuniformly doped cases where $\mathscr{E}_s$ is the electric field at the surface and $N_a$ is the doping concentration at the edge of the depletion layer.) When $\mathscr{E}_s > 10^5$ V/cm, however, the subband spacings become greater than $kT$ and $Q_i^{QM}$ is significantly less than $Q_i$. ***The $Q_i^{QM}$–$\psi_s$ curve [Eq. (4.53)] exhibits a positive parallel shift with respect to the classical $Q_i$–$\psi_s$ curve [Eq. (4.54)] on a semilogarithmic scale, which means that additional band bending is required to achieve the same inversion charge per unit area as the classical value***. The classical threshold condition, $\psi_s = 2\psi_B$, should therefore be modified to $\psi_s = 2\psi_B + \Delta\psi_s^{QM}$, where $Q_i^{QM}(\psi_s = 2\psi_B + \Delta\psi_s^{QM}) = Q_i(\psi_s = 2\psi_B)$. Based on the last expression,

$$\Delta\psi_s^{QM} = \frac{kT}{q}\ln\left(\frac{Q_i(\psi_s = 0)}{Q_i^{QM}(\psi_s = 0)}\right) \tag{4.55}$$

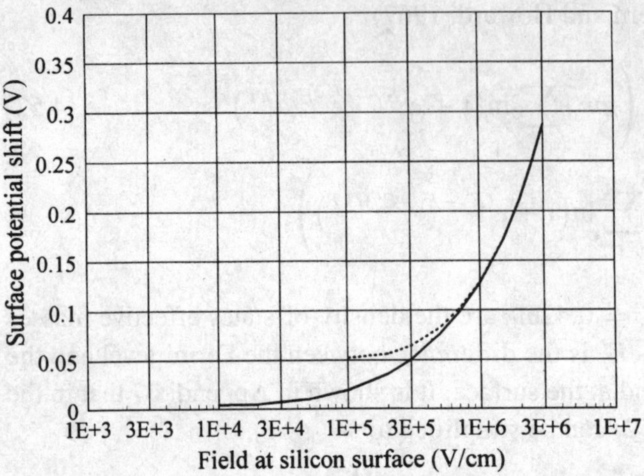


**FIGURE 4.17.** Additional band bending $\Delta\psi_s^{QM}$ (over the classical $2\psi_B$ value) required for reaching the threshold condition as a function of surface electric field. The dotted curve is calculated by keeping only the lowest term (twofold degeneracy) in Eq. (4.53.)

can be evaluated from the preexponential factors in Eqs. (4.54) and (4.53). Figure 4.17 shows the calculated $\Delta\psi_s^{QM}$ as a function of $\mathscr{E}_s$. Beyond $10^6$ V/cm, only the lowest subband is occupied by electrons, and

$$\Delta\psi_s^{QM} \approx \frac{E_0}{q} - \frac{kT}{q}\ln\left(\frac{8\pi q m_d \mathscr{E}_s}{h^2 N_c}\right), \tag{4.56}$$

as indicated by the dotted curve in Fig. 4.17. Knowing $\Delta\psi_s^{QM}$, one can easily calculate the threshold voltage shift due to the quantum effect:

$$\Delta V_t^{QM} = \frac{dV_g}{d\psi_s}\,\Delta\psi_s^{QM} = m\,\Delta\psi_s^{QM}, \tag{4.57}$$

where $m = 1 + (3t_{ox}/W_{dm}^0)$ as before. For the 0.1-μm MOSFET example discussed in Section 4.2.3, $N_a = 10^{18}$ cm$^{-3}$ and $m = 1.3$. If the channel is uniformly doped, $\mathscr{E}_s = 5.4 \times 10^5$ V/cm, $\Delta\psi_s^{QM} = 0.08$ V, and $\Delta V_t^{QM} = 0.10$ V, which makes the threshold voltage (0.57 V) unacceptably high. With an extreme retrograde doping profile, the surface electric field is reduced by a factor of two to $2.7 \times 10^5$ V/cm, for which $\Delta\psi_s^{QM} = 0.046$ V. This brings $\Delta V_t^{QM} = 0.06$ V and a minimum $V_t$ of 0.25 V. It is not very difficult to adjust the retrograde profile to obtain a $V_t$ higher than the extreme value, e.g., 0.4 V, suitable for the 1.5-V power-supply voltage for such devices (Fig. 4.9).

#### 4.2.4.3 QUANTUM EFFECT ON INVERSION-LAYER DEPTH

After strong inversion, the inversion charge density builds up rapidly and the triangular potential-well model is no longer valid. If the field is high enough that only the lowest subband is populated, a variational approach leads to an approximate

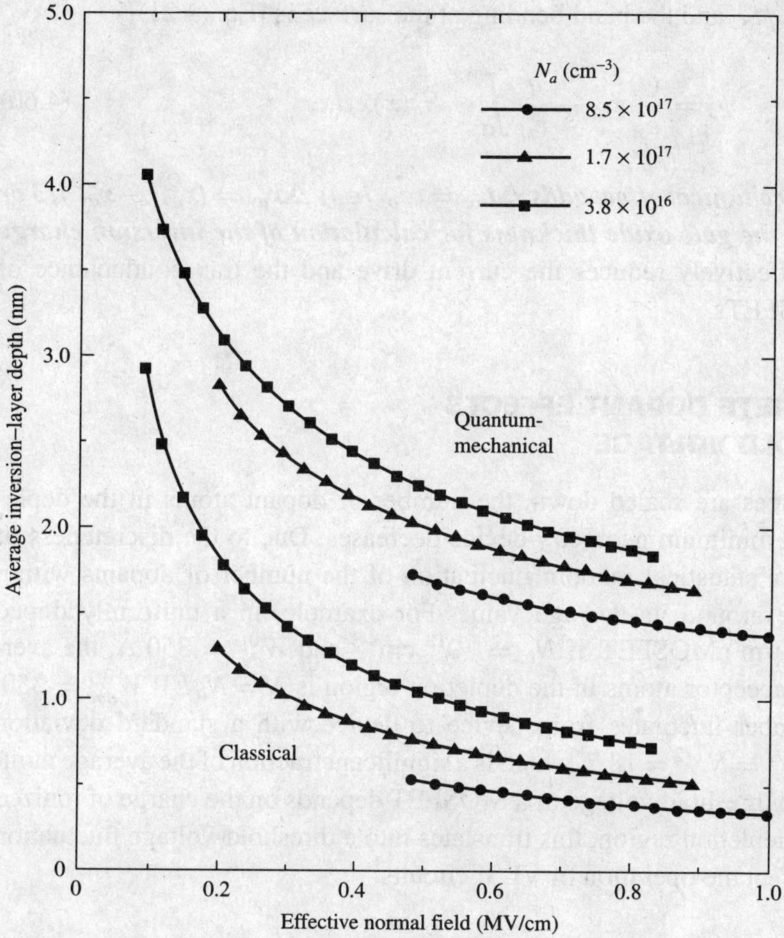


**FIGURE 4.18.** Calculated QM and classical inversion-layer depth versus effective normal field for several uniform doping concentrations. (After Ohkura, 1990.)

expression for the average distance of electrons from the surface (Stern, 1972):

$$x_{av}^{QM} = \left(\frac{9\varepsilon_{si}h^2}{16\pi^2 m_x q Q^*}\right)^{1/3}, \tag{4.58}$$

where $Q^* = Q_d + \frac{11}{32}Q_i$ is a combination of the depletion and inversion charge per unit area in the channel. For intermediate fields, the solution must be obtained numerically. Figure 4.18 shows a comparison of the classical and QM inversion-layer depths versus the effective normal field defined in Eq. (3.47) (Ohkura, 1990). The QM value is consistently larger than the classical value by about 10–12 Å for a wide range of channel doping (uniform) and effective fields. Since

$$V_g = V_{fb} + \psi_s + \frac{Q_d}{C_{ox}} + \frac{Q_i}{C_{ox}}, \tag{4.59}$$

where $C_{ox} = \varepsilon_{ox}/t_{ox}$ and the band bending at the surface is [Eq. (4.21)]

$$\psi_s = \frac{Q_i}{\varepsilon_{si}} x_{av} + \frac{q}{\varepsilon_{si}} \int_0^{W_d} N_a(x) x \, dx, \tag{4.60}$$

***the quantum-mechanical effect adds*** $\Delta t_{ox} = (\varepsilon_{ox}/\varepsilon_{si})\, \Delta x_{av} = (x_{av}^{QM} - x_{av}^{CL})/3$ ***or about 3–4 Å to the gate oxide thickness for calculation of the inversion charge density***. This effectively reduces the current drive and the transconductance of thin-oxide MOSFETs.

### 4.2.5 DISCRETE DOPANT EFFECTS ON THRESHOLD VOLTAGE

As CMOS devices are scaled down, the number of dopant atoms in the depletion region of a minimum geometry device decreases. Due to the discreteness of atoms, there is a statistical random fluctuation of the number of dopants within a given volume around its average value. For example, in a uniformly doped $W = L = 0.1$-μm nMOSFET, if $N_a = 10^{18}$ cm$^{-3}$ and $W_{dm}^0 = 350$ Å, the average number of acceptor atoms in the depletion region is $N = N_a L W W_{dm}^0 = 350$. The actual number fluctuates from device to device with a standard deviation $\sigma_N = \langle(\Delta N)^2\rangle^{1/2} = N^{1/2} = 18.7$, which is a significant fraction of the average number $N$. Since the threshold voltage of a MOSFET depends on the charge of ionized dopants in the depletion region, this translates into a threshold-voltage fluctuation which could affect the operation of VLSI circuits.

#### 4.2.5.1 A SIMPLE FIRST-ORDER MODEL

To estimate the effect of depletion charge fluctuation on threshold voltage, we consider a small volume $dx\,dy\,dz$ at a point $(x, y, z)$ in the depletion region of a uniformly doped ($N_a$) MOSFET. The $x$-axis is in the depth direction, the $y$-axis in the length direction, and the $z$-axis in the width direction. The average number of dopant atoms in this small volume is $N_a\,dx\,dy\,dz$. The actual number fluctuates around this value with a standard deviation of $\sigma_{dN} = (N_a\,dx\,dy\,dz)^{1/2}$. This fluctuation can be thought of as a small delta function of nonuniform doping (either positive or negative) at $(x, y, z)$ superimposed on a uniformly doped background $N_a$. Here we focus on the linearly extrapolated threshold voltage $V_{on}$, as defined in Fig. 3.16. When there is a slight local nonuniformity of doping in either the channel-width or the channel-length direction, the first-order influence on the linear threshold voltage is through its effect on the depletion charge density averaged over the entire channel area (Nguyen, 1984). This is similar to the assumption made in the charge-sharing model for short-channel effects in Section 3.2.1. The effect of the above doping fluctuation on the linear threshold voltage is then equivalent to that of a uniform delta-function implant of dose (number of ions per unit area) $\Delta D$ and depth $x$, where $\langle(\Delta D)^2\rangle^{1/2} = \sigma_{dN}/WL = (N_a\,dx\,dy\,dz)^{1/2}/WL$. The threshold-voltage shift is obtained by substituting $D_I = \Delta D$ and $x_c = x$ in Eq. (4.36) and retaining only the first-order terms in $\Delta D$:

$$\Delta V_{on} = \frac{q\,\Delta D}{C_{ox}}\left(1 - x\sqrt{\frac{qN_a}{2\varepsilon_{si}(2\psi_B)}}\right) = \frac{q\,\Delta D}{C_{ox}}\left(1 - \frac{x}{W_{dm}^0}\right). \tag{4.61}$$

The last expression is quite general and is applicable to a nonuniformly doped background as well. It follows directly from Eq. (4.21) or can be seen from the graphical representation in Fig. 4.5(b). The mean square deviation (variance) of threshold voltage due to the depletion charge fluctuation in $dx\,dy\,dz$ is then

$$\langle \Delta V_{on}^2 \rangle_{x,y,z} = \frac{q^2 N_a}{C_{ox}^2 L^2 W^2}\left(1 - \frac{x}{W_{dm}^0}\right)^2 dx\,dy\,dz. \tag{4.62}$$

Since dopant number fluctuations at various points are completely random and uncorrelated, the total mean square fluctuation of the threshold voltage is obtained by integrating Eq. (4.62) over the entire depletion region:

$$\sigma_{V_{on}}^2 = \frac{q^2 N_a}{C_{ox}^2 L^2 W^2}\int_0^W \int_0^L \int_0^{W_{dm}^0}\left(1 - \frac{x}{W_{dm}^0}\right)^2 dx\,dy\,dz. \tag{4.63}$$

It is straightforward to carry out the integration and obtain

$$\sigma_{V_{on}} = \frac{q}{C_{ox}}\sqrt{\frac{N_a W_{dm}^0}{3LW}}. \tag{4.64}$$

In the above 0.1-μm example, $\sigma_{V_{on}} = 17.5$ mV if $t_{ox} = 35$ Å. This is small compared with the worst-case short-channel threshold roll-off in Section 4.2.1, but can be significant in minimum-geometry devices, for example, in an SRAM cell.

In the above analysis, it was assumed that the surface potential is uniform in both the length and the width directions of the device. In other words, all the lumpiness due to local fluctuations of the depletion charge is smoothed out and the surface potential depends only on the average (or total) depletion charge of the device. This assumption is not valid in the subthreshold region, where current injection is dominated by the highest potential barrier in the channel rather than by the average value (Nguyen, 1984). In general, the problem needs to be solved by 3-D numerical simulations (Wong and Taur, 1993). The results indicate that in addition to the threshold fluctuations of a similar magnitude to that expected from Eq. (4.64), there is also a negative shift of the average threshold voltage, especially in the subthreshold region. This is believed to be due to the inhomogeneity of surface potential resulting from the microscopic random distribution of discrete dopant atoms in the channel. For the same reason, the source–drain current may exhibit some statistical asymmetry under high-drain-bias conditions (Wong and Taur, 1993).

#### 4.2.5.2 DISCRETE DOPANT EFFECTS IN A RETROGRADE-DOPED CHANNEL

Threshold voltage fluctuations due to discrete dopants are greatly reduced in a retrograde-doped channel. Consider the profile in Fig. 4.11 with $N_s = 0$, i.e., the channel is undoped within $0 < x < x_s$. The average threshold voltage and the maximum depletion width $W_{dm}^0$ are given by Eq. (4.43) and Eq. (4.44), respectively. For a small volume of dopants at $(x, y, z)$ where $x_s < x < W_{dm}^0$, Eq. (4.62) still holds. The $x$-integral in Eq. (4.63), however, is carried out from $x_s$ to $W_{dm}^0$, which results in

$$\sigma_{V_{on}} = \frac{q}{C_{ox}}\sqrt{\frac{N_a W_{dm}^0}{3LW}\left(1 - \frac{x_s}{W_{dm}^0}\right)^{3/2}} \tag{4.65}$$

for a retrograde-doped channel. ***In the extreme retrograde or ground-plane limit shown in Fig. 4.12(b), $x_s = W_{dm}^0$, and the threshold voltage fluctuation goes to zero***. This is also clear from Eq. (4.45), where the threshold voltage is essentially independent of $N_a$ (or $N_a'$). Of course, the technological challenge is then to control the tolerance of the undoped-layer thickness $x_s$ so that it does not introduce a different kind of threshold voltage variations.

## 4.3 MOSFET CHANNEL LENGTH

Channel length is a key parameter in CMOS technology used for performance projection (circuit models), short-channel design, and model–hardware correlation. This section focuses on MOSFET channel length: its definition, extraction, and physical interpretation.

### 4.3.1 VARIOUS DEFINITIONS OF CHANNEL LENGTH

A number of quantities, e.g., *mask length* ($L_{mask}$), *gate length* ($L_{gate}$), *metallurgical channel length* ($L_{met}$), and *effective channel length* ($L_{eff}$), have been used to describe the length of a MOSFET. Even though they are all related to each other, their relationships are strongly process-dependent.

Figure 4.19 shows schematically how various channel lengths are defined. $L_{mask}$ is the design length on the polysilicon etch mask. It is reproduced on the wafer as $L_{gate}$ through lithography and etching processes. Depending on the lithography and etching biases, $L_{gate}$ can be either longer or shorter than $L_{mask}$. There are also process tolerances associated with $L_{gate}$. For the same $L_{mask}$ design, $L_{gate}$ may vary

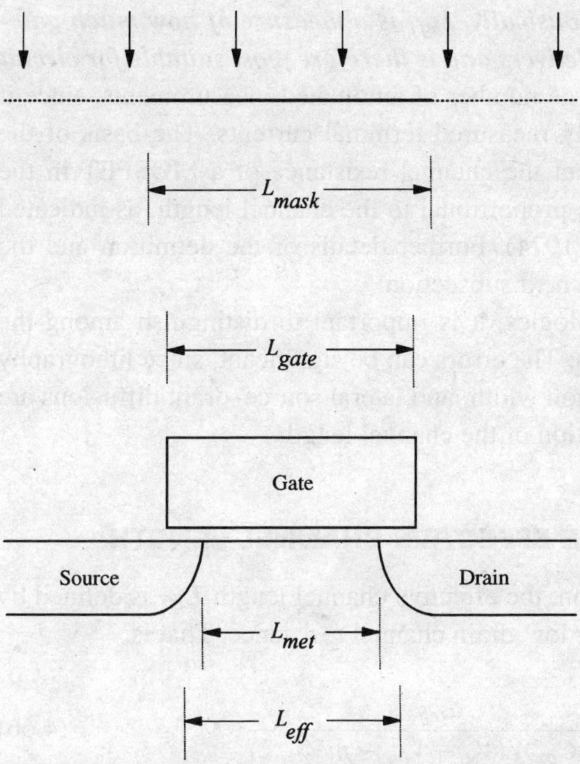


**FIGURE 4.19.** Schematic diagram showing the definitions of and relationship among the various notions of channel length. The physical interpretation of $L_{eff}$ is examined in Section 4.3.3.

from chip to chip, wafer to wafer, and run to run. Although $L_{gate}$ is an important parameter for process control and monitoring, there is no simple way of making a large number of measurements of it. Usually, $L_{gate}$ is measured with a scanning electron microscope (SEM) and only sporadically across the wafer. There is also an uncertainty in the precise definition of $L_{gate}$ when the polysilicon etch profile is not vertical, as to whether $L_{gate}$ refers to the top or to the bottom dimension of the gate.

$L_{met}$ is defined as the distance between the metallurgical junctions of the source and drain diffusions at the silicon surface. In a modern CMOS process, the source and drain regions are self-aligned to the polysilicon gate by performing the source–drain implant after gate patterning (Kerwin *et al.*, 1969). As a result, there is a close correlation between $L_{met}$ and $L_{gate}$. Usually, $L_{met}$ is shorter than $L_{gate}$ by a certain amount due to the lateral implant straggle and the lateral source–drain diffusion in the process. Accurate physical measurement of $L_{met}$ in actual hardware is very difficult. Normally, $L_{met}$ is used only in 2-D models for short-channel device design. Even for that purpose, difficulties arise in defining $L_{met}$ when dealing with a buried-channel device or a retrograde channel profile with zero surface doping, where there are no metallurgical junctions at the silicon surface.

The parameter $L_{eff}$ is different from all other channel lengths discussed above in that it is defined through some electrical characteristics of the MOSFET device and is not a physical parameter. ***Basically, $L_{eff}$ is a measure of how much gate-controlled current a MOSFET delivers and is therefore most suitable for circuit models***. $L_{eff}$ also allows for a large number of automated measurements, since it can be extracted from electrically measured terminal currents. The basis of the $L_{eff}$ definition lies in the fact that the channel resistance of a MOSFET in the linear or low-drain bias region is proportional to the channel length, as indicated by Eq. (3.100) (Dennard *et al.*, 1974). Further details of the definition and the extraction of $L_{eff}$ are given in the next subsection.

For submicron CMOS technologies, it is important to distinguish among the various notions of channel length. The errors can be significant, since lithography and etching bias, junction depletion width, and lateral source–drain diffusions are all becoming an appreciable fraction of the channel length.

### 4.3.2 EXTRACTION OF THE EFFECTIVE CHANNEL LENGTH

As discussed in the last subsection, the effective channel length $L_{eff}$ is defined by its proportionality to the linear or low-drain channel resistance. That is,

$$R_{ch} \equiv \frac{V_{ds}}{I_{ds}} = \frac{L_{eff}}{\mu'_{eff} C_{ox} W (V_g - V_{on} - m V_{ds}/2)} \tag{4.66}$$

from Eq. (3.61), where $V_{on}$ is the linearly extrapolated threshold voltage and $\mu'_{eff}$ is the modified effective mobility, which contains the inversion-layer capacitance effect. $\mu'_{eff}$ is a weak function of $V_g$. For different $L_{mask}$, $L_{eff}$ differs but is assumed to be related to $L_{mask}$ by a constant *channel length bias* $\Delta L$:

$$L_{eff} = L_{mask} - \Delta L. \tag{4.67}$$

All the lithography and etch biases as well as the lateral source–drain implant straggle and diffusion are lumped into $\Delta L$. The assumption that the channel length bias is constant is a reasonable one when the channel length is not too short. However, $\Delta L$ can be linewidth-dependent when $L_{mask}$ approaches the resolution limit of the lithography tool used in the process. This issue will be addressed later.

In the simplest scheme of channel-length extraction (Dennard *et al.*, 1974), $R_{ch}$ is measured for a set of devices with different $L_{mask}$. Based on Eq. (4.66) and Eq. (4.67), a plot of $R_{ch}$ for a given $V_g$ versus $L_{mask}$ should yield a straight line whose intercept with the $x$-axis gives $\Delta L$ and therefore $L_{eff}$. In practice, however, two issues must be addressed for short-channel devices. The first one is the source–drain series resistance. The second one is the short-channel effect (SCE), which causes $V_{on}$ in Eq. (4.66) to depend on $L_{mask}$.

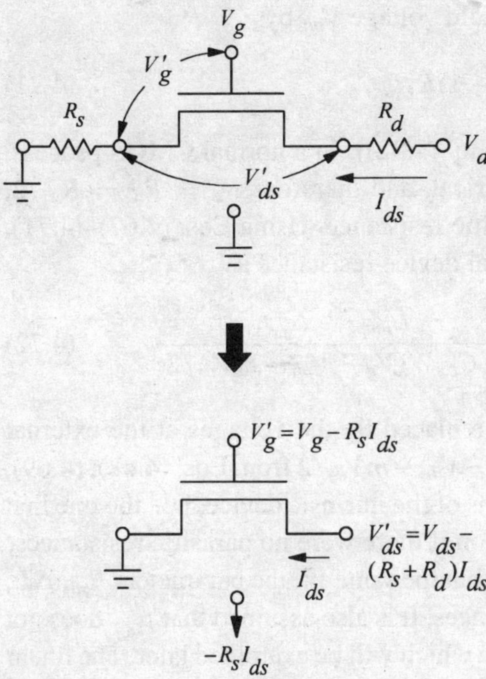


**FIGURE 4.20.** Equivalent circuit of MOSFET with source and drain series resistance. The intrinsic part of the top circuit is equivalent to the bottom circuit with redefined terminal voltages.

#### 4.3.2.1 CHANNEL-RESISTANCE METHOD

The effect of source–drain resistance is examined using the equivalent circuit in Fig. 4.20. A source resistance $R_s$ and a drain resistance $R_d$ are assumed to connect an *intrinsic MOSFET* to the external terminals where voltages $V_{ds}$ and $V_g$ are applied. The internal voltages are $V'_{ds}$ and $V'_g$ for the intrinsic MOSFET. One can write the following relations:

$$V'_{ds} = V_{ds} - (R_s + R_d)I_{ds} \tag{4.68}$$

and

$$V'_g = V_g - R_s I_{ds}. \tag{4.69}$$

As shown in Fig. 4.20, the intrinsic part of an actual device with parasitic resistance is equivalent to an intrinsic MOSFET with a grounded source, with $V'_g$ and $V'_{ds}$ at the gate and the drain terminals, and with a reverse bias $-R_s I_{ds}$ on the substrate. Based on Eq. (4.66), but with redefined voltage symbols on the intrinsic nodes, the channel resistance of the intrinsic device is given by

$$R_{ch} \equiv \frac{V'_{ds}}{I_{ds}} = \frac{L_{eff}}{\mu'_{eff} C_{ox} W\left(V'_g - V'_{on} - mV'_{ds}/2\right)}, \tag{4.70}$$

where $V'_{on}$ is the linear threshold voltage with the reverse bias on the substrate. It is related to the zero-substrate-bias threshold voltage $V_{on}$ by

$$V'_{on} = V_{on} + (m-1)R_s I_{ds}, \tag{4.71}$$

where $m-1$ is the substrate sensitivity [Eq. (4.32)]. In a normal CMOS process, the source and drain regions are symmetrical, and therefore $R_s = R_d = R_{sd}/2$, where $R_{sd}$ is the total source–drain parasitic resistance. Using Eqs. (4.67)–(4.71), one can write the externally measured total device resistance as

$$R_{tot} \equiv \frac{V_{ds}}{I_{ds}} = R_{sd} + R_{ch} = R_{sd} + \frac{L_{mask} - \Delta L}{\mu'_{eff} C_{ox} W (V_g - V_{on} - m V_{ds}/2)}. \tag{4.72}$$

Here all the internal voltages have been replaced by the voltages at the external terminals, since $V'_g - V'_{on} - mV'_{ds}/2 = V_g - V_{on} - mV_{ds}/2$ from Eqs. (4.68), (4.69), and (4.71). Note that $V_{on}$ is defined in terms of the intrinsic device, i.e., the one that would be obtained from linear extrapolation if there were no parasitic resistances.

For a set of devices with different $L_{mask}$ but the same $W$, the parameters $R_{sd}$, $\Delta L$, and $C_{ox}$ are the same within process tolerances. It is also assumed that $\mu'_{eff}$ does not change with channel length, an assumption which will be examined later. The linear threshold voltage $V_{on}$, however, does depend on channel length because of short-channel effects. When comparing $R_{tot}$ of devices with different $L_{mask}$, therefore, it is important to measure $V_{on}$ for each device and adjust $V_g$ so that the gate overdrive $V_g - V_{on}$ is the same from device to device. ***A plot of $R_{tot}$ (at small $V_{ds}$) versus $L_{mask}$ for a given $V_g - V_{on}$ will then yield a straight line that passes through the point ($\Delta L$, $R_{sd}$)***. An example is shown in Fig. 4.21. The slope of the line depends on the specific value of the gate overdrive. $\Delta L$ and $R_{sd}$ are determined by the common intercept of several lines, each for a different $V_g - V_{on}$ (Chern *et al.*, 1980).

#### 4.3.2.2 SHIFT-AND-RATIO METHOD

Despite the simplicity of the channel-resistance method described above, two main issues remain. First, it is not always straightforward to find the intrinsic $V_{on}$ of short-channel devices. The presence of $R_{sd}$ adds considerable difficulty in the usual linear extrapolation of $V_{on}$ from the measured $I_{ds}$–$V_g$ curve (Sun *et al.*, 1986). Typically, one tends to underestimate $V_{on}$ in short-channel devices, as the degradation of $I_{ds}$ by $R_{sd}$ is more severe at higher currents. This introduces errors in channel-length extraction. The problem is further aggravated by a strong dependence of mobility on gate voltage, for example, in low-temperature and/or 0.1-μm MOSFETs. The second problem with the resistance method is that the $R_{tot}$-versus-$L_{mask}$ lines for different gate overdrives may not intersect at a common point. Significant errors may result if only a limited number of $V_g - V_{on}$ are investigated.

An improved channel-length extraction algorithm, called the *shift-and-ratio* (S&R) method, is able to circumvent the above problems (Taur *et al.*, 1992). This method is based on the same channel-resistance concept described above. It starts

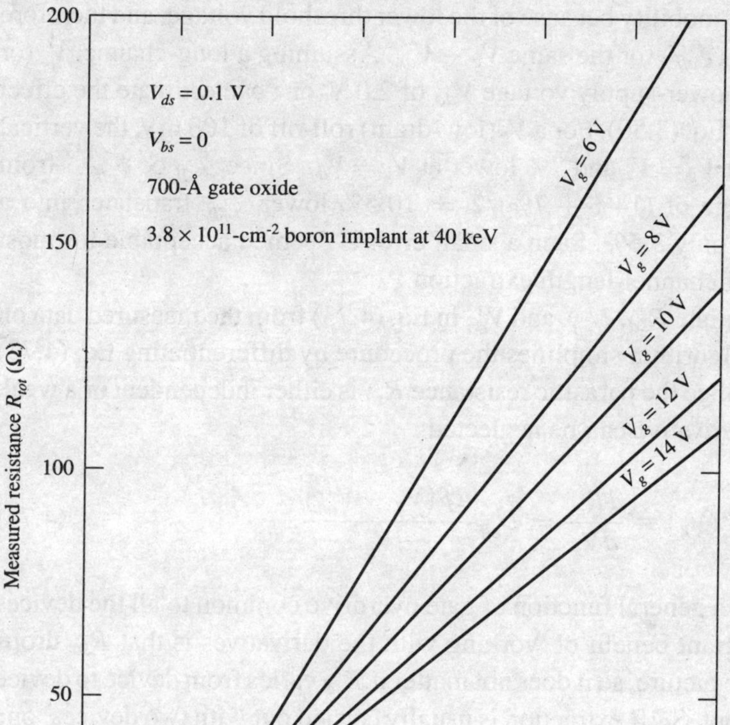


**FIGURE 4.21.** Measured $R_{tot}$ at a low drain voltage versus $L_{mask}$ for several different values of $V_g - V_{on}$. The common intercept determines both $\Delta L$ and $R_{sd}$. (After Chern *et al.*, 1980.)

with a generalization of Eq. (4.72) to the form

$$R^i_{tot}(V_g) = R_{sd} + L^i_{eff} f\left(V_g - V^i_{on}\right), \tag{4.73}$$

where $f$ is a general function of gate overdrive common to all the measured devices. The superscript $i$ denotes the $i$th device, with an unknown effective channel length $L^i_{eff} = L^i_{mask} - \Delta L$ and linear threshold voltage $V^i_{on}$. The key assumption behind Eq. (4.73) is that the modified effective mobility $\mu'_{eff}$ is a common function of $V_g - V_{on}$ for all the measured devices. This is a reasonable assumption in view of Eq. (3.50), Eq. (3.51), Eq. (3.58), and Eq. (3.60). Note that for this argument, $V_t$ and $V_{on}$ are interchangeable as long as their difference is a constant [≈ (2–3)$kT/q$], independent of channel length. Strictly speaking, short-channel devices have a slightly higher mobility because of the lower threshold voltage, and therefore a lower vertical field ($\mathscr{E}_{eff}$) for the same $V_g - V_{on}$. Assuming a long-channel $V_t$ (or $V_{on}$) of 0.5 V and a power-supply voltage $V_{dd}$ of 2.0 V, one can estimate the effect of SCE on $\mathscr{E}_{eff}$ using Eq. (3.50). For a $V_t$ (low-drain) roll-off of 100 mV, the vertical field is 14% lower at $V_g = V_t$ and 7% lower at $V_g = V_{dd}$. Since $\mu_{eff} \propto \mathscr{E}_{eff}^{-1/3}$ from Eq. (3.51), an average of $(14\% + 7\%)/2 = 10.5\%$ lower $\mathscr{E}_{eff}$ translates into a mobility increase of only 3.5%. Such a small error is deemed acceptable for most practical purposes of channel-length extraction.

The task is to calculate $R_{sd}$, $L^i_{eff}$, and $V^i_{on}$ in Eq. (4.73) from the measured data on $R^i_{tot}(V_g)$. The S&R algorithm simplifies the procedure by differentiating Eq. (4.73) with respect to $V_g$. Since the parasitic resistance $R_{sd}$ is either independent or a weak function of $V_g$, its derivative can be neglected:

$$S^i(V_g) \equiv \frac{dR^i_{tot}}{dV_g} = L^i_{eff} \frac{df\left(V_g - V^i_{on}\right)}{dV_g}. \tag{4.74}$$

Here $df/dV_g$ is also a general function of gate overdrive common to all the devices measured. An important benefit of working with the derivatives is that $R_{sd}$ drops completely out of the picture, so it does not matter if $R_{sd}$ varies from device to device as long as it is constant. S&R extraction is usually carried out with two devices: one long-channel and one short-channel. Equation (4.74) with superscript $i$ represents the short-channel device, while superscript 0 refers to the long-channel device:

$$S^0(V_g) \equiv \frac{dR^0_{tot}}{dV_g} = L^0_{eff} \frac{df\left(V_g - V^0_{on}\right)}{dV_g}. \tag{4.75}$$

An example is shown in Fig. 4.22 for two devices: $S^0(V_g)$ for $L^0_{mask} = 10\ \mu\text{m}$, and $S^i(V_g)$ for $L^i_{mask} = 0.25\ \mu\text{m}$.

It would have been easy if $V^i_{on} = V^0_{on}$, in which case $S^i$ and $S^0$ would be similar functions of $V_g$ and $L^i_{eff}$ would be simply obtained from the ratio $S^i/S^0 = L^i_{eff}/L^0_{eff} \approx L^i_{eff}/L^0_{mask}$. In general, however, $V^i_{on} \neq V^0_{on}$, and the two $S$-functions must be shifted with respect to each other before the ratio is taken. For example, in Fig. 4.22, we can shift one curve ($S^i$) horizontally to the right by a varying amount $\delta$ and compute the ratio between the two curves,

$$r(\delta, V_g) \equiv \frac{S^0(V_g)}{S^i(V_g - \delta)}, \tag{4.76}$$

as a function of $V_g$. The purpose here is to find the $\delta$-value for which the ratio $r$ is a constant, independent of $V_g$. If $\delta$ is zero or too small, $r$ is a monotonically decreasing function of $V_g$. On the other hand, if $\delta$ is too large, $r$ becomes a monotonically increasing function of $V_g$. These are shown in Fig. 4.23. ***Only when $S^i$ is shifted by an amount $\delta$ equal to the threshold voltage difference between the two devices,***

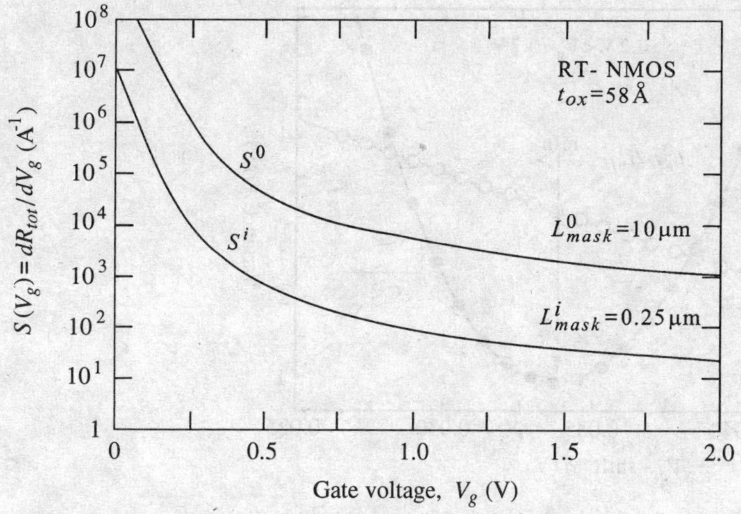


**FIGURE 4.22.** Examples of $S(V_g) = dR_{tot}(V_g)/dV_g$ curves measured from long-channel ($L_{mask}$ = 10 μm) and short-channel ($L_{mask}$ = 0.25 μm) devices. (After Taur *et al.*, 1992.)

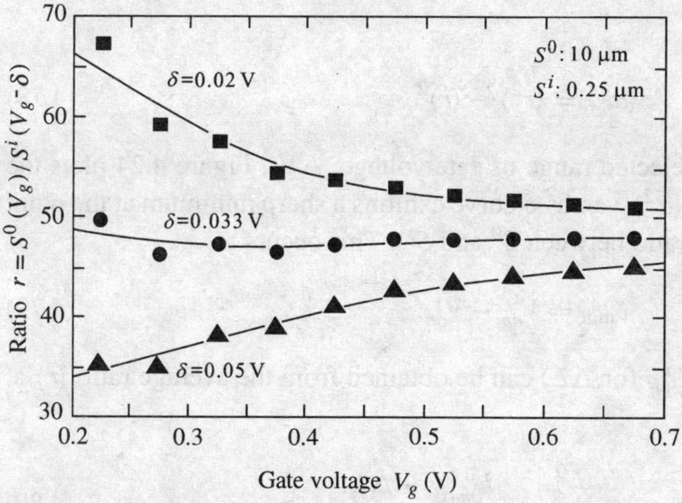


**FIGURE 4.23.** Curves of $r(\delta, V_g)$ (ratio of $S$-functions after shift) versus $V_g$ for three different amounts of shift $\delta$. The data are taken from the device examples in Fig. 4.22.

***$V^0_{on} - V^i_{on}$, does r become nearly independent of $V_g$.*** Once the correct shift is found, it is a simple matter to find $L^i_{eff}$ from the ratio $r$ evaluated at that shift.

The above procedure can be automated by computing the average $r$ and the mean square deviation of $r$ from its average value, i.e.,

$$\langle r \rangle = \frac{\int_{\Delta V_g} r(\delta, V_g)\, dV_g}{\int_{\Delta V_g} dV_g} \tag{4.77}$$

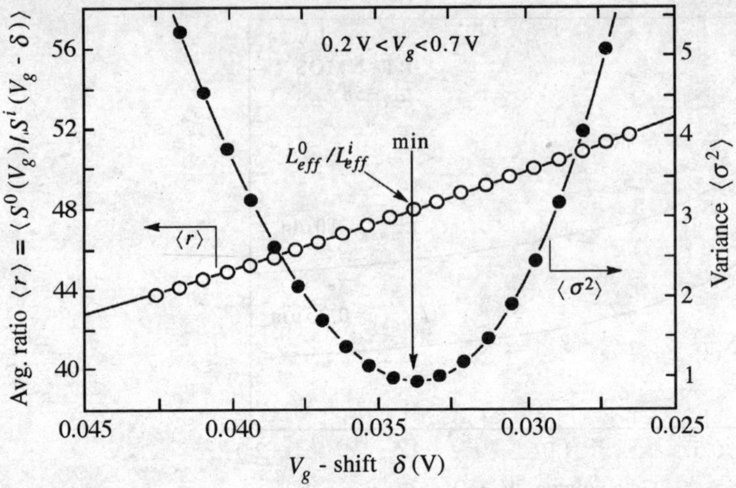


**FIGURE 4.24.** Average ratio $\langle r \rangle$ (open circles) and variance $\langle \sigma^2 \rangle$ (solid dots) versus shift $\delta$. The data are taken from the device examples in Fig. 4.22 and Fig. 4.23. (After Taur *et al.*, 1992.)

and

$$\langle \sigma^2 \rangle = \langle r^2 \rangle - \langle r \rangle^2, \tag{4.78}$$

as functions of $\delta$ for a selected range of gate voltage, $\Delta V_g$. Figure 4.24 plots the example case where the $\langle \sigma^2 \rangle$-versus-$\delta$ curve exhibits a sharp minimum at the point of best match (constant ratio between $S^0$ and $S^i$). This occurs at

$$\delta_{\min} = V_{on}^0 - V_{on}^i, \tag{4.79}$$

and the channel length $L_{eff}^i$ (or $\Delta L$) can be obtained from the average ratio $\langle r \rangle$ at this point:

$$\langle r \rangle_{\delta_{\min}} = \frac{L_{eff}^0}{L_{eff}^i} = \frac{L_{mask}^0 - \Delta L}{L_{mask}^i - \Delta L}. \tag{4.80}$$

Note that even if $\Delta L$ is different between the long-channel and the short-channel devices, very little error is introduced, as $\Delta L \ll L_{mask}^0$ and hence $L_{eff}^0 \approx L_{mask}^0$, insensitive to $\Delta L$.

Once $\delta_{\min}$ and $\langle r \rangle_{\delta_{\min}}$ are found, $R_{sd}$ can be calculated from the measured resistances by using Eq. (4.73) and the corresponding equation for the long-channel device (superscript 0):

$$R_{sd} = \frac{\langle r \rangle_{\delta_{\min}} R_{tot}^i(V_g - \delta_{\min}) - R_{tot}^0(V_g)}{\langle r \rangle_{\delta_{\min}} - 1}. \tag{4.81}$$

For a given long-channel reference, the above extraction procedure can be repeated for a number of short-channel devices with different $L^i_{mask}$. A by-product of the process is the low-drain short-channel threshold-voltage roll-off given by $\delta_{\min}$. The S&R algorithm has been shown to yield consistent results for effective channel lengths down to the 0.1-μm range (Taur *et al.*, 1992). Although it is not necessary to assume a constant $\Delta L$ and $R_{sd}$ for the S&R algorithm to work, comparable $\Delta L$ and $R_{sd}$ results among different devices are a good indication that the channel-length extraction has been carried out properly.

It is important in the S&R algorithm to choose a proper gate voltage range $\Delta V_g$ for $\langle r \rangle$ and $\langle \sigma^2 \rangle$ calculations.The subthreshold region should be avoided, where the current conduction mechanism is different from that in the linear region (Nguyen, 1984). ***A good choice for $\Delta V_g$ is from about $V_{on}$ + 0.2 V all the way up to $V_{dd}$, which covers most regions of interest in the $I_{ds}$–$V_g$ curve***. The sensitivity of the extracted $L_{eff}$ to $\Delta V_g$ depends on how abrupt the lateral source–drain profile is. This is discussed in more detail in the next subsection.

### 4.3.3 PHYSICAL MEANING OF EFFECTIVE CHANNEL LENGTH

This subsection examines the physical meaning of $L_{eff}$ extracted from electrically measured terminal currents. The effective channel length is defined through the linear channel resistance by Eq. (4.66). This equation is derived for long-channel devices and is not strictly valid for short-channel devices. By its definition, $L_{eff}$ represents a measure of the effective current-carrying capability of the device and is not associated with any fixed physical quantity. When the channel profile is reasonably uniform and the source–drain doping is not too graded, $L_{eff}$ is approximately equal to $L_{met}$ (Laux, 1984). In general, however, one cannot take $L_{eff} = L_{met}$ for granted, especially in very short-channel MOSFETs.

An example is illustrated in Fig. 4.25, where $L_{eff}$, extracted (by the S&R algorithm) from currents calculated using a 2-D device simulator, is plotted versus $L_{met}$ for a variety of source–drain and channel doping conditions (Taur *et al.*, 1995b). A 2-D Gaussian profile is used to simulate the falloff of the source and drain doping concentration near the gate edge:

$$N_d(x, y) = N_0 e^{-(x-x_0)^2/2\sigma_V^2} e^{-(y-y_0)^2/2\sigma_L^2}, \tag{4.82}$$

where $x$ is in the vertical direction and $y$ is in the lateral direction. The junction depth $x_j$ is mainly determined by the vertical straggle $\sigma_V$. A key parameter for $L_{eff}$ is the lateral source–drain doping gradient characterized by the lateral straggle $\sigma_L$. For each of the doping cases in Fig. 4.25, $L_{eff}$ varies with $L_{met}$ linearly with a slope of one. In other words, $L_{eff} - L_{met}$ is essentially independent of $L_{met}$, indicating that the linear relationship between $R_{tot}$ and $L_{mask}$ assumed in Fig. 4.21 and Eq. (4.73) can be extended to short-channel devices. However, $L_{eff} - L_{met}$ varies considerably with the doping profile. For an infinitely abrupt source–drain

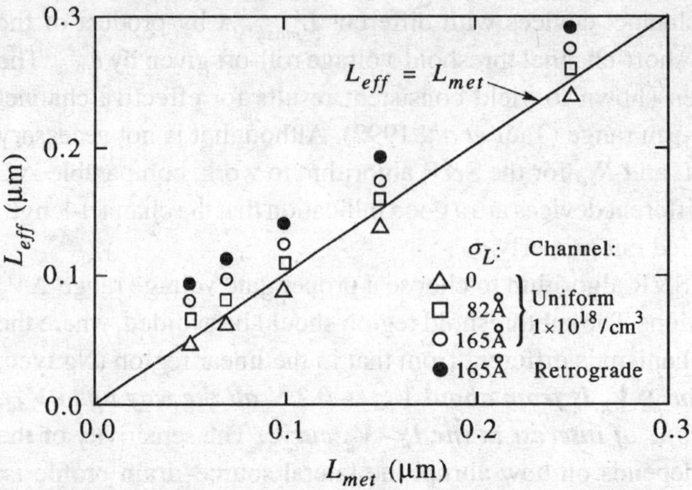


**FIGURE 4.25.** $L_{eff}$ extracted from simulated currents versus $L_{met}$ for four doping cases. Open symbols are for a uniformly doped channel with different lateral source–drain gradients. Solid dots are for a retrograde-doped channel. (After Taur *et al.*, 1995b.)

profile ($\sigma_L = 0$), $L_{eff} - L_{met}$ is slightly negative. As the lateral straggle $\sigma_L$ increases, $L_{eff} - L_{met}$ becomes increasingly more positive. The difference grows even larger in the retrograde channel case where $L_{eff}$ is significantly longer than $L_{met}$. Such a deviation can be understood in terms of the spatial dependence of channel sheet resistivity as discussed below.

#### 4.3.3.1 SHEET RESISTIVITY IN SHORT-CHANNEL DEVICES

Equation (4.66) implicitly assumes that the sheet resistivity, $\rho_{ch}$ given by Eq. (3.101), is uniform in both the MOSFET width and length directions. If the device is wide enough, $\rho_{ch}$ can be considered uniform in that direction. However, the variation of $\rho_{ch}$ in the length direction cannot be ignored in a short-channel device. From Eq. (3.8),

$$I_{ds} = -\mu_{eff} W Q_i(y) \frac{dV}{dy}, \tag{4.83}$$

where $V(y)$ is the quasi-Fermi level at a point $y$ along the channel length direction. $I_{ds}$ is a constant independent of $y$ as required by current continuity. One can define a laterally varying sheet resistivity as

$$\rho_{ch}(y) = \frac{dV/dy}{I_{ds}/W} = \frac{1}{-\mu_{eff} Q_i(y)}. \tag{4.84}$$

Note that $Q_i < 0$ for nMOSFETs. This expression is valid as long as the current flow is largely parallel to the $y$-direction and the equipotential contours are perpendicular to the silicon surface. The total resistance is given by

$$\frac{V_{ds}}{I_{ds}} = \frac{1}{W}\int_{S-D} \rho_{ch}(y)\,dy, \tag{4.85}$$

where the integration is carried out from the heavily doped source region to the heavily doped drain region.

Figure 4.26 plots $\rho_{ch}(y)$ calculated from the 2-D device simulator versus distance from the source to the drain for $L_{met} = 0.10$ μm at three different gate voltages (Taur *et al.*, 1995b). The area under each curve gives the total source-to-drain resistance

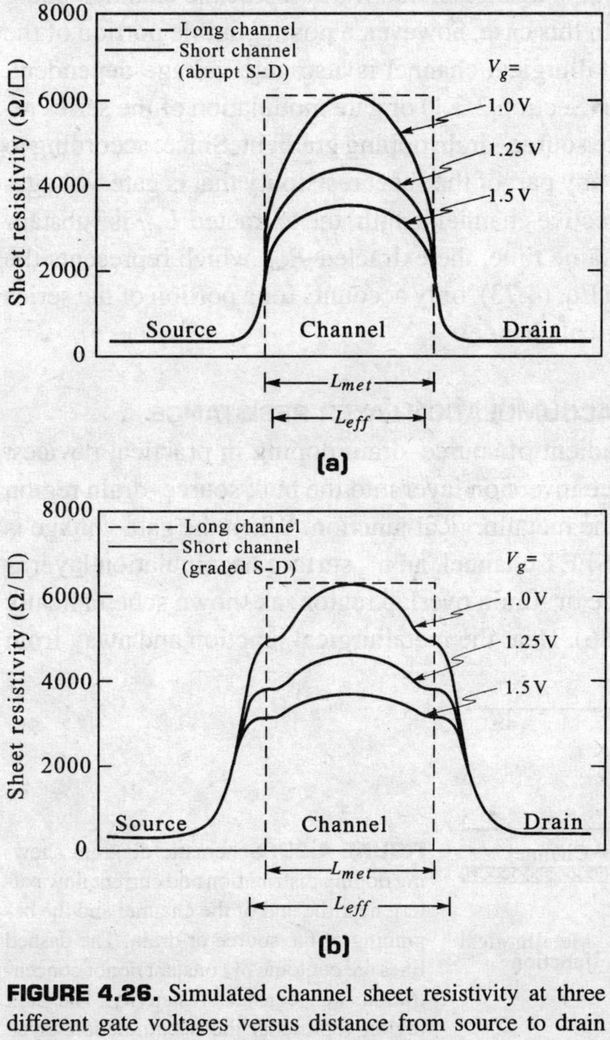


**FIGURE 4.26.** Simulated channel sheet resistivity at three different gate voltages versus distance from source to drain of an $L_{met} = 0.10$-μm MOSFET. The curves in (a) are for an infinitely abrupt (laterally) source–drain in which $L_{eff} = 0.091$ μm. The curves in (b) are for a graded ($\sigma_L = 165$ Å) source–drain in which $L_{eff} = 0.124$ μm. In both cases, the dashed lines represent the ideal, uniform-sheet resistivity of a scaled long-channel device. (After Taur *et al.*, 1995b.)

as indicated by Eq. (4.85). In Fig. 4.26(a) for an infinitely abrupt (laterally) source–drain junction, the sheet resistivity is modulated by gate voltage inside the (metallurgical) channel and independent of gate voltage outside the (metallurgical) channel. However, in contrast to a long-channel device, $\rho_{ch}(y)$ is highly nonuniform, with a peak near the middle of the channel and decreasing toward the edges. This is due to SCEs from the source–drain fields, which help lower the potential barrier near the junctions and raise the local inversion charge density (Wordeman *et al.*, 1985). This effect is more pronounced at low gate voltages near threshold. The resulting $L_{eff}$ extracted by the S&R method is slightly shorter than $L_{met}$.

Figure 4.26(b) shows similar plots for the same $L_{met} = 0.10$ μm, but with a finite lateral source–drain gradient. $\rho_{ch}(y)$ again is nonuniform inside the channel, being modulated by the gate voltage. In this case, however, a nonnegligible portion of the sheet resistivity outside the metallurgical channel is also gate-voltage-dependent. This is because of accumulation (Section 2.3.1) or gate modulation of the series resistance associated with the finite source–drain doping gradient. Since, according to the $L_{eff}$ definition in Eq. (4.73), any part of the sheet resistivity that is gate-voltage-dependent contributes to the effective channel length, the extracted $L_{eff}$ is substantially longer than $L_{met}$. At the same time, the extracted $R_{sd}$, which represents the constant part of the resistance in Eq. (4.73), only accounts for a portion of the series resistance outside the metallurgical channel.

#### 4.3.3.2 GATE-MODULATED ACCUMULATION-LAYER RESISTANCE

Because of the finite lateral gradient of source–drain doping in practical devices, current injection from the surface inversion layer into the bulk source–drain region does not occur immediately at the metallurgical junction. When the gate voltage is high enough to turn on the MOSFET channel, an n$^+$ surface accumulation layer is also formed in the gate-to-source or -drain overlap region, as shown schematically in Fig. 4.27 (Ng and Lynch, 1986). Near the metallurgical junction and away from

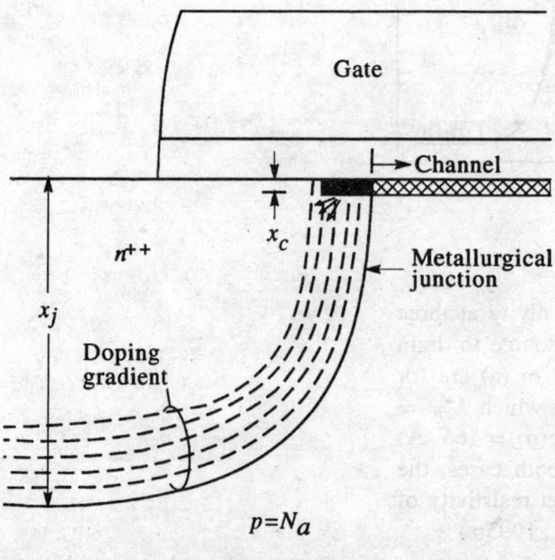


**FIGURE 4.27.** Schematic diagram showing doping distribution and current flow pattern near the end of the channel and the beginning of the source or drain. The dashed lines are contours of constant donor concentration, i.e., constant resistivity. The dark region represents the accumulation layer. (After Ng and Lynch, 1986.)

the surface, the donor concentration (also compensated by the p-type background) is low and the conductivity of the accumulation layer is higher than that of the bulk source–drain. As a result, current flow stays in the accumulation layer near the surface. This continues until the source–drain doping becomes high enough that the bulk conductance exceeds that of the accumulation layer. The point or region of current injection into the bulk depends on the lateral source–drain doping gradient. The more graded the profile is, the farther away the injection point is from the metallurgical junction.

The sheet resistivity of the accumulation layer can be estimated by applying Eq. (2.180) to the gate-to-source–drain overlap region:

$$V_g = V_{fb} + \psi_s - \frac{Q_{ac}}{C_{ox}}, \tag{4.86}$$

where $Q_{ac} < 0$ is the accumulation charge (electrons) per unit area induced by the gate field, $\psi_s$ is the band bending at the surface with respect to the bulk n-type region, and $V_{fb}$ is the flat-band voltage largely determined by the work-function difference between the gate electrode and the n-type silicon. For an $n^+$-polysilicon-gated nMOSFET, $V_{fb} = -E_g/2q + \psi_B$, where $\psi_B$ is given by Eq. (2.37) in terms of the local n-type doping concentration. The band bending in accumulation is approximately given by the distance between the n-type Fermi level and the conduction-band edge, i.e., $\psi_s \approx E_g/2q - \psi_B$. Therefore, $V_{fb}$ and $\psi_s$ in Eq. (4.86) nearly cancel each other and one obtains $V_g \approx -Q_{ac}/C_{ox}$. The sheet resistivity of the accumulation layer is then

$$\rho_{ac} = \frac{1}{\mu_{ac}|Q_{ac}|} = \frac{1}{\mu_{ac} C_{ox} V_g}, \tag{4.87}$$

where $\mu_{ac}$ is the average electron mobility in the accumulation layer. If, for process reasons (gate reoxidation), the oxide thickness in the gate to source–drain overlap region is different from $t_{ox}$ in the channel region, a different $C_{ox}$ should be used in Eq. (4.87). The electron mobility in the accumulation layer has a similar field dependence to that in the inversion layer (Sun and Plummer, 1980). From Gauss's law, the average electric field in the accumulation layer is $\mathscr{E}_{eff} = |Q_{ac}|/2\varepsilon_{si} = C_{ox}V_g/2\varepsilon_{si}$. Knowing $\mathscr{E}_{eff}$, one can look up $\mu_{ac}$ from Fig. 3.13. Like the channel mobility, $\mu_{ac}$ is not limited by the impurity scattering which usually dominates the bulk mobility (Fig. 2.7) for the moderately high doping levels ($N_d \approx 10^{18}$ cm$^{-3}$) at the surface. This is because of the screening of Coulomb scattering when the carrier concentration in the accumulation layer greatly exceeds the donor concentration at that point.

#### 4.3.3.3 INTERPRETATION OF $L_{eff}$ IN TERMS OF CURRENT INJECTION POINTS

The dependence of $\rho_{ac}$ on $V_g$ in Eq. (4.87) is too similar to that of $\rho_{ch}$ in Eq. (3.101) to allow separation of the accumulation-layer resistance from the channel resistance.

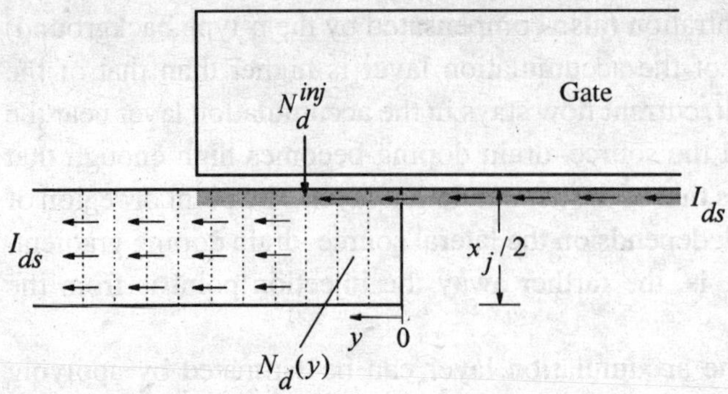


**FIGURE 4.28.** Schematic diagram of a simple 1-D model for estimating the source–drain doping concentration at the point of current injection from the surface into the bulk. As far as the resistance is concerned, the source or drain region is modeled as uniform stripes of doping concentration $N_d(y)$ and width $x_j/2$.

***The region where the current flows predominantly in the accumulation layer is therefore considered as a part of $L_{eff}$.*** To estimate the source–drain doping concentration at which current injection into the bulk takes place, we consider a simple 1-D current model shown in Fig. 4.28. The donor concentration at the surface is assumed to be $N_d(y)$, which varies along the direction of current flow as dictated by the lateral doping gradient. If the bulk resistivity corresponding to $N_d$ is $\rho(N_d)$ and the source–drain junction depth is $x_j$, the average sheet resistivity of a thin stripe perpendicular to the current flow is approximately $\rho(N_d)/(x_j/2)$. Here the effective depth of uniform current flow is taken as $x_j/2$, since for any given stripe the n-type conductivity is highest at the surface and drops to zero at $x_j$. As $N_d$ increases toward the heavily doped source–drain region, $\rho(N_d)/(x_j/2)$ decreases accordingly. Current injection into the bulk takes place at $N_d = N_d^{inj}$, where the sheet resistivity $\rho(N_d^{inj})/(x_j/2)$ equals $\rho_{ac}$ of Eq. (4.87). Since $\rho_{ac}$ depends on the gate voltage, so does the point of injection (Hu *et al.*, 1987). At low gate overdrives, the injection point is closer to the metallurgical junction edge. As the gate voltage increases, the injection point moves out toward the more heavily doped source–drain region. For reasonably abrupt source–drain doping profiles, $N_d(y)$ is a strong (exponential) function of $y$ and the injection points do not spread too far apart. If a $V_g$-range from slightly above $V_t$ to $V_{dd}$ is used in the S&R algorithm, the resulting $L_{eff}$ is, to the first-order approximation, given by the injection point corresponding to $V_g \approx V_{dd}/2$ in the middle of the $V_g$-range. Setting $\rho(N_d^{inj})/(x_j/2)$ equal to $\rho_{ac}$ of Eq. (4.87) with $V_g = V_{dd}/2$, one can then find $N_d^{inj}$ from

$$\rho(N_d^{inj}) = \frac{x_j}{\mu_{ac} C_{ox} V_{dd}}, \tag{4.88}$$

where $\mu_{ac}$ is evaluated at an average normal field of $\mathscr{E}_{eff} = C_{ox}V_{dd}/4\varepsilon_{si}$.

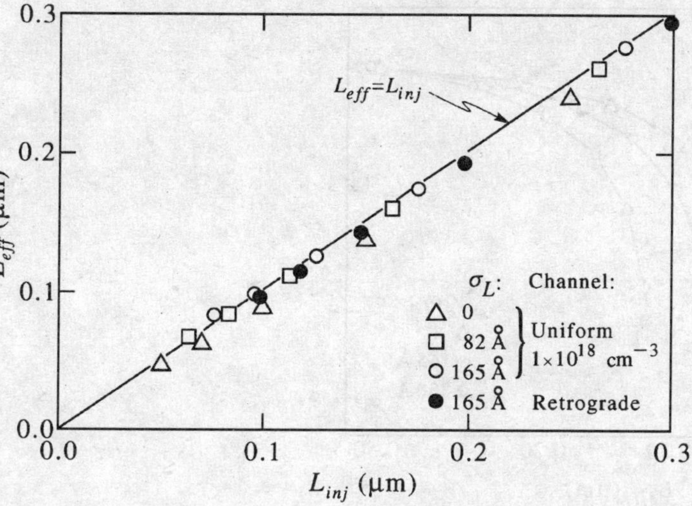


**FIGURE 4.29.** Same $L_{eff}$ data as in Fig. 4.25, but plotted versus $L_{inj}$ in terms of current injection points. Open symbols are for a uniformly doped channel with different lateral source–drain gradients. Solid dots are for a retrograde-doped channel.

In general, $N_d^{inj}$ increases as the device dimensions are scaled down. For 1-μm CMOS technology, $N_d^{inj} \approx 10^{17}$ cm$^{-3}$. For the 0.1-μm devices in Fig. 4.25, with $V_{dd} = 1.5$ V, $t_{ox} = 30$ Å, $x_j = 500$ Å, one obtains $\mathscr{E}_{eff} = 4.2 \times 10^5$ V/cm, $\mu_{ac} =$ 420 cm$^2$/V-s (Fig. 3.13), and $\rho(N_d^{inj}) = 0.007$ Ω-cm. The last number corresponds to $N_d^{inj} \approx 8 \times 10^{18}$ cm$^{-3}$ from the n-type resistivity curve in Fig. 2.8. Figure 4.29 replots the simulated $L_{eff}$ data in Fig. 4.25 against $L_{inj}$, defined as the distance between the points where the source–drain doping equals $N_d^{inj}$. All the points lie within 100 Å of the $L_{eff} = L_{inj}$ line, independent of the lateral doping gradient and channel profile. This supports the physical interpretation of $L_{eff}$ in terms of the current injection points from the accumulation layer.

#### 4.3.3.4 IMPLICATIONS FOR SHORT-CHANNEL EFFECTS

The fact that $L_{eff}$ can be much longer than $L_{met}$ has significant implications for the short-channel $V_t$ roll-off curves. Figure 4.30 shows the low-drain threshold voltage roll-off versus $L_{eff}$ for several different source–drain doping gradients. The abrupt doping profile has the best short-channel effect. As the lateral straggle increases, the short-channel effect becomes progressively worse. This can be understood from Fig. 4.31, where the net doping concentration $N_d - N_a$ at the surface is plotted along the channel length direction for the various source–drain profiles studied. For a given $L_{eff}$ or $L_{inj}$ (=0.1 μm), the distance between the points where the doping concentration falls to $N_d^{inj}$ (=$8 \times 10^{18}$ cm$^{-3}$ in this case) is fixed. It is clear that the more graded the source–drain profile is, the deeper the n-type doping tail penetrates into the channel and compensates or reverses the p-type doping inside

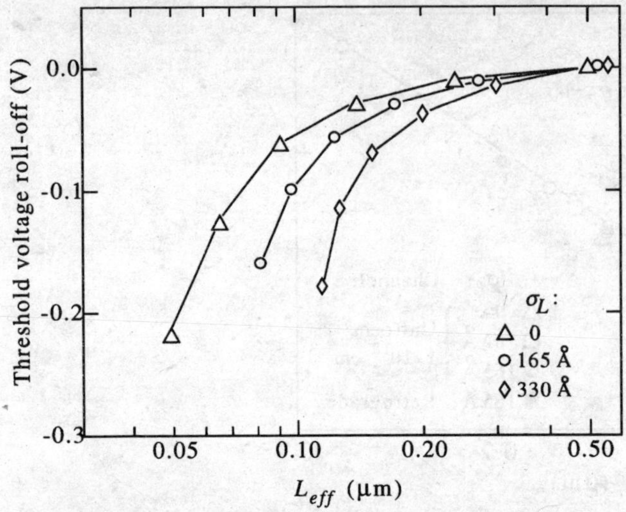


**FIGURE 4.30.** Simulated short-channel threshold roll-off versus $L_{eff}$ for three different lateral source–drain doping gradients. On each curve, the points are for $L_{met} =$ 0.05, 0.07, 0.10, 0.15, 0.25, and 0.50 μm. (After Taur *et al.*, 1995b.)

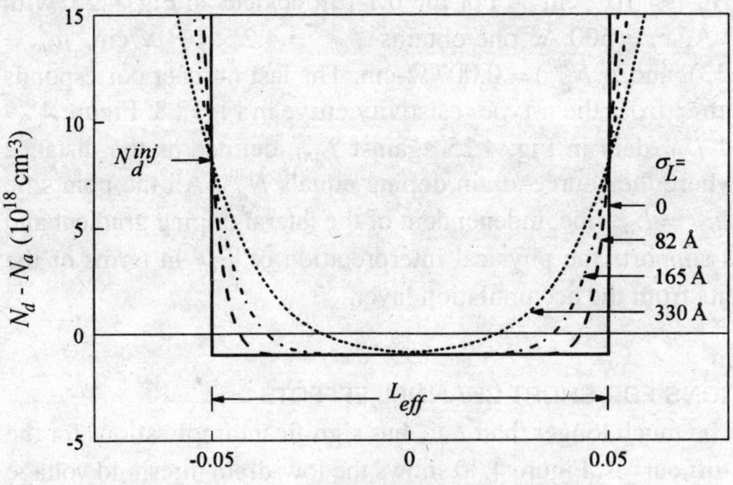


**FIGURE 4.31.** Net n-type concentration versus distance from source to drain at the surface of a 0.1-μm ($L_{inj}$ or $L_{eff}$) nMOSFET. The injection points ($N_d^{inj}$) are kept the same for different lateral doping gradients.

the channel. This is detrimental to the short-channel effect, as the edge regions become more easily depleted and inverted by the source–drain fields (opposite to the halo effect). ***It is therefore very important to reduce the width of the (laterally) graded source–drain region as the channel length is scaled down.***

#### 4.3.3.5 EFFECTIVE CHANNEL LENGTH OF LDD DEVICES

Lightly-doped-drain (LDD) MOSFETs (Ogura *et al.*, 1982) are designed with an extended moderately doped ($10^{17}$–$10^{18}$-$cm^{-3}$ range) source–drain region to relieve high electric fields and related hot-electron effects. The presence of such a lightly doped region poses great difficulties in the interpretation of effective channel length. Current injection from the surface accumulation layer into the bulk region takes place over an extended distance with a location dependent on gate voltage. No matter what kind of $L_{eff}$ extraction method is used, $L_{eff}$ is significantly longer than $L_{met}$, especially at high gate overdrives (Sun *et al.*, 1986).

In one approach, channel resistance extraction (Fig. 4.21) is carried out for a pair of closely spaced $V_g$-values (Hu *et al.*, 1987). The procedure is repeated throughout the entire voltage range, allowing both $L_{eff}$ and $R_{sd}$ to be gate-voltage-dependent. Similar methods can be implemented in the S&R algorithm as well by breaking up the full voltage range used in Eq. (4.77) into small intervals of 0.5 or 0.25 V each. A set of $(L_{eff}, R_{sd})$ can be extracted from the current data in each interval by repeating the S&R algorithm. However, most circuit models do not accommodate gate-voltage-dependent channel lengths. Furthermore, once $L_{eff}$ and $R_{sd}$ are allowed to be gate-voltage-dependent, there is no unique way of breaking up $R_{ch}$ and $R_{sd}$ from the measured $R_{tot}$, and the solutions to Eq. (4.72) or Eq. (4.73) become ambiguous and method-dependent.

There is no consensus on how to define the $L_{eff}$ of an LDD MOSFET. If the entire $I_{ds}$–$V_g$ characteristics can be fitted within acceptable tolerances by a constant $L_{eff}$ and a constant $R_{sd}$ based on Eq. (4.73), a circuit model can be formulated with these parameters. Otherwise, one needs to define a constant $L_{eff}$, perhaps using the figure extracted from the current data at low gate overdrives (Sun *et al.*, 1986), and leave the rest of the resistance as a gate-voltage-dependent series resistance.

#### 4.3.3.6 EXTRACTION OF CHANNEL LENGTH BY *C–V* MEASUREMENTS

In an entirely different approach, another type of channel length has been extracted from the measured *C–V* data of a series of MOSFETs with different $L_{mask}$ (Sheu and Ko, 1984). Capacitance measurements in general are more difficult to perform, as they require specially designed test sites. It is by no means straightforward to interpret the capacitively measured channel length and apply it to circuit models for current calculations.

The capacitive extraction of channel length is based on the fact that when a MOSFET is turned on, the intrinsic gate-to-channel capacitance is proportional to the channel length:

$$C_{gc} = C_{ox} W L_{cap} \tag{4.89}$$

Here $L_{cap}$ is the capacitively defined channel length, which may or may not be

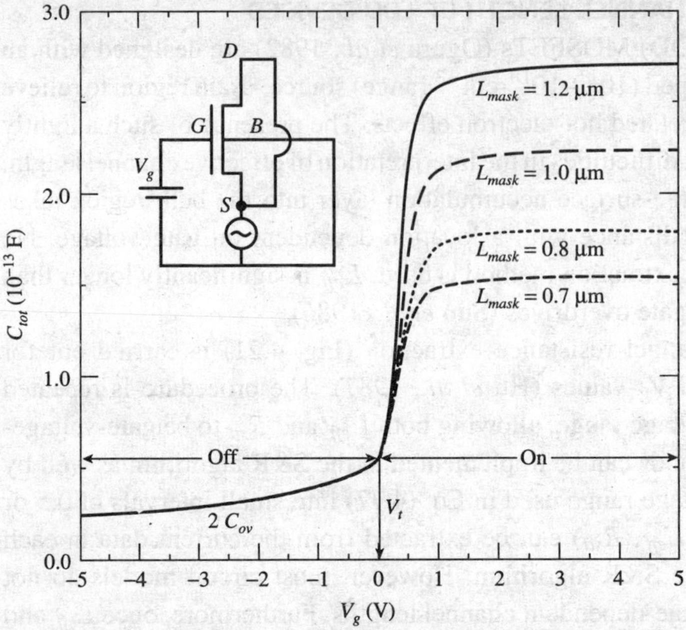


**FIGURE 4.32.** Example of measured capacitance from gate to source–drain versus gate voltage for MOSFETs of different mask lengths. The inset shows the split *C–V* measurement setup. (After Guo *et al.*, 1994.)

the same as $L_{eff}$ or $L_{met}$. The gate-to-channel capacitance is usually measured in a split *C–V* setup that separates the majority-carrier response from the minority-carrier response, as shown in the inset of Fig. 4.32. The total measured capacitance consists of both the intrinsic gate-to-channel capacitance and a parasitic overlap capacitance from the gate to source–drain which is independent of channel length:

$$C_{tot} = C_{gc} + 2C_{ov} = C_{ox}WL_{cap} + 2C_{ov}. \tag{4.90}$$

Here $C_{ov}$ is the overlap capacitance per gate edge (see Fig. 5.16). Typical examples of $C_{tot}$–$V_g$ curves are shown in Fig. 4.32 (Guo *et al.*, 1994). Using a large-area MOS capacitor, one can easily calibrate $C_{ox}$, taking all the polysilicon depletion and inversion–layer quantum effects into account. To find out $L_{cap}$, it is critical to determine what $2C_{ov}$ to subtract from the measured $C_{tot}$. In principle, $2C_{ov}$ in Eq. (4.90) is the parasitic capacitance at a gate voltage when the MOSFET is on and $C_{gc}$ is given by Eq. (4.89). In practice, $2C_{ov}$ cannot be separated from $C_{gc}$, since, unlike channel resistance, channel capacitance does not vary significantly with gate voltage once the device is turned on. What is usually done is to take $2C_{ov}$ as the measured capacitance when the MOSFET is off. However, from Fig. 4.32 it is clear that $2C_{ov}$ varies with the gate voltage (Oh *et al.*, 1990). There is no guarantee that $2C_{ov}$ in the off state is the same as $2C_{ov}$ in the on state. If $2C_{ov}$ is taken as the capacitance right below the threshold voltage, it will contain an unwanted inner-fringe term that is absent when the conducting channel is formed. If $2C_{ov}$ is taken at a negative gate voltage where the substrate is accumulated to eliminate the inner-fringe component, the lightly doped source–drain in the direct overlap region will be depleted (Sheu and Ko, 1984). Any such error in $2C_{ov}$ translates into a large error in $L_{cap}$ when dealing with short-channel devices having small intrinsic capacitances.

A better interpretation of the capacitively extracted channel length is in terms of the gate length, $L_{gate}$ (Fig. 4.19), since as the gate voltage varies, the same amount of charge per unit area is induced at the silicon surface whether it is in the inversion channel or in the source–drain overlap region under the gate. In other words, as far as the capacitance is concerned, the direct overlap length should be lumped into the channel length. This also circumvents the problem with the inner-fringe component mentioned above. One still needs to estimate the outer fringe capacitance and subtract it from the measured capacitance. But this can be done using a simple formula (Section 5.2.2) and therefore should have less error associated with it.

## EXERCISES

**4.1** Apply constant-field scaling rules to the long-channel currents [Eq. (3.19) for the linear region and Eq. (3.23) for the saturation region], and show that they behave as indicated in Table 4.1.

**4.2** Apply constant-field scaling rules to the subthreshold current, Eq. (3.36), and show that instead of decreasing with scaling ($1/\kappa$), it actually increases with scaling (note that $V_g < V_t$ in subthreshold). What if the temperature is also scaled down by the same factor ($T \to T/\kappa$)?

**4.3** Apply constant-field scaling rules to the saturation current from the $n = 1$ velocity saturation model [Eq. (3.78)] and the fully saturation-velocity limited current [Eq. (3.80)], and show that they behave as indicated in Table 4.1.

**4.4** Apply generalized scaling rules to the saturation current from the $n = 1$ velocity saturation model [Eq. (3.78)], and show that it behaves as indicated in Table 4.3 (between the two limits).

**4.5** In Eqs. (4.59) and (4.60), how should $x_{av}$ be defined in terms of $n(x)$, the electron volume concentration as a function of depth? The same definition applies to $x_j$ in Eq. (4.50) and $x_{av}^{QM}$ in Eq. (4.58).

**4.6** *Nonuniform $V_t$ in the width direction.* A MOSFET is nonuniformly doped in the width direction. Part of the width ($W_1$) has a linear threshold voltage $V_{on1}$. The other part of the width ($W_2$) has a linear threshold voltage $V_{on2}$. Show that as far as the linear region characteristics [Eq. (3.61)] are concerned, this device is equivalent to a uniform MOSFET of width $W_1 + W_2$ with a linear threshold voltage $V_{on} = (W_1 V_{on1} + W_2 V_{on2})/(W_1 + W_2)$. Ignore any fringing fields that may exist near the boundary between the two regions.

**4.7** *Nonuniform $V_t$ in the length direction.* A MOSFET is nonuniformly doped in the length direction. Part of the length ($L_1$) has a linear threshold voltage $V_{on1}$. The other part of the length ($L_2$) has a linear threshold voltage $V_{on2}$. Assume $V_{on1} \approx V_{on2}$, and consider only the first-order terms of $V_{on1} - V_{on2}$. Show that as far as the linear region characteristics [Eq. (3.61)] are concerned, this device is equivalent to a uniform MOSFET of length $L_1 + L_2$ with a linear threshold voltage $V_{on} = (L_1 V_{on1} + L_2 V_{on2})/(L_1 + L_2)$. Ignore any fringing fields that may exist near the boundary between the two regions.

**4.8** In the top equivalent circuit of Fig. 4.20, the source–drain current can be considered either as a function of the internal voltages: $I_{ds}(V_g', V_{ds}')$, or as a function of the external voltages: $I_{ds}(V_g, V_{ds})$. The internal voltages are related to the external voltages by Eqs. (4.68) and (4.69). Show that the transconductance of the intrinsic MOSFET can be expressed as

$$g_m' \equiv \left(\frac{\partial I_{ds}}{\partial V_g'}\right)_{V_{ds}'} = \frac{g_m}{1 - g_m R_s - g_{ds}(R_s + R_d)},$$

where

$$g_m \equiv \left(\frac{\partial I_{ds}}{\partial V_g}\right)_{V_{ds}}$$

is the extrinsic transconductance, and

$$g_{ds} \equiv \left(\frac{\partial I_{ds}}{\partial V_{ds}}\right)_{V_g}$$

is the extrinsic output conductance.

**4.9** Show that in the subthreshold region and when the drain bias is low, Eq. (3.12) leads to Eq. (4.54):

$$Q_i = \frac{kTn_i^2}{\mathscr{E}_s N_a} e^{q\psi_s/kT},$$

where $\psi_s$ is the surface potential and $\mathscr{E}_s$ is the surface electric field. This equation is more general than Eq. (3.31) since it is valid for nonuniform (vertically) dopings with $N_a$ being the p-type concentration at the edge of the depletion layer. (Note that the factor $N_a$ merely reflects the fact in Fig. 2.24 that the band bending $\psi_s$ is defined with respect to the bands of the neutral bulk region of doping $N_a$.)

**4.10** In a short-channel device or in a nonuniformly doped (laterally) MOSFET, $\psi_s$ may vary along the channel length direction from the source to drain. Generalize the expression in Exercise 4.9 and show that

$$\frac{V_{ds}}{I_{ds}} = \frac{1}{\mu_{eff} W}\int_0^L \frac{dy}{Q_i(y)} = \frac{N_a}{\mu_{eff} WkTn_i^2}\int_0^L \mathscr{E}_s(y)e^{-q\psi_s(y)/kT}\,dy$$

for the subthreshold region at low drain biases. Since $\mathscr{E}_s(y) \approx [V_g - V_{fb} - \psi_s(y)]/3t_{ox}$ is not a strong function of $\psi_s$, the exponential factor dominates. This implies that the subthreshold current is controlled by the point of highest barrier (lowest $\psi_s$) in the channel. It also implies that the *channel length* factor entering the subthreshold current expression is different from the *effective channel length* defined by the linear region characteristics, Eq. (4.66).

# 5 CMOS PERFORMANCE FACTORS

The performance of a CMOS VLSI chip is measured by its integration density, switching speed, and power dissipation. CMOS circuits have the unique characteristic of practically zero standby power, which enables higher integration levels and makes them the technology of choice for most VLSI applications. This chapter examines the various factors that determine the switching speed of basic CMOS circuit elements.

## 5.1 BASIC CMOS CIRCUIT ELEMENTS

In a modern CMOS VLSI chip, the most important function components are CMOS static gates. In gate array circuits, CMOS static gates are used almost exclusively. In microprocessors and supporting circuits of memory chips, most of the control interface logic is implemented using CMOS static gates. Static logic gates are the most widely used CMOS circuit because of their simplicity and noise immunity. This section describes basic static CMOS circuit elements and their switching characteristics.

Circuit symbols for nMOSFETs and pMOSFETs are defined in Fig. 5.1. A MOSFET is a four-terminal device, although usually only three are shown. Unless specified, the body (p-substrate) terminal of an nMOSFET is connected to the ground (lowest voltage), while the body terminal (n-well) of a pMOSFET is connected to the power supply $V_{dd}$ (highest voltage).

### 5.1.1 CMOS INVERTERS

The most basic element of digital static CMOS circuits is a CMOS inverter. A CMOS inverter is a combination of an nMOSFET and a pMOSFET, as shown in Fig. 5.2 (Burns, 1964). The source terminal of the nMOSFET is connected to the ground, while the source of the pMOSFET is connected to $V_{dd}$. The gates of the two MOSFETs are tied together as the input node. The two drains are tied together as the output node. In such an arrangement, the complementary nature of n- and pMOSFETs allows one and only one transistor to be conducting in one of the two

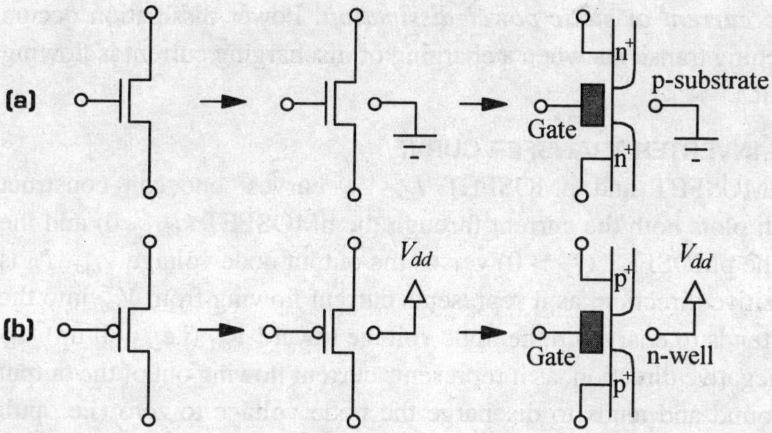


**FIGURE 5.1.** Circuit symbols and voltage terminals of (a) nMOSFET and (b) pMOSFET.

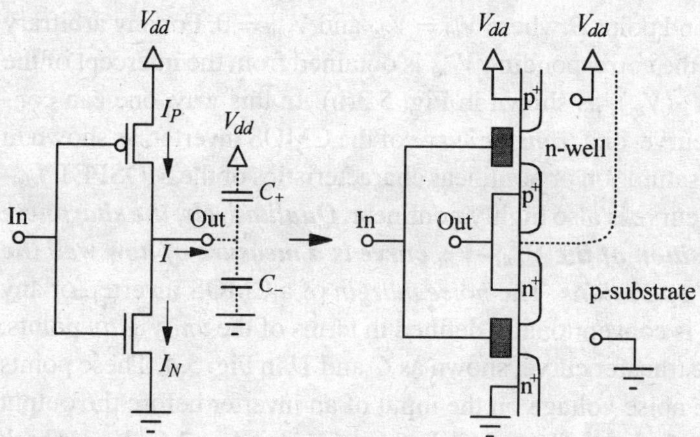


**FIGURE 5.2.** Circuit diagram and schematic cross section of a CMOS inverter.

stable states. For example, when the input voltage is high or when $V_{in} = V_{dd}$, the gate-to-source voltage of the nMOSFET equals $V_{dd}$, which turns it on. At the same time, the gate-to-source voltage of the pMOSFET is zero, so the pMOSFET is off. The output node is then pulled down to the ground potential by currents through the conducting nMOSFET, which is referred to as the *pull-down* transistor. On the other hand, when the input voltage is low or when $V_{in} = 0$, the nMOSFET is off, since its gate-to-source voltage is zero. The gate-to-source voltage of the pMOSFET, however, is $-V_{dd}$, which turns it on (a negative gate voltage turns on a pMOSFET). The output node is now pulled up to $V_{dd}$ by the conducting pMOSFET, which is referred to as the *pull-up* transistor. Since the output voltage is always opposite to the input voltage ($V_{out}$ is high when $V_{in}$ is low and vice versa), this circuit is called an inverter. Notice that ***since only one of the transistors is on in the steady state,***

***there is no static current or static power dissipation***. Power dissipation occurs only during switching transients when a charging or discharging current is flowing through the circuit.

#### 5.1.1.1 CMOS INVERTER TRANSFER CURVE

Based on the nMOSFET and pMOSFET $I_{ds}$–$V_{ds}$ curves, one can construct Fig. 5.3(a), which plots both the current through the nMOSFET ($I_N > 0$) and the current through the pMOSFET ($I_P > 0$) versus the output node voltage $V_{out}$. $I_P$ is plotted in the positive direction, as it represents current flowing from $V_{dd}$ into the output node and tends to charge up the node voltage toward $V_{dd}$ (i.e., pull up). $I_N$ is plotted in the negative direction, as it represents current flowing out of the output node into the ground and tends to discharge the node voltage to zero (i.e., pull down). The net current is given by $I = I_P - I_N$. The output node voltage increases or decreases depending on whether $I > 0$ or $I < 0$. The directions of the currents are depicted in Fig. 5.2.

In the steady state, $I = 0$, there are two stable points of operation: point $A$ where $V_{in} = 0$ and $V_{out} = V_{dd}$, and point $B$ where $V_{in} = V_{dd}$ and $V_{out} = 0$. For any arbitrary value of $V_{in}$ in between, the corresponding $V_{out}$ is obtained from the intercept of the two curves, $I_N(V_{in}) = I_P(V_{in})$, as shown in Fig. 5.3(b). In this way, one can construct a $V_{out}$-versus-$V_{in}$ curve, or a *transfer curve* of the CMOS inverter, as shown in Fig. 5.4. Because of the saturation or nonlinear characteristics of the MOSFET $I_{ds}$–$V_{ds}$ curves, the $V_{out}$–$V_{in}$ curve is also highly nonlinear. ***Qualitatively, the sharpness of the high-to-low transition of the $V_{out}$–$V_{in}$ curve is a measure of how well the circuit performs digital operations***. The *noise margin* of a CMOS inverter, or any digital inverting circuit, is conventionally defined in terms of the *unity-gain* points, $dV_{out}/dV_{in} = -1$, in the transfer curve, shown as $C$ and $D$ in Fig. 5.4. These points determine the allowable noise voltage on the input of an inverter before the output is affected. For example, for any value of $V_{in}$ between points $A$ and $C$, $V_{out}$ is high and the nMOSFET is biased in saturation while the pMOSFET is biased in the linear region, as can be seen from Fig. 5.3(b). For $V_{in}$ between $B$ and $D$, $V_{out}$ is low and the nMOSFET is in the linear region while the pMOSFET is in saturation. Between $C$ and $D$, both the nMOSFET and the pMOSFET are biased in saturation. This is an unstable region where the small-signal voltage gain, $|dV_{out}/dV_{in}|$, is exceedingly high. Ideally, the noise margin is high when both $\Delta V_{in}(A \rightarrow C)$ and $\Delta V_{in}(B \rightarrow D)$ are close to $V_{dd}/2$.

In order for the high-to-low transition of the transfer curve to occur close to the midpoint, $V_{in} = V_{dd}/2$, it is desired for $I_P$ and $I_N$ to be nearly symmetrical, as illustrated in the example in Fig. 5.3(b). However, nMOSFET and pMOSFET are not exactly symmetric devices. Due to the higher electron mobility, an nMOSFET inherently has a higher drive current than a pMOSFET of the same width. To compensate for the difference, a CMOS inverter is usually designed with a wider pMOSFET than nMOSFET. In other words, the device width ($W_n$, $W_p$) should be

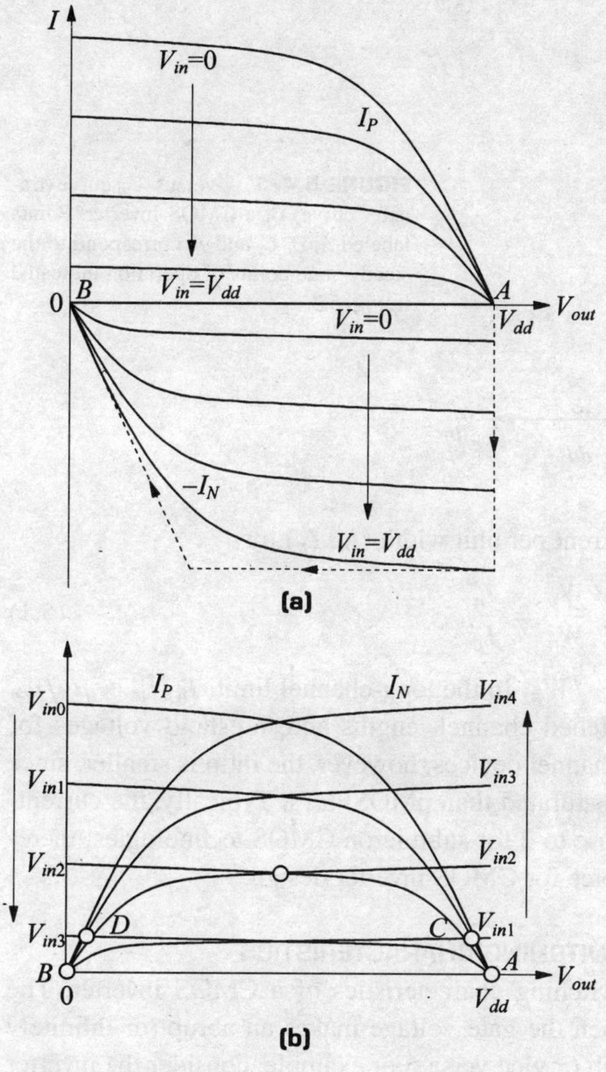


**FIGURE 5.3.** nMOSFET current ($I_N$) and pMOSFET current ($I_P$) in a CMOS inverter versus output node (drain) voltage for a series of input node (gate) voltages from 0 to $V_{dd}$. (a) $I_P$ is plotted in the positive direction for pull-up and $I_N$ in the negative direction for pull-down. The dashed lines depict an approximate bias point trajectory of a pull-down transition from $A$ to $B$ following an abrupt switching of $V_{in}$ from 0 to $V_{dd}$. This approximate trajectory is used later in the text for estimating the switching delay. (b) $I_P$ and $I_N$ are both plotted in the positive direction in order to find the steady-state points of operation (circles) given by the intercepts where $I_P = I_N$ for the same $V_{in}$ and $V_{out}$. The curves are labeled by the input voltages: $0 = V_{in0} < V_{in1} < V_{in2} < V_{in3} < V_{in4} = V_{dd}$.

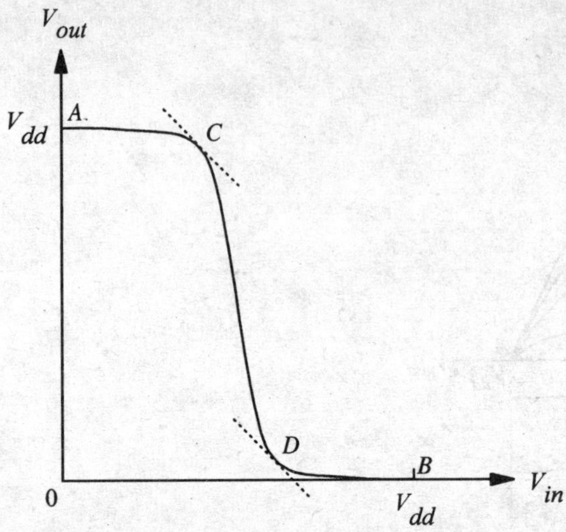


**FIGURE 5.4.** $V_{out}$-versus-$V_{in}$ curve (transfer curve) of a CMOS inverter. Points labeled *A, B, C,* and *D* correspond to the steady-state points of operation indicated in Fig. 5.3.

inversely proportional to the current per unit width ($I_n$, $I_p$), or

$$\frac{W_p}{W_n} = \frac{I_n}{I_p}, \tag{5.1}$$

where $I_n = I_N/W_n$ and $I_p = I_P/W_p$. In the long-channel limit, $I_n/I_p \propto \mu_n/\mu_p$ from Eq. (3.23), assuming matched channel lengths and threshold voltages for n- and pMOSFETs. For short-channel devices, however, the ratio is smaller, since nMOSFETs are more velocity-saturated than pMOSFETs. Typically, the current-per-unit-width ratio $I_n/I_p$ is close to 2 for submicron CMOS technologies; therefore, $W_p/W_n = 2$ is a good choice for CMOS inverter design.

#### 5.1.1.2 CMOS INVERTER SWITCHING CHARACTERISTICS

We now consider the basic switching characteristics of a CMOS inverter. The simplest input waveform is when the gate voltage makes an abrupt or infinitely sharp transition from low to high or vice versa. For example, consider the inverter biased at point *A* in Fig. 5.3(a) when $V_{in}$ makes a step transition from 0 to $V_{dd}$. Before the transition, the nMOSFET is off and the pMOSFET is on. After the transition, the nMOSFET is on and the pMOSFET is off. The trajectory of $V_{out}$ from point *A* to point *B* follows the $V_{in} = V_{dd}$ curve of the nMOSFET as shown in Fig. 5.3(a). If the total capacitance of the output node (including both the output capacitance of the switching inverter and the input capacitance of the next stage or stages it drives) is represented by two capacitors – one ($C_-$) to the ground and one ($C_+$) to the $V_{dd}$ rail, as illustrated in Fig. 5.2 – then the pull-down switching characteristics are described by

$$C_- \frac{d(V_{out} - 0)}{dt} + C_+ \frac{d(V_{out} - V_{dd})}{dt} = -I_N(V_{in} = V_{dd}),$$

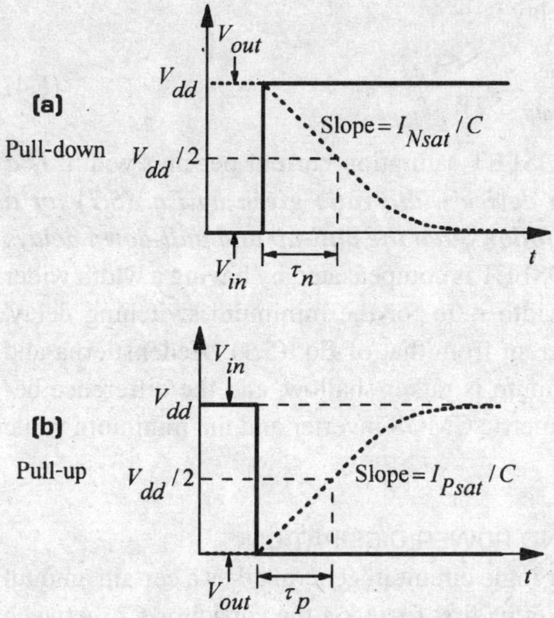


**FIGURE 5.5.** Waveforms of the output node voltage (dotted) of a CMOS inverter. (a) Pull-down transition after an abrupt rise of input voltage (solid). (b) Pull-up transition after an abrupt fall of input voltage (solid).

or

$$(C_- + C_+)\frac{dV_{out}}{dt} = C\frac{dV_{out}}{dt} = -I_N(V_{in} = V_{dd}), \tag{5.2}$$

with the initial condition $V_{out}(t=0) = V_{dd}$. Here $C = C_- + C_+$ includes both the capacitance to ground and the capacitance to $V_{dd}$. For simplicity, we approximate the $I_N(V_{in} = V_{dd})$ curve by two piecewise continuous lines. In the saturation region ($V_{out} > V_{dsat}$), $I_N = I_{Nsat}$ is a constant. In the linear region ($V_{out} < V_{dsat}$), $I_N = (I_{Nsat}/V_{dsat})V_{out}$, like a resistor with a resistance $V_{dsat}/I_{Nsat}$. These are shown as dashed lines in Fig. 5.3(a). The solution $V_{out}(t)$ is depicted in Fig. 5.5(a). Right after $V_{in}$ switches from 0 to $V_{dd}$, $V_{out}$ decreases linearly with time at a rate given by $I_{Nsat}/C$ until $V_{out} = V_{dsat}$ is reached. Below that, $V_{out}$ decreases exponentially toward zero with a time constant $CV_{dsat}/I_{Nsat}$. Similar switching characteristics are shown in Fig. 5.5(b) for pMOSFET pull-up when $V_{in}$ switches abruptly from $V_{dd}$ to 0 [point $B$ to point $A$ in Fig. 5.3(a)]. In this case, $V_{out}$ follows the $V_{in} = 0$ curve of the pMOSFET, and the initial rate of increase with time is given by $I_{Psat}/C$.

While it takes a significantly longer time for the output voltage to approach zero, it is conventional to define an nMOSFET pull-down delay $\tau_n$ as the time it takes for the output node voltage to reach $V_{dd}/2$. From Fig. 5.5(a), it is clear that

$$\tau_n = \frac{CV_{dd}}{2I_{Nsat}} = \frac{CV_{dd}}{2W_n I_{nsat}}, \tag{5.3}$$

where $I_{nsat} \equiv I_{Nsat}/W_n$ is the nMOSFET saturation current per unit width.

Similarly, the pMOSFET pull-up delay is

$$\tau_p = \frac{CV_{dd}}{2I_{Psat}} = \frac{CV_{dd}}{2W_p I_{psat}}, \tag{5.4}$$

where $I_{psat} \equiv I_{Psat}/W_p$ is the pMOSFET saturation current per unit width. ***If a CMOS inverter is designed with a device width ratio given by Eq. (5.1) for a symmetrical transfer curve, it also follows that the pull-up and pull-down delays are equal.*** The less conductive pMOSFET is compensated by having a width wider than that of the nMOSFET. The width ratio for the minimum switching delay, $\tau = (\tau_n + \tau_p)/2$, is generally different from that of Eq. (5.1) (Hedenstierna and Jeppson, 1987). However, the minimum is rather shallow, and the difference between the switching delay of a symmetric CMOS inverter and the minimum value is usually no more than 5%.

#### 5.1.1.3 SWITCHING ENERGY AND POWER DISSIPATION

Switching a CMOS inverter or other logic circuit in general takes a certain amount of energy from the power supply. Let us first focus on the capacitor $C_-$ between the output node and ground in Fig. 5.2. During the pull-up transition of a CMOS inverter, the charge on $C_-$ changes from zero to $\Delta Q_- = C_- V_{dd}$. This means that there is an energy of $V_{dd}\Delta Q_- = C_- V_{dd}^2$ flowing out of the power supply in the pull-up transition. Half of this energy, or $C_- V_{dd}^2/2$ is dissipated by the charging current in the pMOSFET resistance. The other half, another $C_- V_{dd}^2/2$, is stored in the capacitor $C_-$. This energy stays in the capacitor until the next pull-down takes place. Then the charge on $C_-$ drops to zero and the stored energy is dissipated by the discharging current through the nMOSFET resistance. Likewise, for the capacitor $C_+$ between the output node and $V_{dd}$ in Fig. 5.2, an amount of energy $C_+ V_{dd}^2$ is supplied by the power source during the pull-down transition; half of which is dissipated by the discharging current in the nMOSFET resistance, while the other half $(C_+ V_{dd}^2/2)$ is stored in the capacitor $C_+$. The stored energy is later dissipated in the pMOSFET resistance during the next pull-up transition.

From the above discussion, it is clear that for any capacitor $C$ (either to the ground or to $V_{dd}$) to be charged or discharged, an energy of $CV_{dd}^2/2$ is dissipated irreversibly. It is often conventional to consider a complete cycle consisting of a pair of transitions, either up–down ($0 \to V_{dd} \to 0$) or down–up ($V_{dd} \to 0 \to V_{dd}$). In that case, we say an energy of $CV_{dd}^2$ is dissipated per cycle. (An exception is Miller capacitances between two switching nodes, to be discussed in Section 5.3.4.)

Since dc power dissipation is negligible in CMOS circuits, the only power consumption comes from switching. (Standby power dissipation of low-$V_t$ devices is discussed in Section 5.3.3.) While the peak power dissipation in a CMOS inverter can reach $V_{dd}I_{Nsat}$ or $V_{dd}I_{Psat}$, the average power dissipation depends on how often it switches. In a CMOS processor, the switching of logic gates is controlled by a clock generator of frequency $f$. If on the average a total equivalent capacitance $C$ is charged and discharged within a clock cycle of period $T = 1/f$, the average power dissipation is

$$P = \frac{CV_{dd}^2}{T} = CV_{dd}^2 f. \tag{5.5}$$

Note here that each up or down transition of a capacitor within the period $T$ contributes half of that capacitance to $C$. If, for example, a capacitor is switched four times (goes through the up–down cycle twice) within the clock period, its capacitance is counted twice in $C$. Equation (5.5) will be used in the discussion of the power–delay tradeoff in Section 5.3.

The above simplified delay and power analysis assumes abrupt switching of $V_{in}$. In general, $V_{in}$ is fed from a previous logic stage and has a finite rise or fall time associated with it. The switching trajectory from $A$ to $B$ or from $B$ to $A$ in Fig. 5.3(a) then becomes much more complicated. Instead of staying on one constant-$V_{in}$ curve, the bias point moves through different curves as $V_{in}$ ramps up or down. Furthermore, both $I_N$ and $I_P$ must be considered during either a pull-up or a pull-down transition, since the other transistor is not switched off completely as one transistor is turned on. This also means that there is a *crossover*, or *short-circuit*, current that flows momentarily between the power-supply terminal and the ground in a switching event, which adds another power dissipation component to Eq. (5.5). One last complication is that the output node capacitances $C_-$ and $C_+$ are generally voltage-dependent rather than being constant as assumed above. More extensive numerical analysis of the general case will be given in Section 5.3.

#### 5.1.1.4 QUASISTATIC ASSUMPTION

In the above discussion of CMOS switching characteristics, it was implicitly assumed that the device response time, i.e., the time required for charge redistribution, is fast compared with the rate the terminal voltage is changed. This is called the *quasistatic* assumption. In other words, the device current responds instantaneously to an external voltage change. This assumption is valid if the input rise or fall time is much longer than the carrier transit time across the channel. Since the average carrier drift velocity is given by $\mu_{eff} V_{dd}/L$ ($V_{dd}/L$ is the average field along the channel) until it is limited by the saturation velocity $v_{sat}$, the transit time is of the order of $L^2/\mu_{eff} V_{dd}$ or $L/v_{sat}$, whichever is greater. For short-channel devices used in digital circuits, the transit time approaches $L/v_{sat}$, which is of the order of 10 ps for 1-μm MOSFETs and 1 ps for 0.1-μm MOSFETs. These numbers are at least an order of magnitude shorter than the delay of an unloaded CMOS inverter made in the corresponding technology (Taur *et al.*, 1985, 1993c). This indicates that the switching time is limited by the parasitic capacitances rather than by the time required for charge redistribution within the transistor itself and thus validates the quasistatic approach.

### 5.1.2 CMOS NAND AND NOR GATES

CMOS inverters described in the last subsection are used to invert a logic signal, to act as a buffer or output driver, or to form a latch (two inverters connected back to back). However, they cannot perform logic computation, since there is only one input voltage. In the static CMOS logic family, the most widely used circuits with multiple inputs are *NAND* and *NOR* gates as shown in Fig. 5.6. In a NAND gate, a number of nMOSFETs are connected in series between the output node and the ground. The same number of pMOSFETs are connected in parallel between $V_{dd}$ and the output node. Each input signal is connected to the gates of a pair of n- and pMOSFETs as in the inverter case. In this configuration, the output node is pulled to ground only if all the nMOSFETs are turned on, i.e., only if all the input voltages are high ($V_{dd}$). If one of the input signals is low (zero voltage), the low-resistance path between the output node and ground is broken, but one of the pMOSFETs is turned on, which pulls the output node to $V_{dd}$. On the contrary, the NOR circuit in Fig. 5.6(b) consists of parallel-connected nMOSFETs between the output node and ground, but serially connected pMOSFETs between $V_{dd}$ and the output node. The output voltage is high only if all the input voltages are low, i.e., all the pMOSFETs are on and all the nMOSFETs are off. Otherwise, the output is low.

Due to the complementary nature of n- and pMOSFETs and the serial-versus-parallel connections, there is no direct low-resistance path between $V_{dd}$ and ground except during switching. In other words, just like CMOS inverters, ***there is no static current or standby power dissipation for any combination of inputs in either the CMOS NAND or NOR circuits***. The circuit output resistance is low, however, because of the conducting transistor(s).

In CMOS technology, NAND circuits are much more frequently used than NOR. This is because it is preferable to put the transistors with the higher resistance in parallel and those with the lower resistance in series. Since pMOSFETs have a higher resistance due to the lower hole mobility, they are rarely used in series (stacked). By connecting low-resistance nMOSFETs in series and high-resistance pMOSFETs in parallel, a NAND gate is more balanced in terms of the pull-up and the pull-down operations and achieves better noise immunity as well as a higher overall circuit speed.

#### 5.1.2.1 TWO-INPUT CMOS NAND GATE

As an example, we will examine the transfer curve and the switching characteristics of a two-input NAND gate, also referred to as a two-way NAND, or NAND with a fan-in of two, shown in Fig. 5.7. With the two pMOSFETs connected in parallel between $V_{dd}$ and the output node, the pull-up operation of a two-way NAND is similar to that of an inverter. If either one of the transistors is turned on while the other one is turned off, the charging current is identical to that of the pMOSFET

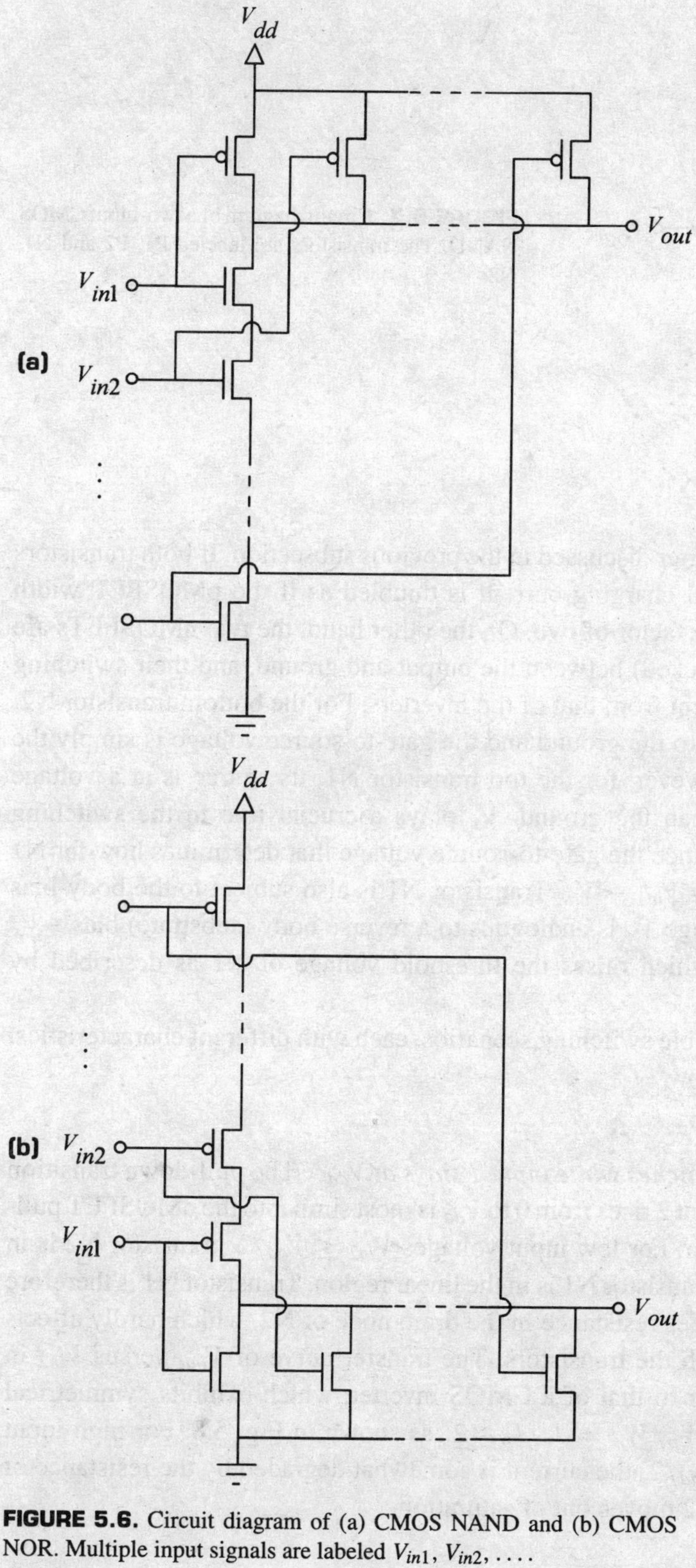


**FIGURE 5.6.** Circuit diagram of (a) CMOS NAND and (b) CMOS NOR. Multiple input signals are labeled $V_{in1}$, $V_{in2}$, ....

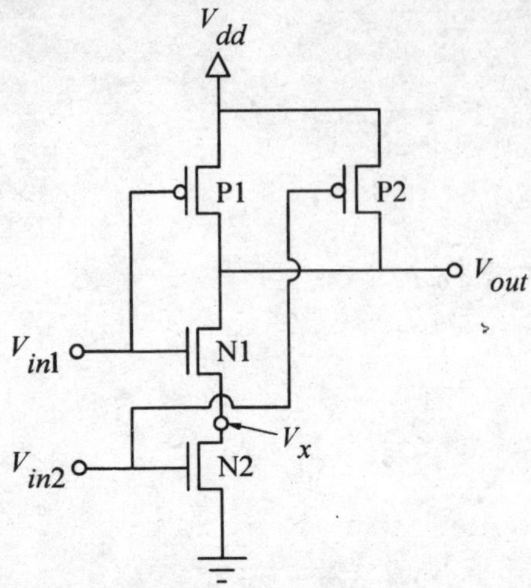


**FIGURE 5.7.** Circuit diagram of a two-input CMOS NAND. The transistors are labeled P1, P2 and N1, N2.

pull-up in a CMOS inverter discussed in the previous subsection. If both transistors are pulling up, the total charging current is doubled as if the pMOSFET width had been increased by a factor of two. On the other hand, the two nMOSFETs are connected in series (stacked) between the output and ground, and their switching behavior is quite different from that of the inverters. For the bottom transistor N2, its source is connected to the ground and the gate-to-source voltage is simply the input voltage $V_{in2}$. However, for the top transistor N1, its source is at a voltage $V_x$ (Fig. 5.7) higher than the ground. $V_x$ plays a crucial role in the switching characteristics of N1, since the gate-to-source voltage that determines how far N1 is turned on is given by $V_{in1} - V_x$. Transistor N1 is also subject to the body-bias effect, as a source voltage $V_x$ is analogous to a reverse body (substrate) bias $-V_x$ ($-V_{bs}$ in Fig. 3.11), which raises the threshold voltage of N1 as described by Eq. (3.40).

There are three possible switching scenarios, each with different characteristics. They are described below.

- *Case A. Input 2 switches while input 1 stays at $V_{dd}$.* The pull-down transition in case A when input 2 rises from 0 to $V_{dd}$ is most similar to the nMOSFET pull-down in an inverter. For low input voltages $V_{in2} \leq V_{dd}/2$, transistor N2 is in saturation, while transistor N1 is in the linear region. Transistor N1 is therefore equivalent to a series resistance in the drain node of N2, which hardly affects the current through the transistors. The transfer curve of $V_{out}$ versus $V_{in2}$ in this case is similar to that of a CMOS inverter, which exhibits symmetrical characteristics if $W_p/W_n = I_n/I_p \approx 2$, as shown in Fig. 5.8. For high input voltages $V_{in2} > V_{dd}/2$, the current is somewhat degraded by the resistance of N1 as transistor N2 moves out of saturation.

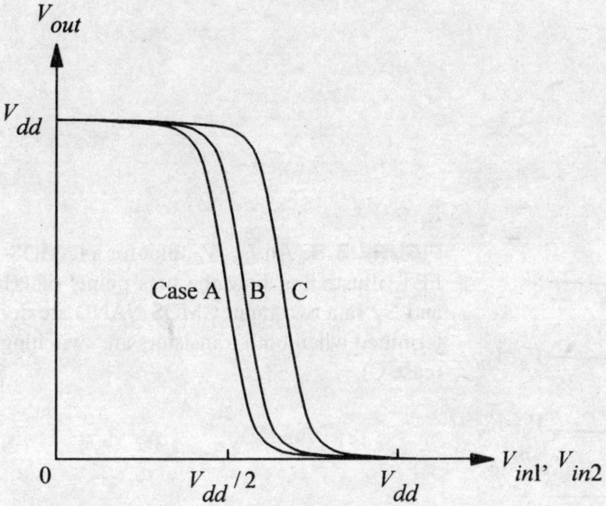


**FIGURE 5.8.** Transfer curves of a two-input CMOS NAND for different cases of switching discussed in the text. The device width ratio $W_p/W_n$ is taken to be 2 in this illustration.

- *Case B. Input 1 switches while input 2 stays at $V_{dd}$.* For the pull-down transition in case B, transistor N1 is in saturation while N2 is in linear mode during most part of the switching cycle. Transistor N2 therefore acts like a series resistance connected in the source terminal of N1. The voltage $V_x$ between the two transistors rises slightly above ground, depending on the current level. This degrades the pull-down current as the gate-to-source voltage of N1 is reduced to $V_{in1} - V_x$ and its threshold voltage is increased by $(m-1)V_x$ due to the body effect. As a result, a slightly higher input voltage $V_{in1}$ is needed to reach the high-to-low transition of the transfer curve in Fig. 5.8. Even though the pull-down current in case B is slightly less than in case A, the switching time in case B is comparable to that in A if the output is not too heavily loaded. This is because of the additional capacitance in case A associated with the top transistor N1 that needs to be discharged from $V_{dd}$ to ground when the bottom device is switching. These factors are further discussed in detail in Section 5.3.5.
- *Case C. Both input 1 and input 2 switch simultaneously.* The worst case for pull-down in a two-input CMOS NAND is case C, in which both inputs rise from 0 to $V_{dd}$. The bias point in the steady state can be found using the method illustrated in Fig. 5.9. For a given $V_{in}$ (both inputs 1 and 2), the bottom transistor (N2) follows the $I$–$V$ curve labeled $V_g = V_{in}$. If we assume a $V_x$-value as shown, the bias point of the bottom transistor is easily determined. As far as the gate overdrive, $V_g - V_t$, is concerned, the top transistor (N1) is biased at an equivalent gate voltage of $V_{in} - mV_x$ because the gate-to-source voltage

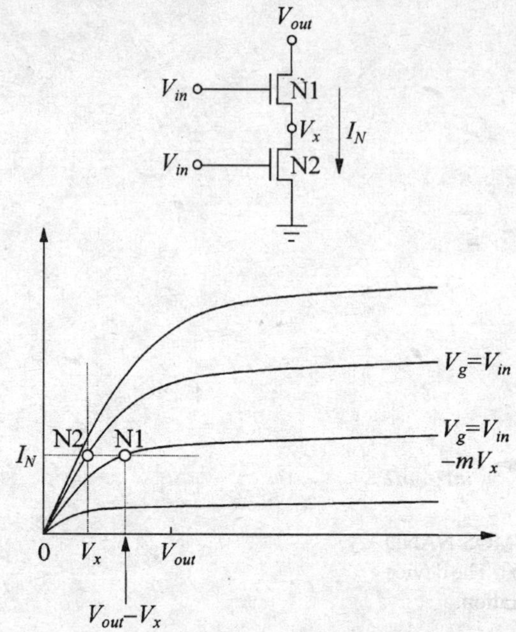


**FIGURE 5.9.** An $I_{ds}$–$V_{ds}$ plot for an nMOSFET, illustrating how the bias points of N1 and N2 in a two-input CMOS NAND are determined when both transistors are switching (case C).

is $V_{in} - V_x$ and the threshold voltage is higher than that of N2 by $(m-1)V_x$ due to the body effect. Since N1 and N2 are connected in series, the current through both transistors is the same ($I_N$). This determines the bias point for the top transistor N1 as shown in Fig. 5.9. Once both bias points are known, the output node voltage is simply the sum of the two source-to-drain voltages, $V_x$ and $V_{out} - V_x$. This yields one point for the $I_N$-versus-$V_{out}$ curve as indicated in Fig. 5.9. It is easy to see that transistor N2 is always biased in the linear region (otherwise the solution does not exist), while transistor N1 is in the linear region for small values of $V_{out}$ and in saturation for large $V_{out}$. In principle, the above procedure can be repeated to generate a family of $I_N$–$V_{out}$ curves with $V_{in}$ as a parameter. A $V_{out}$-versus-$V_{in}$ transfer curve can then be constructed using the method described in Fig. 5.3 given the $I_P$–$V_{out}$ curves of the simultaneously switching pMOSFETs. It can be seen from Fig. 5.9 that the nMOSFET pull-down current is reduced by approximately a factor of two from the inverter case because of the serial connection. The pull-up current, on the other hand, is twice that of the inverter case due to the parallel connection of pMOSFETs. This moves the high-to-low transition in the transfer curve to a $V_{in}$ significantly higher than $V_{dd}/2$, as shown in case C of Fig. 5.8.

Because of the spread of transfer curves under different switching conditions, the noise margin of a CMOS NAND gate is inferior to that of a CMOS inverter. When the device width ratio is chosen following the symmetric-inverter case, $W_p/W_n = I_n/I_p \approx 2$, as in Fig. 5.8, the noise margin for input low is intact, but the noise margin for input high is eroded because of case C. To shift the high-to-low transition edge in the transfer curve of case C to the midpoint, $V_{dd}/2$, a transistor width ratio such that $I_N/I_P = 4$ is needed, where $I_P$ and $I_N$ represent the currents of individual devices. This means $W_n/W_p = 4I_p/I_n \approx 2$; in other words, the nMOSFET width is twice the pMOSFET width. However, this would cause the transitions in cases A and B to shift to input voltages lower than $V_{dd}/2$ and degrade the noise margin for input low. A compromise would be to use $W_p = W_n$ for a two-input CMOS NAND such that the transition edges of case A and case C are more or less symmetric with respect to the midpoint of the input range, $V_{dd}/2$.

## 5.1.3 INVERTER AND NAND LAYOUTS

### 5.1.3.1 LAYOUT OF A SINGLE DEVICE

Both the CMOS circuit density and the delay performance are determined by the layout ground rules of the particular technology. Figure 5.10 shows a typical layout of an isolated MOSFET and its corresponding cross section. Only three major masking levels are shown: active region (isolation), polysilicon gate, and contact hole. To complete a CMOS process, several additional implant blockout masks are needed for doping the channel and the source–drain regions of nMOSFETs and pMOSFETs, respectively (Appendix 1). After the device or the front-end-of-line (FEOL) process, a number of metal levels are laid down in the back-end-of-

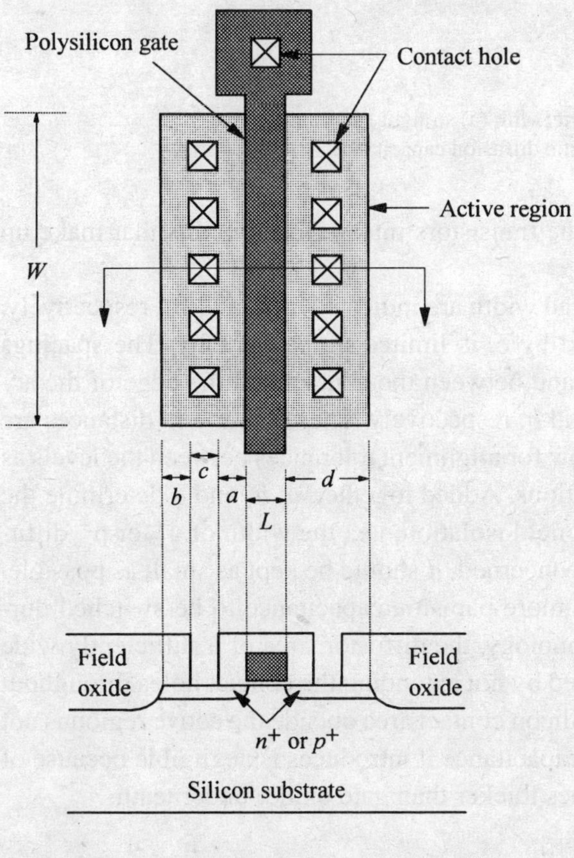


**FIGURE 5.10.** Basic layout and corresponding cross-section of a single MOSFET, illustrating several key layout ground rules.

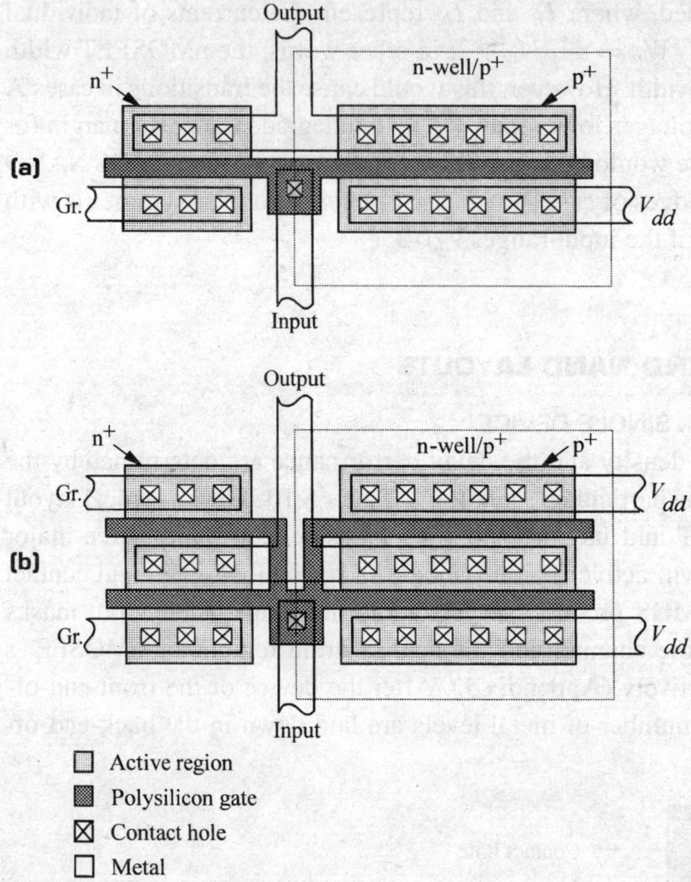


**FIGURE 5.11.** Layout of a CMOS inverter with (a) straight gates and (b) folded gates for minimizing the parasitic diffusion capacitance.

line (BEOL) process to connect the transistors into various circuits that make up the chip.

In Fig. 5.10, the device length and width are indicated by $L$ and $W$, respectively. The contact-hole size, represented by $c$, is limited by lithography. The spacings between the contact and the gate and between the contact and the edge of the active region are represented by $a$ and $b$, respectively. These minimum distances are required in the ground rules to allow for alignment tolerances between the levels as well as linewidth biases and variations. Added together, $a$, $b$, and $c$ determine the distance between the gate and the field isolation, i.e., the width of $n^+$ or $p^+$ diffusion, $d$. As far as CMOS delay is concerned, $d$ should be kept as small as possible, since a larger diffusion area adds more parasitic capacitance to be switched during a transition. In a silicided technology, the diffusion area of a sufficiently wide MOSFET can be somewhat reduced by not extending the contact holes throughout the entire device width. The polysilicon contact area outside the active region is not a critical factor, as the additional capacitance it introduces is negligible because of the thick field oxide (about 50 times thicker than gate oxide) underneath.

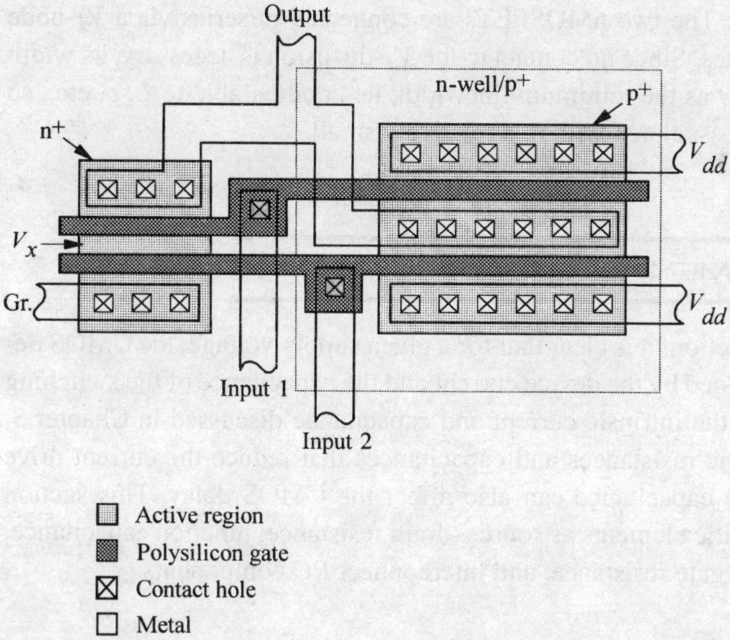


**FIGURE 5.12.** Layout example of a two-input CMOS NAND with the equivalent circuit in Fig. 5.7.

#### 5.1.3.2 LAYOUT OF A CMOS INVERTER

Figure 5.11(a) shows a simple layout of a CMOS inverter with $W_p/W_n \approx 2$. Four metal wires are shown, leading to $V_{dd}$, ground, input, and output. The pMOSFET receives n-well and $p^+$ source–drain implants with the use of two block-out masks. The nMOSFET receives a p-type channel implant and an $n^+$ source–drain implant with block-out masks of the complementary polarity. The intrinsic delay of a CMOS inverter, defined in terms of one stage driving an identical stage (fan-out = 1), is independent of the device width except for some parasitic effects at the ends. This is because both the current and the capacitance (gate and diffusion) are proportional to the device width in the straight-gate layout in Fig. 5.11(a). ***A dramatic reduction in the junction contributions to the parasitic capacitance can be achieved using the folded layout shown in Fig. 5.11(b).*** By sandwiching the drain node between two symmetric source regions with a fork-shaped polysilicon gate, the device width and therefore the current is effectively doubled without increasing the diffusion area. In other words, the junction capacitance per effective device width in layout (b) is about half of that in layout (a), assuming $a + b + c$ is comparable to $2a + c$ in Fig. 5.10. Note that the area of the source regions is of no importance to the delay, since the source is not being switched.

#### 5.1.3.3 LAYOUT OF A TWO-INPUT CMOS NAND

A typical layout for a two-input CMOS NAND is shown in Fig. 5.12. The two parallel-connected pMOSFETs are arranged as in the folded inverter, with the switching node sandwiched between the two input gates. This again minimizes junction capacitance. The two nMOSFETs are connected in series via a $V_x$-node between the input gates. Since no contact to the $V_x$-diffusion is necessary, its width can be kept as narrow as the minimum linewidth, i.e., comparable to $L$, $c$, etc., so that the capacitance associated with it is relatively small.

## 5.2 PARASITIC ELEMENTS

From the previous section, it is clear that for a given supply voltage, the CMOS delay is mainly determined by the device current and the capacitance of the switching node. In addition to the intrinsic current and capacitance discussed in Chapter 3, however, any parasitic resistances and capacitances that reduce the current drive or increase the node capacitance can also affect the CMOS delay. This section examines such parasitic elements as source–drain resistance, junction capacitance, overlap capacitance, gate resistance, and interconnect $RC$ components.

### 5.2.1 SOURCE–DRAIN RESISTANCE

It was discussed in Section 3.2.4 that source–drain series resistance degrades the current of a short-channel MOSFET whose intrinsic resistance is low. Resistance on the source side is particularly troublesome, as it degrades the gate drive as well.

A schematic diagram of the current-flow pattern in the source or drain region of a MOSFET is shown in Fig. 5.13 (Ng and Lynch, 1986). The total source or drain resistance can be divided into several parts: $R_{ac}$ is the accumulation-layer resistance in the gate–source (or –drain) overlap region where the current mainly stays at the surface; $R_{sp}$ is associated with current spreading from the surface layer into a uniform pattern across the depth of the source–drain; $R_{sh}$ is the sheet resistance of the source–drain region where the current flows uniformly; and $R_{co}$ is the contact resistance (including the spreading resistance in silicon under the contact) in the region where the current flows into a metal line. Once the current flows into an aluminum line, there is very little additional resistance, since the resistivity of aluminum is very low, $\rho_{Al} \approx 3 \times 10^{-6}$ Ω-cm. In VLSI interconnects, the aluminum thickness is typically 0.5–1.0 μm. From Eq. (2.26), the sheet resistivity is on the order of 0.05 Ω/□. This is negligible compared with the channel sheet resistivity, $\rho_{ch} \approx$ 2000–7000 Ω/□, except when a long, thin wire is connected to a wide MOSFET. Figure 5.13 shows only the series resistance on one side of the device. The total source–drain series resistance per device is, of course, twice of that shown in Fig. 5.13, assuming that the source and drain are symmetrical. Below we examine the various components of the source–drain resistance.

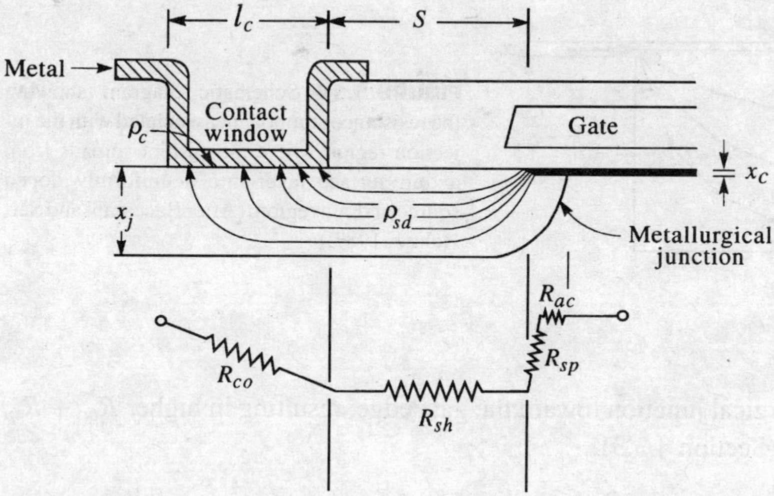


**FIGURE 5.13.** A schematic cross section showing the pattern of current flow from a MOSFET channel through the source or drain region to the aluminum contact. The diagram below identifies various contributions to the series resistance. The device width in the $z$-direction is assumed to be $W$. (After Ng and Lynch, 1986.)

#### 5.2.1.1 ACCUMULATION-LAYER RESISTANCE AND SPREADING RESISTANCE

The accumulation-layer resistance $R_{ac}$ depends on the gate voltage. Since it is not easily separable from the active channel resistance, $R_{ac}$ is considered as a part of $L_{eff}$ as discussed in Section 4.3.3.

Next we consider the spreading resistance component, $R_{sp}$. An analytical expression has been derived for $R_{sp}$ assuming an idealized case shown in Fig. 5.14 where the current spreading takes place in a uniformly doped medium with resistivity $\rho_j$ (Baccarani and Sai-Halasz, 1983):

$$R_{sp} = \frac{2\rho_j}{\pi W} \ln\left(0.75 \frac{x_j}{x_c}\right). \tag{5.6}$$

Here $W$ is the device width, and $x_j$ and $x_c$ are the junction depth and the inversion (or accumulation) layer thickness, respectively. For typical values of $x_j/x_c \approx 40$, we have $R_{sp} \approx 2\rho_j/W$. In practice, however, it is difficult to apply Eq. (5.6), since current spreading usually takes place in a region where the local resistivity is highly nonuniform due to the lateral source–drain doping gradient. In general, a 2-D numerical simulation is needed to evaluate $R_{ac}$ and $R_{sp}$. Qualitatively, current injection from the surface into the bulk takes place such that the sum of the resistances, $R_{ac} + R_{sp}$, is a minimum (Ng and Lynch, 1986). For an abrupt source–drain profile, the injection point is close to the metallurgical end of the channel; both $R_{ac}$ and $R_{sp}$ are low, and $L_{eff} \approx L_{met}$. For a graded profile, the injection point moves away

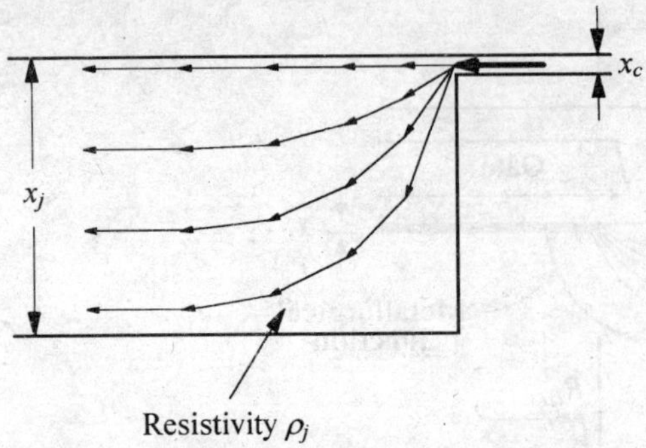


**FIGURE 5.14.** Schematic diagram showing the resistance component associated with the injection region where the current spreads from a thin surface layer into a uniformly doped source or drain region. (After Baccarani and Sai-Halasz, 1983.)

from the metallurgical junction toward the gate edge, resulting in higher $R_{ac} + R_{sp}$ and $L_{eff} > L_{met}$ (Section 4.3.3).

#### 5.2.1.2 SHEET RESISTANCE

Next, we examine $R_{sh}$ and $R_{co}$. In Fig. 5.13, the sheet resistance of the source–drain diffusion region is simply,

$$R_{sh} = \rho_{sd}\frac{S}{W}, \tag{5.7}$$

where $W$ is the device width, $S$ is the spacing between the gate edge and the contact edge, and $\rho_{sd}$ is the sheet resistivity of the source–drain diffusion, typically of the order of 50–500 $\Omega/\square$. Since $\rho_{sd} \ll \rho_{ch}$ of the device, this term is usually negligible if $S$ is kept to a minimum limited by the overlay tolerance between the contact and the gate lithography levels. In a nonsilicided technology, $S = a$ in Fig. 5.10, provided that most of the device width dimension is covered by contacts.

#### 5.2.1.3 CONTACT RESISTANCE

Based on a transmission-line model (Berger, 1972), the contact resistance can be expressed as

$$R_{co} = \frac{\sqrt{\rho_{sd}\rho_c}}{W}\coth\left(l_c\sqrt{\frac{\rho_{sd}}{\rho_c}}\right), \tag{5.8}$$

where $l_c$ is the width of the contact window (Fig. 5.13), and $\rho_c$ is the interfacial *contact resistivity* (in $\Omega$-cm$^2$) of the ohmic contact between the metal and silicon. $R_{co}$ includes the resistance of the current crowding region in silicon underneath the contact. In a nonsilicided technology, $l_c = c$ in Fig. 5.10. Equation (5.8) has two limiting cases: *short contact* and *long contact*. In the short-contact limit, $l_c \ll (\rho_c/\rho_{sd})^{1/2}$, and

$$R_{co} = \frac{\rho_c}{Wl_c} \tag{5.9}$$

is dominated by the interfacial contact resistance. The current flows more or less uniformly across the entire contact. In the long-contact limit, $l_c \gg (\rho_c/\rho_{sd})^{1/2}$, and

$$R_{co} = \frac{\sqrt{\rho_{sd}\rho_c}}{W}. \tag{5.10}$$

This is independent of the contact width $l_c$, since most of the current flows into the front edge of the contact. Once in the long-contact regime, there is no advantage increasing the contact width.

For ohmic contacts between metal and heavily doped silicon, current conduction is dominated by tunneling or field emission. The contact resistivity $\rho_c$ depends exponentially on the barrier height $\phi_B$ and the surface doping concentration $N_d$ (Yu, 1970):

$$\rho_c \propto \exp\left(\frac{4\pi\phi_B}{qh}\sqrt{\frac{m^*\varepsilon_{si}}{N_d}}\right), \tag{5.11}$$

where $h$ is Planck's constant and $m^*$ is the electron effective mass. Depending on the doping concentration and contact metallurgy, $\rho_c$ is typically in the range of $10^{-6}$–$10^{-7}$ Ω-cm$^2$.

#### 5.2.1.4 RESISTANCE IN A SELF-ALIGNED SILICIDE TECHNOLOGY

***Both $R_{sh}$ and $R_{co}$ are greatly reduced in advanced CMOS technologies with self-aligned silicide*** (Ting *et al.*, 1982). As shown schematically in Fig. 5.15, a highly conductive (≈2–10 Ω/□) silicide film is formed on all the gate and source–drain surfaces separated by dielectric spacers in a self-aligned process. Since the sheet resistivity of silicide is 1–2 orders of magnitude lower than that of the source–drain, the silicide layer practically shunts all the currents, and the only significant contribution to $R_{sh}$ is from the nonsilicided region under the spacer. This reduces the length $S$ in Eq. (5.7) to 0.1–0.2 μm, which means that $R_{sh}W$ should be no

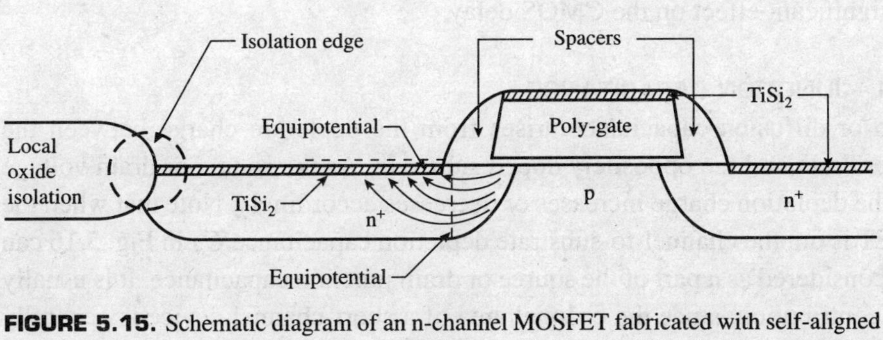


**FIGURE 5.15.** Schematic diagram of an n-channel MOSFET fabricated with self-aligned $TiSi_2$, showing the current flow pattern between the channel and the silicide. (After Taur *et al.*, 1987.)

more than 50 Ω-μm. At the same time, $R_{co}$ between the source–drain and silicide is also reduced, since now the contact area is the entire diffusion. In other words, the diffusion width $d$ in Fig. 5.10 becomes the contact length $l_c$ in Eq. (5.8). Current flow in this case is almost always in the long-contact limit, so that Eq. (5.10) applies. However, the parameters $\rho_{sd}$ and $\rho_c$ in Eq. (5.10) should be replaced by $\rho'_{sd}$ and $\rho'_c$: the sheet resistivity of the source–drain region under the silicide and the contact resistivity between the silicide and silicon. $\rho'_{sd}$ is higher than the nonsilicided sheet resistivity $\rho_{sd}$, since a surface layer of heavily doped silicon is consumed in the silicidation process (Taur *et al.*, 1987). $\rho'_c$ is also higher than $\rho_c$ if the interface doping concentration becomes lower due to silicon consumption. This is particularly a concern when a thick silicide film is formed over a shallow source–drain junction. As a rule of thumb, no more than a third of the source–drain depth should be consumed in the silicide process.

In a CMOS process, a silicide material such as $TiSi_2$ with a near-midgap work function is needed to obtain approximately equal barrier heights to $n^+$ and $p^+$ silicon. The experimentally measured $\rho'_c$ between $TiSi_2$ and $n^+$ or $p^+$ silicon is of the order of $10^{-6}$–$10^{-7}$ Ω-cm$^2$ (Hui *et al.*, 1985). Based on Eq. (5.10), therefore, $R_{co}$ for a silicided diffusion is in the range of 50–200 Ω-μm (Taur *et al.*, 1987). The minimum contact width $l_c$ (or diffusion width $d$) required to satisfy the long-contact criterion can be estimated from $(\rho'_c/\rho'_{sd})^{1/2}$ to be about 0.25 μm. Contact resistance between silicide and metal is usually negligible, since the interfacial contact resistivity is of the order of $10^{-7}$–$10^{-8}$ Ω-cm$^2$ in a properly performed process.

### 5.2.2 PARASITIC CAPACITANCES

A schematic diagram of the MOSFET capacitances is shown in Fig. 5.16. In addition to the intrinsic capacitances $C_g$ and $C_d$ discussed in Section 3.1.6, there are also parasitic capacitances: namely, junction capacitance between the source or drain diffusion and the substrate (or n-well in the case of pMOSFETs), and overlap capacitance between the gate and the source or drain region. These capacitances have a significant effect on the CMOS delay.

#### 5.2.2.1 JUNCTION CAPACITANCE

Junction or diffusion capacitance arises from the depletion charge between the source or drain and the oppositely doped substrate. As the source or drain voltage varies, the depletion charge increases or decreases accordingly. Note that when the MOSFET is on, the channel-to-substrate depletion capacitance $C_d$ in Fig. 5.16 can also be considered as a part of the source or drain junction capacitance. It is usually a small contribution, since the channel area of a short-channel device is generally much less than the diffusion area.

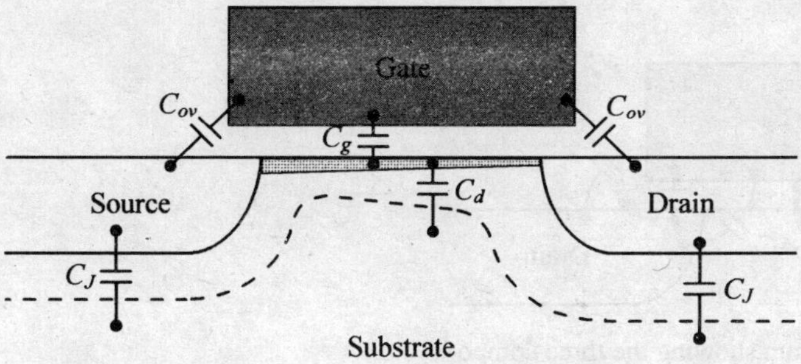


**FIGURE 5.16.** Schematic diagram of a MOSFET showing both the intrinsic capacitances ($C_g$, $C_d$) and the parasitic capacitances ($C_J$, $C_{ov}$). The two $C_J$'s at the source and the drain may have different values depending on the bias voltages.

From Eq. (2.70), the capacitance per unit area of an abrupt p–n junction is

$$C_j = \frac{\varepsilon_{si}}{W_{dj}} = \sqrt{\frac{\varepsilon_{si} q N_a}{2(\psi_{bi} + V_j)}}, \tag{5.12}$$

where $W_{dj}$ is the depletion-layer width, $N_a$ is the impurity concentration of the lightly doped side, $\psi_{bi}$ is the built-in potential, typically around 0.9 V as shown in Fig. 2.12, and $V_j$ is the reverse bias voltage across the junction. Equation (5.12) indicates that the source or drain junction capacitance is voltage-dependent. At a higher drain voltage, the depletion layer widens and the capacitance decreases. Figure 2.13 plots the depletion-layer width and the capacitance at zero bias versus $N_a$. Since the junction capacitance increases with $N_a$, one should avoid doping the substrate (or n-well) regions under the source–drain junctions unnecessarily highly. Too low a doping concentration between the source and drain, however, would cause excessive short-channel effect or lead to punch-through as discussed in Section 3.2.1.

The total diffusion-to-substrate capacitance is simply equal to $C_j$ times the diffusion area in the layout:

$$C_J = WdC_j, \tag{5.13}$$

where $W$ is the device width, and $d$ is the diffusion width in Fig. 5.10. For a noncontacted diffusion, $d$ can be as small as the minimum linewidth of the lithography. The diffusion capacitance of the switching node can be reduced by a factor of two using the folded layout in Fig. 5.11(b). Strictly speaking, there are also perimeter contributions to the diffusion capacitance, since the substrate doping concentration is usually higher at the diffusion boundary due to field implants. The extra contribution can be minimized by careful process design or by using again the folded

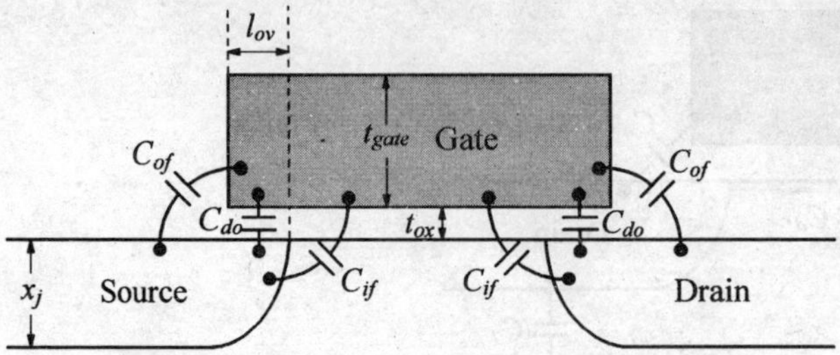


**FIGURE 5.17.** Schematic diagram showing the three components of the gate-to-diffusion overlap capacitance.

layout in which the diffusion is bounded by two gates, thus avoiding diffusion-field boundaries except at the ends.

#### 5.2.2.2 OVERLAP CAPACITANCE

Another parasitic capacitance in a MOSFET is the gate-to-source or gate-to-drain overlap capacitance. It consists of three components: direct overlap, outer fringe, and inner fringe, as shown schematically in Fig. 5.17. The direct overlap component is simply

$$C_{do} = Wl_{ov}C_{ox} = \frac{\varepsilon_{ox} W l_{ov}}{t_{ox}}, \tag{5.14}$$

where $l_{ov}$ is the length of the source or drain region under the gate. In a typical process, the oxide in the overlap region is somewhat thicker than $t_{ox}$ due to bird's-beak near the gate edge resulting from a reoxidation step (Wong *et al.*, 1989). Therefore, $l_{ov}$ should be interpreted as an equivalent overlap length, rather than an actual physical length.

By solving Laplace's equation analytically with proper boundary conditions, the outer and inner fringe components can be expressed as (Shrivastava and Fitzpatrick, 1982)

$$C_{of} = \frac{2\varepsilon_{ox} W}{\pi} \ln\left(1 + \frac{t_{gate}}{t_{ox}}\right), \tag{5.15}$$

and

$$C_{if} = \frac{2\varepsilon_{si} W}{\pi} \ln\left(1 + \frac{x_j}{2t_{ox}}\right), \tag{5.16}$$

where $t_{gate}$ is the height of the polysilicon gate, and $x_j$ is the depth of the source or drain junction. Equations (5.15) and (5.16) assume ideal shapes of the polysilicon gate and source–drain regions with square corners. For typical values of $t_{gate}/t_{ox} \approx 40$ and $x_j/t_{ox} \approx 20$, one obtains $C_{of}/W \approx 2.3\varepsilon_{ox} \approx 0.08$ fF/μm and $C_{if}/W \approx 1.5\varepsilon_{si} \approx 0.16$ fF/μm. Even though the inner fringe component is larger due to the higher dielectric constant of silicon, it is present only when $V_g < V_t$ and the region under the gate is depleted. Once $V_g > V_t$, the inversion layer forms, which effectively shields any electrostatic coupling between the gate and the inner edges of the source or drain junction. Similar shielding of the inner fringe capacitance also takes place when the gate voltage is negative (for nMOSFETs) and the surface is accumulated. Under these conditions, the overlap capacitance consists of only the direct overlap and the outer fringe components.

From the above numerical estimates, one can write the total overlap capacitance at $V_g = 0$ (silicon is depleted under the gate) as

$$C_{ov}(V_g = 0) = C_{do} + C_{of} + C_{if} \approx \varepsilon_{ox} W\left(\frac{l_{ov}}{t_{ox}} + 7\right). \tag{5.17}$$

Note that Eq. (5.17) is the maximum overlap capacitance per edge. It assumes perfectly conducting source and drain regions. In reality, because of the lateral source–drain doping gradient at the surface, the overlap capacitance depends on the drain voltage. When the drain voltage increases in an nMOSFET (with the same gate-to-substrate voltage), the overlap capacitance tends to decrease slightly because the reverse bias widens the depletion region at the surface and therefore reduces the effective overlap length (Oh *et al.*, 1990). This is especially the case with LDD MOSFETs.

It has been reported that a minimum length of direct overlap region of the order of $l_{ov} \approx (2\text{–}3)t_{ox}$ is needed to avoid reliability problems arising from hot-carrier injection into the ungated region (Chan *et al.*, 1987b). In other words, such a margin is required to avoid "underlap" of the gate and the source–drain. Combining this requirement with Eq. (5.17), one obtains $C_{ov}/W \approx 10\varepsilon_{ox} \approx 0.3$ fF/μm at zero gate voltage, independent of technology generation.

### 5.2.3 GATE RESISTANCE

In modern CMOS technologies, silicides are formed over polysilicon gates to lower the resistance and provide ohmic contacts to both $n^+$ and $p^+$ gates. The sheet resistivity of silicides is of the order of 2–10 Ω/□, which is generally adequate for 0.5-μm CMOS technology and above. For 0.25-μm CMOS technology and below, however, the device delay improves and gate *RC* delays may not be negligible. Compounding the problem is a tendency for silicide resistivity to increase in fine-line structures. This is due to either agglomeration or lack of nucleation sites to initiate the phase transformation in the case of $TiSi_2$. Gate *RC* delay is an ac effect not observable in dc *I–V* curves. It shows up as an additional delay component in ring oscillators, delay chains, and other logic circuits.

Gate *RC* delay can be analyzed with a distributed network shown in Fig. 5.18 for a MOSFET device of width *W* and length *L*. The resistance per unit length *R*

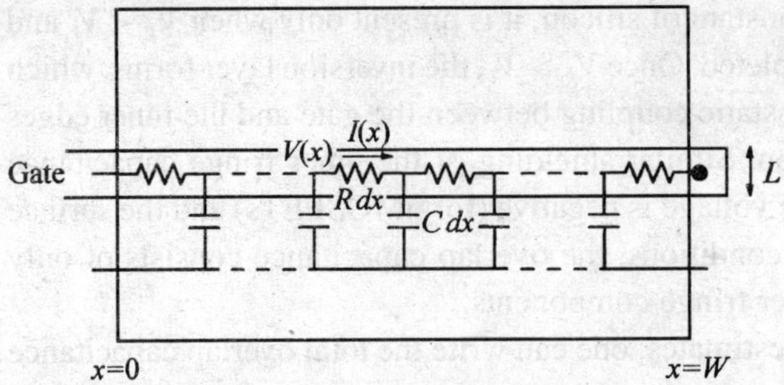


**FIGURE 5.18.** A distributed network for gate $RC$ delay analysis. The lower rail represents the MOSFET channel, which is connected to the source–drain. The input step voltage is applied at the left.

is related to the silicide sheet resistivity $\rho_g$ ($\Omega/\square$) by

$$R = \frac{\rho_g}{L}. \tag{5.18}$$

The capacitance per unit length, $C$, mainly arises from the inversion charge that must be supplied (or taken away) when the voltage at a particular point along the gate increases (or decreases). To a good approximation, $C$ is given by the gate oxide capacitance,

$$C = C_{ox}L = \frac{\varepsilon_{ox}L}{t_{ox}}. \tag{5.19}$$

For a higher accuracy, one should also include the overlap capacitance per unit gate width in $C$.

At any point $x$ along the gate, one can write

$$V(x+dx) - V(x) = \frac{\partial V}{\partial x}\,dx = -I(x)R\,dx, \tag{5.20}$$

and

$$I(x+dx) - I(x) = \frac{\partial I}{\partial x}\,dx = -C\,dx\,\frac{\partial V}{\partial t}. \tag{5.21}$$

Eliminating $I(x)$ from the above equations, one obtains

$$\frac{\partial^2 V}{\partial x^2} = RC\frac{\partial V}{\partial t}. \tag{5.22}$$

The differential equation that governs the $RC$ delay of a distributed network therefore resembles the diffusion equation with a diffusion coefficient $D = 1/RC$. If a step voltage from 0 to $V_{dd}$ is applied at $x = 0$, the boundary conditions are $V(0,t) = V_{dd}$

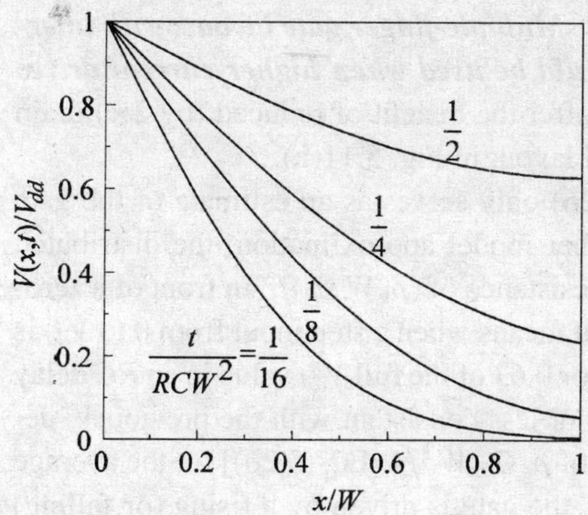


**FIGURE 5.19.** Local gate voltage versus distance along the width of the device at different time intervals after an input voltage $V_{dd}$ is applied at $x = 0$.

and $I(W, t) = 0$, which is analogous to constant-source diffusion into a finite-width medium. The numerical solution for this case is plotted in Fig. 5.19 (Sakurai, 1983). For $t \ll RCW^2/8$, the solution can be approximated by a *complementary error function*,

$$V(x, t) = V_{dd}\,\mathrm{erfc}\left(\frac{x}{\sqrt{4t/RC}}\right), \tag{5.23}$$

where

$$\mathrm{erfc}(y) \equiv \frac{2}{\sqrt{\pi}}\int_y^{\infty} e^{-z^2}\,dz. \tag{5.24}$$

For $t \gg RCW^2/4$, the approximate solution is given by

$$V(x, t) = V_{dd}\left[1 - \frac{4}{\pi}\sin\left(\frac{\pi x}{2W}\right)\exp\left(-\frac{\pi^2 t}{4RCW^2}\right)\right]. \tag{5.25}$$

It can be seen from Fig. 5.19 that the average value of $V(x, t)$ within $0 < x < W$ reaches $V_{dd}/2$ when $t \approx RCW^2/4$. If one takes this value as the effective $RC$ delay $\tau_g$ due to the gate resistance and substitutes Eqs. (5.18) and (5.19) for $R$ and $C$, one obtains

$$\tau_g = \frac{\rho_g C_{ox} W^2}{4}. \tag{5.26}$$

Note that $\tau_g$ is independent of device length but is proportional to the square of the device width. Clearly, the gate $RC$ delay has a more significant effect percentage-wise in fast-switching unloaded inverters than in heavily loaded circuits. In order to limit $\tau_g$ to less than 1 ps, assuming $\rho_g = 10\ \Omega/\square$ and $t_{ox} = 50$ Å, the device width

$W$ must be restricted to 7.6 μm or below. ***Multiple-finger gate layouts with interdigitated source and drain regions should be used when higher-current drives are needed***. Such types of layouts also offer the benefit of reduced (by 2×) drain junction capacitance, just like the folded layout in Fig. 5.11(b).

It should be pointed out that Eq. (5.26) only serves as an estimate of the gate $RC$ delay for a particular case. In another model approximation, the distributed gate resistance is replaced by a lumped resistance of $(\rho_g W/L)/3$ in front of a zero-resistance gate (Razavi *et al.*, 1994). That means when a step input from 0 to $V_{dd}$ is applied, the gate voltage rises to $1 - e^{-1}$ or 0.63 of the full $V_{dd}$ value in an $RC$ delay time of $\rho_g C_{ox} W^2/3$. This result is more or less consistent with the previously described model, which gives a delay time of $\rho_g C_{ox} W^2/4$ [Eq. (5.26)] for the average gate voltage to reach $V_{dd}/2$. In practice, the gate is driven by a rising (or falling) signal with a finite ramp rate. Also, partial current conduction takes place early in the near end ($x = 0$) of the device, which constitutes the leading edge of signal propagation. Generally speaking, the gate $RC$ delay depends on the drive condition of the previous stage and the capacitive loading of the following stage. Quantitative results should be obtained numerically from appropriate circuit models.

### 5.2.4 INTERCONNECT *R* AND *C*

Unlike other parasitic elements discussed above, interconnect capacitance and resistance have negligible effects on the delay of local circuits such as CMOS inverters or NAND gates discussed in Section 5.1. On a VLSI chip or system level, however, interconnect $R$ and $C$ can play a major role in system performance, especially in standard-cell designs where wire capacitance dominates circuit delay. We shall discuss interconnect capacitance first, followed by interconnect resistance.

#### 5.2.4.1 INTERCONNECT CAPACITANCE

Because of its geometry, the capacitance of an interconnect line cannot be calculated by the parallel-plate capacitance alone. In general, interconnect capacitance has three components: the parallel-plate (or area) component, fringing-field component, and wire-to-wire component. Figure 5.20 shows schematically electric field lines that constitute the parallel-plate and the fringing-field capacitance of an isolated line (Bakoglu, 1990). The total capacitance per unit length, $C_w$,

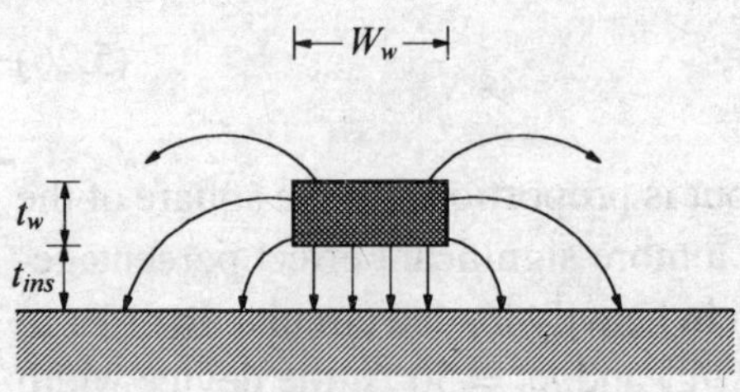


**FIGURE 5.20.** Schematic diagram showing electrostatic coupling between an isolated wire and a conducting plane. The straight field lines underneath the wire represent the parallel-plate component of the capacitance. The field lines emerging from the side and the top of the wire make up the fringing-field component of the capacitance. (After Bakoglu, 1990.)

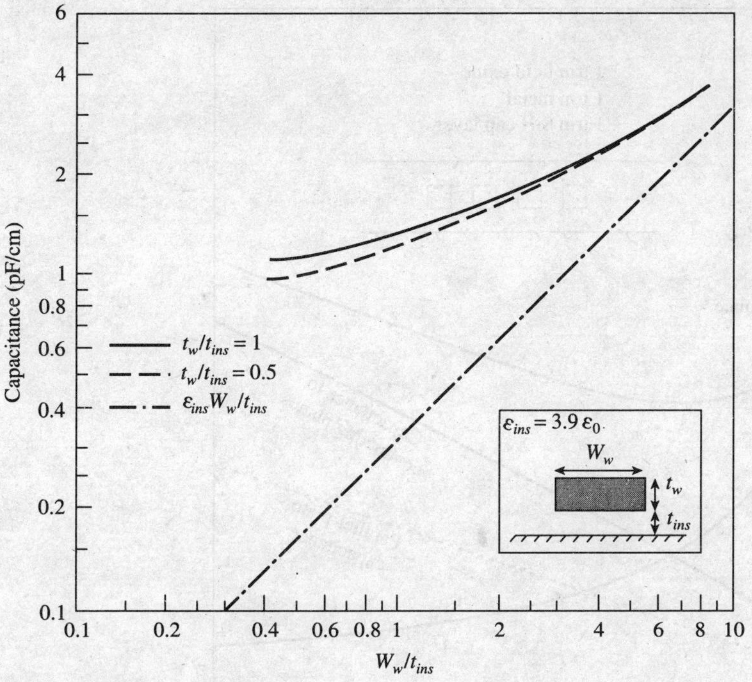


**FIGURE 5.21.** Wire capacitance per unit length as a function of width-to-gap ratio, $W_w/t_{ins}$, for the system in Fig. 5.20. The straight line represents the parallel-plate component of the capacitance. The dielectric medium is assumed to be oxide with a dielectric constant of 3.9. (After Schaper and Amey, 1983.)

calculated numerically by solving a 2-D Laplace's equation, is shown in Fig. 5.21 versus the ratio of wire width to insulator thickness, $W_w/t_{ins}$ (Schaper and Amey, 1983). Only when $W_w \gg t_{ins}$ can the total capacitance per unit length be approximated by the parallel-plate component, $\varepsilon_{ins} W_w/t_{ins}$ (the straight line in Fig. 5.21). As $W_w/t_{ins} \rightarrow 1$, the fringing-field component becomes important and the total capacitance can be much higher than the parallel-plate component. In fact, a minimum capacitance of about 1 pF/cm (for silicon dioxide as the interlevel dielectric) is reached even if $W_w \ll t_{ins}$. This shows that reducing the wire capacitance by increasing the insulator thickness becomes ineffective when the insulator thickness becomes comparable to the width of the wire. Decreasing the wire thickness $t_w$ does not help much either, as is evident in Fig. 5.21.

To improve the packing density in today's VLSI chips, wires of minimum pitch with nearly equal lines and spaces are frequently used. This causes a still higher wiring capacitance due to contributions from the neighboring lines. Figure 5.22 shows the calculated total capacitance as a sum of two components for an array of wires with equal line and space sandwiched between two conducting planes. As shown in the inset, the thickness of metal lines and the thickness of insulators below (oxide) and above (nitride) them are all assumed to be 1 μm. The capacitances are

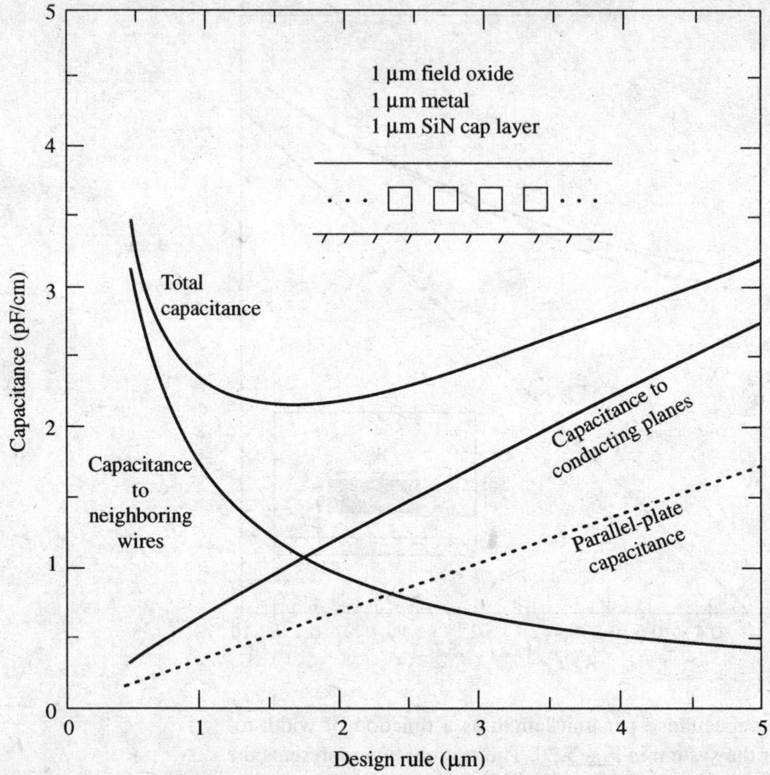


**FIGURE 5.22.** Capacitance per unit length as a function of design rules for an array of wires with equal line and space sandwiched between two conducting planes shown in the inset. The total capacitance of each wire is made up of two components: capacitance to the conducting planes, and capacitance to the neighboring wires. Both the metal and insulator thicknesses are held constant as the design rule is varied. The parallel-plate capacitance is also shown (dotted line) for reference. (After Schaper and Amey, 1983.)

calculated as a function of the metal line or space dimension. When the metal line and space are much larger than the thicknesses, the capacitance is dominated by the component (parallel-plate plus fringing) to the conducting planes above and below. When the metal pitch is much smaller than the thicknesses, however, wire-to-wire capacitance dominates. ***The total capacitance exhibits a broad minimum value of about 2 pF/cm when the metal line or space dimension is approximately equal to the insulator (and wire) thickness***. This conclusion is more general than the specific dimensions assumed. If all the line, space, metal thickness, and insulator thickness are scaled by the same factor, the result remains unchanged. The number 2 pF/cm can be understood from the capacitance per unit length between two concentric cylinders of radii $a$ and $b$:

$$C_w = \frac{2\pi\varepsilon_{ins}}{\ln(b/a)}. \tag{5.27}$$

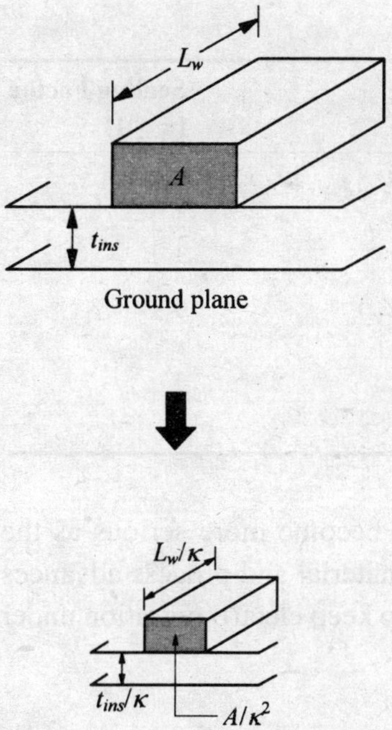


**FIGURE 5.23.** Scaling of interconnect lines and insulator thicknesses. (After Dennard, 1986.)

If one takes $\varepsilon_{ins} = \varepsilon_{ox}$ and $b/a = 2$, then $C_w \approx 2\pi\,\varepsilon_{ox} \approx$ 2 pF/cm. If an alternative insulator with a lower dielectric constant than that of oxide is used, $C_w$ will decrease proportionally.

#### 5.2.4.2 INTERCONNECT SCALING

Based on the above discussions, one can easily set a strategy for interconnect scaling similar to that for MOSFET scaling described in Section 4.1.1. This is shown schematically in Fig. 5.23 (Dennard *et al.*, 1974). ***All linear dimensions – wire length, width, thickness, spacing, and insulator thickness – are scaled down by the same factor, κ, as the device scaling factor***. Wire lengths ($L_w$) are reduced by $\kappa$ because the linear dimension of the devices and circuits that they connect to is reduced by $\kappa$. Both the wire and the insulator thicknesses are scaled down along with the lateral dimension, for otherwise the fringe capacitance and wire-to-wire coupling (crosstalk) would increase disproportionally, as illustrated in Fig. 5.22. Table 5.1 summarizes the rules for interconnect scaling. All material parameters, such as the metal resistivity $\rho_w$ and dielectric constant $\varepsilon_{ins}$, are assumed to remain the same. The wire capacitance then scales down by $\kappa$, the same way as the device capacitance (Table 4.1), while the wire capacitance per unit length, $C_w$, remains unchanged (approximately 2 pF/cm for silicon dioxide insulation, as mentioned above). The wire resistance, on the other hand, scales up by $\kappa$, in contrast to the device resistance, which does not change with scaling (Table 4.1). The wire resistance per unit length, $R_w$, then scales up by $\kappa^2$, as indicated in Table 5.1. It is also noted that the current density of interconnects increases with $\kappa$, which implies

**TABLE 5.1** Scaling of Local Interconnect Parameters

| | Interconnect Parameters | Scaling Factor ($\kappa \geq 1$) |
|---|---|---|
| Scaling assumptions | Interconnect dimensions ($t_w, L_w, W_w, t_{ins}, W_{sp}$) | $1/\kappa$ |
| | Resistivity of conductor ($\rho_w$) | 1 |
| | Insulator permittivity ($\varepsilon_{ins}$) | 1 |
| Derived wire scaling behavior | Wire capacitance per unit length ($C_w$) | 1 |
| | Wire resistance per unit length ($R_w$) | $\kappa^2$ |
| | Wire $RC$ delay ($\tau_w$) | 1 |
| | Wire current density ($I/W_w t_w$) | $\kappa$ |

that reliability issues such as electromigration may become more serious as the wire dimension is scaled down. Fortunately, a few material and process advances in metallurgy have taken place over the generations to keep electromigration under control in VLSI technologies.

#### 5.2.4.3 INTERCONNECT RESISTANCE

The interconnect $RC$ delay can be examined using the same distributed $RC$ network model introduced in Section 5.2.3. From Fig. 5.19 or Eq. (5.25), the voltage at the receiving end of an interconnect line rises to $1 - e^{-1} \approx 63\%$ of the source voltage after a delay of $t = RCW^2/2$. If one takes this value as the equivalent $RC$ delay ($\tau_w$) of an interconnect line and substitutes $R_w, C_w, L_w$ for $R, C, W$, one obtains

$$\tau_w = \frac{1}{2} R_w C_w L_w^2. \tag{5.28}$$

Using $R_w = \rho_w / W_w t_w$ and Eq. (5.27) for $C_w$ with $\ln(b/a) \approx 1$, one can express Eq. (5.28) as

$$\tau_w \approx \pi \varepsilon_{ins} \rho_w \frac{L_w^2}{W_w t_w}, \tag{5.29}$$

where $W_w$ and $t_w$ are the wire width and thickness, respectively. One of the key conclusions of interconnect scaling is that the wire $RC$ delay $\tau_w$ does not change as the device dimension and intrinsic delay are scaled down. Eventually, this will impose a limit on VLSI performance. Fortunately, for conventional aluminum metallurgy with silicon dioxide insulation, $\rho_w \approx 3 \times 10^{-6}$ $\Omega$-cm and

$$\tau_w \approx (3 \times 10^{-18}\ \text{s}) \frac{L_w^2}{W_w t_w}. \tag{5.30}$$

It is easy to see that the $RC$ delay of local wires is negligible as long as $L_w^2 / W_w t_w < 3 \times 10^5$. For example, a 0.25 μm × 0.25 μm wire 100 μm long has an $RC$ delay of 0.5 ps, which is quite negligible even when compared with the intrinsic delay (≈20 ps) of a 0.1-μm CMOS inverter (Taur *et al.*, 1993c). Therefore, ***a local circuit macro can be scaled down with all $W_w$, $t_w$, and $L_w$ reduced by the same factor without running into serious RC problems.***

#### 5.2.4.4 *RC* DELAY OF GLOBAL INTERCONNECTS

Based on the above discussion, the *RC* delay of local wires will not limit the circuit speed even though it cannot be reduced through scaling. The *RC* delay of global wires, on the other hand, is an entirely different matter. Unlike local wires, the length of global wires, on the order of the chip dimension, does not scale down, since the chip size actually increases slightly for advanced technologies with better yield and defect density to accommodate a much larger number of circuit counts. Even if we assume the chip size does not change, the *RC* delay of global wires scales up by $\kappa^2$ from Eq. (5.30). It is clear that one quickly runs into trouble if the cross-sectional area of global wires is scaled down the same way as the local wires. For example, for today's 0.25-μm CMOS technology, $L_w^2/W_w t_w \sim 10^8$–$10^9$ and $\tau_w \sim 1$ ns, severely degrading the system performance. The use of copper wires instead of aluminum would reduce the numerical factor in Eq. (5.30) by a factor of about 1.5 and provide some relief.

A number of solutions have been proposed to deal with the problem. The most obvious one is to minimize the number of cross-chip global interconnects in the critical paths through custom layout/design and use of sophisticated design tools. One can also use repeaters to reduce the dependence of *RC* delay on wire length from a quadratic one to a linear one (Bakoglu, 1990). A more fundamental solution is to increase or not to scale the cross-sectional area of global wires. However, just increasing the width and thickness of global wires is not enough, since the wire capacitance will then increase significantly, which degrades both performance and power. The intermetal dielectric thickness must be increased in proportion to keep the wire capacitance per unit length constant. Of course, there is a technology price to pay in building such low-*RC* global wires. It also means more levels of interconnects, since one still needs several levels of thin, dense local wires to make the chip wirable.

***The best strategy for interconnect scaling is then to scale down the size and spacing of lower levels in step with device scaling for local wiring, and to use unscaled or even scaled-up levels on top for global wiring***, as shown schematically in Fig. 5.24 (Sai-Halasz, 1995). Unscaled wires allow the global *RC* delay to remain essentially unchanged, as seen from Eq. (5.30). Scaled-up (together with the insulator thickness) wires allow the global *RC* delay to scale down together with the device delay. This is even more necessary if the chip size increases with every generation. Ultimately, the scaled-up global wires will approach the transmission-line limit when the inductive effect becomes more important than the resistive effect.

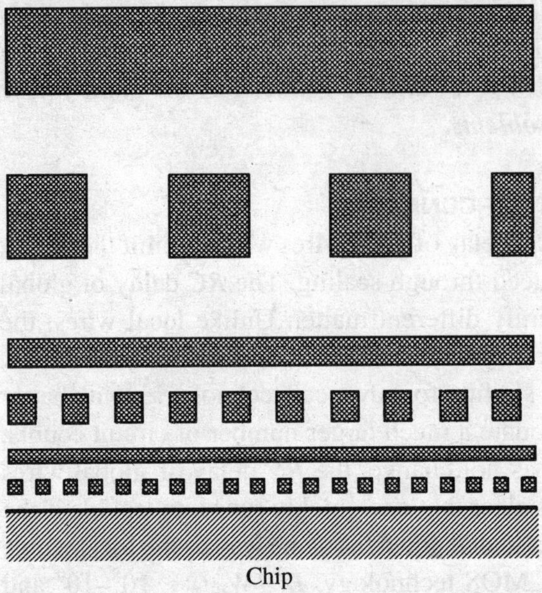


**FIGURE 5.24.** Schematic cross section of a wiring hierarchy that addresses both the density and the global *RC* delay in a high-performance CMOS processor. (After Sai-Halasz, 1995.)

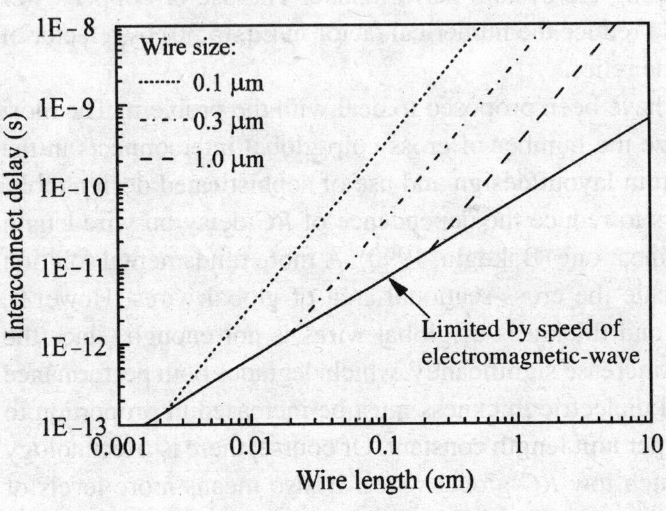


**FIGURE 5.25.** *RC* delay versus wire length for three different wire sizes (assuming square wire cross sections). Wires become limited by electromagnetic-wave propagation when the *RC* delay equals the time of flight over the line length. An oxide insulator is assumed here.

This happens when the signal rise time is shorter than the time of flight over the length of the line. Signal propagation is then limited by the speed of electromagnetic waves, $c/(\varepsilon_{ins}/\varepsilon_0)^{1/2}$, instead of by *RC* delay. Here $c = 3 \times 10^{10}$ cm/s is the velocity of light in vacuum. For oxide insulators, $(\varepsilon_{ins}/\varepsilon_0)^{1/2} \approx 2$, the time of flight is approximately 70 ps/cm. Figure 5.25 shows the interconnect delay versus wire length calculated from Eq. (5.30) for three different wire cross sections. For a longer global wire to reach the speed-of-light limit, a larger wire cross section is needed. The transmission-line situation is more often encountered in packaging wires (Bakoglu, 1990).

## 5.3 SENSITIVITY OF CMOS DELAY TO DEVICE PARAMETERS

This section focuses on the performance factors of basic CMOS circuit elements and their sensitivities to both the intrinsic device parameters and the parasitic resistances and capacitances. Using 2.5-V, 0.25-μm CMOS devices as an example, we first define the propagation delay of an inverter chain and discuss the loading effect due to fan-out and wiring capacitances. Three performance factors – the switching resistance $R_{sw}$, input capacitance $C_{in}$, and output capacitance $C_{out}$ – are introduced in terms of a delay equation, followed by several subsections detailing their sensitivity to various device parameters. The last subsection deals with the performance factors of two-way NAND circuits.

### 5.3.1 PROPAGATION DELAY AND DELAY EQUATION

In this subsection, we define the propagation delay and the delay equation of a static CMOS gate. While CMOS inverters are used as an example to build the basic framework, most of the formulation and performance factors are equally applicable to other NAND and NOR circuits that perform more general logic operations.

#### 5.3.1.1 PROPAGATION DELAY OF A CMOS INVERTER CHAIN

The basic switching characteristics of a CMOS inverter with a step input waveform have been briefly touched upon in Section 5.1.1. In a practical logic circuit, a CMOS inverter is driven by the output from a previous stage whose waveform has a finite rise or fall time associated with it. One way to characterize the switching delay or the performance of an inverter is to construct a cascaded chain of identical inverters as shown in Fig. 5.26, and consider the propagation delay of a logic signal going through them. Load capacitors can be added to the output node of each inverter to simulate the wiring capacitance it may drive in addition to the next inverter.

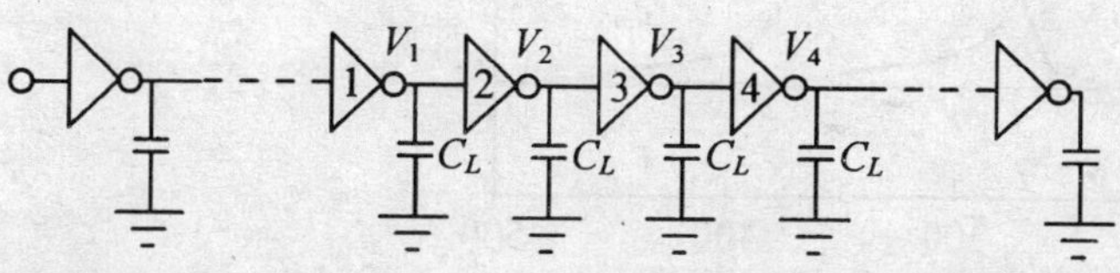


**FIGURE 5.26.** A linear chain of CMOS inverters (fan-out = 1). Each triangular symbol represents a CMOS inverter consisting of an nMOSFET and a pMOSFET as shown in Fig. 5.2. Power-supply connections are not shown.

For a given CMOS technology, the propagation delay is experimentally determined by constructing a *ring oscillator* with a large, odd number of CMOS inverters connected head to tail and measuring the oscillating frequency of the signal at any given point on the ring when the power-supply voltage is applied. The sustained oscillation is a result of strong positive feedback due to the voltage gain of each inverter stage ($|dV_{out}/dV_{in}| > 1$ in Fig. 5.4). The period of the oscillation is given by $n(\tau_n + \tau_p)$, where $n$ is the number of stages (an odd number) and $\tau_n$, $\tau_p$ are inverter delays per stage for rising and falling inputs, respectively. In other words, in one period the logic signal propagates around the ring twice.

Because of the complexity of the current expressions for short-channel MOSFETs and the voltage dependence of both intrinsic and extrinsic capacitances, a circuit model such as SPICE is needed to solve the propagation delay numerically. In order to gain insight into how the voltage and current waveforms look during a switching event, we consider the example of a 0.25-μm CMOS inverter with the device parameters listed in Table 5.2. All lithography dimensions and contact borders, e.g., $a$, $b$, and $c$ in Fig. 5.10, are assumed to be 0.35 μm (nonfolded). The power-supply voltage is 2.5 V (Davari *et al.*, 1988a).

The propagation delay is evaluated by introducing a step voltage signal at the input of the linear inverter chain in Fig. 5.26. After a few stages, the signal waveform has become a *standardized signal*, i.e., one that has stablized and remains a constant shape independent of the number of stages of propagation. There are also a few stages following the ones of interest for maintaining the same capacitive loading of each stage. Figure 5.27 shows an example of waveforms at four successive stages, $V_1$, $V_2$, $V_3$, $V_4$, for the unloaded case, $C_L = 0$. As $V_1$ rises, the nMOSFET

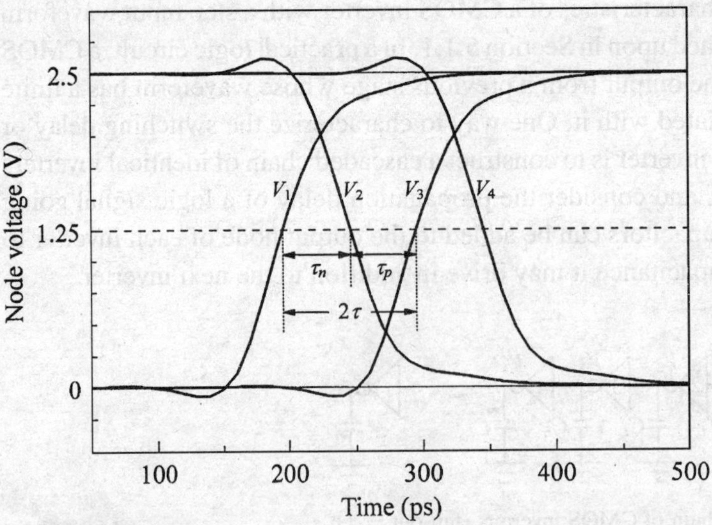


**FIGURE 5.27.** Successive voltage waveforms of the CMOS inverter chain in Fig. 5.26 ($C_L = 0$). The delay is measured by their intersections with the $V_{dd}/2$ (dashed) line as shown.

**TABLE 5.2** 0.25 μm CMOS Parameters for Circuit Modeling

| Source | Parameter | Value | |
|---|---|---|---|
| Assumed | Power supply voltage $V_{dd}$ (V) | 2.5 | |
| | Channel length $L$ (μm) | 0.25 | |
| | Lithography ground rules $a, b, c$ (μm) | 0.35 | |
| | Gate oxide thickness $t_{ox}$ (nm) | 7 | |
| | Linearly extrapolated threshold voltage $V_{on}$ (V) | ±0.5 | |
| | | nMOSFET | pMOSFET |
| | Source-and-drain series resistance $R_{sd}$ (Ω-μm) | 1000 | 2000 |
| | Saturation velocity $v_{sat}$ (cm/s) | $7 \times 10^6$ | $7 \times 10^6$ |
| | Substrate and well doping concentrations $N_a$, $N_d$ (cm$^{-3}$) | $10^{17}$ | $10^{17}$ |
| | Gate-to-source or -drain (per edge) overlap capacitance $C_{ov}$ (fF/μm) | 0.3 | 0.3 |
| | Drain-induced barrier lowering $\Delta V_t$ between $V_{ds} = 0$ and $V_{ds} = V_{dd}$ (V) | 0.03 | 0.03 |
| | Body-effect coefficient $m$ | 1.13 | 1.13 |
| | Device widths $W_n$, $W_p$ (μm) | 10 | 25 |
| Derived | Intrinsic channel capacitance per unit width $\varepsilon_{ox}L/t_{ox}$ (fF/μm) | 1.23 | 1.23 |
| | Source and drain junction capacitance (at $V_j = V_{dd}/2$, based on $N_a$, $N_d$), $C_j$ (fF/μm$^2$) | 0.64 | 0.64 |
| | Effective mobility (at $V_g = V_{dd}/2$), $\mu_{eff}$ (cm$^2$/V-s) | 400 | 100 |
| | On currents (at $V_g = V_{ds} = V_{dd}$), $I_n$, $I_p$ (mA/μm) | 0.414 | 0.17 |

of inverter 2 is turned on, which pulls $V_2$ to ground. The fall of $V_2$ then turns on the pMOSFET of inverter 3, which causes $V_3$ to rise, and so on. If one draws a straight line through the midpoints of all the waveforms at $V = V_{dd}/2$, one can define the pull-down propagation delay $\tau_n$ as the time interval between $V_1$ and $V_2$ along that line. Similarly, the pull-up propagation delay $\tau_p$ is the time interval between $V_2$ and $V_3$ along $V = V_{dd}/2$. The definitions here are consistent with those in Fig. 5.5 for step inputs. A better-defined quantity is the CMOS propagation delay, $\tau = (\tau_n + \tau_p)/2$, which is one-half the time delay between the parallel waveforms $V_1$ and $V_3$ or $V_2$ and $V_4$. The time $\tau$ is also the delay measured experimentally from CMOS ring oscillators. It is equal to the oscillation period divided by two times the number of stages as stated before. In this specific example, the device width ratio is chosen to be $W_p/W_n = 2.5$ so that the pull-down delay equals the pull-up delay, i.e., $\tau_n = \tau_p = \tau$. In general, $\tau_n$ and $\tau_p$ may be different from each other, and the CMOS delay may be dominated by either the nMOSFET or the pMOSFET.

#### 5.3.1.2 BIAS-POINT TRAJECTORIES IN A SWITCHING EVENT

As the logic signal arrives at the input gate of an inverter, a transient current flows in either the nMOSFET or the pMOSFET of that inverter until the output node completes its high-to-low or low-to-high transition. It is instructive to examine the bias-point trajectories through a family of $I_{ds}$–$V_{ds}$ curves during a pull-down or pull-up switching event. Figure 5.28(a) plots the trajectories of nMOSFET (solid dots) and pMOSFET (open circles) currents of inverter 2 in Fig. 5.26 versus the output node voltage $V_2$, as $V_2$ is pulled down from $V_{dd}$ to ground. The points are plotted in constant 10-ps time intervals over a background of nMOSFET $I_{ds}$–$V_{ds}$ curves in the negative direction (discharging currents) and pMOSFET $I_{ds}$–$V_{ds}$ curves in the positive direction (charging currents). The pMOSFET current is very low throughout the transition, indicating negligible power dissipation from crossover currents between the power-supply terminal and the ground. The output node, initially at $V_{dd}$, is discharged by the nMOSFET current, which reaches its highest value midway during switching when $V_2 \approx V_{dd}/2$ and $V_{in} \approx 0.9\, V_{dd}$. This point is also where the voltage waveform $V_2$ in Fig. 5.27 exhibits the maximum downward slope. The peak current is typically 80–90% of the maximum on current at $V_g = V_{ds} = V_{dd}$. The exact percentage depends on the detailed device parameters such as mobility, velocity saturation, and series resistance. Likewise, Fig. 5.28(b) shows the bias-point trajectories of both transistors in inverter 3 when the output node $V_3$ is pulled up from ground to $V_{dd}$. In this case, the nMOSFET current (open circles) is negligible, while the pMOSFET current (solid dots) reaches its peak value when $V_3 \approx V_{dd}/2$, as in the pull-down case. The two bias trajectories are basically similar to each other and are insensitive to loading conditions. At larger $C_L$, the delay time between the points increases, but the shape of the curve remains essentially the same.

#### 5.3.1.3 DELAY EQUATION: SWITCHING RESISTANCE, INPUT AND OUTPUT CAPACITANCE

The simulations plotted in Fig. 5.27 and Fig. 5.28 are for the unloaded case with fan-out = 1. In general, $C_L \neq 0$, and the output of an inverter may drive more than one stage. In the latter case the fan-out is 2, 3, . . . , which means that each inverter in the chain is driving 2, 3, . . . stages in parallel. Each receiving stage is assumed to have the same widths as the sending stage. There are also situations where an inverter is driving another stage wider than its own widths. Such cases can be covered mathematically by generalizing the definition of "fan-out" to include nonintegral numbers, provided that the same n- to p-width ratio is always maintained. Fan-outs greater than 3 are rarely used in CMOS logic circuits, as they lead to significantly longer delays.

Figure 5.29 plots the inverter delay $\tau$ versus the load capacitance $C_L$ for fan-out = 1, 2, and 3, simulated with the device parameters in Table 5.2. For each fan-out, the delay increases linearly with $C_L$ with a constant slope independent of fan-out.

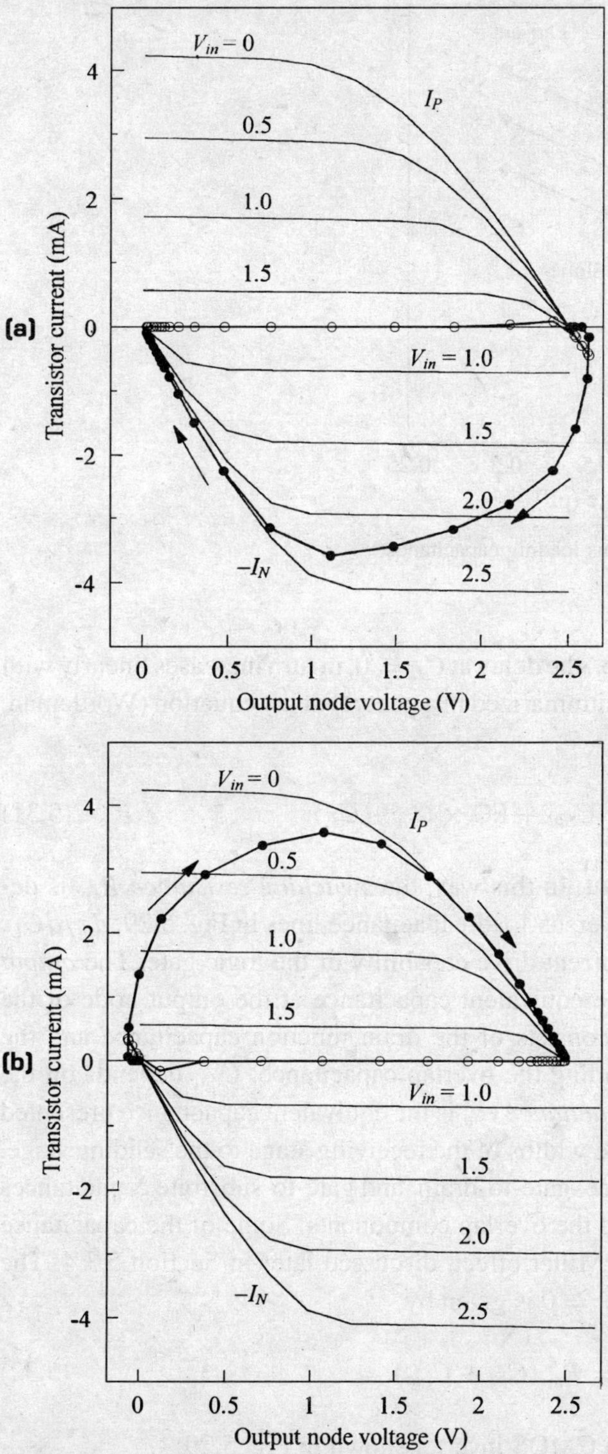


**FIGURE 5.28.** Bias-point switching trajectories of n- and pMOSFETs for (a) pull-down transition of node $V_2$ in inverter 2, and (b) pull-up transition of node $V_3$ in inverter 3. The bias points are plotted in equal time intervals, which for the unloaded case ($C_L = 0$) are 10 ps apart.

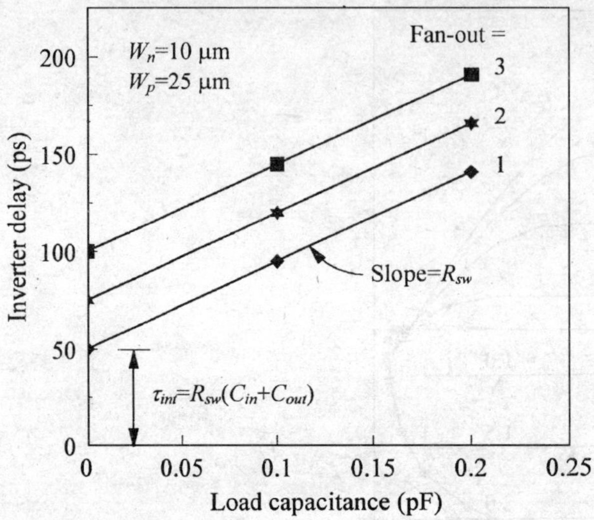


**FIGURE 5.29.** Inverter delay $\tau$ versus loading capacitance $C_L$ for fan-out of 1, 2, and 3.

The intercept with the $y$-axis, i.e., the delay at $C_L = 0$, in turn increases linearly with the fan-out. These facts can be summarized in a general delay equation (Wordeman, 1989),

$$\tau = R_{sw} \times (C_{out} + \text{FO} \times C_{in} + C_L), \tag{5.31}$$

where FO represents the fan-out. In this way, the *switching resistance* $R_{sw}$ is defined as the slope of the delay-versus-load-capacitance lines in Fig. 5.29, $d\tau/dC_L$. It is a direct indicator of the current drive capability of the logic gate. The *output capacitance* $C_{out}$ represents the equivalent capacitance at the output node of the sending stage, which usually consists of the drain junction capacitance and the drain-to-gate capacitance including the overlap capacitance. $C_{out}$ depends on the layout geometry. The *input capacitance* $C_{in}$ is the equivalent capacitance presented by one-unit (FO = 1) input-gate widths of the receiving stage to the sending stage. $C_{in}$ consists of the gate-to-source, gate-to-drain, and gate-to-substrate capacitances including both the intrinsic and the overlap components. Some of the capacitance components are subject to the Miller effect, discussed later in Section 5.3.4. The minimum unloaded delay at $C_L = 0$ is given by

$$\tau_{int} = R_{sw}(C_{in} + C_{out}), \tag{5.32}$$

which is 50 ps for the 0.25-μm CMOS inverter shown in Fig. 5.29.

***The delay equation (5.31) not only allows the delay to be calculated for any fan-out and loading conditions but also decouples the two important factors that govern CMOS performance: current and capacitance***. Current drive capability is represented by $R_{sw}$, which is inversely proportional to the large-signal transconductance appropriate for digital circuits (Solomon, 1982). Strictly speaking, n- and pMOSFETs are coupled together in a CMOS inverter and there is no unique way of separating their individual switching factors. In an effort to illustrate how CMOS performance is related to the device currents, however, we decompose the switching resistance $R_{sw}$ of an inverter into $R_{swn}$ and $R_{swp}$ in terms of the pull-down and pull-up delays $\tau_n$ and $\tau_p$ defined in Fig. 5.27, i.e., $R_{swn} \equiv d\tau_n/dC_L$ and $R_{swp} \equiv d\tau_p/dC_L$. Since $\tau = (\tau_n + \tau_p)/2$, it follows that $R_{sw} = (R_{swn} + R_{swp})/2$. One can write the following empirical expressions for $R_{swn}$ and $R_{swp}$:

$$R_{swn} = k_n \frac{V_{dd}}{W_n I_n} \tag{5.33}$$

and

$$R_{swp} = k_p \frac{V_{dd}}{W_p I_p}. \tag{5.34}$$

Here $W_n$ and $W_p$ are the device widths, $I_n$ and $I_p$ are the maximum on currents per unit width at $V_g = V_{ds} = V_{dd}$ as listed in Table 5.2, and $k_n$ and $k_p$ are numerical fitting parameters, which are 0.74 and 0.78 for the 0.25-μm devices in Table 5.2. These values are significantly higher than those in Eqs. (5.3) and (5.4), where $k_n = k_p = 0.5$ for the case of step inputs. The difference is due to the finite rise time or fall time of the standardized input signal.

The switching resistances extracted from the above specific example are listed in Table 5.3. For the CMOS inverters, $W_p/W_n$ was chosen to be 2.5 to compensate for the difference between $I_n$ and $I_p$, so that $R_{swn} \approx R_{swp} \approx R_{sw}$ and $\tau_n \approx \tau_p \approx \tau$.

Both the input and the output capacitances, $C_{in}$ and $C_{out}$, in Eq. (5.31) are approximately proportional to $W_n + W_p$, since both nMOSFET and pMOSFET contribute more or less equally per unit width to the node capacitance whether they are being turned on or being turned off. This assumes that all the capacitances per unit width are symmetrical between the n- and p-devices, as is the case in Table 5.2. The specific numbers for the case in Fig. 5.29 are listed in Table 5.3. Note that $(C_{in} + C_{out})/(W_n + W_p)$ is substantially larger than the intrinsic channel capacitance per unit width, 1.23 fF/μm, listed in Table 5.2.

**TABLE 5.3** Extracted Performance Factors of the 0.25-μm CMOS in Table 5.2

| | |
|---|---|
| nMOSFET switching resistance, $W_n R_{swn}$ (Ω-μm) | 4450 |
| pMOSFET switching resistance, $W_p R_{swp}$ (Ω-μm) | 11500 |
| Input capacitance, $C_{in}/(W_n + W_p)$ (fF/μm) | 1.55 |
| Output capacitance, $C_{out}/(W_n + W_p)$ (fF/μm) | 1.61 |

#### 5.3.1.4 CMOS DELAY SCALING

It is instructive to reexamine, from the delay-equation point of view, how CMOS performance improves under the rules of constant-field scaling outlined in Section 4.1.1. Let us assume that the first five parameters in Table 5.2 are scaled down by a factor of two, i.e., $V_{dd} = 1.25$ V, $L = 0.125$ μm, $t_{ox} = 3.5$ nm, $V_{on} = \pm 0.25$ V, and $a, b, c = 0.175$ μm (lithography ground rules). If the source and drain series resistances in the scaled CMOS are also reduced by a factor of two, i.e., $R_{sdn} = 500$ Ω-μm and $R_{sdp} = 1000$ Ω-μm, the on currents per unit device width will remain essentially unchanged, i.e., $I_n = 0.414$ mA/μm and $I_p = 0.17$ mA/μm (both the mobility and the saturation velocity are the same as before). Since $V_{dd}$ is reduced by a factor of two, both n- and p-switching resistances normalized to unit device width, $W_n R_{swn}$ and $W_p R_{swp}$, improve by a factor of two. At the same time, all the capacitances per unit width should be kept the same. These include the gate capacitance $\varepsilon_{ox} L/t_{ox}$, the overlap capacitance (0.3 fF/μm), and the junction capacitance. Note that the junction capacitance per unit area, $C_j$, may go up by a factor of two due to the higher doping needed to control the short-channel effect, but the junction capacitance per unit device width is proportional to $(a + b + c)C_j$ and therefore remains unchanged. Combining all the above factors, one obtains that both $C_{in}/(W_n + W_p)$ and $C_{out}/(W_n + W_p)$ are unchanged and the intrinsic delay given by Eq. (5.32) improves by a factor of two to 25 ps.

In practice, one cannot follow the above ideal scaling for various reasons. The most important one is that the threshold voltage cannot be reduced without a substantial increase in the off current, as discussed extensively in Section 4.2. A more detailed tradeoff among CMOS performance, active power, and standby power will be considered in Section 5.3.3.

### 5.3.2 DELAY SENSITIVITY TO CHANNEL WIDTH, LENGTH, AND GATE OXIDE THICKNESS

The next few subsections examine CMOS delay sensitivity to various device parameters, both intrinsic and parasitic, as listed in Table 5.2. To begin with, this subsection discusses the effect of device width, channel length, and gate oxide thickness on CMOS performance.

#### 5.3.2.1 CMOS DELAY SENSITIVITY TO pMOSFET/nMOSFET WIDTH RATIO

When the p- to n-device width ratio $W_p/W_n$ is varied in a CMOS inverter, the relative current drive capabilities $R_{swn}$ and $R_{swp}$, and therefore $\tau_n$ and $\tau_p$, also vary. Figure 5.30 plots the intrinsic delay (FO = 1, $C_L = 0$) of CMOS inverters as a function of the device width ratio. The rest of the device parameters are the same as in Table 5.2. As $W_p/W_n$ increases, $\tau_p$ decreases but $\tau_n$ increases. At $W_p/W_n \approx 2.5$, the pull-up time becomes equal to the pull-down time, which gives the best noise

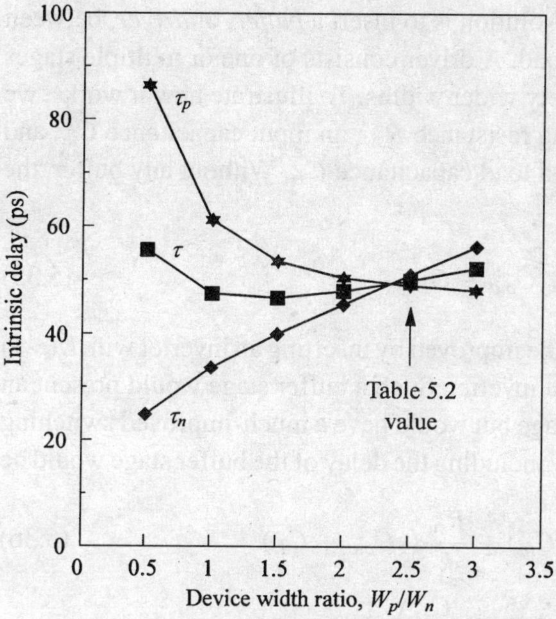


**FIGURE 5.30.** Intrinsic CMOS inverter delays $\tau_n$, $\tau_p$, and $\tau$ for FO = 1 and $C_L = 0$ versus p- to n-device width ratio.

margin, as discussed in Section 5.1.1. The overall delay, $\tau = (\tau_n + \tau_p)/2$, on the other hand, is rather insensitive to the width ratio, showing a shallow minimum at $W_p/W_n \approx 1.5$. The specific example in the last subsection used $W_p/W_n = 2.5$, so that $\tau_n \approx \tau_p \approx \tau = 50$ ps, which is within 10% of the minimum delay at $W_p/W_n = 1.5$. It should be noted that only the intrinsic or unloaded delay exhibits a minimum at $W_p/W_n = 1.5$. The minimum delay for wire-loaded circuits tends to occur at a larger $W_p/W_n$ ratio.

#### 5.3.2.2 DEVICE WIDTH EFFECT WITH RESPECT TO LOAD CAPACITANCE

From the discussions in Section 5.3.1, it is clear that if $W_n$ and $W_p$ are scaled up by the same factor without changing the ratio $W_p/W_n$, the intrinsic delay remains the same. The switching resistance, $R_{sw} = d\tau/dC_L$, however, is reduced by that same factor. So for a given capacitive load $C_L$, the delay improves. In fact, it has been argued that for high-performance purposes, one can scale up the device size until the circuit delays are mostly device-limited, i.e., approaching intrinsic delays (Sai-Halasz, 1995). This can be accomplished, if necessary, by increasing the chip size, because the capacitance due to wire loading increases only as the linear dimension of the chip (2 pF/cm in Section 5.2.4), while the effective device width can increase as the area of the chip if one uses corrugated (folded) gate structures. Of course, delays of global interconnects, as well as chip power and cost, will go up as a result.

In practical CMOS circuits, one tries to avoid the situation where a device drives a capacitive load much greater than its own capacitance, as that results in delays much longer than the intrinsic delay. One solution is to insert a *buffer*, or *driver*, between the original sending stage and the load. A driver consists of one or multiple stages of CMOS inverters with progressively wider widths. To illustrate how it works, we consider an inverter with a switching resistance $R_{sw}$, an input capacitance $C_{in}$, and an output capacitance $C_{out}$, driving a load capacitance $C_L$. Without any buffer, the single-stage delay is

$$\tau = R_{sw}(C_{out} + C_L). \tag{5.35}$$

If $C_L \gg C_{in}$ and $C_{out}$, the delay may be improved by inserting an inverter with $k$ (>1) times wider widths than the original inverter. Such a buffer stage would present an equivalent FO $= k$ to the sending stage but would have a much-improved switching resistance, $R_{sw}/k$. The overall delay including the delay of the buffer stage would be

$$\tau_b = R_{sw}(C_{out} + kC_{in}) + \frac{R_{sw}}{k}(kC_{out} + C_L) \tag{5.36}$$

$$= R_{sw}\left(2C_{out} + kC_{in} + \frac{C_L}{k}\right).$$

It is easy to see that the best choice of the buffer width is $k = (C_L/C_{in})^{1/2}$, which yields a minimum delay of

$$\tau_{b\,\min} = R_{sw}(2C_{out} + 2\sqrt{C_{in}C_L}). \tag{5.37}$$

For heavy loads ($C_L \gg C_{in}$, $C_{out}$), $\tau_{b\,\min}$ can be substantially shorter than the unbuffered delay $\tau$. To drive even heavier loads, multiple-stage buffers can be designed for best results.

#### 5.3.2.3 SENSITIVITY OF DELAY TO CHANNEL LENGTH

Channel length offers the biggest lever for CMOS performance improvement. Higher on currents are obtained at shorter channel lengths, which means lower switching resistance. Figure 5.31(a) shows the variation of $R_{swn}$ and $R_{swp}$ with channel length assuming the rest of the device parameters are given by Table 5.2 (no short-channel $V_t$-roll-off is assumed). Note that the pMOSFET tends to improve more than the nMOSFET, since the latter is more velocity-saturated, rendering $I_n$ less sensitive to channel length. Not only does the switching resistance improve with shorter channel lengths, intrinsic capacitances also decrease with channel length, as shown in Fig. 5.31(b). Here $C_{in}$ is more sensitive to the channel length than $C_{out}$ since a larger part of $C_{in}$ comes from the intrinsic channel capacitance. Combining all of the above effects, one can plot the intrinsic delay given by Eq. (5.32) as a function of channel length, as shown in Fig. 5.31(c). In general, the inverter delay improves approximately linearly with channel length near the design point of interest.

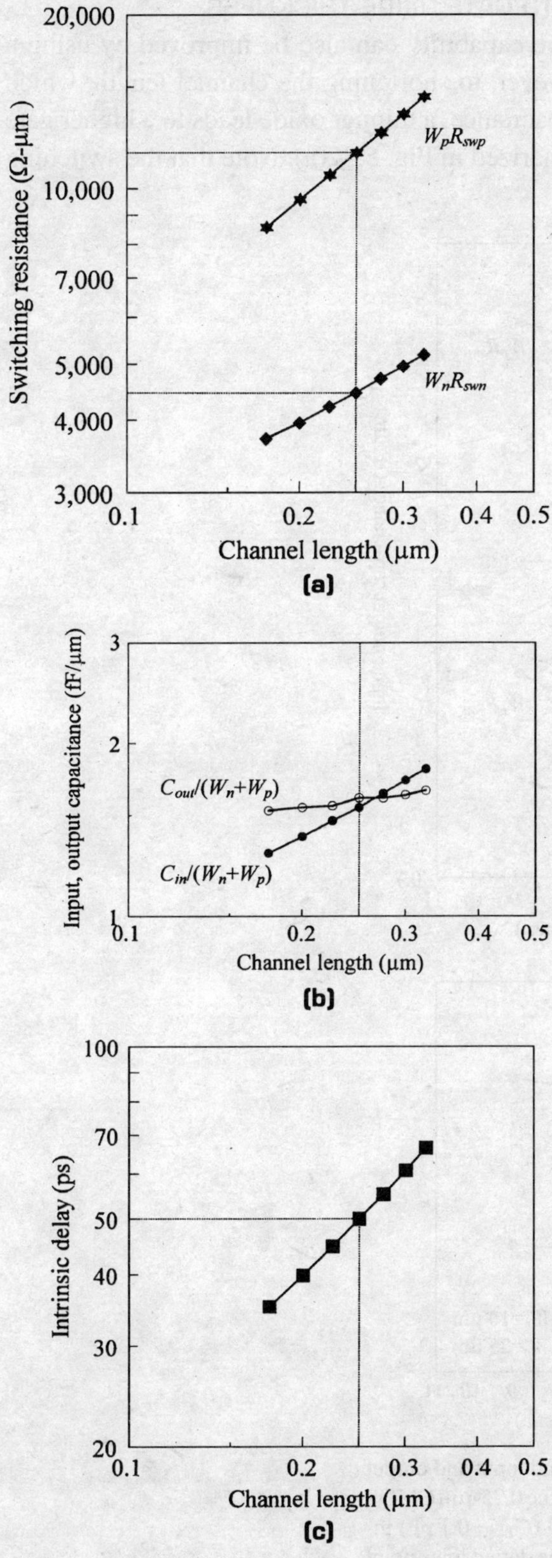


**FIGURE 5.31.** (a) Switching resistances, (b) input and output capacitances, and (c) intrinsic CMOS inverter delay versus channel length for the devices listed in Table 5.2. Both n- and pMOSFETs are assumed to have the same channel length. The dotted lines point to the values for the standard 0.25-μm channel length.

#### 5.3.2.4 SENSITIVITY OF DELAY TO GATE OXIDE THICKNESS

Switching resistance or current drive capability can also be improved by using a thinner gate oxide. In contrast, however, to shortening the channel length, which helps both the resistance and the capacitance, a thinner oxide leads to a higher gate capacitance. These effects are summarized in Fig. 5.32(a). Note that the switching

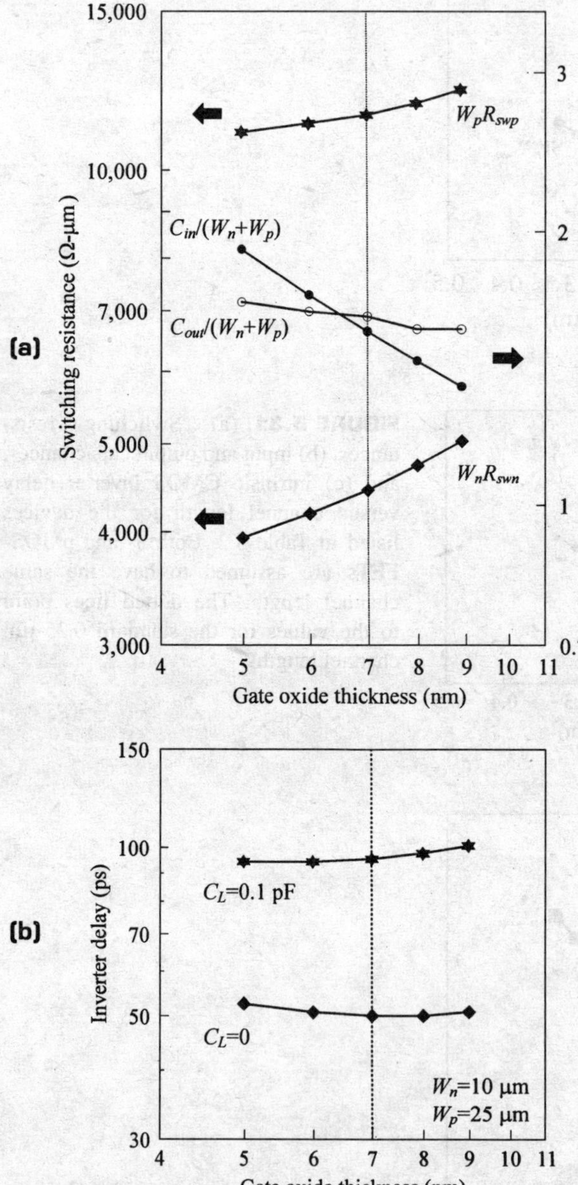


**FIGURE 5.32.** (a) Switching resistances and input and output capacitances versus gate oxide thickness for the 0.25-μm CMOS listed in Table 5.2. (b) Unloaded and loaded ($C_L = 0.1$ pF) inverter delays versus gate oxide thickness. The dotted lines point to the standard thickness of 7 nm.

resistances do not improve linearly with $t_{ox}$, due to two effects: series resistance and field-dependent mobility. Series resistance tends to dilute any reduction in the intrinsic channel resistance gained from thinner oxides. For a given gate voltage, thinner oxides also cause a higher vertical field in the inversion layer, which results in mobility degradation as discussed in Section 3.1.5. This is more apparent in the less velocity-saturated pMOSFETs, whose on current suffers more from mobility degradation than nMOSFETs. It can also be seen in Fig. 5.32(a) that $C_{in}$ is more sensitive to $t_{ox}$ than is $C_{out}$, for the same reason as their dependence on channel length stated in the previous paragraph.

If one combines all the factors listed in Fig. 5.32(a), the net result is that the intrinsic delay is nearly independent of the gate oxide thickness, as shown in Fig. 5.32(b). Loaded delays, on the other hand, do improve with thinner oxides because of the higher transconductances or on currents. It should be pointed out that the above sensitivity study only considers $t_{ox}$ variations at the level of the circuit model, while keeping all other parameters unchanged. In other words, the interdependence between $t_{ox}$ and $V_t$ or $L$ at the process or device level is not taken into account. From a device-design point of view, thinner oxides would allow shorter channel lengths and therefore potentially higher circuit performance.

### 5.3.3 SENSITIVITY OF DELAY TO POWER-SUPPLY VOLTAGE AND THRESHOLD VOLTAGE

This subsection addresses the dependence of CMOS delay on power-supply voltage and threshold voltage. The effect is mainly through the switching-resistance factor, as both the input and output capacitances are relatively insensitive to $V_{dd}$ and $V_{on}$. If one reduces the power-supply voltage for given hardware with a fixed threshold voltage, $V_{on} = 0.5$ V (linearly extrapolated), the intrinsic delay increases rapidly as shown by the solid curve in Fig. 5.33. Similar performance degradation occurs to the loaded delay as well, since the curve basically reflects the behavior of the switching resistance. This is mainly due to the loss of large-signal transconductance, $I_{on}/V_{dd}$, as $V_{dd}$ is reduced. In addition, the empirical $k$-factors defined by Eqs. (5.33) and (5.34) also degrade at lower voltages. For example, they increase from about 0.76 at $V_{dd} = 2.5$ V to about 0.95 at $V_{dd} = 1.0$ V following the $V_{on} = 0.5$ V curve. The delays of two-way NAND gates exhibit a very similar $V_{dd}$-dependence to the inverter delay. More discussions on two-way NAND delays can be found in Section 5.3.5.

It is possible to retain most of the performance at a lower supply voltage if $V_{on}$ is reduced in proportion to $V_{dd}$. This is shown by the dashed curve in Fig. 5.33 (assuming different hardware), where the intrinsic delay is computed as a function of power-supply voltage while keeping a constant $V_{on}/V_{dd}$ ratio of 0.2. The delay still increases with the reduction of $V_{dd}$ because of the lower field and therefore lower carrier velocity, but not nearly as steeply as for the constant-$V_{on}$ curve. The

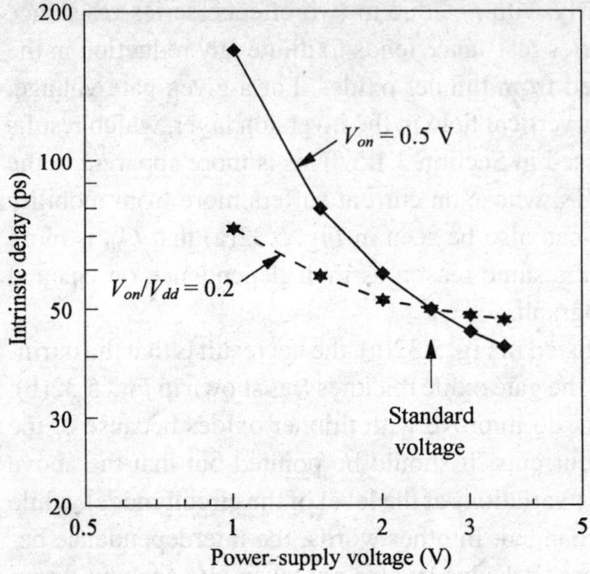


**FIGURE 5.33.** CMOS intrinsic delay versus power-supply voltage. The solid curve is for a constant (linearly extrapolated) threshold voltage, $V_{on} = 0.5$ V (same hardware). The dashed curve is for a constant $V_{on}/V_{dd}$ ratio of 0.2; in other words, $V_{on} = 0.2$, 0.3, 0.4, 0.5, 0.6, 0.7 V (different hardware) for $V_{dd} = 1.0$, 1.5, 2.0, 2.5, 3.0, 3.5 V, respectively.

simulation results in Fig. 5.33 can be summarized in an empirical expression for the switching resistance (Wordeman, 1989):

$$R_{sw} = \frac{f(V_{dd})}{K - (V_{on}/V_{dd})}, \tag{5.38}$$

applicable for $V_{on}/V_{dd} \leq 0.5$. Here $f(V_{dd})$ is a slowly varying function of $V_{dd}$ proportional to the dashed curve in Fig. 5.33, and $K \approx 0.7$ is a numerical fitting parameter. The denominator in Eq. (5.38) can be interpreted as proportional to the factor $V_g - V_t - mV_{dsat}$ in the MOSFET saturation-current expression, Eq. (3.82), with $\langle V_g - mV_{dsat}\rangle \approx 0.7V_{dd}$.

### 5.3.3.1 POWER AND DELAY TRADEOFF IN A $V_{dd}$–$V_t$ DESIGN PLANE

It was discussed in Section 5.1.1 that the active or switching power of a CMOS circuit is given by (neglecting crossover currents)

$$P_{ac} = CV_{dd}^2 f, \tag{5.39}$$

where $C$ is the total equivalent capacitance being charged and discharged in a clock cycle, and $f$ is the clock frequency. The maximum clock frequency is inversely proportional to the device delay $\tau$. Clearly, the most effective way of reducing active power at the device level is to lower the supply voltage. This works fine for low-power electronics in which power dissipation is of primary concern. For high-performance logic chips, however, threshold voltage needs to be reduced as well; otherwise the delay rises sharply as discussed above. Lower threshold voltage, on the other hand, leads to exponentially increasing standby power dissipation.

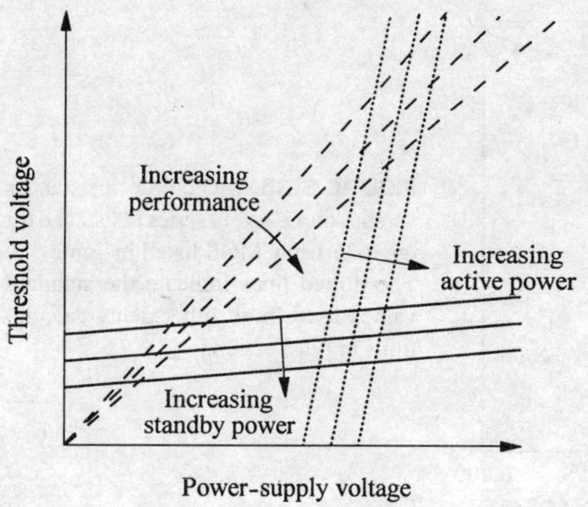


**FIGURE 5.34.** Tradeoff of CMOS delay, active power, and standby power in a $V_{dd}$–$V_t$ design plane. Solid curves represent constant-standby-power contours, which are mainly controlled by $V_t$. Dotted curves represent constant-active-power contours, which are largely controlled by $V_{dd}$. Dashed curves represent constant-delay contours, which are primarily controlled by the ratio $V_t/V_{dd}$.

Using Eq. (4.13), one can express the worst-case standby power dissipation as

$$P_{off} = WV_{dd}I_{off} = WV_{dd}I_0 e^{-qV_t/mkT}, \tag{5.40}$$

where $I_{off}$ is the worst-case off current per unit width at the worst-case temperature $T$, $V_t$ is the worst-case threshold voltage, $I_0$ is the MOSFET current per unit width at threshold voltage, and $W$ is the effective total device width having the worst-case channel length and with a $V_{dd}$ drop across the source and drain.

Qualitatively, the above considerations can be represented by a power-versus-delay tradeoff in a $V_{dd}$–$V_t$ design plane shown in Fig. 5.34 (Mii *et al.*, 1994). A triangular design space is bounded by performance, active power, and standby power constraints. ***Depending on specific requirements of the application, CMOS technologies can be tailored by choosing an appropriate set of power-supply and threshold voltages***. High-performance CMOS usually operates at the lower right-hand corner of the design space and pushes both power limits. In addition to active-power considerations, device reliability issues may also impose a limit on how high $V_{dd}$ can be. A practical limit for the nominal threshold voltage for room-temperature CMOS is about 0.3 V, below which the chip standby power quickly becomes unmanageable. It is possible, in fact not uncommon in DRAM technologies, to provide multiple threshold voltages on a chip to allow the design flexibility of using different devices for different purposes, e.g., in memory and logic circuits (Taur *et al.*, 1995a). This comes, of course, at the expense of additional process complexity and cost.

Low-power CMOS operates at lower supply voltages and possibly at a higher threshold voltage if the standby power is of primary concern. Conservative circuit design must be exercised, however, if $V_{on}/V_{dd} \geq 0.5$, in which case the device delay is very sensitive to $V_{dd}$ or $V_{on}$ tolerances, as is evident in Fig. 5.33.

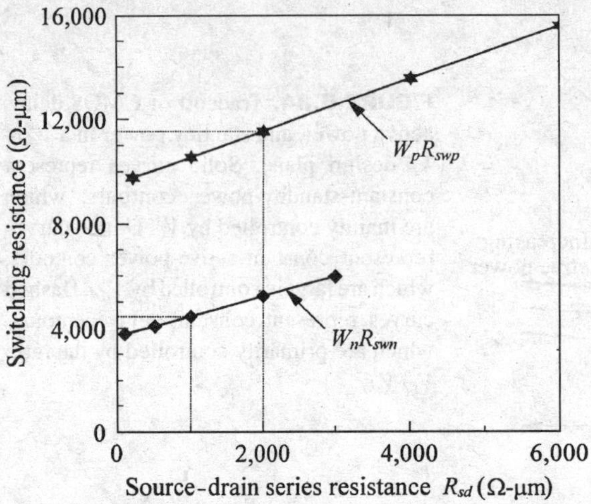


**FIGURE 5.35.** Switching resistances versus source–drain series resistance for the 0.25-μm CMOS listed in Table 5.2. The dotted lines indicate the standard values used in all other delay calculations in this section.

### 5.3.4 SENSITIVITY OF DELAY TO PARASITIC RESISTANCE AND CAPACITANCE

This subsection examines the sensitivity of CMOS delay to parasitic source–drain series resistance, overlap capacitance, and junction capacitance, using the 0.25-μm devices listed in Table 5.2 as an example.

#### 5.3.4.1 SENSITIVITY OF DELAY TO SERIES RESISTANCE

The effect of source–drain series resistance on CMOS delay comes through n- and pMOSFET currents and therefore their switching resistances. Figure 5.35 shows the sensitivity of n- and p-switching resistances to the series resistance $R_{sd}$. Since pMOSFETs have a lower current per unit width, they can tolerate a higher $R_{sd}$ for the same percentage of degradation. For the default values assumed in Table 5.2, $R_{sd} = 1000$ Ω-μm for n and 2000 Ω-μm for p, both devices are degraded by about 20% in terms of their current drive capability. A simple rule of thumb for estimating the performance loss due to series resistance is to add 75–100% of $R_{sd}$ to the intrinsic switching resistance, i.e., $\Delta R_{swn} \approx (0.75\text{–}1.0)R_{sdn}$ and $\Delta R_{swp} \approx (0.75\text{–}1.0)R_{sdp}$.

#### 5.3.4.2 SENSITIVITY OF DELAY TO OVERLAP CAPACITANCE

Gate-to-drain overlap capacitance is a serious performance detractor in lightly loaded CMOS circuits. It not only enters the input capacitance but is also a component of the output capacitance, sometimes further amplified by feedback effects. Figure 5.36 shows both the input and the output capacitances as a function of the overlap capacitance $C_{ov}$ (per edge). The dotted line points to the value assumed in Table 5.2, 0.3 fF/μm. This is about the lowest value of $C_{ov}$ that can be achieved in practice, as discussed in Section 5.2.2. Both the gate-to-source and the gate-to-drain capacitances contribute to the input capacitance. Only the gate-to-drain component enters the output capacitance. However, its contribution is nearly doubled from its

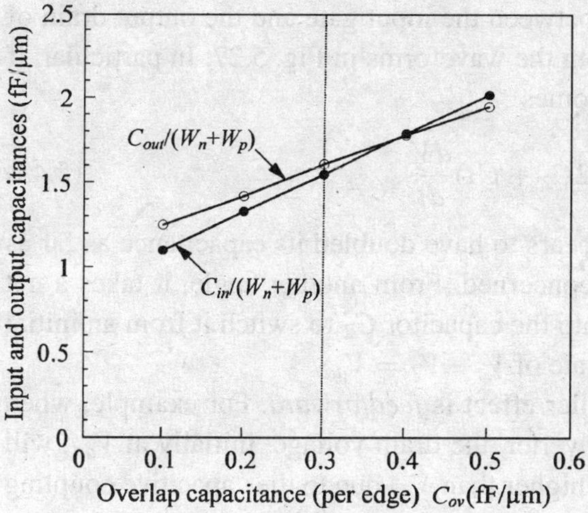


**FIGURE 5.36.** Input and output capacitances versus overlap capacitance. Both n- and pMOSFETs are assumed to have the same $C_{ov}$ per edge. The dotted line indicates the value listed in Table 5.2.

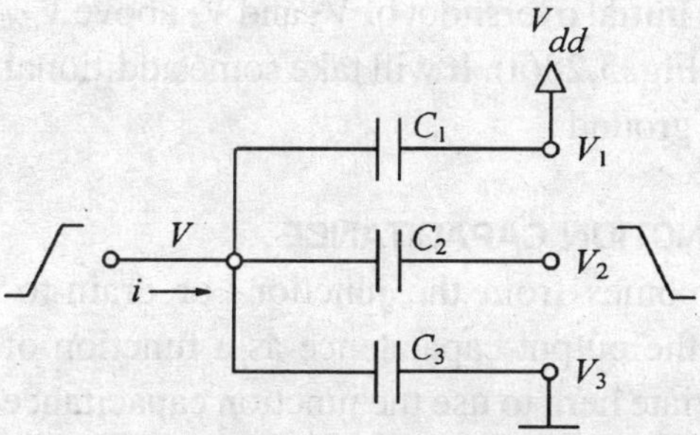


**FIGURE 5.37.** A circuit example illustrating the Miller effect.

original value due to the *Miller effect* explained below. It is estimated that overall an overlap capacitance of 0.3 fF/μm accounts for nearly 40% of the intrinsic delay.

#### 5.3.4.3 MILLER EFFECT

The Miller effect arises when the voltages on both sides of a capacitor being charged or discharged vary with time. Figure 5.37 shows an example of three capacitors connected to a node of voltage $V$ being charged. Each capacitor is connected to a different voltage level on the other side. One can express the charging current $i$ as

$$i = C_1\frac{d(V - V_1)}{dt} + C_2\frac{d(V - V_2)}{dt} + C_3\frac{d(V - V_3)}{dt}. \tag{5.41}$$

Since both $V_1 = V_{dd}$ and $V_3 = 0$ are fixed voltages, one obtains

$$i = C_1\frac{dV}{dt} + C_2\frac{dV}{dt} - C_2\frac{dV_2}{dt} + C_3\frac{dV}{dt}, \tag{5.42}$$

and there is no Miller effect on $C_1$ and $C_3$. However, $dV_2/dt \neq 0$, since $V_2$ varies with time. If $V_2$ varies with time in a direction opposite to that of $V$, it will take more time (and charge) to charge up the node voltage $V$ to a certain level than it otherwise would. This happens, for example, between the input gate and the output drain of a CMOS inverter, as can be seen from the waveforms in Fig. 5.27. In particular, if $dV_2/dt = -dV/dt$, Eq. (5.42) becomes

$$i = (C_1 + 2C_2 + C_3)\frac{dV}{dt}. \tag{5.43}$$

In other words, the capacitor $C_2$ appears to have doubled its capacitance as far as the charging of node voltage $V$ is concerned. From another angle, it takes a net flow of charge of $\Delta Q_2 = 2C_2 V_{dd}$ into the capacitor $C_2$ to switch it from an initial state of $V - V_2 = -V_{dd}$ to a final state of $V - V_2 = V_{dd}$.

Another manifestation of the Miller effect is *feedforward*. For example, when the gate voltage rises in a CMOS inverter, the drain voltage, initially at $V_{dd}$, will momentarily rise to a value slightly higher than $V_{dd}$ due to the capacitive coupling to the gate. This happens before the nMOSFET current starts to flow and brings the drain voltage down, as can be seen from the initial overshoot of $V_2$ and $V_4$ above $V_{dd}$ in Fig. 5.27 and from the $I$–$V$ trajectory in Fig. 5.28(a). It will take some additional amount of charge to pull the drain node to ground.

#### 5.3.4.4 SENSITIVITY OF DELAY TO JUNCTION CAPACITANCE

A major part of the output capacitance comes from the junction- or drain-to-substrate capacitance. Figure 5.38 shows the output capacitance as a function of the junction capacitance. It is most appropriate here to use the junction capacitance per unit device width, $C_J/W$, which equals $(a+b+c)C_j$ for a nonfolded layout. Here $C_j$ is the junction capacitance per unit area given by Eq. (5.12), and $a$, $b$, $c$ are the layout dimensions defined in Fig. 5.10. The point corresponding to the value in Table 5.2 is indicated by the dotted lines. It can be seen that in this case the junction capacitance accounts for about 50% of the output capacitance. The junction

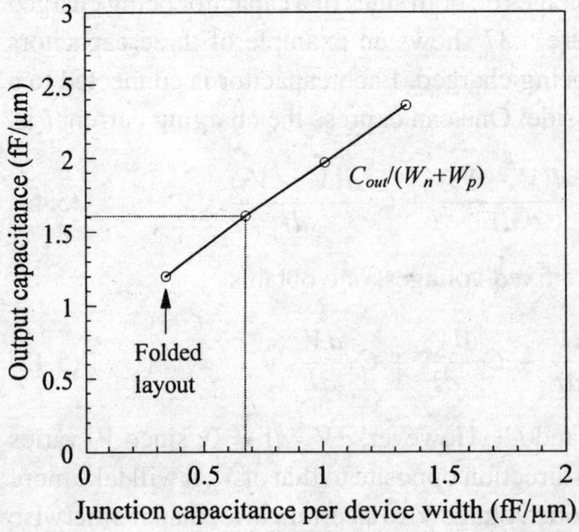


**FIGURE 5.38.** Output capacitance per unit width versus junction capacitance per unit width. Both n- and pMOSFETs are assumed to have the same $C_j$. The dotted lines indicate the value listed in Table 5.2.

**TABLE 5.4** Components of $C_{in}$ and $C_{out}$

| Component | Input Capacitance (%) | Output Capacitance (%) |
|---|---|---|
| Intrinsic gate oxide capacitance | 57 | 14 |
| Overlap capacitance | 43 | 35 |
| Junction capacitance (nonfolded) | — | 51 |

capacitance contribution can be cut in half using the folded layout in Fig. 5.11. This is indicated by the arrow in Fig. 5.38. The folded layout improves the intrinsic inverter delay by about 15%.

It is instructive to break $C_{in}$ and $C_{out}$ for the 0.25-μm CMOS devices listed in Table 5.2 into various components: intrinsic gate capacitance, overlap capacitance, and junction capacitance. This can be done by extrapolating the simulation results in Fig. 5.31(b), Fig. 5.36, and Fig. 5.38. The results are given in Table 5.4. Note that the values of $C_{in}$ and $C_{out}$ are given in Table 5.3. The unloaded delay is proportional to $C_{in} + C_{out}$, in which only about 35% comes from the intrinsic gate oxide capacitance.

### 5.3.5 DELAY OF TWO-WAY NAND AND BODY EFFECT

So far we have been using CMOS inverters, i.e., with fan-in of 1, for studying the performance factors. Many of the basic characteristics also apply to more general CMOS circuits. There are, however, a few other factors associated with the multiple-fan-in NAND gates in which more than one nMOSFET is stacked between the output node and the power-supply ground. This subsection examines these factors using a two-way NAND (Fig. 5.7) as an example.

#### 5.3.5.1 TOP AND BOTTOM SWITCHING OF A TWO-WAY NAND GATE

The simulation is set up with the layout shown in Fig. 5.12 and with the same 0.25-μm CMOS devices listed in Table 5.2, except that a width ratio of $W_p/W_n = 2$ is used instead. In this configuration, the two parallel pMOSFETs are naturally folded. The nMOSFETs are nonfolded. The width of the diffusion region ($V_x$-node) between the two stacked nMOSFETs is assumed to be the minimum lithography dimension, 0.35 μm in this case. To construct a linear chain of two-way NAND gates, one must distinguish between two cases: *top switching* and *bottom switching*, as was first outlined in Section 5.1.2. Referring to Fig. 5.7, top switching means that transistors N1 and P1 are driven by a logic transition propagated through input 1. Input 2 is tied to $V_{dd}$ in this case, so that N2 is always on and P2 is always off. On the other hand, in bottom switching transistors N2 and P2 are driven by a logic signal from the output of the previous stage through input 2, while input 1 is tied to $V_{dd}$. These two switching modes have somewhat different delay characteristics as discussed below.

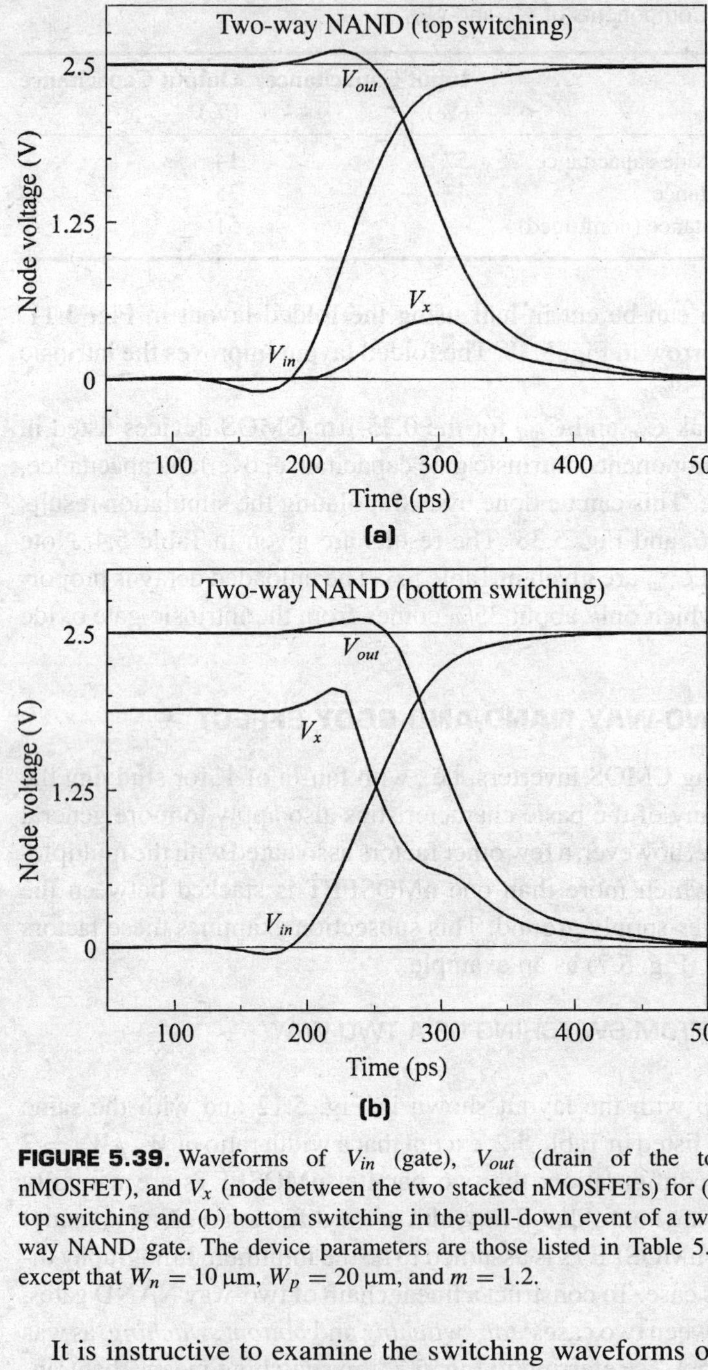


**FIGURE 5.39.** Waveforms of $V_{in}$ (gate), $V_{out}$ (drain of the top nMOSFET), and $V_x$ (node between the two stacked nMOSFETs) for (a) top switching and (b) bottom switching in the pull-down event of a two-way NAND gate. The device parameters are those listed in Table 5.2, except that $W_n = 10\ \mu m$, $W_p = 20\ \mu m$, and $m = 1.2$.

It is instructive to examine the switching waveforms of various node voltages in a two-way NAND. Figure 5.39 plots the input, output, and $V_x$-node voltages versus time during an nMOSFET pull-down event. In the top-switching case in Fig. 5.39(a), the $V_x$-node voltage starts at zero, rises momentarily to a peak about 20% of $V_{dd}$, then falls back to zero together with $V_{out}$. The rise of $V_x$ is necessary to accommodate the discharging current when the top transistor is turned on. In the

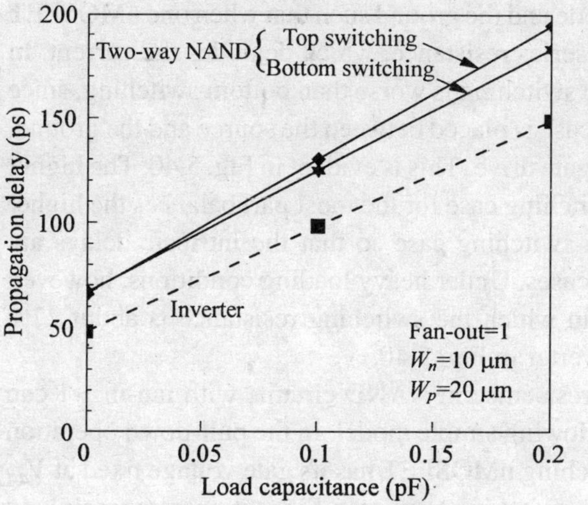


**FIGURE 5.40.** Propagation delay versus load capacitance. The two solid lines are for the top-switching and bottom-switching cases of a two-way NAND gate. The dashed line shows the delay of a CMOS inverter of the same device widths for comparison.

bottom-switching case in Fig. 5.39(b), the $V_x$-node voltage starts at a high value, but quite a bit lower than $V_{dd}$. Even though its gate is tied to $V_{dd}$, the top transistor is initially biased in the subthreshold region since $V_{dd} - mV_x - V_t < 0$ (the factor $m$ comes from the body effect; see the discussion in Section 5.1.2). The exact starting value of $V_x$ depends on a detailed matching of the subthreshold currents in the top and bottom nMOSFETs. When the bottom nMOSFET is turned on, the $V_x$-node is pulled down to ground, followed by $V_{out}$. One can easily figure out the bias point of each transistor, e.g., in the linear or saturation region, from the values of $V_{in}$, $V_x$, and $V_{out} - V_x$ at any given instant.

Figure 5.40 plots the propagation delay of a two-way NAND gate (solid lines), as described above, versus the load capacitance $C_L$. The dashed line shows the delay of an inverter of the same widths for comparison. A delay equation of the same form as Eq. (5.31) also applies to the two-way NAND, but with different values of $R_{sw}$, $C_{in}$, and $C_{out}$. The intrinsic delay ($C_L = 0$) of the two-way NAND is about 40% (1.38×) longer than that of the inverter, for the following reasons. First, let us consider the capacitances. The input capacitance of a two-way NAND stage is essentially the same as that of an inverter. However, a two-way NAND has a higher output capacitance. In the top-switching case, there is an additional gate-to-drain overlap capacitance $C_{ov}$ (no Miller effect) on the pMOSFET side of the two-way NAND layout in Fig. 5.12, compared with the inverter layout in Fig. 5.11(a). In the bottom-switching case, the output capacitance is further increased by additional components on the nMOSFET side. These include the gate capacitance of N1, some overlap capacitance associated with the gate of N1, and a small junction capacitance of the $V_x$-node. In addition to the higher capacitances, the switching resistances, i.e., the slopes of the lines in Fig. 5.40, of the two-way NAND are also higher than that of the inverter. This primarily stems from the stacking of the two nMOSFETs between the output node and the ground such that when one nMOSFET is switching, the other acts like a series resistance, which degrades the current. In terms of switching resistances, top switching is worse than bottom switching, since the series resistance in the former case is placed between the source and the ground, which results in additional loss of gate drive. This is evident in Fig. 5.40. The higher switching resistance in the top-switching case for the most part balances the higher output capacitance in the bottom-switching case so that the intrinsic delays are almost the same between the two cases. Under heavy loading conditions, however, top switching is the worst case, in which the switching resistance is about 27% (1.27×) worse than that of the inverter in Fig. 5.40.

The degradation of switching resistance in NAND circuits with fan-in >1 can be roughly estimated using the following simple model. In the pull-down operation of a two-way NAND, the nonswitching nMOSFET has its gate voltage fixed at $V_{dd}$ and acts like a series resistor to the switching transistor. Since it operates mainly in the linear region during a switching event (see the discussion in Section 5.1.2.), its effective resistance can be approximated by $V_{dsat}/W_n I_n$, where $V_{dsat}$ and $I_n$ are the saturation voltage and current (per unit width) at $V_g = V_{dd}$. In the top-switching case, this resistance is added at the source terminal, which increases the nMOSFET switching resistance by roughly the same amount, i.e., $\Delta R_{swn} = V_{dsat}/W_n I_n$, based on the discussion following Fig. 5.35. Using $R_{sw} = (R_{swn} + \Delta R_{swn} + R_{swp})/2$ and Eqs. (5.33) and (5.34), one can write the switching resistance of a two-way NAND gate as

$$R_{sw}(\text{FI} = 2) \approx \frac{k_n V_{dd} + V_{dsat}}{2W_n I_n} + \frac{k_p V_{dd}}{2W_p I_p}. \tag{5.44}$$

For the above 0.25-μm CMOS example, $W_n I_n \approx W_p I_p$, $k_n \approx k_p \approx 0.75$, and $V_{dsat} \approx 0.4 V_{dd}$ from Fig. 5.28(a). Substitution of these numbers in Eq. (5.44) yields a switching resistance about 1.27 times larger than that of the inverter, consistent with the numerical results stated above. Equation (5.44) can be generalized to larger fan-ins by inserting a multiplying factor of FI − 1 in front of the $V_{dsat}$ term. Since $R_{sw}$ degrades rapidly with increasing fan-in, fan-ins higher than 3 are rarely used in CMOS circuits.

#### 5.3.5.2 DELAY SENSITIVITY TO BODY EFFECT

The delays in Fig. 5.40 are computed based on the set of device parameters in Table 5.2, in which the body-effect coefficient is $m = 1.13$. With all other device parameters being equal, the delay increases with $m$ for two reasons. ***First, the device saturation current decreases with increasing m due to body effect at the drain***. This can be seen from the saturation current expression, Eq. (3.78), for the $n = 1$ case discussed in Section 3.2.2. Other values of $n$ lead to qualitatively similar results. The dependence of the saturation current on $m$ is stronger for less velocity-saturated devices, e.g., pMOSFETs. The fully velocity-saturated current,

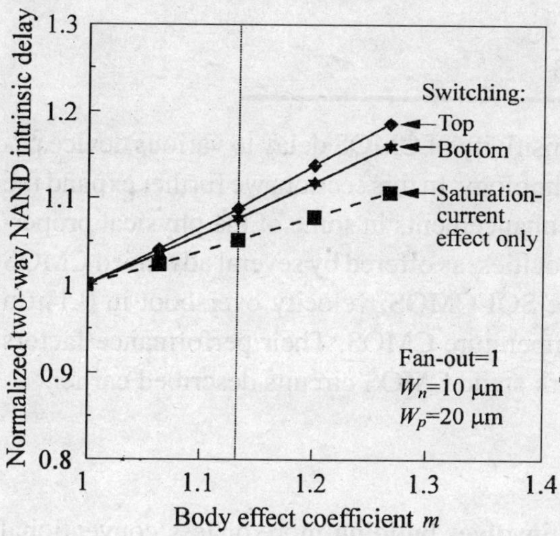


**FIGURE 5.41.** Normalized intrinsic delay of two-way NAND versus body-effect coefficient. The dashed line is for either switching mode, and takes only the body effect on saturation current into account. The solid lines show the overall sensitivity to body effect, including also the effect of source-to-substrate reverse bias.

Eq. (3.80), is independent of $m$. ***The second factor is more applicable to stacked nMOSFETs in NAND gates: when the source potential is higher than the body potential, as in transistor N1 of Fig. 5.7, the threshold voltage increases because of body effect and the current decreases.*** The source-to-body potential of N1 is given by the $V_x$ voltage shown in Fig. 5.39(a) and (b). To a lesser degree, this effect also occurs in a CMOS inverter due to the presence of series resistances at the n- and p-source terminals.

Both of these factors are examined in terms of the intrinsic delay of a two-way NAND gate as a function of the body-effect coefficient in Fig. 5.41. In each case, the delay is normalized with respect to the value at $m = 1$, where there is no body effect. The solid lines show the top switching and bottom switching delays versus $m$, taking both the saturation-current and the source-potential effects into account. At $m = 1.13$, as indicated by the dotted line, an overall delay degradation of $\approx 8\%$ can be attributed to the body effect. The dashed line in Fig. 5.41 shows the delay sensitivity to body effect due to the saturation current factor alone, for which both top and bottom switchings behave nearly the same. The difference between one or the other solid line and the dashed line is a measure of the delay sensitivity to body effect due to the source-to-body potential only. As expected, the top-switching case exhibits a higher sensitivity to such an effect. The effect is not zero in the bottom-switching case, since any source-to-body potential in the nonswitching top transistor can raise its threshold voltage through body effect and therefore increase its effective resistance in the circuit. Roughly speaking, a fraction equal to $m - 1$ of the total switching-resistance degradation in a two-way NAND, which can be estimated from Eq. (5.44), is due to body effect at the source. These factors will be further addressed when we consider the performance of silicon-on-insulator (SOI) CMOS in Section 5.4.1.

## 5.4 PERFORMANCE FACTORS OF ADVANCED CMOS DEVICES

In the last section, we discussed the sensitivity of CMOS delay to various device parameters in a standard bulk CMOS technology. In this section, we further expand the parameter space to include possible enhancements in some of the physical properties, e.g., mobilities and saturation velocities, as offered by several advanced CMOS devices and structures. These include SOI CMOS, velocity overshoot in 0.1-μm CMOS, SiGe MOSFET, and low-temperature CMOS. Their performance factors are assessed in terms of the benchmark static CMOS circuits described earlier.

### 5.4.1 SOI CMOS

Silicon-on-insulator, or SOI, CMOS involves building more or less conventional MOSFETs on very thin layers of crystalline silicon, as illustrated in Fig. 5.42. The thin layer of silicon is separated from the substrate by a thick layer (typically 1000 Å or more) of buried $SiO_2$ film, thus electrically isolating the devices from the underlying silicon substrate and from each other. SOI substrates fabricated by oxygen-ion implantation (SIMOX) and wafer bonding are particularly suitable for VLSI applications due to recent progress in material quality and their compatibility with established CMOS processing technology.

The inherent advantages of SOI devices over bulk CMOS are listed below.

- *Very low junction capacitance.* The source and drain junction capacitance is almost entirely eliminated in SOI MOSFETs. The capacitance through the thick buried oxide layer to the substrate is very small.
- *No body effect.* The threshold voltage of stacked devices in SOI, e.g., transistor N1 in Fig. 5.7, is not degraded by the body effect, since their body potential is not tied to the ground or $V_{dd}$ but can rise to the same potential as the source (node $V_x$ in Fig. 5.7).

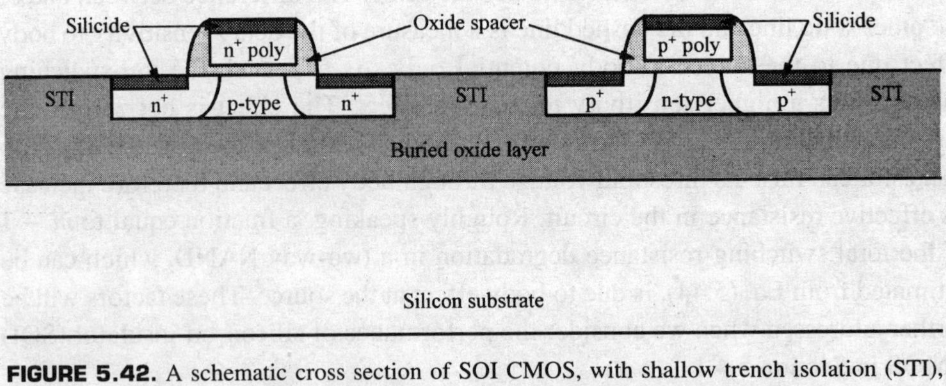


**FIGURE 5.42.** A schematic cross section of SOI CMOS, with shallow trench isolation (STI), dual polysilicon gates, and self-aligned silicide.

- *Soft-error immunity.* In bulk devices, minority carriers are generated along the track of any high-energy particle or ionizing radiation that strikes through the silicon. If the collected charge of a junction node exceeds a certain threshold, it may cause an upset of the stored logic state. This is commonly referred to as a *soft error*. SOI devices offer a potential improvement in the soft-error rate, since the presence of the buried oxide greatly reduces the volume susceptible to ionizing radiation.

#### 5.4.1.1 PARTIALLY DEPLETED AND FULLY DEPLETED SOI MOSFETs

SOI MOSFETs are often distinguished as *partially depleted* (PD) when the silicon film is thicker than the maximum gate depletion width and the devices exhibit a floating-body effect (Yoshimi *et al.*, 1989), and *fully depleted* (FD) when the silicon film is thin enough that the entire film is depleted before the threshold condition is reached. The subthreshold slope of a long-channel FD SOI MOSFET can be near the ideal 60 mV/decade at 300 K. This is because the effective gate depletion width is very large in FD SOI and the body-effect coefficient $m = 1 + 3t_{ox}/W_{dm} \approx 1$. The steeper subthreshold slope permits a lower $V_t$ for the same off current, which in turn allows the devices to be used at lower supply voltages, thereby attracting attention for low-power operation. However, the nearly ideal subthreshold slope (60 mV/decade) occurs only in long-channel devices. ***In short-channel FD SOI MOSFETs, the thick buried oxide acts like a wide gate depletion region, which leaves them vulnerable to source–drain field penetration and results in poor short-channel effects*** (Yan *et al.*, 1991; Su *et al.*, 1994). This can be understood from the $t_{ox}$–$W_{dm}$ design-space plot in Fig. 4.8, where FD SOI tends to operate in a region toward the right.

In PD SOI MOSFETs, there is a neutral body region below the gate depletion boundary. The floating-body or kink effect occurs when carriers of the same type as the body, generated by impact ionization near the drain, are stored in the floating body, which alters the body potential and hence the threshold voltage (Yoshimi *et al.*, 1989). This effect is especially strong in nMOSFETs, due to the higher impact ionization rate of electrons (Fig. 2.41). Floating-body effects become rather complex in dynamic conditions. In a practical switching event, the charging-up of the floating body by impact ionization or junction leakage usually takes a longer time than the input transition. In fast gate ramps, the body potential tends to rise with the gate potential, which reduces $V_t$ and increases the transient current. This phenomenon is referred to as the *drain current overshoot*. ***Even though floating-body effects tend to enhance circuit speed in certain conditions, the drain current overshoot (or undershoot) is history-dependent*** (Gautier *et al.*, 1995; Sherony *et al.*, 1996). The floating-body potential depends on how recently and how often the device has been switched through its high impact ionization conditions. The worst case is when the device (usually the nMOSFET) has been in the on state for a long time and is then turned off but is turned back on again before the body charge reaches the equilibrium state. This poses great difficulties in circuit design. Another undesirable consequence of the floating-body effect in a PD SOI is the higher off current due to a forward-biased body-to-source junction when the drain voltage is high. Using body contacts can restore the device characteristics of SOI MOSFETs to the bulk-MOSFET-like characteristics (Chen *et al.*, 1996). This, however, gives away the body-effect advantage of SOI devices.

#### 5.4.1.2 PERFORMANCE FACTOR OF SOI CMOS

With a major part of the SOI advantage coming from its inherently low junction capacitance (Fig. 5.42), most of the speed improvement occurs in unloaded circuits. Using the 0.25-μm CMOS devices (Table 5.2) analyzed in the previous section as an example, one can estimate the performance advantage of SOI devices over bulk CMOS in terms of the inverter and two-way NAND delays. The layouts assumed are Fig. 5.11(a) (nonfolded) for the inverter stage and Fig. 5.12 for the two-way NAND gate. In general, the propagation delay is expressed by Eq. (5.31):

$$\tau = R_{sw}(C_{out} + \text{FO} \times C_{in} + C_L). \tag{5.45}$$

It is clear from Fig. 5.38 that junction capacitance is a major part of $C_{out}$ in bulk CMOS. If one assumes that the junction capacitance is zero in SOI devices (actually, only the areal component is zero; a small perimeter component is still present), the delay improvement can be estimated from the relative contributions of the various components of $C_{in}$ and $C_{out}$ listed in Tables 5.3 and 5.4. The results are summarized in Table 5.5, where the performance factors for intrinsic ($C_L = 0$) and heavily loaded ($C_L \gg C_{in}, C_{out}$) circuits are listed separately. The percentage of improvement shown is defined as [delay(bulk)/delay(SOI) – 1] × 100%.

For the intrinsic delay of unloaded inverters, $C_L = 0$, there is a substantial improvement of SOI over bulk at fan-out of 1. The improvement factor decreases at larger fan-outs, as the relative weight of $C_{out}$ in $C_{out} + \text{FO} \times C_{in}$ is diluted. For heavily loaded circuits ($C_L \gg C_{in}, C_{out}$), the only important factor as far as the

**TABLE 5.5** SOI Performance Advantage Over Bulk CMOS

| | | | Advantage (%) | |
|---|---|---|---|---|
| **Circuit** | **Fan-in** | **Fan-out** | **Intrinsic** | **Heavily loaded** |
| Inverter | 1 | 1 | 35 | ~0[a] |
| | | 2 | 21 | ~0[a] |
| | | 3 | 15 | ~0[a] |
| Two-way NAND | 2 | 1 | 32 | ~6[b] |
| | | 2 | 22 | ~6[b] |

[a] Assumes floating-body effect is either absent or a neutral factor.
[b] Due to body effect in bulk CMOS with a body-effect coefficient $m = 1.2$.

delay is concerned is $R_{sw}$. If we assume that the floating-body effect is either absent or a neutral factor, there is no improvement from SOI, as it uses the same channel length and oxide thickness as the bulk CMOS. In two-way NAND circuits, there is an additional benefit of SOI due to the absence of body effect in stacked nMOSFETs. This factor enters the switching resistance $R_{sw}$ and therefore helps the heavily loaded delay as well. It should be noted that only the body effect arising from the source-to-body potential is eliminated in SOI, assuming that the floating body is tied to the source. The body effect associated with the drain potential still exists in PD SOI and may affect the saturation current as discussed in Section 5.3.5. From the difference between the top switching curve (worst case at heavy loading) and the "saturation current effect only" curve in Fig. 5.41, one can estimate that the improvement from SOI due to this effect is about 6% for a body-effect coefficient $m \approx 1.2$.

***Since the improvement factor of SOI over bulk varies widely depending on the circuit loading condition, extensive redesign is desirable to take full advantage of the SOI technology***. For example, the device widths should be increased to lessen the effect of wire loading on circuit delays. In general, it is expected that an improvement factor somewhere between the intrinsic and the heavily loaded figures can be achieved at the chip level.

### 5.4.2 VELOCITY OVERSHOOT EFFECT

It was discussed in Section 3.2.2 that for nMOSFETs below 0.2-μm channel length, significant velocity overshoot takes place near the drain, such that the device current may exceed the value governed by the velocity saturation model [e.g., Eq. (3.78) or even Eq. (3.80)] with a $v_{sat}$ valid for longer devices. In this subsection, we examine what benefit this may have on CMOS delays. To serve as an example, we consider a 1.5-V, 0.1-μm CMOS with representative device parameters listed in Table 5.6 (Taur *et al.*, 1993c). The lithography dimension assumed is 0.2 μm. Both the electron and the hole mobilities are somewhat lower than the 0.25-μm case in Table 5.2 because of the higher vertical fields. The parasitic source and drain resistances are reduced by a factor of two from the 0.25-μm values owing to the use of rapid thermal anneal (RTA) process, which produces shallower and more abrupt source–drain extensions. Both the junction capacitance per unit area and the body-effect coefficient are higher because of the higher doping concentration required to control the short-channel effect (i.e., $W_{dm}$ decreases more than $t_{ox}$, as was discussed in Section 4.2.3). The intrinsic inverter delay obtained from the 0.1-μm CMOS with these parameters is 25 ps, which is a factor of two faster than the 0.25-μm devices in the previous section. The two-way NAND delay is about 1.4 times the inverter delay.

To study the primary effect of velocity overshoot on circuit performance, we let the saturation velocity parameter in the circuit model take on values larger than the default value, $7 \times 10^6$ cm/s. Figure 5.43 shows the simulated inverter and two-way NAND delays versus the saturation velocity assumed for both electrons and

**TABLE 5.6** Parameters of the 0.1-μm CMOS Devices

| Parameter | nMOSFET | pMOSFET |
|---|---|---|
| Power-supply voltage $V_{dd}$ (V) | 1.5 | |
| Channel length $L$ (μm) | 0.1 | |
| Lithography ground rules $a, b, c$ (μm) | 0.2 | |
| Gate oxide thickness $t_{ox}$ (nm) | 3 | |
| Linearly extrapolated threshold voltage $V_{on}$ (V) | ±0.4 | |
| Source and drain series resistance $R_{sd}$ (Ω-μm) | 500 | 1000 |
| Saturation velocity $v_{sat}$ (cm/s) | $7 \times 10^6$ | $7 \times 10^6$ |
| Effective mobility (at $V_g = V_{dd}/2$), $\mu_{eff}$ (cm²/V-s) | 330 | 85 |
| Gate-to-source or -drain (per edge) overlap capacitance $C_{ov}$ (fF/μm) | 0.3 | 0.3 |
| Source and drain junction capacitance (at $V_j = V_{dd}/2$, based on $N_a = N_d = 10^{18}$ cm$^{-3}$), $C_j$ (fF/μm²) | 1.75 | 1.75 |
| Body-effect coefficient $m$ | 1.26 | 1.26 |
| Device widths $W_n$, $W_p$ (μm) | 10 | 20 |
| On currents (at $V_g = V_{ds} = V_{dd}$), $I_n$, $I_p$ (mA/μm) | 0.54 | 0.22 |

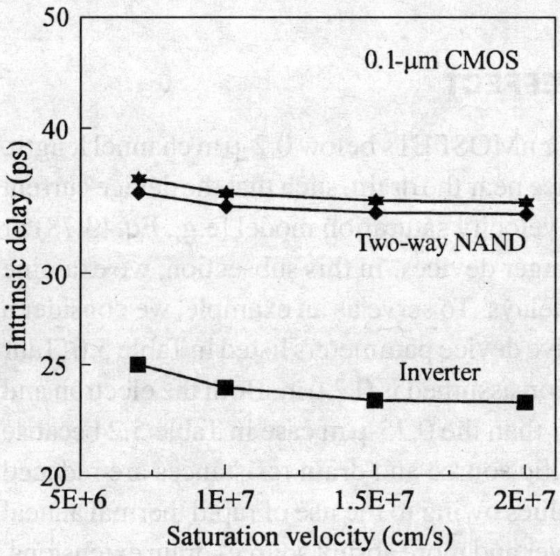


**FIGURE 5.43.** Intrinsic delays ($C_L = 0$) of 0.1-μm CMOS inverter and two-way NAND circuits versus saturation velocity. Electron and hole saturation velocities are varied together.

holes. Physically, the parameter $v_{sat}$ can be interpreted as the carrier velocity at the drain or the thermal velocity at the source, whichever is lower, as was discussed in Section 3.2.2.5. The curves in Fig. 5.43 basically reflect the behavior of the switching resistances, as the capacitances to the first order do not depend on the saturation velocity. It can be seen that both delays are rather insensitive to the saturation velocity, improving by less than 10% for a 3× increase in $v_{sat}$. This is due to several factors discussed below.

First, because of the lower mobilities resulting from the higher vertical fields (Figs. 3.13 and 3.14), the 0.1-μm devices are not as severely velocity-saturated as would be suggested, e.g., by the mobility figures in the 1-μm generation. In fact, since $\mu_p V_{dd}/L \approx 10^7$ cm/s, the pMOSFETs are not greatly velocity-saturated at all, hence their on-current $I_p$ does not increase significantly when $v_{sat}$ is increased from $7 \times 10^6$ to $2 \times 10^7$ cm/s.

Second, even though the nMOSFET on current $I_n$ at $V_g = V_{ds} = V_{dd}$ does improve by some 25% when $v_{sat}$ increases by a factor of two, such a percentage does not translate directly into switching-resistance ($R_{swn}$) improvement. This is because the degree of velocity saturation, as measured by the factor $\mu_{eff}(V_g - V_t)/(m v_{sat} L)$ in Eq. (3.78), lessens at $V_g < V_{dd}$. In other words, currents at lower gate voltages saturate mainly by pinch-off at low drain voltages, $V_{dsat} \approx (V_g - V_t)/m$, and therefore do not improve with $v_{sat}$ by as much as the current at $V_g = V_{dd}$. A manifestation of such effects is that the factor $k_n$ defined by Eq. (5.33) increases as $v_{sat}$ becomes higher and the device becomes more "long-channel-like."

The third factor for the delay insensitivity to $v_{sat}$ comes from the finite series resistance in both n- and pMOSFETs. Taking the source-end parasitic resistance $R_s$ into account using Eqs. (4.69) and (4.71), one can rewrite the saturation-velocity-limited current, Eq. (3.80), as

$$I_{dsat}(L \to 0) = \frac{C_{ox} W v_{sat}(V_g - V_t)}{1 + m R_s C_{ox} W v_{sat}}. \tag{5.46}$$

For the 0.1-μm device parameters listed in Table 5.6, $m R_s C_{ox} W v_{sat} \approx 0.25$ for nMOSFETs and $\approx 0.5$ for pMOSFETs at $v_{sat} = 7 \times 10^6$ cm/s. This means that the second term in the denominator is not negligible compared with unity, which tends to slow the increase of $I_{dsat}(L \to 0)$ with $v_{sat}$.

### 5.4.3 SiGe OR STRAINED Si MOSFET

Carrier mobilities higher than those of conventional silicon MOSFETs have been reported in SiGe or strained silicon channel MOSFETs (Kesan *et al.*, 1991; Ismail, 1995). Theoretically, the hole mobility increases in silicon under either tensile or compressive strain, due to breaking of the valence-band degeneracy and reduction of the conductivity mass (Fischetti and Laux, 1996). The electron mobility is also enhanced in silicon under tensile strain because of increased electron populations in the two lower energy valleys with a lower conductivity mass. Tensile strain can be created by growing an epitaxial silicon layer on a relaxed SiGe film, whose lattice constant is slightly larger than that of the bulk silicon. Alloy scattering in SiGe, however, has a negative effect on both the electron and the hole mobilities (Fischetti and Laux, 1996).

The benefit of increased mobilities on CMOS delay can be investigated using the same circuit model as before. The base device case is that of a 0.1-μm CMOS with the parameters listed in Table 5.6. Two possible scenarios are considered here: either both the electron and the hole mobilities are enhanced by the same factor, or only

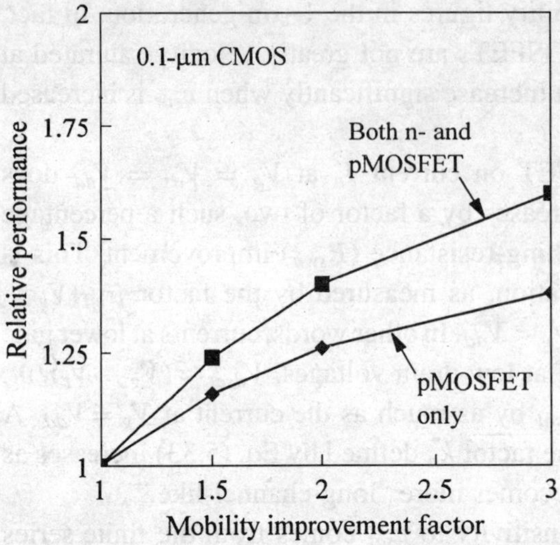


**FIGURE 5.44.** Relative performance factor, defined as inversely proportional to the intrinsic inverter delay, obtained from improvement in either the hole mobility alone or both the electron and hole mobilities. The $W_p/W_n$ ratio is kept at 2. The rest of the device parameters are given in Table 5.6.

the hole mobility is enhanced. The simulated performance factors in terms of the intrinsic delay of CMOS inverters are shown in Fig. 5.44. Significant performance gains of 25% or more can be obtained if both mobilities improve by 1.5× or if the hole mobility alone improves by 2×. The performance gain due to higher mobilities comes in through the switching-resistance factor and is therefore independent of fan-out and wire loading conditions. Similar improvement factors are also obtained in the delays of two-way NAND circuits. It can be seen from Fig. 5.44 that for the same mobility improvement factor, a higher hole mobility offers more performance gain than a higher electron mobility. This is because the pMOSFET is not as velocity-saturation-limited as the nMOSFET. It should be pointed out that both the n- and p-source–drain series resistances are kept constant when the mobility improvement factor is applied. This is one of the contributors to the fact in Fig. 5.44 that ***the initial mobility improvement is most beneficial to CMOS performance***. The parasitic resistances, together with the velocity saturation effect (or limitation by the thermal velocity at the source; see Section 3.2.2.5), put a damper on the performance improvement and eventually limit the total gain attainable from higher mobilities.

Even though high-mobility material has the potential of delivering higher CMOS performance at the same channel length, there remain many technological challenges in integrating such a strained film or films in an advanced CMOS process. In particular, one may be constrained to low-temperature processing ($T < 800°C$) to avoid strain relaxation and Ge segregation problems. That means growing or depositing gate oxide layers at $T < 800°C$, and annealing doped polysilicon gates and source–drain implants at $T < 800°C$ while trying to achieve high dopant activation and low junction leakage. It is clear from the above discussion that the performance advantage quickly erodes away if the source–drain series resistances cannot be kept at least as low as in bulk silicon devices.

### 5.4.4 LOW-TEMPERATURE CMOS

The performance advantage of low-temperature operation of MOSFETs has been recognized for some time (Gaensslen *et al.*, 1977; Sun *et al.*, 1987). ***The benefit is mainly derived from two aspects of the MOSFET characteristics at low temperature: higher carrier mobilities and steeper subthreshold slope***. Field-dependent electron and hole mobilities at 300 and 77 K are shown in Figs. 3.13 and 3.14. In this temperature range, the electron mobility improves by a factor of 2–5, depending on the magnitude of the vertical field. This is because of the much reduced electron–phonon scattering at low temperatures. Strictly speaking, for a given threshold voltage, low-temperature mobilities should be evaluated at a slightly lower field than at room temperature, since the $\psi_B$-term in Eq. (3.49) increases as temperature decreases (Fig. 2.6). But this is a relatively minor effect. Similarly, hole mobility also improves from 300 to 77 K, although by a more moderate factor of 1.7–4. The improvement factors of both electron and hole mobilities decrease at higher vertical fields where surface roughness scattering, which is largely insensitive to temperature, becomes important. In addition to the mobilities, the saturation velocities of carriers in bulk silicon also improve slightly at low temperatures. There are no extensive experimental data on the saturation velocities in a MOSFET channel as a function of temperature and field. In general, it is expected that $v_{sat}$ improves by some 10–30% from 300 to 77 K (Taur *et al.*, 1993a).

Another important aspect of the MOSFET characteristics at low temperatures is that the subthreshold slope steepens by a factor proportional to the absolute temperature (Section 3.1.3), making it much easier to turn off a MOSFET than at room temperature. An example is shown in Fig. 5.45 (Gaensslen *et al.*, 1977). This allows the threshold voltage $V_t$, and therefore the power-supply voltage $V_{dd}$, to scale down further below their permissible values at room temperature. For example, a subthreshold slope of 25 mV/decade at 80 K (Taur *et al.*, 1993b) would allow a $V_t$ of 0.1–0.2 V and a $V_{dd}$ of 0.4–0.8 V, provided that the threshold voltage tolerances from short-channel effects can be tightened as well through the use of optimized channel doping profiles (Taur *et al.*, 1997). Another factor worth mentioning is that, for the same reason as with the subthreshold slope, scaling down $kT/q$ also helps reduce the inversion-layer thickness (classical model) and therefore the inversion-layer capacitance effects depicted in Eq. (3.58) (Baccarani and Wordeman, 1983). This is beneficial to thin-oxide devices operated at low voltages.

To estimate the performance gain of CMOS circuits at low temperatures, we consider the same set of 0.1-μm channel devices listed in Table 5.6 and evaluate the intrinsic inverter delay as a function of temperature. At each temperature, the electron and the hole mobilities are adjusted according to the published data, e.g., in Figs. 3.13 and 3.14. A slight temperature dependence of the saturation velocities is also included in the model. Threshold voltages are adjusted following various strategies described below. In Fig. 5.46, the relative performance factor, defined

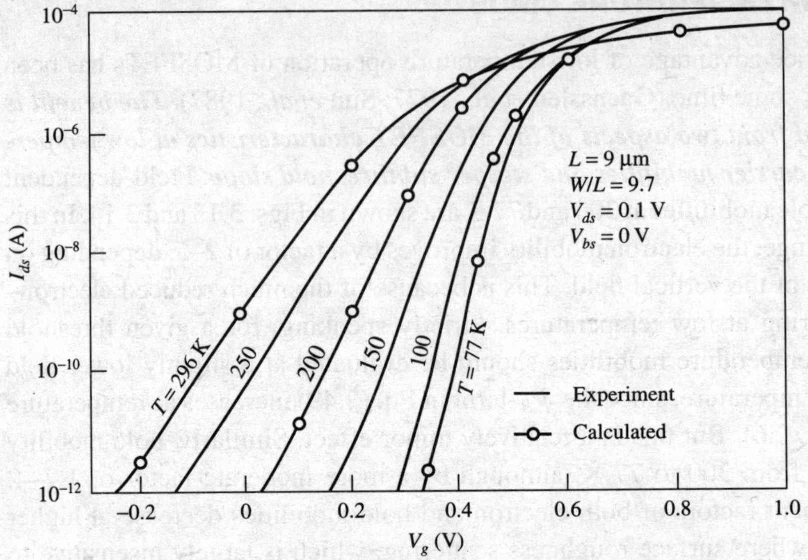


**FIGURE 5.45.** Subthreshold $I$–$V$ characteristics of nMOSFETs as a function of temperature. The gate oxide is 200 Å thick. (After Gaensslen *et al.*, 1977).

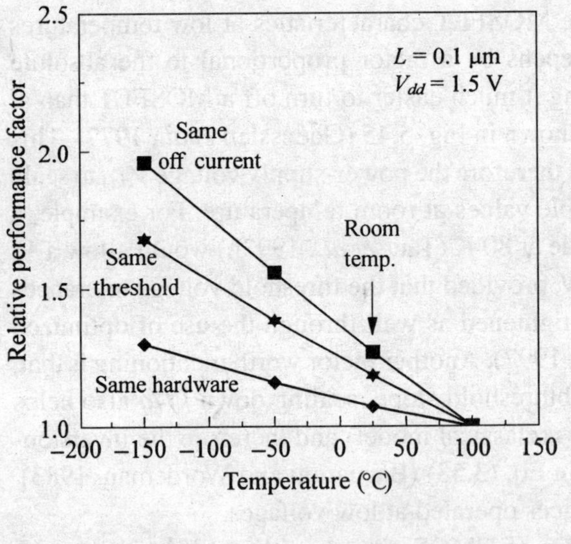


**FIGURE 5.46.** Relative performance factor of 0.1-μm CMOS as a function of temperature. Threshold voltages are adjusted differently with temperature in each of the three scenarios as described in the text. All the performance factors are normalized to the value at 100°C, where $V_{on} = \pm 0.33$ V.

as inversely proportional to the inverter delay, is plotted versus temperature. Since the capacitances to the first order are independent of temperature, the performance factor mainly reflects the reciprocal of the switching resistance in Eq. (5.31) and should be applicable to various static CMOS circuits with different fan-out and loading conditions.

Three different scenarios are considered in Fig. 5.46, depending on the assumption about the threshold voltage. In each case, the performance factor is normalized to the value at 100°C, which is the temperature specified for most of the IC products. In the *same-hardware* case, the magnitude of threshold voltage increases toward lower temperatures, governed by the $\approx -0.8$-mV/°C coefficient discussed in Section 3.1.4. The linearly extrapolated threshold voltage at 100°C is $V_{on} = \pm 0.33$ V, based on the $V_{on} = \pm 0.4$-V figure at room temperature listed in Table 5.6. The threshold behavior versus temperature is also evident in Fig. 5.45. The rise of threshold voltage offsets some of the performance gained from the higher mobilities such that a lesser net improvement is obtained at low temperatures, as shown by the bottom curve in Fig. 5.46. In the *same-threshold* case, the threshold voltages are held constant at $V_{on} = \pm 0.33$ V as the temperature is varied. The middle curve in Fig. 5.46 therefore represents the performance gained from the higher mobilities and, to a lesser extent, from the slightly higher saturation velocities. ***To gain the most performance out of low-temperature CMOS, one should turn the threshold-voltage trend around and take advantage of the steeper subthreshold slope***. This is represented by the *same-off-current* case in Fig. 5.46, in which the threshold voltages are adjusted to lower values as temperature decreases such that the off current is maintained at the same level as the product specification at 100°C (e.g., 50-nA/μm worst case). In principle, this can be accomplished using the retrograde channel doping concept outlined in Section 4.2.3 without degrading the short-channel effect. Up to a factor of two in performance gain can be achieved at –150°C, as indicated by the top curve in Fig. 5.46. The threshold voltages at that temperature are adjusted to $V_{on} = \pm 0.18$ V, which leaves plenty of gate overdrive for $V_{dd} = 1.5$ V. It may be desirable at this point to trade performance for lower power by operating the CMOS devices at a lower supply voltage, e.g., at 0.8 V. In fact, with the steep field dependence of the low-temperature mobilities (Figs. 3.13 and 3.14), a lower voltage allows the devices to operate in a regime of significantly higher mobilities. Following the design principles outlined in Section 5.3.3, one should be able to achieve a 4× power reduction with only a slight loss in performance. This is particularly worthwhile because it is rather expensive to cool a high-power chip to low temperatures.

In addition to the device improvement depicted in Fig. 5.46, which amounts to about 0.3%/°C in the best case, the conductivity of metal interconnects (either aluminum or copper) also improves at low temperatures. Depending on the material purity, the improvement factor lies in the range of 0.3–0.6%/°C. In other words, interconnect *RC* delays will improve by at least as much as the devices. This means that the performance factors projected in Fig. 5.46 at the device level should translate directly to the chip level without extensive design modifications. Apart from packaging issues and system costs, one key challenge for low-temperature CMOS is to be able to tighten the short-channel threshold tolerances through the use of optimized channel doping profiles while following the low-$V_t$ strategy in Fig. 5.46 for best performance.

## EXERCISES

**5.1** Consider the CMOS switching delay, $\tau = (\tau_n + \tau_p)/2$, where $\tau_n$ and $\tau_p$ are given by Eqs. (5.3) and (5.4). If the inverter is driving another stage with the same n- to p-width ratio and if both the n- and p-devices have the same capacitance per unit width, the load capacitance $C$ is proportional to $W_n + W_p$. Show that the minimum delay $\tau$ occurs for a width ratio of $W_p/W_n = (I_{nsat}/I_{psat})^{1/2}$, which is different from Eq. (5.1) for best noise margin where $\tau_n = \tau_p$.

**5.2** For an $RC$ circuit with a capacitor $C$ connected in series with a resistor $R$ and a switchable voltage source, solve for the waveform of the voltage across the capacitor, $V(t)$, when the voltage source is abruptly switched from 0 to $V_{dd}$ with the initial condition $V(t = 0) = 0$. Show that when the equilibrium condition is established, an energy of $CV_{dd}^2/2$ has been dissipated in the resistor $R$ and the same amount of energy is stored in $C$. Since the energy dissipated and the energy stored are independent of $R$, the same results hold even if $R = 0$. What happens if the voltage source is now switched off from $V_{dd}$ to 0 with the initial condition $V(t = 0) = V_{dd}$?

**5.3** The carrier transit time is defined as $\tau_{tr} \equiv Q/I$, where $Q$ is the total inversion charge and $I$ is the total conduction current of the device. For a MOSFET device biased in the linear region (low drain voltage), use Eq. (3.19) and the inversion charge expression above Eq. (3.54) to derive an expression for $\tau_{tr}$. Similarly, use Eq. (3.23) and the expression above Eq. (3.56) to derive $\tau_{tr}$ for a long-channel MOSFET biased in saturation.

**5.4** Use Eq. (3.78) and the inversion-charge expression in Exercise 3.10 to find the carrier transit time $\tau_{tr}$ for a short-channel MOSFET biased in saturation. What is the limiting value of $\tau_{tr}$ when the device becomes fully velocity-saturated as $L \to 0$?

**5.5** A similar distributed network to the one in Fig. 5.18 can be used to formulate the *transmission-line model* of contact resistance in a planar geometry (Berger, 1972). Here we consider the current flow from a thin resistive film (diffusion with a sheet resistivity $\rho_{sd}$) into a ground plane (metal) with an interfacial contact resistivity $\rho_c$ between them (Fig. 5.13). Thus, in Fig. 5.18, $R\,dx$ corresponds to $(\rho_{sd}/W)\,dx$, and $C\,dx$ is replaced by a shunt conductance $G\,dx$, which corresponds to $(W/\rho_c)\,dx$. Show that both the current and voltage along the current flow direction satisfy the following differential equation:

$$\frac{d^2 f}{dx^2} = RGf = \frac{\rho_{sd}}{\rho_c} f,$$

where $f(x) = V(x)$ or $I(x)$ defined in Fig. 5.18.

**5.6** Following the above transmission-line model, with the boundary condition $I(x = l_c) = 0$ where $x = 0$ is the leading edge and $x = l_c$ is the far end of the contact window (Fig. 5.13), solve for $V(x)$ and $I(x)$ within a multiplying factor and show that the total contact resistance, $R_{co} = V(x=0)/I(x=0)$, is given by Eq.(5.8).

**5.7** The insertion of a buffer stage (Section 5.3.2) between the inverter and the load is beneficial only if the load capacitance is higher than certain value. Find, in terms of $C_{in}$ and $C_{out}$, the minimum load capacitance $C_L$ above which the single-stage buffered delay given by Eq. (5.37) is shorter than the unbuffered delay given by Eq. (5.35)

**5.8** Generalize Eq. (5.36) for one-stage buffered delay to $n$ stages: If the width ratios of the successive buffer stages are $k_1, k_2, k_3, \ldots, k_n$ (all >1), show that the $n$-stage buffered delay is

$$\tau_b(n) = R_{sw}\left((n+1)C_{out} + (k_1 + k_2 + \cdots + k_n)C_{in} + \frac{C_L}{k_1 k_2 \cdots k_n}\right).$$

**5.9** Following the previous exercise, show that for a given $n$, the $n$-stage buffered delay is a minimum,

$$\tau_{b\,\min}(n) = R_{sw}\left[(n+1)C_{out} + (n+1)C_{in}\left(\frac{C_L}{C_{in}}\right)^{1/(n+1)}\right],$$

when $k_1 = k_2 = \cdots = k_n = (C_L/C_{in})^{1/(n+1)}$. Here $\tau_{b\,\min}(n)$, as expected, is reduced to Eq. (5.37) if $n = 1$.

**5.10** If one plots the minimum $n$-stage buffered delay from the previous exercise versus $n$, it will first decrease and then increase with $n$. In other words, depending on the ratios of $C_L/C_{in}$ and $C_{out}/C_{in}$, there is an optimum number of buffer stages for which the overall delay is the shortest. Show that this optimum $n$ is given by the closest integer to

$$n = \frac{\ln(C_L/C_{in})}{\ln k} - 1,$$

where $k$ is a solution of

$$k(\ln k - 1) = \frac{C_{out}}{C_{in}}.$$

For typical $C_{out}/C_{in}$ ratios not too different from unity, $k$ is in the range of 3–5. Note that $k$ also gives the optimum width ratio between the successive buffer stages, i.e., $k_1 = k_2 = \cdots = k_n = k$. Also show that the minimum buffered delay is given by

$$\tau_{b\,\min} \approx k R_{sw} C_{in} \ln(C_L/C_{in}),$$

which only increases logarithmically with load capacitance.

# 6 BIPOLAR DEVICES

Although most microelectronics products are now made of CMOS transistors, bipolar transistors remain important in microelectronics because of their inherently high speed for digital circuit applications and superior characteristics for analog circuit applications. There are two types of bipolar devices: the n–p–n type, which has a p-type base and n-type emitter and collector, and the p–n–p type, which has an n-type base and p-type emitter and collector. Commonly used bipolar devices are either *lateral transistors*, where the active device regions are arranged horizontally adjacent to one another and the active currents flow laterally, or *vertical transistors*, where the active device regions are arranged vertically one on top of another and the active currents flow vertically. ***Practically all high-speed bipolar transistors used in digital circuits are of the vertical n–p–n type.***

For simplicity, only vertical n–p–n bipolar transistors will be considered explicitly here. The equations derived for vertical transistors apply to horizontal transistors as well, provided that the device parameter values are adjusted accordingly. Also, the equations for an n–p–n transistor can be extended to a p–n–p transistor simply by reversing the voltage and dopant polarities and using the appropriate device parameter values.

## 6.1 n–p–n TRANSISTORS

Figure 6.1(a) shows a one-dimensional representation of a vertical n–p–n transistor. The transistor consists of an $n^+$-type emitter region and an n-type collector region, with a p-type base region sandwiched in between. The collector sits on an $n^+$-type subcollector region. Figure 6.1(b) shows a cross-sectional schematic of the transistor. The $n^+$ subcollector is brought to the top surface for electrical contact by a vertical $n^+$-type *reach-through* region.

The starting substrate material for fabricating a vertical n–p–n transistor is usually a p-type silicon wafer. The subcollector is formed in the substrate, usually by ion implantation and diffusion. Then the n-type collector is formed on top of the subcollector by an epitaxial growth process. An $n^+$-type vertical reach-through region is formed for electrical connection to the subcollector. After that, the p-type

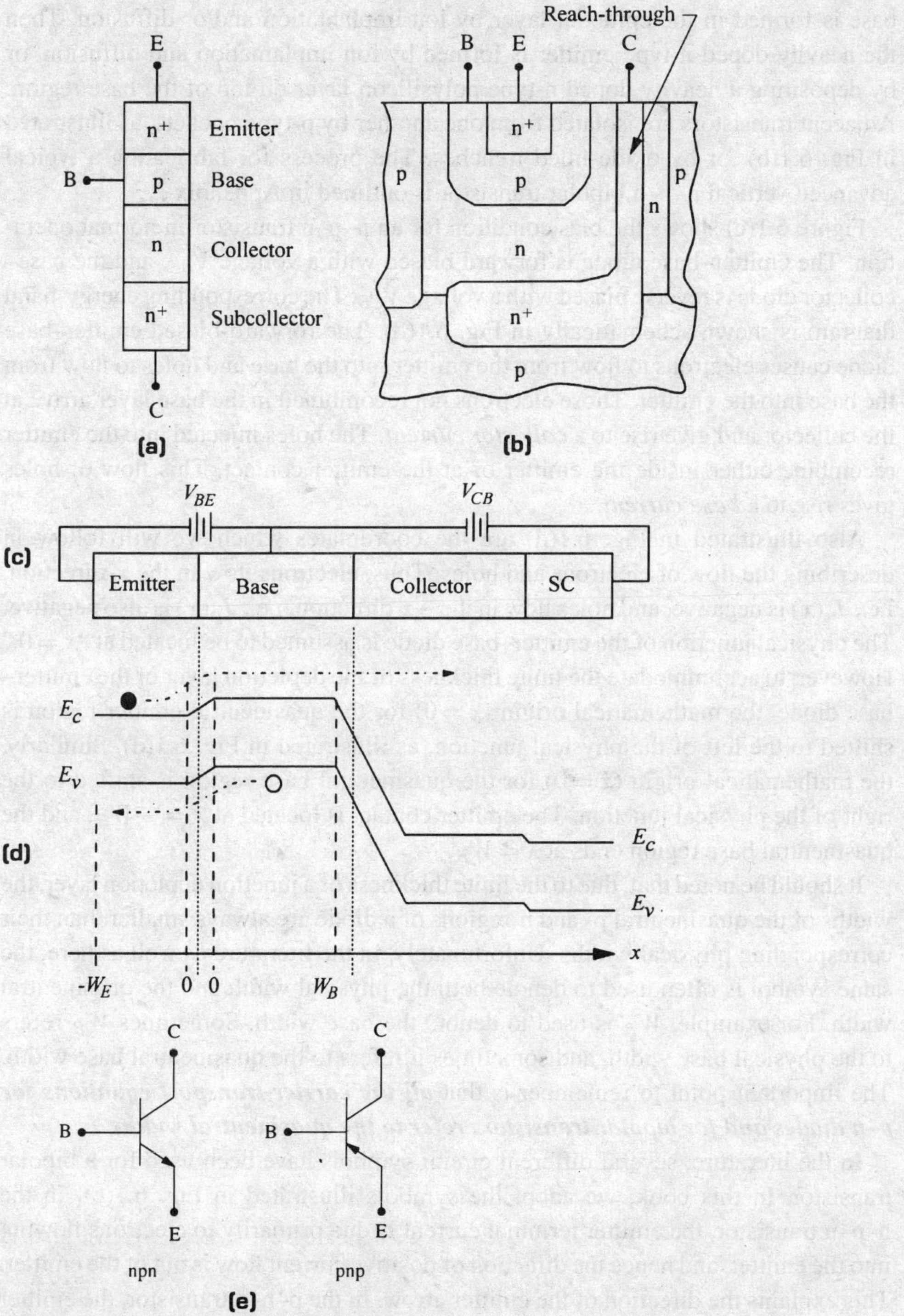


**FIGURE 6.1.** (a) One-dimensional representation of an n–p–n transistor, (b) its cross-sectional schematic, (c) schematic illustrating the applied voltages in normal operation, (d) schematics illustrating the energy-band diagram, carrier flows, and locations of the boundaries of the emitter and base quasineutral regions, and (e) circuit symbols for an n–p–n transistor and a p–n–p transistor.

base is formed in the epitaxial layer by ion implantation and/or diffusion. Then the heavily doped n-type emitter is formed by ion implantation and diffusion, or by depositing a heavily doped n-type polysilicon layer on top of the base region. Adjacent transistors are isolated from one another by p-type pockets, as illustrated in Fig. 6.1(b), or by oxide-filled trenches. The process for fabricating a typical advanced vertical n–p–n bipolar transistor is outlined in Appendix 2.

Figure 6.1(c) shows the bias condition for an n–p–n transistor in normal operation. The emitter–base diode is forward biased with a voltage $V_{BE}$, and the base–collector diode is reverse biased with a voltage $V_{CB}$. The corresponding energy-band diagram is shown schematically in Fig. 6.1(d). The forward-biased emitter–base diode causes electrons to flow from the emitter into the base and holes to flow from the base into the emitter. Those electrons not recombined in the base layer arrive at the collector and give rise to a *collector current*. The holes injected into the emitter recombine either inside the emitter or at the emitter contact. This flow of holes gives rise to a *base current*.

Also illustrated in Fig. 6.1(d) are the coordinates which we will follow in describing the flow of electrons and holes. Thus, electrons flow in the $x$-direction, i.e., $J_n(x)$ is negative, and holes flow in the $-x$ direction, i.e., $J_p(x)$ is also negative. The physical junction of the emitter–base diode is assumed to be located at "$x = 0$." However, to accommodate the finite thickness of the depletion layer of the emitter–base diode, the mathematical origin ($x = 0$) for the quasineutral emitter region is shifted to the left of the physical junction, as illustrated in Fig. 6.1(d). Similarly, the mathematical origin ($x = 0$) for the quasineutral base region is shifted to the right of the physical junction. The emitter contact is located at $x = -W_E$, and the quasineutral base region ends at $x = W_B$.

It should be noted that, due to the finite thickness of a junction depletion layer, the widths of the quasineutral p- and n-regions of a diode are always smaller than their corresponding physical widths. Unfortunately, in the literature as well as here, the same symbol is often used to denote both the physical width and the quasineutral width. For example, $W_B$ is used to denote the base width. Sometimes $W_B$ refers to the physical base width, and sometimes it refers to the quasineutral base width. The important point to remember is that ***all the carrier-transport equations for p–n diodes and for bipolar transistors refer to the quasineutral widths.***

In the literature, several different circuit symbols have been used for a bipolar transistor. In this book, we adopt the symbols illustrated in Fig. 6.1(e). In the n–p–n transistor, the emitter terminal current is due primarily to electrons flowing into the emitter, and hence the direction of positive current flow is out of the emitter. This explains the direction of the emitter arrow. In the p–n–p transistor, the emitter terminal current is due to holes flowing into the emitter. The emitter arrow indicates the direction of positive current flow.

Figure 6.2(a) illustrates the vertical doping profile of an n–p–n transistor with a diffused, or implanted and then diffused, emitter. The emitter junction depth $x_{jE}$ is

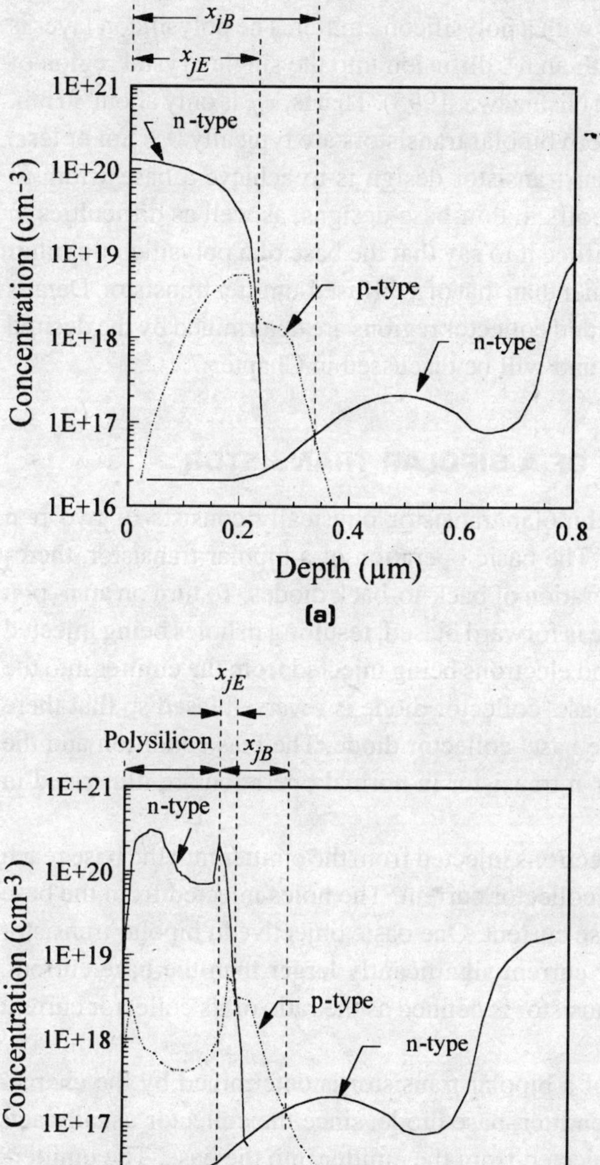


**FIGURE 6.2.** Vertical doping profiles of typical n–p–n transistors: (a) with implanted and/or diffused emitter, and (b) with polysilicon emitter.

typically 0.2 μm or larger (Ning and Isaac, 1980). The base junction depth is $x_{jB}$, and the physical base width is equal to $x_{jB} - x_{jE}$. Figure 6.2(b) illustrates the vertical doping profile of an n–p–n transistor with a polysilicon emitter. The polysilicon layer is typically about 0.2 μm thick, with an $n^+$ diffusion into the single-crystal region of only about 30 nm (Nakamura and Nishizawa, 1995). That is, $x_{jE}$ is only about 30 nm.

The base widths of most modern bipolar transistors are typically 0.1 μm or less. While one of the goals in bipolar transistor design is to achieve a base width as small as possible, there are tradeoffs in thin-base designs, as well as difficulties in fabricating thin-base devices. Suffice it to say that the base of a polysilicon-emitter transistor can be made much thinner than that of a diffused-emitter transistor. Details of the doping profiles of the base and collector regions are determined by the desired device dc and ac characteristics and will be discussed in Chapter 7.

### 6.1.1 BASIC OPERATION OF A BIPOLAR TRANSISTOR

As illustrated in Fig. 6.1(a), a bipolar transistor physically consists of two p–n diodes connected back to back. The basic operation of a bipolar transistor, therefore, can be described by the operation of back-to-back diodes. To turn on an n–p–n transistor, the emitter–base diode is forward biased, resulting in holes being injected from the base into the emitter, and electrons being injected from the emitter into the base. In normal operation, the base–collector diode is reverse biased so that there is no forward current flow in the base–collector diode. The bias condition and the energy-band diagram of an n–p–n transistor in normal operation are illustrated in Fig. 6.1(c) and (d).

As described earlier, as the electrons injected from the emitter into the base reach the collector, they give rise to a collector current. The holes injected from the base into the emitter give rise to a base current. One basic objective in bipolar transistor design is to achieve a collector current significantly larger than the base current. The *current gain* of a bipolar transistor is defined as the ratio of its collector current to its base current.

To first order, the behavior of a bipolar transistor is determined by the characteristics of the forward-biased emitter–base diode, since the collector usually acts only as a sink for the carriers injected from the emitter into the base. The emitter–base diode behaves like a thin-base diode. Thus, qualitatively, the current–voltage characteristics of a thin-base diode discussed in Section 2.2.4 can be applied to describe the current–voltage characteristics of a bipolar transistor.

### 6.1.2 MODIFYING THE SIMPLE DIODE THEORY FOR DESCRIBING BIPOLAR TRANSISTORS

In order to extend the simple diode theory discussed in Section 2.2 to describe the behavior of a bipolar transistor quantitatively, three important effects ignored in it must be included. These are the effects of finite electric field in a quasineutral region, heavy doping, and nonuniform energy bandgap. These effects are discussed below.

#### 6.1.2.1 ELECTRIC FIELD IN A QUASINEUTRAL REGION WITH A UNIFORM ENERGY BANDGAP

In Section 2.2.4, the current–voltage characteristics of a p–n diode were derived for the case of zero electric field in the p- and n-type quasineutral regions. As will be shown below, the zero-field approximation is valid only where the majority-carrier current is zero and concentration is uniform. For bipolar transistors, as shown in Fig. 6.2(a) and (b), the doping profiles are rather nonuniform. A nonuniform doping profile means that the majority-carrier concentration is also nonuniform. Furthermore, at large emitter–base forward biases, to maintain quasineutrality the high concentration of injected minority carriers can cause significant nonuniformity in the majority-carrier concentration as well. Therefore, the effect of nonuniform majority-carrier concentration in a quasineutral region cannot be ignored in determining the current–voltage characteristics of a bipolar transistor.

For a p-type region, Eq. (2.48) gives

$$\phi_p = \psi_i + \frac{kT}{q}\ln\left(\frac{p_p}{n_i}\right), \tag{6.1}$$

where $\phi_p$ is the hole quasi-Fermi potential and $\psi_i$ is the intrinsic potential. (Note that $p_p$ is equal to $N_a$ only for the case of low electron injection, i.e., only at low currents.) The electric field is given by Eq. (2.33), namely

$$\begin{aligned}\mathscr{E} \equiv -\frac{d\psi_i}{dx} &= \frac{kT}{q}\frac{1}{p_p}\frac{dp_p}{dx} - \frac{d\phi_p}{dx} \\ &= \frac{kT}{q}\frac{1}{p_p}\frac{dp_p}{dx} + \frac{J_p}{qp_p\mu_p},\end{aligned} \tag{6.2}$$

where we have used Eq. (2.46), which relates $d\phi_p/dx$ to $J_p$. In Eq. (6.2), the intrinsic-carrier concentration is assumed to be independent of $x$. This is accurate only when the energy bandgap is not a function of $x$. The dependence of energy bandgap on $x$ will be discussed later in connection with heavy-doping effects.

Let us apply Eq. (6.2) to the intrinsic-base region of an n–p–n transistor with a typical current gain of 100. At a typical but high collector current density of 1 mA/μm$^2$, the base current density is 0.01 mA/μm$^2$, i.e., $J_p = 0.01$ mA/μm$^2$ in the base layer. As can be seen from Fig. 6.2, the base doping concentration is typically on the order of $10^{18}$ cm$^{-3}$, and the corresponding hole mobility is about 150 cm$^2$/V-s (Fig. 2.7). That is, $p_p \approx 10^{18}$ cm$^{-3}$ and $\mu_p \approx 150$ cm$^2$/V-s, and $J_p/qp_p\mu_p \approx 40$ V/cm, which is a negligibly small electric field in normal device operation. (This is consistent with the arguments presented in Section 2.2.3 that $d\phi_p/dx$ is often negligibly small in a p-type region.) Therefore, for a p-type region

Eq. (6.2) gives

$$\mathscr{E}(\text{p}-\text{region}) \approx \frac{kT}{q}\frac{1}{p_p}\frac{dp_p}{dx}. \tag{6.3}$$

Similarly, for an n-type region,

$$\mathscr{E}(\text{n}-\text{region}) \approx -\frac{kT}{q}\frac{1}{n_n}\frac{dn_n}{dx}. \tag{6.4}$$

Equations (6.3) and (6.4) show that ***the electric field is negligible in a region of uniform majority-carrier concentration.***

To include the effect of finite electric field, the current-density equations (2.43) and (2.44), which include both the drift and the diffusion components, should be used. These are repeated here:

$$J_n(x) = qn\mu_n\mathscr{E} + qD_n\frac{dn}{dx}, \tag{6.5}$$

and

$$J_p(x) = qp\mu_p\mathscr{E} - qD_p\frac{dp}{dx}. \tag{6.6}$$

It should be noted that if Eq. (6.4) is substituted into Eq. (6.5), the RHS of Eq. (6.5) is equal to zero. Similarly, if Eq. (6.3) is substituted into Eq. (6.6), the RHS of Eq. (6.6) is equal to zero. What this means is that the approximations for the electric fields represented by ***Eqs. (6.3) and (6.4) are good approximations only for describing minority-carrier currents***. The $d\phi_p/dx$ term, although very small in a p-region, is entirely responsible for the majority-carrier current in a p-region. In fact, from Eq. (2.46), the hole current density in a p-region is $J_p = -qp\mu_p\, d\phi_p/dx$. Thus, for describing hole current in a p-region, Eq. (6.2), instead of Eq. (6.3), should be used for the electric field. The electron current in a p-region due to the $d\phi_p/dx$ term, on the other hand, is negligible. Therefore, Eqs. (6.3) and (6.4) are good approximations for describing minority-carrier currents, i.e., for electron current in a p-region and hole current in an n-region. That is, these approximations are applicable to currents in a diode or in a bipolar transistor.

- *Built-in electric field in a nonuniformly doped base region.* Consider the electron current in the p-type base of a forward-biased emitter–base diode. Let $N_B(x)$ be the doping concentration in the base, and, for simplicity, all the dopants are assumed to be ionized. Quasineutrality requires that

$$p_p(x) = N_B(x) + n_p(x). \tag{6.7}$$

Therefore,

$$\frac{dp_p}{dx} = \frac{dN_B}{dx} + \frac{dn_p}{dx}. \tag{6.8}$$

The *built-in electric field* $\mathscr{E}_0$ is defined as the electric field from the nonuniform base dopant distribution alone, ignoring any effect of injected minority carriers. It can be obtained by substituting $N_B$ for $p_p$ in Eq. (6.3), namely

$$\mathscr{E}_0 \equiv \mathscr{E}(n_p = 0) = \frac{kT}{q}\frac{1}{N_B}\frac{dN_B}{dx}. \tag{6.9}$$

Substituting Eq. (6.3) into Eq. (6.5), and using Eqs. (6.8) and (6.9) and the Einstein relationship, we have, for electron current in a nonuniformly doped p-type base region,

$$J_n(x) = qn_p\mu_n\mathscr{E}_0\frac{N_B}{n_p + N_B} + qD_n\left(\frac{2n_p + N_B}{n_p + N_B}\right)\frac{dn_p}{dx}. \tag{6.10}$$

Equation (6.10) suggests that the *effective electric field* $\mathscr{E}_{\mathit{eff}}$ in the p-type base can be written as

$$\mathscr{E}_{\mathit{eff}} \equiv \mathscr{E}_0\frac{N_B}{n_p + N_B}. \tag{6.11}$$

It should be pointed out that Eqs. (6.10) and (6.11) are valid for all levels of electron injection from the emitter, i.e., for all values of $n_p$.

- *Electric field and current density in the low-injection limit.* At low levels of electron injection from the emitter, i.e., for $n_p \ll N_B$, $\mathscr{E}_{\mathit{eff}}$ reduces to $\mathscr{E}_0$ and Eq. (6.10) reduces to

$$J_n(x) \approx qn_p\mu_n\mathscr{E}_0 + qD_n\frac{dn_p}{dx}, \tag{6.12}$$

which simply says that the electron current flowing in the base consists of a drift component due to the built-in field from the nonuniform base dopant distribution, and a diffusion component from the electron concentration gradient in the base.

- *Electric field and current density in the high-injection limit.* When the electron injection level is very high, i.e., when $n_p \gg N_B$, $\mathscr{E}_{\mathit{eff}}$ becomes very small. The built-in electric field is screened out by the large concentration of injected minority carriers. Therefore, the electron current component associated with the built-in field becomes negligible, and the electron current density approaches

$$J_n(x)|_{n_p \gg N_B} \approx q2D_n\frac{dn_p}{dx}. \tag{6.13}$$

That is, at the high-injection limit, the minority-carrier current behaves as if it were purely a diffusion current, but with a diffusion coefficient twice its low-injection value. This is known as the *Webster effect* (Webster, 1954).

#### 6.1.2.2 HEAVY-DOPING EFFECT

As discussed in Section 2.1.2, the effective ionization energy for impurities in a heavily doped semiconductor decreases with its doping concentration, resulting in a decrease in its effective energy bandgap. For a lightly doped silicon region at thermal equilibrium, Eq. (2.12) gives the relationship

$$p_0 n_0 = n_i^2 = N_c N_v \exp(-E_g/kT), \tag{6.14}$$

where $N_c$ and $N_v$ are the effective densities of states in the conduction and valence bands, respectively, and $E_g$ is the energy bandgap. As the energy gap changes and/or as the densities of states change due the effect of heavy doping, the $p_0 n_0$ product will also change. For modeling purposes, it is convenient to define an *effective intrinsic-carrier concentration* $n_{ie}$ and ***lump all the heavy-doping effects into a parameter called the apparent bandgap narrowing,*** $\boldsymbol{\Delta E_g}$, given by the equation

$$p_0(\Delta E_g) n_0(\Delta E_g) \equiv n_{ie}^2 = n_i^2 \exp(\Delta E_g/kT). \tag{6.15}$$

The heavy-doping effect increases the effective intrinsic carrier concentration. ***To include the heavy-doping effect,*** $\boldsymbol{n_i}$ ***should be replaced by*** $\boldsymbol{n_{ie}}$. Thus, including the heavy-doping effect, the product $pn$ in Eq. (2.87) becomes

$$pn = n_{ie}^2 \exp\left(\frac{q(\phi_p - \phi_n)}{kT}\right), \tag{6.16}$$

where $\phi_p$ and $\phi_n$ are the hole and electron quasi-Fermi potentials, respectively.

It is extremely difficult to determine $\Delta E_g$ experimentally, and there is considerable scatter in the reported data in the literature (del Alamo *et al.*, 1985a). Careful analyses of the reported data suggest the following *empirical* expressions for the apparent bandgap-narrowing parameter:

$$\Delta E_g(N_d) = 18.7 \ln\left(\frac{N_d}{7 \times 10^{17}}\right) \text{ meV} \tag{6.17}$$

for $N_d \geq 7 \times 10^{17}$ cm$^{-3}$, and zero for lower doping levels, for n-type silicon (del Alamo *et al.*, 1985b); and

$$\Delta E_g(N_a) = 9\left(F + \sqrt{F^2 + 0.5}\right) \text{ meV}, \tag{6.18}$$

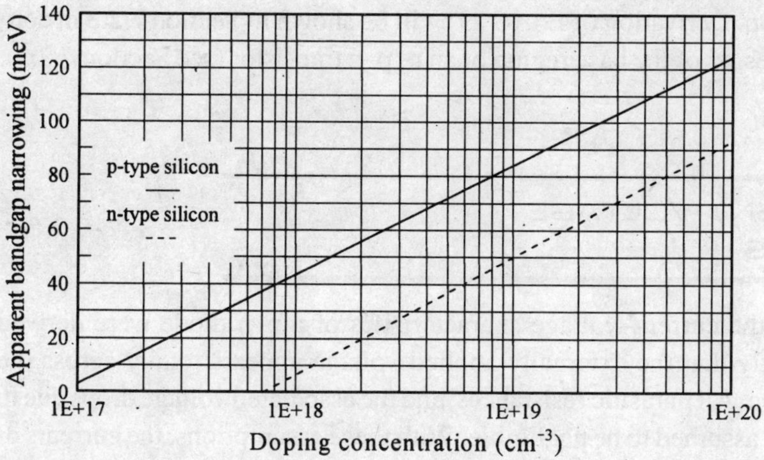


**FIGURE 6.3.** Apparent bandgap narrowing as given by the empirical expressions in Eqs. (6.17) and (6.18).

where $F = \ln(N_a/10^{17})$, for $N_a > 10^{17}$ cm$^{-3}$, and zero for lower doping levels, for p-type silicon (Slotboom and de Graaff, 1976; Swirhun *et al.*, 1986). Figure 6.3 is a plot of $\Delta E_g$ as a function of doping concentration, as given by these two empirical expressions.

#### 6.1.2.3 ELECTRIC FIELD IN A QUASINEUTRAL REGION WITH A NONUNIFORM ENERGY BANDGAP

Aside from the heavy-doping effect, the energy bandgap can also be modified by incorporating a relatively large amount of germanium into silicon. In this case, the bandgap becomes narrower (People, 1986). If both heavy-doping effect and the effect of germanium are included in the parameter $\Delta E_g$ in Eq. (6.15), then the product $pn$ given by Eq. (6.16) can be used to describe transport in heavily doped SiGe alloys.

When the energy bandgap is nonuniform, the electric field is no longer simply given by Eqs. (6.3) and (6.4), which include only the effect of nonuniform dopant distribution. When the effect of nonuniform energy bandgap is included, the electric fields are given by (van Overstraeten *et al.*, 1973)

$$\mathscr{E}(\text{p-region}) = \frac{kT}{q}\left(\frac{1}{p_p}\frac{dp_p}{dx} - \frac{1}{n_{ie}^2}\frac{dn_{ie}^2}{dx}\right) \tag{6.19}$$

for a p-type region, and

$$\mathscr{E}(\text{n-region}) = -\frac{kT}{q}\left(\frac{1}{n_n}\frac{dn_n}{dx} - \frac{1}{n_{ie}^2}\frac{dn_{ie}^2}{dx}\right) \tag{6.20}$$

for an n-type region. Derivation of Eq. (6.19) will be shown in Section 7.2.3 in connection with the design of the base region of an n–p–n transistor (see Section 7.2.3).

## 6.2 IDEAL CURRENT-VOLTAGE CHARACTERISTICS

In Section 2.2.4, the current–voltage characteristics of a p–n diode were derived assuming implicitly that the externally applied voltage appears totally across the immediate junction. All parasitic resistances, and the associated voltage drops due to current flow, were assumed to be negligible. With these assumptions, the currents or current densities in a forward-biased diode increase exponentially with the applied voltage. These are the *ideal current–voltage characteristics*.

In practice, the measured current–voltage characteristics of a bipolar transistor are ideal only over a certain range of applied voltage. At low voltages, the base current is larger than the ideal base current. At large voltages, both the base and the collector currents are significantly smaller than the corresponding ideal currents. In this section, the ideal current–voltage characteristics are discussed. Deviations from the ideal characteristics are discussed in the next section.

It was shown in Section 2.2.5 that, for modern bipolar transistors, ***the base transit time is much smaller than the minority-carrier lifetime in the base, and there is negligible recombination in the base region***. For an n–p–n transistor, neglecting second-order effects, such as avalanche multiplication and generation currents due to defects and/or surface states, the base current is due entirely to the injection of holes from the base into the emitter. Similarly, the collector current is due entirely to the injection of electrons from the emitter into the base. (The effect of avalanche multiplication in the base–collector junction is considered in Section 6.5, where breakdown voltages are discussed. Also, that recombination in the base of modern bipolar transistors is negligible is confirmed in Exercise 6.7.)

Referring to Fig. 6.1(a), we see that the base terminal contact is located at the side of the base region. Therefore, the hole current first flows horizontally from the base terminal into the base region and then bends upward and enter the emitter. The horizontal hole current flow causes a lateral voltage drop within the base region, which in turn causes the forward-bias voltage across the immediate emitter–base junction to vary laterally, with the emitter–base forward bias largest nearest the base contact, and smallest furthest away from the base contact. This is known as the *emitter current-crowding effect*. When emitter current crowding is significant, the base and collector current densities are not just a function of $x$ [Fig. 6.1(d)], but also a function of distance from the base contact. Fortunately, as shown in Appendix 9, ***emitter current crowding is negligible in modern bipolar devices*** because of their narrow emitter stripe widths. Therefore, we shall ignore the emitter current-crowding effect and assume both the base and collector current densities to be uniform over the entire emitter–base junction area.

### CURRENT-DENSITY EQUATION FOR ELECTRONS IN A p-TYPE BASE

Let us consider the electrons injected from the emitter into the p-type base region of an n–p–n transistor. Instead of starting with Eq. (6.10), it is often convenient to reformulate the electron current density in terms of carrier concentrations (Moll and Ross, 1956). To this end, we start with the electron current density given by Eq. (2.45), namely

$$J_n(x) = -qn_p\mu_n \frac{d\phi_n}{dx}, \tag{6.21}$$

where $\phi_n$ is the electron quasi-Fermi potential. As discussed in Section 2.2.3, the hole quasi-Fermi potential is approximately constant in a p-region, i.e.,

$$\frac{d\phi_p}{dx} \approx 0. \tag{6.22}$$

Combining Eqs. (6.21) and (6.22), we have

$$J_n(x) \approx qn_p\mu_n \frac{d(\phi_p - \phi_n)}{dx}. \tag{6.23}$$

Now, Eq. (6.16) gives

$$\phi_p - \phi_n = \frac{kT}{q} \ln\left(\frac{p_p n_p}{n_{ie}^2}\right). \tag{6.24}$$

Substituting Eq. (6.24) into Eq. (6.23) and rearranging the terms, we have

$$J_n(x) = qD_n \frac{n_{ie}^2}{p_p} \frac{d}{dx}\left(\frac{n_p p_p}{n_{ie}^2}\right) \tag{6.25}$$

for the electron current in the base. It gives the electron current density in terms of the electron and hole concentrations in the base.

### CURRENT-DENSITY EQUATION FOR HOLES IN AN n-TYPE EMITTER

The hole current density due to holes injected from the p-type base into the n-type emitter can be derived in a similar manner. The result is

$$J_p(x) = -qD_p \frac{n_{ie}^2}{n_n} \frac{d}{dx}\left(\frac{n_n p_n}{n_{ie}^2}\right). \tag{6.26}$$

Equations (6.25) and (6.26) can be used to calculate the collector and base currents for arbitrary doping profiles, arbitrary energy bandgap grading, and arbitrary injection current levels (Moll and Ross, 1956).

### 6.2.1 COLLECTOR CURRENT

Consider the electrons injected from the emitter into the base. As these electrons reach the collector region, they give rise to a collector current. Referring to Fig. 6.1(d), let $x = 0$ denote the depletion-layer edge on the base side of the emitter–base junction, and $x = W_B$ denote the depletion-layer edge on the base side of the base–collector junction. That is, the width of the quasineutral base region is $W_B$. Since there is negligible recombination in this thin base layer (see Exercise 6.7), the electron current density in steady state in the base is independent of $x$. Therefore, Eq. (6.25) can be integrated to give

$$J_n \int_0^{W_B} \frac{p_p}{qD_n n_{ie}^2}\, dx = \left.\frac{n_p p_p}{n_{ie}^2}\right|_{x=W_B} - \left.\frac{n_p p_p}{n_{ie}^2}\right|_{x=0}. \tag{6.27}$$

At $x = 0$, the electron concentration is given by Eq. (2.89), namely,

$$n_p(0) = n_{p0}(0) \exp(qV_{BE}/kT), \tag{6.28}$$

where $V_{BE}$ is the base–emitter forward-bias voltage. At $x = W_B$, the base–collector junction depletion region acts as a sink for the excess electrons in the base region, i.e., $n_p(W_B) = n_{p0}(W_B)$, which is negligible compared to $n_p(0)$. Therefore, the first term on the RHS of Eq. (6.27) can be neglected, and Eq. (6.27) can be rewritten as

$$J_n \int_0^{W_B} \frac{p_p}{qD_n n_{ie}^2}\, dx = -\frac{n_{p0}(0)\, p_p(0)}{n_{ie}^2(0)} \exp\left(\frac{qV_{BE}}{kT}\right). \tag{6.29}$$

Notice that $J_n$ is negative. This is due to the fact that electrons flowing in the $x$-direction give rise to a negative current.

Most modern bipolar transistors have a base doping concentration that peaks at or near the emitter–base junction. As long as this peak concentration is large compared to the injected minority-carrier concentration, the majority-carrier concentration near this peak-doping region is about the same as its thermal-equilibrium value. Again referring to the coordinates illustrated in Fig. 6.1(d), this means $p_p(0) \approx p_{p0}(0)$, and Eq. (6.29) is reduced to

$$J_n = -\frac{q \exp(qV_{BE}/kT)}{\int_0^{W_B} (p_p/D_n n_{ie}^2)\, dx}, \tag{6.30}$$

where we have used the fact that $n_{p0}(0)p_{p0}(0) = n_{ie}^2(0)$.

This electron current is the source of the collector current. Therefore, the collector current $I_C$ is given by

$$I_C \equiv A_E|J_C| = A_E|J_n| = \frac{qA_E \exp(qV_{BE}/kT)}{\int_0^{W_B} (p_p/D_{nB}n_{ieB}^2)\,dx}, \tag{6.31}$$

where $A_E$ denotes the emitter area, and the subscript $B$ denotes quantities in the base region. [Avalanche multiplication in the base–collector junction will increase the collector current to a value larger than that given by Eq. (6.31). This effect is neglected here but is considered in Section 6.5 in connection with the transistor breakdown voltages.]

The collector current is often written in the form

$$I_C = A_E J_{C0} \exp(qV_{BE}/kT),$$

or

$$I_C = A_E \frac{qn_i^2}{G_B} \exp(qV_{BE}/kT), \tag{6.32}$$

where $J_{C0}$ is the *saturated collector current density* and $G_B$ is the *base Gummel number* (Gummel, 1961), and $n_i$ is the intrinsic carrier concentration. Comparing Eqs. (6.31) and (6.32) gives

$$J_{C0} = \frac{q}{\int_0^{W_B} (p_p/D_{nB}n_{ieB}^2)\,dx} \tag{6.33}$$

and

$$G_B = \int_0^{W_B} \frac{n_i^2}{n_{ieB}^2}\frac{p_p}{D_{nB}}\,dx. \tag{6.34}$$

[In the literature, the base Gummel number is often defined as the total integrated base dose (Gummel, 1961). However, here ***we follow the convention of de Graaff (de Graaff et al., 1977) and define $G_B$ to include both the minority-carrier diffusion coefficient and the effect of heavy doping in the base***. Thus, $J_{C0} = qn_i^2/G_B$. If heavy-doping effect is negligible and $D_{nB}$ is a constant, then $G_B =$ (total integrated base dose)/$D_{nB}$.]

It should be noted that ***the collector current is a function of the base-region parameters only, and is independent of the properties of the emitter***. All the effects in the base region, such as bandgap narrowing, bandgap nonuniformity, and dopant distribution, are contained in the parameter $G_B$. For the special case of a uniformly doped base region at low injection currents, with uniform energy bandgap and negligible heavy-doping effect, the base Gummel number reduces to $N_B W_B/D_{nB}$, and Eq. (6.30) reduces to Eq. (2.112), as expected.

### 6.2.2 BASE CURRENT

Neglecting both base–collector junction avalanche effect and recombination in the base layer, the base current in an n–p–n transistor is equal to the hole current injected from the base into the emitter. Referring to Fig. 6.1(d), let $x = 0$ denote the depletion-layer edge on the emitter side of the emitter–base junction, and $x = -W_E$ denote the location of the ohmic contact to the emitter. $W_E$ is the width of the emitter quasineutral region. Since the emitter is usually so heavily doped that its electron concentration is not affected at all by the hole current level, it is a good approximation to assume $n_n \approx n_{n0} = N_E$, where $N_E$ is the emitter doping concentration. With this approximation, the hole current density in the emitter, i.e., Eq. (6.26), can be rewritten as

$$J_p(x) = -qD_p \frac{n_{ie}^2}{n_{n0}} \frac{d}{dx} \left( \frac{n_{n0} p_n}{n_{ie}^2} \right). \tag{6.35}$$

Equation (6.35) gives the hole current density at any point in the emitter, and $J_p(0)$ is equal to the base current density. It should be noted that ***the base current is a function of the emitter-region parameters only and is independent of the properties of the base region***. Thus, the base current density changes as the emitter structure and design are changed. In this subsection, we shall use Eq. (6.35) to derive the base current in terms of the more familiar emitter parameters.

#### 6.2.2.1 SHALLOW OR TRANSPARENT EMITTER

An emitter is considered *shallow* or *transparent* when its width is small compared to its minority-carrier diffusion length. For a shallow emitter, there is negligible recombination in the emitter region except at the emitter contact at $x = -W_E$, and the minority-carrier current density in the emitter is independent of $x$.

For an n–p–n transistor, the hole current density at the emitter contact is usually written in terms of the surface recombination velocity for holes, $S_p$, defined by

$$J_p(x = -W_E) \equiv -q(p_n - p_{n0})_{x=-W_E} S_p. \tag{6.36}$$

Notice that $J_p$ is negative because holes flowing in the $-x$ direction give rise to a negative current. Since $J_p$ is independent of $x$ for a transparent emitter, Eq. (6.35) can be rearranged and integrated to give

$$J_p \int_{-W_E}^{0} \frac{n_{n0}}{qD_p n_{ie}^2} dx = - \left. \frac{n_{n0} p_n}{n_{ie}^2} \right|_{x=0} + \left. \frac{n_{n0} p_n}{n_{ie}^2} \right|_{x=-W_E} \tag{6.37}$$

At $x = 0$, the relation between the hole concentration and the emitter–base voltage is given by Eq. (2.90), namely

$$p_n(0) = p_{n0}(0) \exp(qV_{BE}/kT), \tag{6.38}$$

where $V_{BE}$ is the base–emitter bias voltage. Substituting Eqs. (6.36) and (6.38) into Eq. (6.37), and using the relation $n_{ie}^2 = p_{n0}n_{n0}$, we obtain

$$J_p \int_{-W_E}^{0} \frac{n_{n0}}{qD_p n_{ie}^2}\, dx = -\exp\left(\frac{qV_{BE}}{kT}\right) - 1 - \frac{n_{n0}(-W_E)}{n_{ie}^2(-W_E)qS_p} J_p \tag{6.39}$$

$$\approx -\exp\left(\frac{qV_{BE}}{kT}\right) - \frac{n_{n0}(-W_E)}{n_{ie}^2(-W_E)qS_p} J_p,$$

or

$$J_p \approx \frac{-q\exp\left(\frac{qV_{BE}}{kT}\right)}{\int_{-W_E}^{0} \frac{n_{n0}}{D_p n_{ie}^2}\, dx + \frac{n_{n0}(-W_E)}{n_{ie}^2(-W_E)S_p}}, \tag{6.40}$$

Equation (6.40) is valid for a transparent emitter of arbitrary doping profile and arbitrary surface recombination velocity at the emitter contact (Shibib *et al.*, 1979).

Equation (6.40) gives the hole current density entering the emitter. The base current is therefore

$$I_B \equiv A_E|J_B| = A_E|J_p| = \frac{qA_E\exp\left(\frac{qV_{BE}}{kT}\right)}{\int_{-W_E}^{0} \frac{N_E}{D_{pE} n_{ieE}^2}\, dx + \frac{N_E(-W_E)}{n_{ieE}^2(-W_E)S_p}}, \tag{6.41}$$

where $A_E$ is the emitter area, $N_E$ is the emitter doping concentration, and the subscript $E$ denotes parameters in the emitter region.

The base current is often written in the form

$$I_B = A_E J_{B0} \exp(qV_{BE}/kT),$$

or

$$I_B = A_E \frac{qn_i^2}{G_E} \exp\left(\frac{qV_{BE}}{kT}\right), \tag{6.42}$$

where $J_{B0}$ is the *saturated base current density*, and $G_E$ is the *emitter Gummel number* (de Graaff *et al.*, 1977). For a shallow or transparent emitter, Eq. (6.41) gives

$$J_{B0} = \frac{q}{\int_{-W_E}^{0} \frac{N_E}{D_{pE} n_{ieE}^2}\, dx + \frac{N_E(-W_E)}{n_{ieE}^2(-W_E)S_p}} \tag{6.43}$$

and

$$G_E = \int_{-W_E}^{0} \frac{n_i^2}{n_{ieE}^2} \frac{N_E}{D_{pE}}\, dx + \frac{n_i^2 N_E(-W_E)}{n_{ieE}^2(-W_E)S_p}. \tag{6.44}$$

- *Transparent emitter with uniform doping concentration and uniform energy bandgap.* Let us consider an n–p–n bipolar transistor with an emitter doping profile as indicated in Fig. 6.2(a). The emitter doping profile is not really uniform or boxlike. Even if we assume the most heavily doped region to be uniform, there is still a transition region where the emitter doping concentration drops from about $10^{20}$ to about $10^{18}$ cm$^{-3}$ at the emitter–base junction. This transition region plays an important role in determining the emitter–base junction capacitance and the emitter–base junction breakdown voltage. However, as far as the base current is concerned, the effect of this transition region is relatively small (Roulston, 1990). This is due to the fact that the hole diffusion length in this relatively lightly doped transition region is very large compared to the thickness of the region. As a result, the transition region is almost completely transparent to the holes entering the emitter. Therefore, at least for purposes of modeling the base current, it is common to ignore this transition region and simply assume the emitter region to be uniformly doped and boxlike. Besides, such an approximation makes modeling the emitter region relatively simple. For such a uniformly doped transparent emitter with uniform energy bandgap, Eq. (6.43) reduces to

$$J_{B0} = \frac{qD_{pE}n_{ieE}^2}{N_E W_E(1 + D_{pE}/W_E S_P)}, \tag{6.45}$$

and Eq. (6.44) reduces to

$$G_E = N_E\left(\frac{n_i^2}{n_{ieE}^2}\right)\left(\frac{W_E}{D_{pE}} + \frac{1}{S_p}\right). \tag{6.46}$$

For an ohmic emitter contact, $S_p$ is infinite, and Eq. (6.45) becomes proportional to $1/W_E$, as expected from the properties of a narrow-base diode [cf. Eqs. (2.112) and (2.114)]. The base current increases rapidly as the emitter width, or depth, is reduced.

- *Polysilicon emitter.* The simplest model for describing a polysilicon emitter is to treat the polysilicon–silicon interface located at $x = -W_E$ as a contact with finite surface recombination velocity. In this case, Eq. (6.43) or Eq. (6.45) can be used, depending on whether the single-crystal emitter region is uniformly doped or not. Under certain conditions, a model for the polysilicon emitter can be developed which allows the surface recombination to be evaluated in terms of the properties of the polysilicon layer (Exercise 6.3). In practice, the surface recombination velocity is often used as a fitting parameter to the measured base current. The detailed physics of transport in a polysilicon emitter is very complicated and is dependent on the polysilicon-emitter fabrication process. Therefore, the surface recombination velocity obtained by fitting to the measured base current is also dependent on the polysilicon-emitter fabrication process. The reader is referred to the vast published literature on polysilicon-emitter physics and technology (Ashburn, 1988; Kapoor and Roulston, 1989).

#### 6.2.2.2 DEEP EMITTER WITH UNIFORM DOPING CONCENTRATION AND UNIFORM ENERGY BANDGAP

An emitter is *deep*, or *nontransparent*, when its width is large compared to its minority-carrier diffusion length. For a deep emitter, most or all of the injected minority carriers recombine before they reach the emitter contact, and the minority-carrier current is a function of $x$. The minority-carrier current density given by Eq. (6.35) becomes rather simple if the emitter is assumed to be uniformly doped, has a uniform energy bandgap, and has an ohmic contact at $x = -W_E$. With these assumptions, Eq. (6.35) reduces to

$$J_p(x) = -qD_p\frac{dp_n}{dx}, \tag{6.47}$$

which is simply the hole diffusion current density in the uniformly doped n-side of a diode under low injection, with the hole density given by the equivalent of Eq. (2.103).

The base current density is simply this hole current density at $x = 0$. The base current, therefore, can be obtained from the hole equivalent of Eq. (2.104), namely

$$I_B = A_E|J_p(x=0)| = \frac{qA_E D_{pE} n_{ieE}^2 \exp(qV_{BE}/kT)}{N_E L_{pE}\tanh(W_E/L_{pE})}, \tag{6.48}$$

where $A_E$ is the emitter area, $N_E$ is the emitter doping concentration, and the subscript $E$ denotes parameters in the emitter region. The corresponding base saturation current density and emitter Gummel number are

$$J_{B0} = \frac{qD_{pE}n_{ieE}^2}{N_E L_{pE}\tanh(W_E/L_{pE})}, \tag{6.49}$$

and

$$G_E = \left(\frac{n_i^2}{n_{ieE}^2}\right)\frac{N_E L_{pE}\tanh(W_E/L_{pE})}{D_{pE}}. \tag{6.50}$$

### 6.2.3 CURRENT GAINS

The static *common-emitter* current gain $\beta_0$ is defined by

$$\begin{aligned}\beta_0 &\equiv \frac{\partial I_C}{\partial I_B} \\ &= \frac{I_C}{I_B} = \frac{J_{C0}}{J_{B0}} = \frac{G_E}{G_B}.\end{aligned} \tag{6.51}$$

From Eqs. (6.34) and (6.44),

$$\beta_0 = \frac{\int_{-W_E}^{0} \frac{N_E}{D_{pE} n_{ieE}^2} dx + \frac{N_E(-W_E)}{n_{ieE}^2(-W_E) S_p}}{\int_0^{W_B} \frac{p_p}{D_{nB} n_{ieB}^2} dx} \quad \text{for a transparent emitter,} \tag{6.52}$$

and from Eqs. (6.34) and (6.50),

$$\beta_0 = \frac{N_E L_{pE} \tanh(W_E/L_{pE})}{D_{pE} n_{ieE}^2 \int_0^{W_B} (p_p/D_{nB} n_{ieB}^2)\, dx} \quad \text{for a deep emitter.} \tag{6.53}$$

The static *common-base current gain* $\alpha_0$ is defined by

$$\alpha_0 \equiv \frac{\partial I_C}{\partial(-I_E)} \tag{6.54}$$

$$= \frac{I_C}{-I_E},$$

where $I_E$ is the emitter current. Here we have defined $I_E$ as the current flowing *into* the emitter, so that $-I_E$ is positive. Since $I_E + I_B + I_C = 0$, we have

$$\alpha_0 = \frac{\beta_0}{1 + \beta_0} \tag{6.55}$$

and

$$\beta_0 = \frac{\alpha_0}{1 - \alpha_0}. \tag{6.56}$$

For modern VLSI bipolar transistors, $\beta_0$ is typically about 100. Therefore, $\alpha_0$ is almost unity. In principle, either $\alpha_0$ or $\beta_0$ can be used to describe the current gain of a bipolar transistor. In practice, $\beta_0$ is often used in discussing the device characteristics, device design, and device physics. Throughout this book, we shall use $\beta_0$. (However, we shall use $\alpha_0$ when we consider breakdown voltages in Section 6.5.)

The common-emitter current gain is often quoted as a figure of merit for a bipolar transistor. However, it should be noted that, being the ratio of two currents, the current gain changes as either one of the currents changes. Therefore, to really understand the device design and the device characteristics, both the collector current and the base current, not just the current gain, should be considered. As discussed in the previous subsections, the collector current is a function of only the base parameters, while the base current is a function of only the emitter parameters.

For digital logic circuits, the circuit speed is insensitive to the current gain of the transistors (Ning *et al.*, 1981). However, for many analog circuits, a high current gain is desirable. Most transistors are designed with a current gain of about 100 or larger. For a given bipolar transistor fabrication process, the current gain can be increased or decreased readily by changing the base Gummel number, or the base parameters. Design considerations for the base region will be covered in Section 7.2.

#### 6.2.3.1 CURRENT GAIN FOR UNIFORMLY DOPED DEEP EMITTER AND UNIFORMLY DOPED BASE

For the special case of a uniformly doped emitter with $W_E/L_{pE} \gg 1$, and a uniformly doped base with concentration $N_B$, Eq. (6.53) reduces to

$$\beta_0 = \frac{n_{ieB}^2 D_{nB} N_E L_{pE}}{n_{ieE}^2 D_{pE} \int_0^{W_B} p_p\, dx}. \tag{6.57}$$

If the electron current density injected into the base is *low*, then the electron density in the base is also small compared to the hole density, and the hole density is approximately equal to the base doping concentration $N_B$. In this case, Eq. (6.57) reduces further to

$$\beta_0 = \frac{n_{ieB}^2}{n_{ieE}} \frac{D_{nB}}{D_{pE}} \frac{N_E L_{pE}}{N_B W_B}, \tag{6.58}$$

which is independent of current. (The current gain at high currents can be rather complex and will be discussed in Section 6.3.)

It is instructive to estimate the magnitude of the current gain given by Eq. (6.58). If we assume $N_E = 1 \times 10^{20}\ \text{cm}^{-3}$, $N_B = 1 \times 10^{18}\ \text{cm}^{-3}$, and $W_B = 0.1\ \mu\text{m}$ for a typical deep-emitter thin-base n–p–n transistor, then Fig. 6.3 gives $(n_{ieB}/n_{ieE})^2 = \exp[(\Delta E_{gB} - \Delta E_{gE})/kT] \approx 0.19$ at room temperature, Fig. 2.18(a) gives $D_{nB}/D_{pE} = \mu_{nB}/\mu_{pE} \approx 2.6$, $N_E/N_B = 100$, and Fig. 2.18(c) gives $L_{pE}/W_B \approx 4.6$. Substituting these values into Eq. (6.58) gives $\beta_0 = 230$.

### 6.2.4 IDEAL $I_C$–$V_{CE}$ CHARACTERISTICS

Figure 6.4 illustrates the ideal $I_C$-versus-$V_{CE}$ characteristics of an n–p–n transistor, with $I_B$ as a parameter. Each base current corresponds to a given $V_{BE}$-value. The dashed curve indicates where $V_{CE} = V_{BE}$.

For $V_{CE} < V_{BE}$, the collector–base diode is forward biased and the transistor is said to be in *saturation*. In this case, to first order, the collector current is the

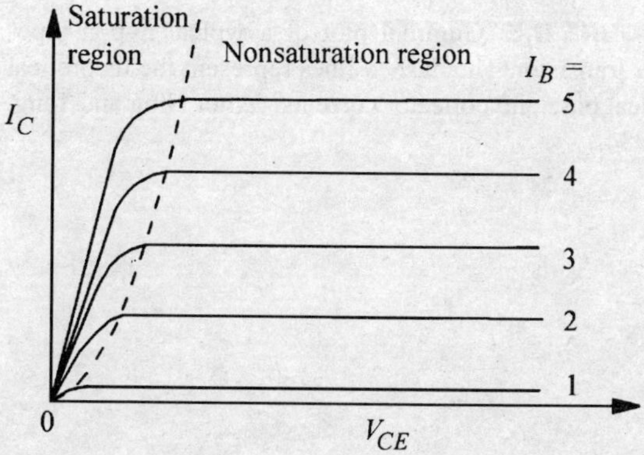


**FIGURE 6.4.** Schematic illustration of the ideal $I_C$-versus-$V_{CE}$ characteristics of an n–p–n transistor. The dashed line is the locus for $V_{CE} = V_{BE}$.

difference of the electron current injected from the emitter into the base and the electron current injected from the collector into the base. As a result, the collector current increases with increasing $V_{CE}$, i.e., as the transistor becomes less saturated.

For $V_{CE} > V_{BE}$, the collector–base diode is reverse-biased and the transistor is said to be in its normal *forward-active* mode of operation. All the electrons injected from the emitter into the base are collected by the collector, and there is no electron injection from the collector into the base. The collector current is therefore constant, independent of $V_{CE}$. The current gain is also constant, and the constant-$I_B$ curves are spaced apart by an amount determined by the base-current step.

***The measured current–voltage characteristics of typical bipolar devices are not ideal.*** The degree of deviation from ideal characteristics depends on the device structure, the device design, the device fabrication process, and on the bias condition of the transistor. The behavior of a typical n–p–n transistor is discussed next.

## 6.3 CHARACTERISTICS OF A TYPICAL n-p-n TRANSISTOR

Figure 6.5 is the so-called *Gummel plot* of a typical n–p–n transistor. It plots both the collector current $I_C$ and the base current $I_B$ on a logarithmic scale as a function of the forward-bias voltage $V_{BE}$ applied to the emitter and base terminals. The theoretical ideal base and collector currents, discussed in Section 6.2, are indicated by the dashed lines. Figure 6.5 shows that the measured collector current is ideal except at large $V_{BE}$, while the measured base current is ideal except at small and at large $V_{BE}$.

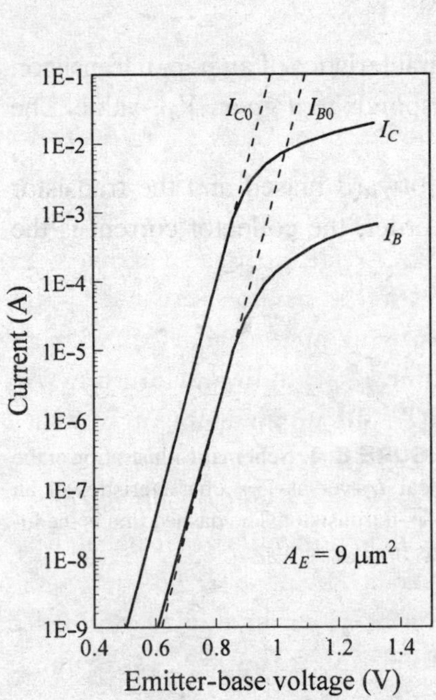


**FIGURE 6.5.** Gummel plot of a typical n–p–n bipolar transistor. The dashed lines represent the theoretical ideal base and collector currents. (After Ning and Tang, 1984.)

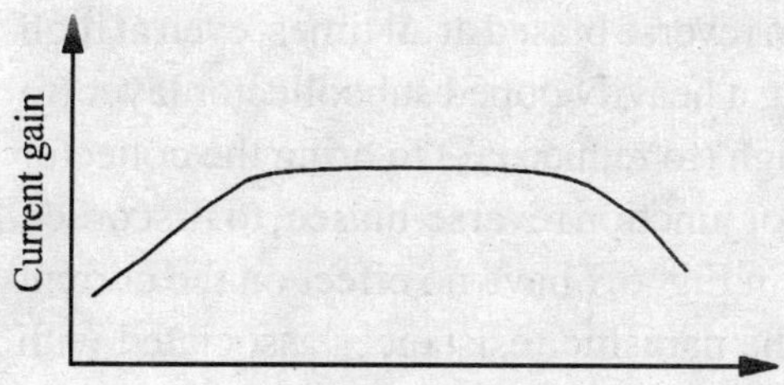


**FIGURE 6.6.** Schematic illustration of the current gain $I_C/I_B$ as a function of collector current for a typical bipolar transistor.

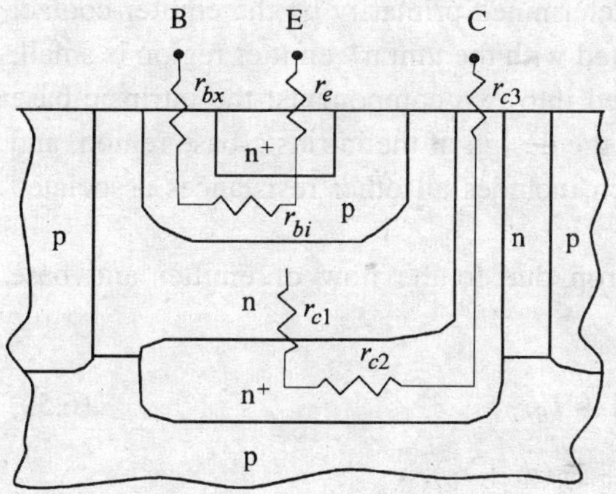


**FIGURE 6.7.** Schematic illustrating the parasitic resistances in a typical modern n–p–n transistor.

Figure 6.6 illustrates the typical measured current gain, $I_C/I_B$, as a function of collector current. For the voltage range where both the base and the collector currents are approximately ideal, the current gain is approximately constant. At low currents, the current gain is less than its ideal value because the base current is larger than its ideal value. At high currents, the current gain rolls off with collector current because the percentage by which the collector current is smaller than its ideal value is larger than the percentage by which the base current is smaller than its ideal value. The dominant physical mechanisms responsible for the nonideal behavior of the base and collector currents are discussed in the subsections below.

### 6.3.1 EFFECT OF EMITTER AND BASE SERIES RESISTANCES

Figure 6.7 shows schematically the physical origins of the parasitic resistances in a typical n–p–n transistor. These resistances are ignored in Section 6.2 in the description of the ideal current–voltage characteristics. As the currents flow through these parasitic resistors, voltage drops are developed, which tend to offset the externally applied voltages. The parasitic resistances can therefore be neglected at low currents but can be very important at large currents.

In normal forward-active operation, the base–collector junction is reverse biased. In most bipolar circuits, particularly those designed for high-speed applications, the collector–base junction is designed to remain reverse biased at all times, even at high currents. This is accomplished by employing a heavily doped subcollector layer (to reduce $r_{c2}$) and a heavily doped reach-through (to reduce $r_{c3}$) to bring the collector contact to the surface. With the base–collector junction reverse-biased, to first order, the collector resistance components shown in Fig. 6.7 have no effect on the current flows in the emitter–base diode, and only the parasitic resistances associated with the emitter and the base need to be considered. (The effect of collector–base voltage on collector current is discussed in the following subsection.)

The emitter series resistance $r_e$ is determined primarily by the emitter contact resistance, since the resistance associated with the thin $n^+$ emitter region is small. The base resistance $r_b$ can be separated into two components: the intrinsic-base resistance $r_{bi}$, which is determined by the design of the intrinsic-base region, and the extrinsic-base resistance $r_{bx}$, which includes all other resistances associated with the base terminal.

The emitter–base diode voltage drop due to the flow of emitter and base currents is

$$\begin{aligned} \Delta V_{BE} &= -I_E r_e + I_B r_b \\ &= I_C r_e + I_B(r_e + r_b), \end{aligned} \tag{6.59}$$

where we have used the fact that $I_E + I_B + I_C = 0$. The relation between the voltage $V_{BE}$ applied to the emitter and base terminals and the voltage $V'_{BE}$ appearing across the immediate emitter–base junction is

$$V'_{BE} = V_{BE} - \Delta V_{BE}. \tag{6.60}$$

To include the effect of the emitter and base series resistances, the equations in Section 6.2 for the ideal collector and base currents should be modified by replacing $V_{BE}$ by $V'_{BE}$. This results in both the measured collector and base currents, when plotted as a function of $V_{BE}$, being significantly smaller than the ideal currents at large $V_{BE}$, as illustrated in Fig. 6.5.

As can be seen from Eq. (6.33), even in the ideal case, the collector saturation current density is a function of the majority-carrier concentration in the base and the base width. Therefore, ***the measured collector current is a function of $\Delta V_{BE}$ as well as a function of the base majority-carrier concentration and the base width, which in turn depend on $V_{BE}$***. The dependence of $I_C$ on $V_{BE}$ is very complex, as can be seen in later subsections.

On the other hand, as can be seen from Eqs. (6.43) and (6.49), the base saturation current density is a function of the emitter parameters only, which, due to the emitter being very heavily doped, do not vary with the minority-carrier injection level. Therefore, at high currents, ***deviation of the base current from its ideal behavior is due to $\Delta V_{BE}$ alone*** (Ning and Tang, 1984). The relation between the ideal base current $I_{B0}$ and the measured base current $I_B$ is therefore

$$I_{B0} = I_B \exp(q\Delta V_{BE}/kT), \tag{6.61}$$

which can be used to evaluate the emitter and base series resistances. This is shown in Appendix 8. Many other methods for determining the emitter and base series resistances have been discussed in the literature (Schroder, 1990). Some of these are discussed in Appendix 8 as well.

### 6.3.2 EFFECT OF BASE–COLLECTOR VOLTAGE ON COLLECTOR CURRENT

In many transistors, particularly in modern high-speed transistors where the base width is very small, the measured collector current, and hence the measured current gain, increases as the base–collector reverse-bias voltage is increased. This is due to two effects, or a combination of them. The first effect is the dependence of the quasineutral base width on collector–base voltage. The second effect is the avalanche multiplication in the base–collector junction. We shall discuss these two effects individually in this subsection.

#### 6.3.2.1 MODULATION OF QUASINEUTRAL BASE WIDTH BY BASE–COLLECTOR VOLTAGE

As the reverse bias across the base–collector junction is increased, the base–collector junction depletion-layer width increases, and hence the quasineutral base width $W_B$ decreases. This in turn causes the collector current to increase, as can be seen from Eq. (6.31). Thus, instead of as illustrated in Fig. 6.4, where the collector current is independent of collector voltage for $V_{CE} > V_{BE}$, the collector current of a typical bipolar transistor increases with collector voltage, as illustrated in Fig. 6.8.

- *Early voltage.* For circuit modeling purposes, the collector current in the nonsaturation region is often assumed to depend linearly on the collector voltage. The collector voltage at which the linearly extrapolated $I_C$ reaches zero is

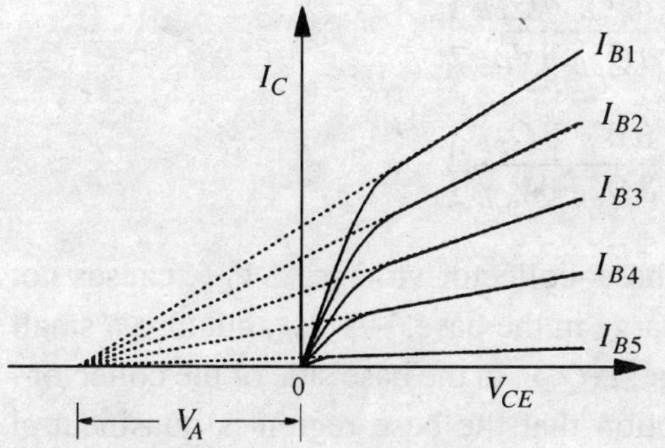


**FIGURE 6.8.** Schematic illustrating the approximately linear dependence of $I_C$ on $V_{CE}$. The linearly extrapolated $I_C$ intersects the $V_{CE}$-axis at $-V_A$.

denoted by $-V_A$. As we shall show later, it is a good and useful approximation to assume that $V_A$ is independent of $V_{BE}$. This is illustrated in Fig. 6.8. $V_A$ is called the *Early voltage* (Early, 1952). It is defined by

$$V_A + V_{CE} \equiv I_C \left( \frac{\partial I_C}{\partial V_{CE}} \right)^{-1} . \tag{6.62}$$

In practice, except for transistors that tend to punch through (to be discussed later), $V_A$ is much larger than the operation range of $V_{CE}$. Therefore, $V_A$ can be approximated by

$$V_A \approx I_C \left( \frac{\partial I_C}{\partial V_{CE}} \right)^{-1} . \tag{6.63}$$

The collector current is given by Eq. (6.32), which can be written as

$$I_C = A_E J_{C0} \exp(qV_{BE}/kT) = \frac{qA_E \exp(qV_{BE}/kT)}{F(W_B)}, \tag{6.64}$$

where, for convenience, a function $F$ has been introduced (Kroemer, 1985) which is defined by

$$F(W_B) \equiv \frac{q}{J_{C0}} = \int_0^{W_B} \frac{p_p(x)}{D_{nB}(x) n_{ieB}^2(x)} \, dx. \tag{6.65}$$

The majority-carrier hole charge per unit area in the base is

$$Q_{pB} = q \int_0^{W_B} p_p(x) \, dx. \tag{6.66}$$

Since $V_{BE}$ is fixed for a given $I_B$, Eq. (6.63) can be rewritten as

$$\begin{aligned} V_A &\approx I_C \left( \frac{-I_C}{F} \frac{\partial F}{\partial V_{CE}} \right)^{-1} \\ &= \left( \frac{-1}{F} \frac{\partial F}{\partial W_B} \frac{\partial W_B}{\partial Q_{pB}} \frac{\partial Q_{pB}}{\partial V_{CE}} \right)^{-1} \\ &= \left( \frac{-1}{F} \frac{\partial F}{\partial W_B} \frac{\partial W_B}{\partial Q_{pB}} \frac{\partial Q_{pB}}{\partial V_{CB}} \right)^{-1} . \end{aligned} \tag{6.67}$$

Notice that a small increase in the base–collector voltage, $\Delta V_{CB}$, causes not only a small decrease in the hole charge in the base, $-\Delta Q_{pB}$, but also a small increase in the depletion-layer charge, $\Delta Q_{dB}$, in the base side of the collector–base depletion region. The assumption that the base region is quasineutral requires that $-\Delta Q_{pB} = \Delta Q_{dB}$. Therefore

$$-\frac{\partial Q_{pB}}{\partial V_{CB}} = \frac{\partial Q_{dB}}{\partial V_{CB}} = C_{dBC}, \tag{6.68}$$

where $C_{dBC}$ is the base–collector junction depletion-layer capacitance [cf. Eq. (2.68)]. The other two derivatives in Eq. (6.67) can be evaluated directly, namely

$$\frac{\partial F}{\partial W_B} = \frac{p_p(W_B)}{D_{nB}(W_B)n_{ieB}^2(W_B)} \tag{6.69}$$

and

$$\frac{\partial W_B}{\partial Q_{pB}} = \left(\frac{\partial Q_{pB}}{\partial W_B}\right)^{-1} = \frac{1}{qp_p(W_B)}. \tag{6.70}$$

Therefore, Eq. (6.67) gives

$$V_A \approx \frac{qD_{nB}(W_B)n_{ieB}^2(W_B)}{C_{dBC}} \int_0^{W_B} \frac{p_p(x)}{D_{nB}(x)n_{ieB}^2(x)}\,dx. \tag{6.71}$$

For a uniformly doped base, Eq. (6.71) reduces to

$$V_A \approx \frac{Q_{pB}}{C_{dBC}}. \tag{6.72}$$

At sufficiently low collector currents such that the base majority-carrier concentration is approximately the same as its equilibrium value, i.e., $p_p \approx p_{p0} = N_B$, Eq. (6.71) gives

$$V_A \approx \frac{qD_{nB}(W_B)n_{ieB}^2(W_B)}{C_{dBC}} \int_0^{W_B} \frac{N_B(x)}{D_{nB}(x)n_{ieB}^2(x)}\,dx. \tag{6.73}$$

Equation (6.73) is independent of base current, so that the slope of the curves in Fig. 6.8 intercept the $V_{CE}$-axis at the same value, namely $V_A$, as illustrated. It is instructive to estimate the magnitude of Eq. (6.73) for a uniformly doped base. In this case, $V_A \approx qW_BN_B/C_{dBC}$. For a base of $W_B = 0.1$ μm and $N_B = 10^{18}$ cm$^{-3}$, we have $qW_BN_B \approx 1.6 \times 10^{-6}$ C/cm$^2$. For a collector of $N_C = 2 \times 10^{16}$ cm$^{-3}$, then, from Fig. 2.13, $C_{dBC} \approx 4 \times 10^{-8}$ F/cm$^2$. Therefore, $V_A \approx 40$ V. In practice, $V_A$ can vary a lot as the transistor design is "optimized." This will be discussed further later in this section and in Chapter 7. As can be seen in Eq. (6.71), $V_A$ is a function of $W_B$, which, as discussed earlier, is a function of the collector voltage. Therefore, strictly speaking, the Early voltage is a function of the collector voltage at which the slope is used for extrapolating to $I_C = 0$. In other words, strictly speaking, $I_C$ does not increase linearly with $V_{CE}$. However, the linear dependence is a good approximation and is a useful approximation for circuit analyses and modeling purposes. The Early voltage is a figure of merit for devices used in analog circuits. The larger the Early voltage, the more independent is the collector current on collector voltage. Another device figure of merit is the product of the current gain and Early voltage. Using Eqs. (6.33), (6.51), and (6.71), this product can be written as (Prinz and Sturm, 1991)

$$\beta_0 V_A = \frac{q^2 D_{nB}(W_B) n_{ieB}^2(W_B)}{C_{dBC} J_{B0}}, \tag{6.74}$$

where the base saturated current density $J_{B0}$ is a function of the emitter parameters. That is, while $V_A$ is a function of the base parameters only, the product $\beta_0 V_A$ is a function of both the emitter and the base parameters.

- *Emitter–collector punch-through.* As shown in Eq. (6.72), the Early voltage is proportional to the majority-carrier charge in the base. As the collector voltage is increased, the width of the quasineutral base region, and hence the majority-carrier charge in the base, is reduced. For a device with a small majority-carrier base charge or small Early voltage to start with, it does not take much increase in collector voltage before all the majority-carrier base charge is depleted, or before the collector punches through to the emitter. At collector–emitter punch-through, the collector current becomes excessively large, being limited only by the emitter and collector series resistances. The collector current at or close to punch-through is no longer controlled adequately by the base voltage for proper device operation. ***Punch-through must be avoided*** under normal device operation, by designing the device to have a sufficiently large majority-carrier base charge.

#### 6.3.2.2 BASE–COLLECTOR JUNCTION AVALANCHE

For a device with a large majority-carrier base charge or large Early voltage to begin with, as the collector voltage is increased, usually the condition of significant base–collector junction avalanche is reached before punch-through is reached. This is certainly the case for transistors where the collector side of the base–collector junction space–charge–layer boundary reaches the subcollector region before punch-through occurs, because as the base-collector junction space-charge-layer boundary reaches the heavily doped subcollector, further increase of the base–collector reverse bias will increase the junction electric field very rapidly.

For an n–p–n transistor, the base–collector junction avalanche process is illustrated in Fig. 6.9(a) (Lu and Chen, 1989). As the electrons injected from the emitter into the base reach the base–collector junction space-charge region, they can cause impact ionization and generate electron–hole pairs. The secondary electrons flow towards the collector terminal, adding to the measured collector current, while the

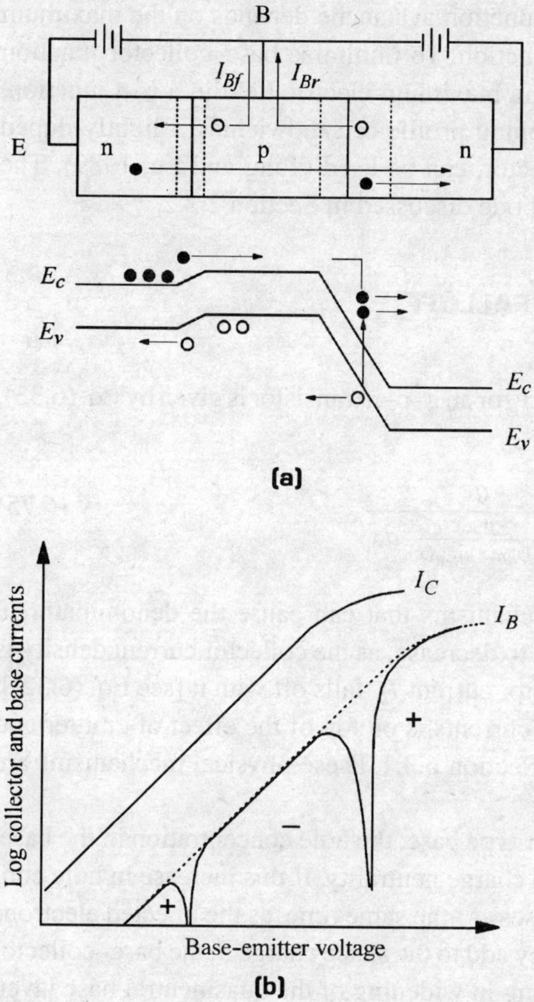


**FIGURE 6.9.** (a) Schematics of an n–p–n transistor operated in the forward-active mode with a large base–collector voltage. As electron–hole pairs are generated in the base–collector junction space-charge region, the secondary hole current, $I_{Br}$, subtracts from the usual forward base current, $I_{Bf}$. The current measured at the base terminal is $I_B = I_{Bf} - I_{Br}$. (b) Typical Gummel plot of an n–p–n transistor where significant avalanche multiplication occurs in the base–collector junction space-charge region. (After Lu and Chen, 1989.)

secondary holes flow towards the base terminal, subtracting from the measured base current. If the secondary hole current is large enough, the current measured at the base terminal could be negative (Lu and Chen, 1989). This is illustrated in Fig. 6.9(b).

At very small emitter–base forward biases, the measured base current is positive as usual. The secondary hole current is not large enough to completely offset the usual base current. As the electron current injected from the emitter into the base–collector space-charge region increases with increased emitter–base forward bias, the secondary hole current increases and may reach a point at which the measured base current turns negative. At sufficiently large emitter–base forward biases, as will be discussed in the next subsection, significant base widening can occur and the electric field in the base–collector junction can be reduced. As a result, avalanche multiplication is reduced and the measured base current returns to positive.

The magnitude of base–collector junction avalanche depends on the maximum electric field in the base–collector junction. To minimize base–collector junction avalanche, techniques for reducing the maximum electric field in a p–n junction, such as retrograding the collector doping profile or sandwiching a lightly doped layer between the base and the collector, can be used (Tang and Lu, 1989). The concept is similar to that of a p–i–n diode discussed in Section 2.2.2.

### 6.3.3 COLLECTOR CURRENT FALLOFF AT HIGH CURRENTS

The collector saturation current density for an n–p–n transistor is given by Eq. (6.33), namely

$$J_{C0} = \frac{q}{\int_0^{W_B} \frac{p_p(x)}{D_{nB}(x) n_{ieB}^2(x)} dx}. \tag{6.75}$$

There are a number of physical mechanisms that can cause the denominator in Eq. (6.75) to increase, and hence $J_{C0}$ to decrease, as the collector current density is increased. As $J_{C0}$ falls off, the collector current $I_C$ falls off with it [see Eq. (6.32)]. This collector current falloff at high currents is on top of the effect of emitter and base series resistances discussed in Section 6.3.1 These physical mechanisms are discussed in this subsection.

As electrons are injected into the p-type base, the hole concentration in the base, $p_p(x)$, increases in order to maintain charge neutrality. If this increase in hole concentration is appreciable, $J_{C0}$ decreases. At the same time, as the injected electrons reach the base–collector junction, they add to the space charge in the base–collector junction space-charge region, resulting in widening of the quasineutral base layer. As $W_B$ increases, $J_{C0}$ also decreases. This is known as the *base-conductivity modulation effect.*

As $W_B$ increases, the collector side of the base–collector space-charge layer also widens into the collector. At sufficiently high collector current densities, base widening can push the "base–collector junction" deep into the collector region. This is known as the *base-widening* or *Kirk effect* (Kirk, 1962), and it also causes $J_{C0}$ to decrease.

Base-conductivity modulation and base-widening effects are not really separate and do not act independently. Dependent on the details of the device design, their combined effect can contribute significantly to the observed collector current saturation in the Gummel plot. In this subsection we discuss base widening in more detail.

#### 6.3.3.1 BASE WIDENING AT LOW CURRENTS

Consider the base–collector junction of an n–p–n transistor. For simplicity, let us assume the base region to have a uniform doping concentration $N_B$, and the collector

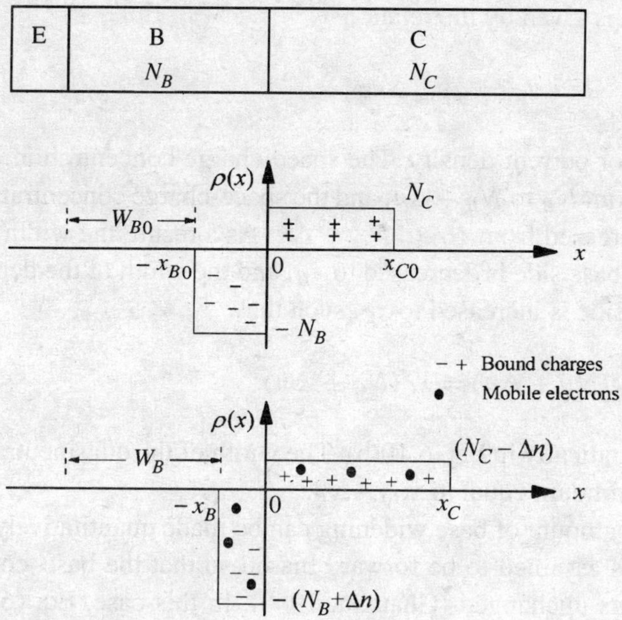


**FIGURE 6.10.** Schematics illustrating the charge distribution in the base–collector junction of an n–p–n transistor. (a) Emitter–base diode is not forward biased, and (b) emitter–base diode is forward biased.

region to have a uniform doping concentration $N_C$. When the transistor is turned off, the charge distribution in the base–collector junction is as shown schematically in Fig. 6.10(a), where $x_{B0}$ and $x_{C0}$ are the widths of the depletion regions on the base side and on the collector side, respectively. The relationship between these widths is given by Eq. (2.63), namely.

$$x_{B0} N_B = x_{C0} N_C. \tag{6.76}$$

The maximum potential drop across the base–collector junction, $\psi_{mBC}$, is given by Eq. (2.64), which can be rewritten as

$$\psi_{mBC} = \frac{q}{2\varepsilon_{si}} \left( N_B x_{B0}^2 + N_C x_{C0}^2 \right). \tag{6.77}$$

When the n–p–n transistor is turned on, electrons are injected into the base and collector regions. These mobile electrons add to the space charge in the base–collector junction region. As long as this additional mobile-electron concentration is small compared with the ionized doping concentrations, the depletion approximation discussed in Section 2.2.2 can be used to estimate its effect. For simplicity, let us assume these mobile electrons traverse the base–collector junction space-charge region at a saturated velocity $v_{sat}$. The mobile electron concentration $\Delta n$ in the space-charge region is given by the relation

$$J_C = qv_{sat}\Delta n, \tag{6.78}$$

where $J_C$ is the collector current density. The space-charge concentration on the base side is increased from $N_B$ to $N_B + \Delta n$, and the space-charge concentration on the collector side is decreased from $N_C$ to $N_C - \Delta n$. As a result, the width of the depletion region on the base side is decreased to $x_B$, and the width of the depletion region on the collector side is increased to $x_C$, such that

$$x_B(N_B + \Delta n) = x_C(N_C - \Delta n). \tag{6.79}$$

This is illustrated schematically in Fig. 6.10(b). The width of the quasineutral base layer is widened by an amount equal to $x_{B0} - x_B$.

An estimation of the amount of base widening can be made quantitatively if the emitter–base junction is assumed to be forward biased so that the base–collector junction voltage remains unchanged (Ghandhi, 1968). In this case, Eq. (6.77) is replaced by

$$\psi_{mBC} = \frac{q}{2\varepsilon_{si}}\left[(N_B + \Delta n)x_B^2 + (N_C - \Delta n)x_C^2\right]. \tag{6.80}$$

Combining Eqs. (6.77) and (6.80), and assuming $\Delta n/N_C \ll 1$, we have

$$x_C = x_{C0}\sqrt{\frac{1 + (\Delta n/N_B)}{1 - (\Delta n/N_C)}} \approx \frac{x_{C0}}{\sqrt{1 - (\Delta n/N_C)}} \tag{6.81}$$

where we have used the fact that $N_B$ is typically much larger than $N_C$, so that $\Delta n/N_B \lll 1$. Similarly

$$x_B = x_{B0}\sqrt{\frac{1 - (\Delta n/N_C)}{1 + (\Delta n/N_B)}} \approx x_{B0}\sqrt{1 - (\Delta n/N_C)}. \tag{6.82}$$

#### 6.3.3.2 BASE WIDENING AT HIGH CURRENTS

At high current densities, the assumption of $\Delta n$ being small compared to $N_C$ is no longer valid, and the above equations cannot be used to estimate the base-widening effect. With the mobile-charge concentration comparable to or larger than the fixed ionized-impurity concentration, the depletion approximation is certainly not valid. Furthermore, the excess electrons in the n-type collector can produce a substantial electric field in the collector, according to Eq. (6.4), and the classical concept of a well-defined junction boundary in the base–collector diode is no longer valid. Also, in order to maintain quasineutrality, the excess electrons induce an excess of holes in the n-type collector. The region of the collector with excess holes becomes an extension of the p-type base. In other words, the base region widens into the collector region, until it reaches the subcollector where the excess electron concentration is small compared with the n-type doping concentration. As a result, the high-field region, originally located at the physical base–collector junction, is relocated to near the collector–subcollector intersection (Poon *et al.*, 1969). The numerical simulation results (Poon *et al.*, 1969) shown in Fig. 6.11 illustrate clearly the effects of base widening at high currents. They show that the relocation of the high-field region is accompanied by a buildup of excess electrons and holes in the collector region.

It is instructive to estimate the collector current density at which substantial base widening occurs. The saturated velocity $v_{sat}$ for electrons in silicon is about $1 \times 10^7$ cm/s, as indicated in Fig. 2.9. At low collector currents, the maximum electron concentration in the n-type collector region is equal to the collector doping concentration $N_C$. The maximum electron current density that can be supported by an electron concentration $N_C$ is $J_{\max} = qv_{sat}N_C$. When the injected electron current density approaches $J_{\max}$, there is an increase of the electron concentration in excess of $N_C$ in order to support the injected electron current flow. As the excess electrons build up, there is a buildup of excess holes in order to maintain quasineutrality, and a relocation of the high-field region. The results shown in Fig. 6.11 suggest that significant base widening starts at a collector current density of approximately $0.3J_{\max}$. This value is consistent with the reported peak cutoff-frequency data for modern VLSI bipolar devices (Crabbé *et al.*, 1993). Thus, ***to avoid significant base widening, a bipolar transistor should not be operated at collector current densities approaching*** $\boldsymbol{J_{\max}}$. For a relatively high $N_C$ of $10^{17}$ cm$^{-3}$, $J_{\max}$ is about 1.6 mA/μm$^2$. To avoid significant base widening, $J_C$ should be less than about 0.5 mA/μm$^2$.

### 6.3.4 NONIDEAL BASE CURRENT AT LOW CURRENTS

As shown in Fig. 6.5, for small emitter–base voltages, the base current is larger than its ideal value. The origins of this excess base current are (a) the generation–recombination current in the emitter–base junction depletion region and (b) the tunneling current in the emitter–base junction (Li *et al.*, 1988). The amount of deviation from ideal behavior depends strongly on the transistor structure, device design, and fabrication process. For most well-designed bipolar transistors and fabrication processes, this excess current is often negligibly small. In any case, since this excess current is often quite noticeable in experimental devices, particularly before the fabrication process has been optimized, its physical origins are discussed here.

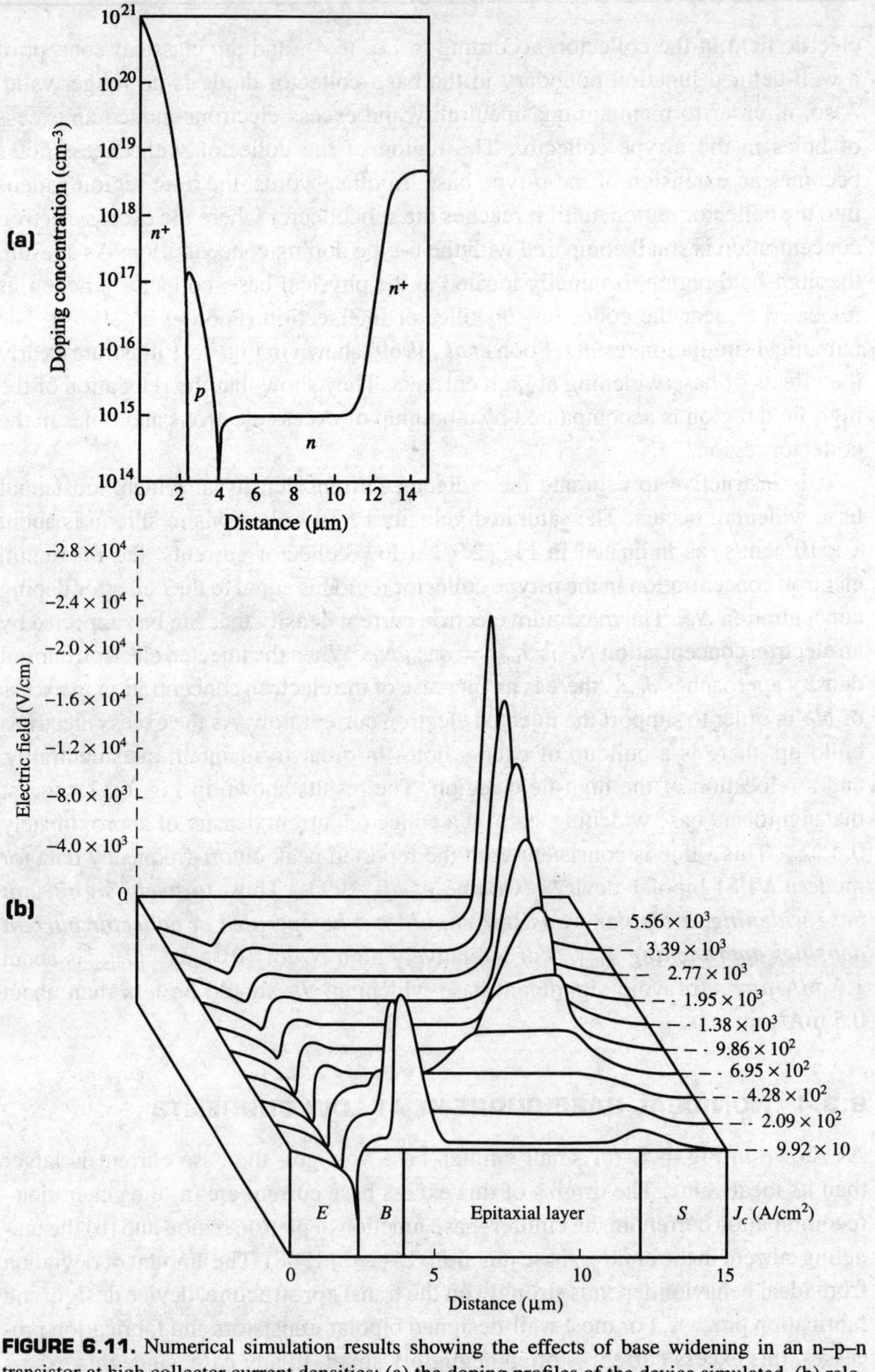


**FIGURE 6.11.** Numerical simulation results showing the effects of base widening in an n–p–n transistor at high collector current densities: (a) the doping profiles of the device simulated, (b) relocation of the high-field region from the physical base–collector junction to the collector–subcollector intersection, (c) buildup of excess holes in the collector, and (d) buildup of excess electrons in the collector. (After Poon *et al.*, 1969).

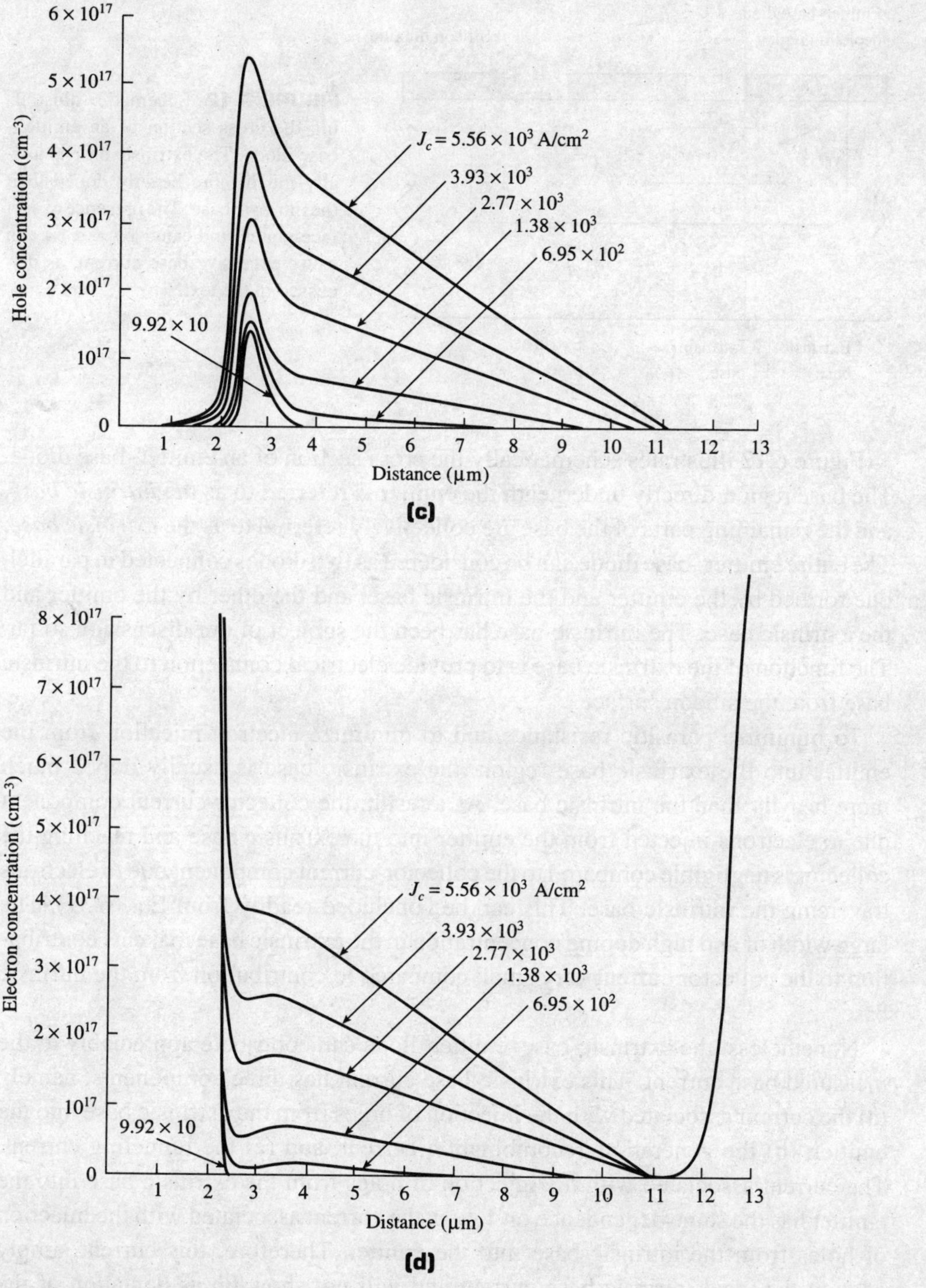


**FIGURE 6.11.** (*Continued*)

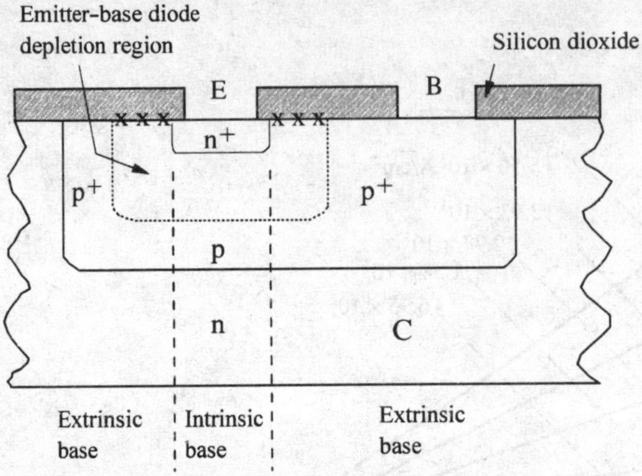


**FIGURE 6.12.** Schematic illustrating the cross section of an emitter–base diode. The extrinsic base is usually much more heavily doped than the intrinsic base. The presence of surface states, indicated by x x x, can cause excessive base current, as discussed in the text.

Figure 6.12 illustrates schematically the cross section of an emitter–base diode. The base region directly underneath the emitter is referred to as the *intrinsic base*, and the remaining parts of the base are collectively referred to as the *extrinsic base*. The entire emitter–base diode can be considered as two diodes connected in parallel, one formed by the emitter and the intrinsic base, and the other by the emitter and the extrinsic base. The intrinsic base has been the subject of our discussion so far. The function of the extrinsic base is to provide electrical connection to the intrinsic base from the silicon surface.

To minimize parasitic resistance and to minimize electron injection from the emitter into the extrinsic base region, the extrinsic base is usually doped much more heavily than the intrinsic base. As a result, the collector-current component due to electrons injected from the emitter into the extrinsic base and reaching the collector is negligible compared to the collector-current component due to electrons traversing the intrinsic base. This can be concluded readily from Eq. (6.31). The large width of and high doping concentration in the extrinsic base make its contribution to the collector current very small compared to contribution from the intrinsic base.

Nonetheless, the extrinsic-base–emitter diode can contribute appreciably to the measured base current. This extrinsic-base current has three components, namely (a) the current associated with the injection of holes from the extrinsic base into the emitter, (b) the generation–recombination current, and (c) the tunneling current. The current associated with the injection of holes from the extrinsic base into the emitter has the same dependence on $V_{BE}$ as the current associated with the injection of holes from the intrinsic base into the emitter. Therefore, this current simply adds to the ideal intrinsic-base current and will not show up as deviation of the measured base current from its ideal behavior. Only the generation–recombination and tunneling currents contribute to the nonideal behavior of the measured base currents. Therefore, only these two components are discussed further here.

#### 6.3.4.1 BASE CURRENT DUE TO GENERATION–RECOMBINATION

Generation-recombination current due to defect centers in silicon is negligible in modern VLSI devices because, unless there is a contamination problem, the concentration of defects that can cause generation–recombination current is negligibly low for all modern VLSI fabrication processes. This may not be true for processes in early development, but it is certainly true by the time a process reaches manufacturing. However, as can be seen from Fig. 6.12, the extrinsic-base–emitter diode has a surface component. The presence of interface states, as indicated in the figure, could give rise to significant surface generation–recombination current, as discussed in Section 2.3.7. The generation–recombination hole current adds to the base current and hence degrades the current gain (Werner, 1976). Surface generation–recombination current, by itself, usually can be recognized by its $\exp(V_{BE}/2kT)$ dependence on $V_{BE}$ (Sah *et al.*, 1957). Fortunately, for properly designed fabrication processes, the density of interface states can be so low that this current component, though usually observable, is not significant in modern bipolar devices.

#### 6.3.4.2 BASE CURRENT DUE TO TUNNELING

The tunneling current in the emitter–base junction, on the other hand, is expected to increase as the transistor dimensions are scaled down (Stork and Isaac, 1983). The emitter–base junction is a fairly abrupt junction. The emitter is very heavily doped, and the base doping concentration is typically in excess of $1 \times 10^{18}$ cm$^{-3}$, as can be seen from Fig. 6.2. Furthermore, as discussed in Chapters 7 and 8, the peak base doping concentration is increased as the physical dimensions of a bipolar transistor are scaled down, resulting in enhanced tunneling current in the emitter–base diode.

Since the extrinsic base is always more heavily doped than the intrinsic base, the observed tunneling current is usually dominated by the component from the extrinsic-base–emitter diode. Furthermore, the interface states in the extrinsic-base–emitter diode can assist in the tunneling process and thus can enhance the tunneling current very significantly (Li *et al.*, 1988). Figure 6.13 illustrates the typical current–voltage characteristics of a bipolar transistor which has excessive base current due to tunneling in the emitter–base diode. When excessive emitter–base tunneling dominates the base current, the base current is usually much larger than suggested by an $\exp(V_{BE}/2kT)$ dependence. Furthermore, as expected from a tunneling process, the excessive emitter–base tunneling current is nearly independent of temperature (Li *et al.*, 1988). Also, the excessive emitter–base tunneling current increases very rapidly with voltage when the emitter–base diode is reverse biased, as can be seen from Fig. 6.13. Fortunately, excessive tunneling current in the emitter–base diode can be suppressed easily by optimizing the emitter–base diode doping profile and the device fabrication process.

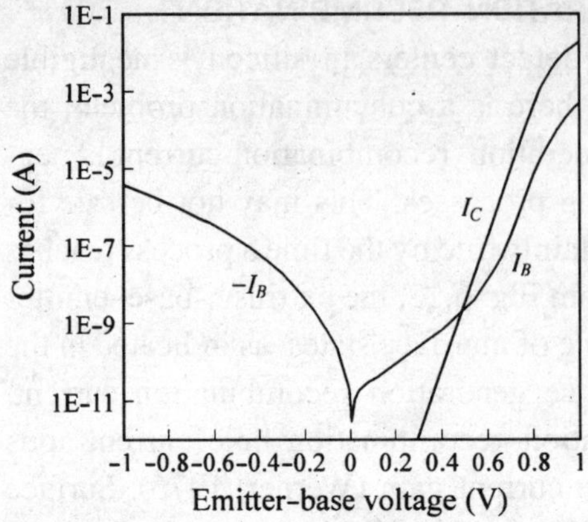


**FIGURE 6.13.** Typical voltage–current characteristics of an n–p–n transistor which has excessive tunneling current in the extrinsic-base–emitter diode. (After Li *et al.*, 1988.)

## 6.4 BIPOLAR DEVICE MODELS FOR CIRCU AND TIME-DEPENDENT ANALYSE

The merits of a bipolar device should be discussed in the context of the circuit in which it is used. For circuit applications, the device electrical characteristics must be first transformed into equivalent circuit parameters. The merits of a device are then interpreted from the behavior of the circuit or from the characteristics of the equivalent-circuit parameters. In this section, the equivalent-circuit models needed for the discussion of bipolar device design, which will be covered in Chapter 7, and device optimization, which will be covered in Chapter 8, are developed. The models suitable for dc or large-signal analyses will be developed first, followed by the models suitable for small-signal analyses. This is followed by the development of the charge-control model, which is suitable for quasistatic time-dependent analyses.

### 6.4.1 BASIC dc MODEL

The Ebers–Moll model (Ebers and Moll, 1954) for an n–p–n transistor is shown in Fig. 6.14. It describes an n–p–n transistor as two diodes in series, arranged in the common-base mode. When a voltage $V_{BE}$ is applied to the emitter–base diode, a forward current $I_F$ flows in the emitter–base diode. This current causes a current $\alpha_F I_F$ to flow in the collector, where $\alpha_F$ is the common-base current gain in the forward direction. Similarly, when a voltage $V_{BC}$ is applied across the base–collector diode, a reverse current $I_R$ flows in the collector–base diode, causing a current $\alpha_R I_R$ to flow in the emitter, where $\alpha_R$ is the common-base current gain in the reverse direction. These currents are indicated in Fig. 6.14. They are related by

$$I_E = \alpha_R I_R - I_F, \tag{6.83}$$

$$I_C = \alpha_F I_F - I_R, \tag{6.84}$$

and

$$I_B = (1 - \alpha_F) I_F + (1 - \alpha_R) I_R. \tag{6.85}$$

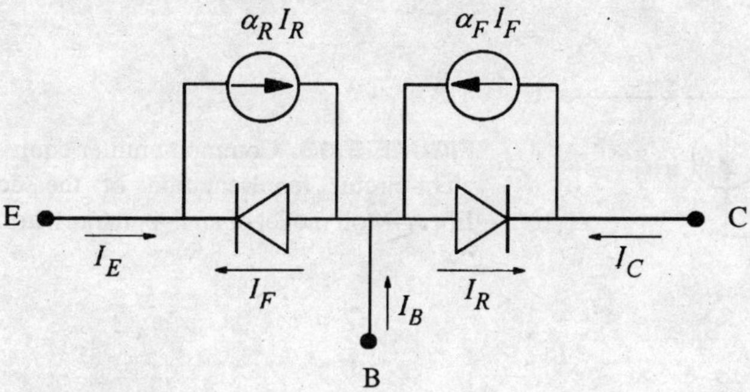


**FIGURE 6.14.** Equivalent-circuit representation of the basic dc Ebers–Moll model of an n–p–n transistor.

The emitter, base, and collector currents are related by $I_E + I_B + I_C = 0$.

From Eq. (2.102) we can write $I_F$ and $I_R$ in the form

$$I_F = I_{F0}[\exp(qV_{BE}/kT) - 1] \tag{6.86}$$

and

$$I_R = I_{R0}[\exp(qV_{BC}/kT) - 1]. \tag{6.87}$$

Therefore, Eqs. (6.83) and (6.84) can be rewritten as

$$I_E = -I_{F0}[\exp(qV_{BE}/kT) - 1] + \alpha_R I_{R0}[\exp(qV_{BC}/kT) - 1] \tag{6.88}$$

and

$$I_C = \alpha_F I_{F0}[\exp(qV_{BE}/kT) - 1] - I_{R0}[\exp(qV_{BC}/kT) - 1]. \tag{6.89}$$

Reciprocity characteristics of the emitter and collector terminals require the off-diagonal coefficients of the equations for $I_E$ and $I_C$ to be equal (Gray *et al.*, 1964; Muller and Kamins, 1977), i.e.,

$$\alpha_R I_{R0} = \alpha_F I_{F0}. \tag{6.90}$$

The common-emitter form of the Ebers–Moll model is often more desirable for circuit analyses. To accomplish this, let us define

$$I_{SF} \equiv \alpha_F I_F, \tag{6.91}$$

$$I_{SR} \equiv \alpha_R I_R, \tag{6.92}$$

$$I_{CT} \equiv I_{SF} - I_{SR}, \tag{6.93}$$

$$\beta_F \equiv \frac{\alpha_F}{1 - \alpha_F}, \tag{6.94}$$

$$\beta_R \equiv \frac{\alpha_R}{1 - \alpha_R}. \tag{6.95}$$

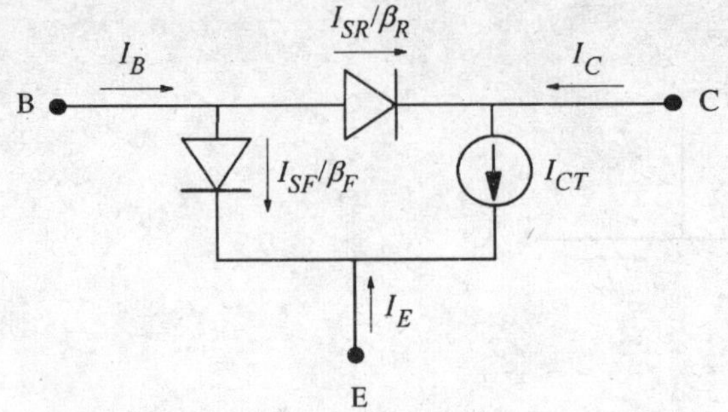


**FIGURE 6.15.** Common-emitter equivalent-circuit representation of the dc Ebers–Moll model of an n–p–n transistor.

Comparison with Eq. (6.56) shows that $\beta_F$ and $\beta_R$ are the common-emitter current gains in the forward and the reverse directions, respectively. Substituting Eqs. (6.90) to (6.95) into Eqs. (6.83) to (6.85) gives

$$I_E = -I_{CT} - \frac{I_{SF}}{\beta_F}, \tag{6.96}$$

$$I_C = I_{CT} - \frac{I_{SR}}{\beta_R}, \tag{6.97}$$

$$I_B = \frac{I_{SF}}{\beta_F} + \frac{I_{SR}}{\beta_R}. \tag{6.98}$$

The equivalent-circuit model for these currents is shown in Fig. 6.15.

### 6.4.2 BASIC ac MODEL

To model the ac behavior of a bipolar transistor, the parasitic internal capacitances and resistances of the transistor must be included. In general, the parasitic resistances can be made rather small by using large device areas and device layout techniques, as well as fabrication process techniques. However, the parasitic capacitances usually can be reduced only by reducing the associated device areas. As a result, the basic behavior of a transistor is determined more by its parasitic capacitances than by its parasitic resistances. For simplicity, we shall first neglect the parasitic resistances and consider only the parasitic capacitances.

As discussed in Section 2.2, there are two components in the capacitance of a p–n diode, namely the depletion-layer capacitance $C_d$ and the diffusion capacitance $C_D$. Let $C_{dBE}$ and $C_{dBC}$ be the depletion-layer capacitances of the emitter–base and collector–base diodes, respectively. Let $C_{DE}$ be the diffusion capacitance associated with forward-biasing the emitter–base diode, and $C_{DC}$ be the diffusion capacitance associated with forward-biasing the collector–base diode. When these capacitances are included, the common-emitter equivalent-circuit model is shown in Fig. 6.16. In Fig. 6.16, the depletion-layer capacitance of the collector–substrate diode, $C_{dCS}$, is also included for completeness.

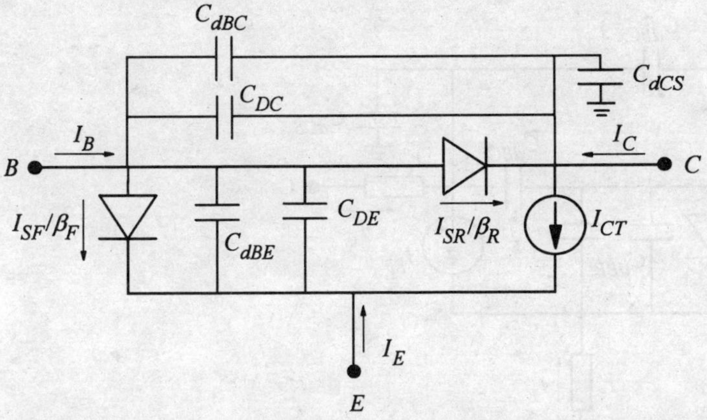


**FIGURE 6.16.** Common-emitter equivalent-circuit representation of the ac Ebers–Moll model of a bipolar transistor. Internal capacitances are included.

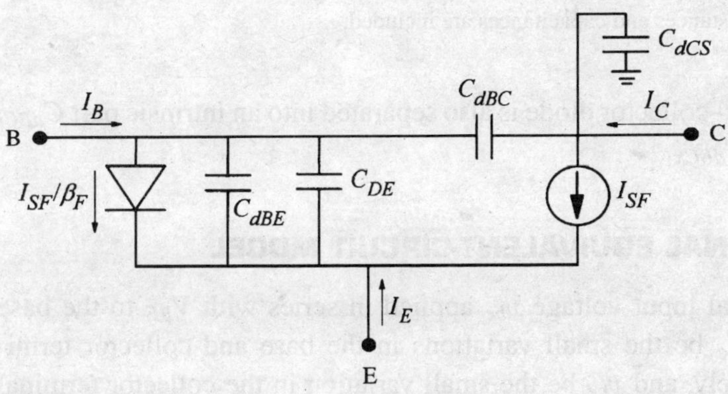


**FIGURE 6.17.** Equivalent-circuit representation of the ac Ebers–Moll model of an n–p–n transistor biased in the forward-active mode of operation. Internal capacitances are included.

#### 6.4.2.1 MODEL FOR A TRANSISTOR BIASED IN THE FORWARD-ACTIVE MODE OF OPERATION

For simplicity, we shall consider only transistors biased in the forward-active mode of operation, i.e., with the emitter–base diodes forward biased and the base–collector diodes reverse biased. (Transistors biased in the reverse-active mode, i.e., with the base–collector diodes forward biased, cannot be switched fast because of the very large diffusion capacitance associated with the forward-biased base–collector diodes. As a result, high-speed circuits usually use transistors biased only in the forward-active mode.) In this case, $I_{SR}$ can be neglected compared to $I_{SF}$, and $C_{DC} = 0$. The model in Fig. 6.16 then simplifies to that shown in Fig. 6.17.

If the internal parasitic resistances indicated in Fig. 6.7 are now included in the equivalent circuit of Fig. 6.17, the resultant equivalent circuit is shown in Fig. 6.18. Here, for purposes of discussion in later chapters, the base resistance is shown as two parts, an intrinsic part $r_{bi}$ and an extrinsic part $r_{bx}$. The depletion-layer

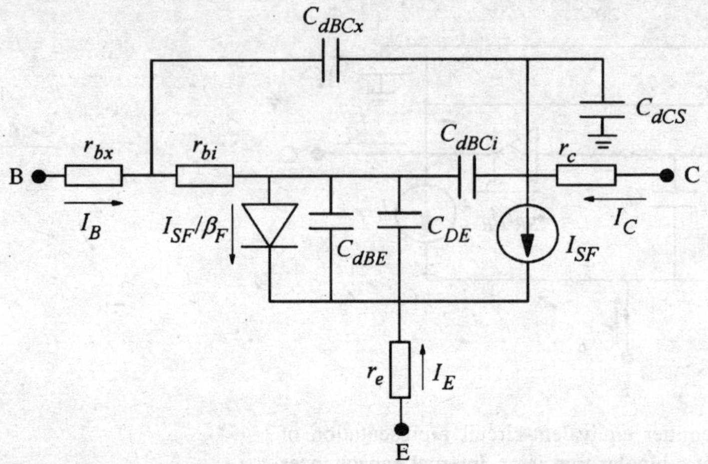


**FIGURE 6.18.** Equivalent-circuit representation of the ac Ebers-Moll model of an n–p–n transistor biased in the forward-active mode of operation. Internal parasitic resistances and capacitances are included.

capacitance of the base–collector diode is also separated into an intrinsic part $C_{dBCi}$ and an extrinsic part $C_{dBCx}$.

### 6.4.3 SMALL-SIGNAL EQUIVALENT-CIRCUIT MODEL

Consider a small-signal input voltage $v_{be}$ applied in series with $V_{BE}$ to the base terminal. Let $i_b$ and $i_c$ be the small variations in the base and collector terminal currents, respectively, and $v_{ce}$ be the small variation in the collector terminal voltage, as a result of $v_{be}$. The small-signal equivalent-circuit model provides a relationship among these parameters. We first develop the small-signal model ignoring the transistor parasitic resistances, and then the model with these resistances included.

- *Small-signal model when parasitic resistances are negligible.* When the device parasitic resistances are negligible, the device terminal voltages $V_C$, $V_E$, and $V_B$ are the same as the internal junction voltages $V'_C$, $V'_E$, and $V'_B$, respectively. The internal junction voltages are the voltages that govern the ideal currents described in Section 6.2. The transconductance $g_m$ relates $i_c$ to $v_{be}$, i.e.,

$$g_m = \frac{\partial I_c}{\partial V'_{BE}} = \frac{qI_C}{kT}, \tag{6.99}$$

where we have used the fact that $I_C$ is proportional to $\exp(qV'_{BE}/kT)$. The input resistance $r_\pi$ relates $v_{be}$ to $i_b$, i.e.,

$$r_\pi = \left(\frac{\partial I_B}{\partial V'_{BE}}\right)^{-1} = \frac{kT}{qI_B} = \frac{\beta_0}{g_m}, \tag{6.100}$$

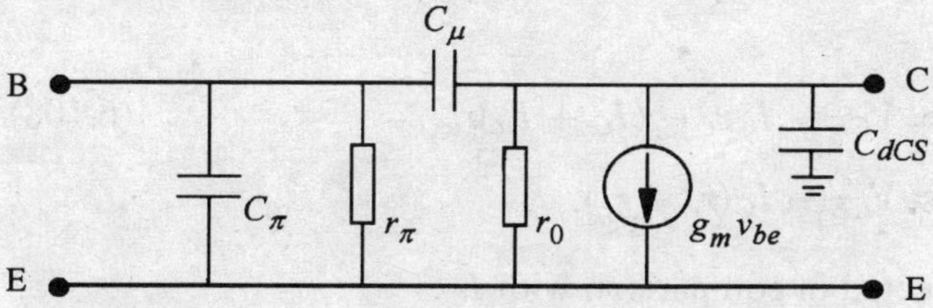


**FIGURE 6.19.** Small-signal hybrid-$\pi$ model of a bipolar transistor when the parasitic resistances are neglected.

where we have used the fact that $\beta_0 = I_C/I_B$. The output resistance $r_0$ relates $i_c$ to $v_{ce}$, i.e.,

$$r_0 = \left(\frac{\partial I_C}{\partial V'_{CE}}\right)^{-1} = \frac{V_A}{I_C}, \tag{6.101}$$

where we have used Eq. (6.63) for the Early voltage $V_A$. The capacitances are designated by

$$C_\mu = C_{dBC} \tag{6.102}$$

and

$$C_\pi = C_{dBE} + C_{DE}. \tag{6.103}$$

The resulting small-signal equivalent circuit is shown in Fig. 6.19. This is the well-known small-signal hybrid-$\pi$ model (Gray *et al.*, 1964).

- *Small-signal model including parasitic resistances.* When the parasitic resistances are included, the device terminal voltages $V_C$, $V_E$, and $V_B$ are no longer the same as the internal junction voltages $V'_C$, $V'_E$, and $V'_B$, respectively, due to the *IR* drops in the parasitic resistors $r_e$, $r_c$, and $r_b$. For simplicity, we have lumped $r_{bi}$ and $r_{bx}$ into $r_b$. The terminal voltages and the internal junction voltages are related by

$$V'_C = V_C - I_C r_c, \tag{6.104}$$

$$V'_B = V_B - I_B r_b, \tag{6.105}$$

and

$$\begin{aligned} V'_E &= V_E - I_E r_e \\ &= V_E + (I_C + I_B) r_e, \end{aligned} \tag{6.106}$$

where we have used the fact that $I_E + I_C + I_B = 0$. Therefore,

$$\begin{aligned} V'_{BE} &= V_{BE} - I_B r_b - (I_C + I_B) r_e \\ &= V_{BE} - I_C r_e - I_B (r_b + r_e). \end{aligned} \tag{6.107}$$

Similarly,

$$V'_{CE} = V_{CE} - I_C r_c - (I_C + I_B) r_e \approx V_{CE} - I_C(r_e + r_c), \tag{6.108}$$

where $I_B$ has been neglected in comparison with $I_C$.

Let us denote the circuit parameters including the effect of the parasitic resistances with a prime ($'$). Thus the transconductance $g'_m$ is

$$\begin{aligned} g'_m &= \frac{\partial I_C}{\partial V_{BE}} = \left(\frac{\partial V_{BE}}{\partial I_C}\right)^{-1} \\ &= \left(\frac{\partial V'_{BE}}{\partial I_C} + r_e + \frac{r_b + r_e}{\beta_0}\right)^{-1} = \left(\frac{1}{g_m} + r_e + \frac{r_b + r_e}{\beta_0}\right)^{-1} \\ &\approx \frac{g_m}{1 + g_m r_e}, \end{aligned} \tag{6.109}$$

where $g_m$ is the intrinsic transconductance given by Eq. (6.99), and $\beta_0 = \partial I_C/\partial I_B$ is the static common-emitter current gain. Since $\beta_0$ is typically about 100, we have also neglected the $r_b/\beta_0$ and $r_e/\beta_0$ terms in Eq. (6.109). Similarly, the input resistance $r'_\pi$ is

$$\begin{aligned} r'_\pi &= \left(\frac{\partial I_B}{\partial V_{BE}}\right)^{-1} \\ &= \frac{\partial V_{BE}}{\partial I_B} = \left(\frac{\partial V'_{BE}}{\partial I_B} + (r_e + r_b) + r_e \beta_0\right) \\ &= r_\pi(1 + g_m r_e) + (r_b + r_e), \end{aligned} \tag{6.110}$$

where the intrinsic input resistance $r_\pi$ is given by Eq. (6.100). The output resistance $r'_0$ is given by

$$\begin{aligned} r'_0 &= \left(\frac{\partial I_C}{\partial V_{CE}}\right)^{-1} \\ &= \frac{\partial V_{CE}}{\partial I_C} = \left(\frac{\partial V'_{CE}}{\partial I_C} + r_e + r_c\right) \\ &= r_0 + (r_e + r_c), \end{aligned} \tag{6.111}$$

where the intrinsic output resistance $r_0$ is given by Eq. (6.101).

The device capacitance components are still the same as before, with $C_\mu$ given by Eq. (6.102) and $C_\pi$ given by Eq. (6.103). It should be noted that $C_\mu$ is determined by $V'_{BC}$, and not $V_{BC}$. Similarly, $C_\pi$ is determined by $V'_{BE}$, and not by $V_{BE}$. The equivalent circuit can be deduced from Eqs. (6.109) to (6.111) and is shown in Fig. 6.20.

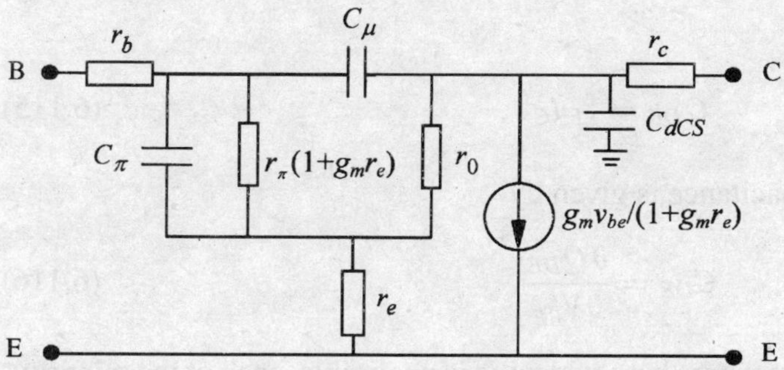


**FIGURE 6.20.** Small-signal hybrid-$\pi$ model of a bipolar transistor including the intrinsic parasitic resistances.

### 6.4.4 EMITTER DIFFUSION CAPACITANCE

The diffusion capacitance $C_{DE}$ is due to all the minority carriers caused by forward-biasing the emitter–base diode. Minority carriers are present in the emitter region and the base region, as well as in the space-charge layers of the emitter–base and base–collector diodes. The total minority-carrier charge can therefore be written as the sum of these individual charges:

$$Q_{DE} = |Q_E| + |Q_B| + |Q_{BE}| + |Q_{BC}|, \tag{6.112}$$

where $Q_E$, $Q_B$, $Q_{BE}$, and $Q_{BC}$ represent the minority-carrier charge in the emitter, the base, the emitter–base space-charge region, and the base–collector space-charge region, respectively. Notice that $Q_{DE}$ is the sum of the *absolute values* of the individual minority-carrier charge components. It is not the sum of the net charge components. For example, for a perfectly symmetrical diode, with the n-region doping concentration equal to the p-region doping concentration, we have $Q_B = -Q_E$. The contributions of $Q_B$ and $Q_E$ to $Q_{DE}$ are $|Q_B|$ and $|Q_E|$, and not $Q_B + Q_E$, which is zero.

As shown in Section 2.2.6, the stored minority-carrier charge can be written as the product of a charging current and a delay or transit time. For modeling purposes, it is convenient to rewrite Eq. (6.112) as

$$Q_{DE} = (t_E + t_B + t_{BE} + t_{BC})I_C, \tag{6.113}$$

where $t_E$ is the emitter delay time, $t_B$ is the base transit time, $t_{BE}$ is the emitter–base space-charge-region transit time, and $t_{BC}$ is the base–collector space-charge-region transit time (Ashburn, 1988). The sum of these times is often referred to as the *forward transit time* $\tau_F$, i.e.,

$$\tau_F = t_E + t_B + t_{BE} + t_{BC}, \tag{6.114}$$

and

$$Q_{DE} = \tau_F I_C. \tag{6.115}$$

The emitter diffusion capacitance is given by

$$C_{DE} = \frac{\partial Q_{DE}}{\partial V'_{BE}}. \tag{6.116}$$

At low current densities where base widening is negligible, each of the minority-charge components in Eq. (6.112) is simply proportional to the charging (collector) current. In this case, $\tau_F$ is constant, independent of $I_C$, and Eq. (6.116) gives

$$C_{DE} = \tau_F \frac{qI_C}{kT} = \tau_F g_m, \tag{6.117}$$

where we have used Eq. (6.99) for $g_m$. The capacitance $C_{DE}$ increases linearly with collector current or transconductance.

However, at sufficiently large current densities, base widening occurs, and, as discussed in Section 6.3.3, $Q_B$ and $Q_{BC}$ increase very rapidly with collector current density. In this case, $t_B$ and $t_{BC}$, and hence $\tau_F$, are no longer constant but increase rapidly with collector current density. Equation (6.116), instead of Eq. (6.117), should be used to determine the emitter diffusion capacitance.

### 6.4.5 CHARGE-CONTROL ANALYSIS

The behavior of a bipolar transistor is often analyzed in a *charge-control* model where the charges within the various regions of the transistor are related to the currents feeding them. The charge-control model is especially useful for time-dependent analyses. It was used in Section 2.2.5 to describe the discharging of a diode that has been switched from forward bias to reverse bias. In this subsection, we describe the time-dependent behavior of an n–p–n transistor using charge-control analysis.

Consider an n–p–n transistor biased in an amplifier mode. Its circuit schematic is shown in Fig. 6.21(a). The input voltage, which is the base terminal voltage, is assumed to be time-dependent. The currents flowing in the transistor are illustrated schematically in Fig. 6.21(b). The time-dependent emitter current, base current, and collector current are denoted by $i_E(t)$, $i_B(t)$, and $i_C(t)$, respectively. The displacement currents in the base–emitter and base–collector junction depletion-layer capacitors are also included.

As the electrons flow through the emitter–base and base–collector junction space-charge regions, they contribute to the minority charges $Q_{BE}$ and $Q_{BC}$ stored in these depletion layers. (As the holes flow from the base into the emitter, they also contribute to a minority-charge component in the emitter–base depletion layer. However, this hole component, which is proportional to the base current, is small

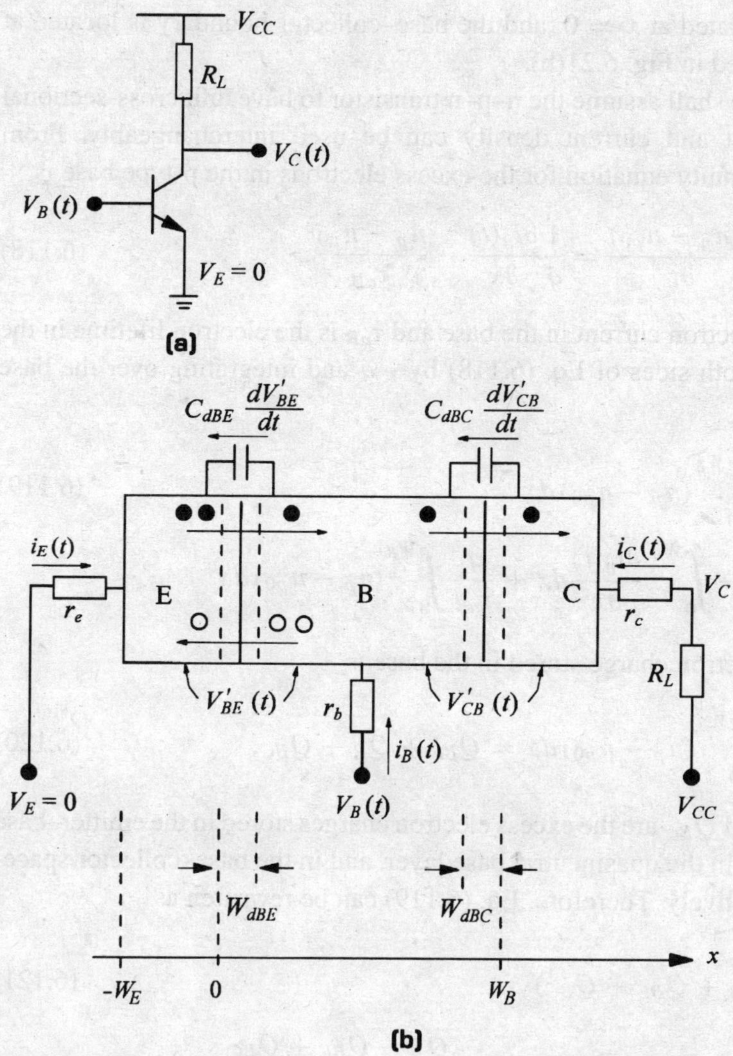


**FIGURE 6.21.** (a) Schematic of an n–p–n transistor biased to operate as an amplifier. The input voltage $V_B$ is assumed to be time-dependent. (b) Schematic illustrating the resistances and the terminal currents in the amplifier. Also illustrated are the displacement currents and the flow of electrons and holes within the transistor. The locations of the emitter contact, the emitter–base boundary, and the base–collector boundary, used in the charge-control model, are also indicated. $W_{dBE}$ and $W_{dBC}$ are the base–emitter and the base–collector junction depletion-layer widths, respectively.

compared to the electron component $Q_{BE}$, which is proportional to the collector current. For simplicity, the hole component of minority charge stored in the emitter–base space-charge layer is ignored.) To facilitate including $Q_{BE}$ and $Q_{BC}$ in the charge-control analysis, we define the *base region* to include the emitter–base space-charge layer and the base–collector space-charge layer. Thus, for our charge-control analysis, the emitter contact is located at $x = -W_E$, the emitter–

base boundary is located at $x = 0$, and the base–collector boundary is located at $x = W_B$, as illustrated in Fig. 6.21(b).

For simplicity, we shall assume the n–p–n transistor to have unit cross-sectional area so that current and current density can be used interchangeably. From Eq. (2.92), the continuity equation for the excess electrons in the p-type base is,

$$\frac{\partial(n_p - n_{p0})}{\partial t} = \frac{1}{q}\frac{\partial i_n(t)}{\partial x} - \frac{n_p - n_{p0}}{\tau_{nB}}, \tag{6.118}$$

where $i_n(t)$ is the electron current in the base and $\tau_{nB}$ is the electron lifetime in the base. Multiplying both sides of Eq. (6.118) by $-q$ and integrating over the base region, we have

$$-q\frac{\partial}{\partial t}\int_0^{W_B}(n_p - n_{p0})\,dx \tag{6.119}$$
$$= -\int_0^{W_B}\frac{\partial i_n(t)}{\partial x}dx + \frac{q}{\tau_{nB}}\int_0^{W_B}(n_p - n_{p0})\,dx.$$

The total excess electron charge stored in the base is

$$-q\int_0^{W_B}(n_p - n_{p0})\,dx = Q_{BE} + Q_B + Q_{BC}, \tag{6.120}$$

where $Q_{BE}$, $Q_B$, and $Q_{BC}$ are the excess electron charges stored in the emitter–base space-charge layer, in the quasineutral base layer, and in the base–collector space-charge layer, respectively. Therefore, Eq. (6.119) can be rewritten as

$$\frac{d}{dt}(Q_B + Q_{BE} + Q_{BC}) \tag{6.121}$$
$$= i_n(0, t) - i_n(W_B, t) - \frac{Q_B + Q_{BE} + Q_{BC}}{\tau_{nB}}.$$

Similarly, integrating the continuity equation for the excess holes over the emitter region, we obtain

$$\frac{dQ_E}{dt} = i_p(-W_E, t) - i_p(0, t) - \frac{Q_E}{\tau_{pE}}, \tag{6.122}$$

where

$$Q_E = q\int_{-W_E}^{0}(p_n - p_{n0})\,dx \tag{6.123}$$

is the excess hole charge stored in the emitter, and $\tau_{pE}$ is the hole lifetime in the emitter.

From the current components illustrated in Fig. 6.21(b), the emitter current is

$$i_E(t) = i_n(0,t) + i_p(0,t) - C_{dBE}\frac{dV'_{BE}(t)}{dt}, \tag{6.124}$$

where $V'_{BE} = V'_B - V'_E$ is the time-dependent voltage across the immediate base–emitter junction, and $C_{dBE}$ is the base–emitter junction depletion-layer capacitance. The collector current is

$$i_C(t) = -i_n(W_B,t) + C_{dBC}\frac{dV'_{CB}(t)}{dt}, \tag{6.125}$$

where $V'_{CB} = V'_C - V'_B$ is the time-dependent voltage across the immediate base–collector junction, and $C_{dBC}$ is the base–collector junction depletion-layer capacitance. The base current is

$$\begin{aligned} i_B(t) &= -i_E(t) - i_C(t) \\ &= -[i_n(0,t) - i_n(W_B,t)] - i_p(0,t) \\ &\quad + C_{dBE}\frac{dV'_{BE}(t)}{dt} - C_{dBC}\frac{dV'_{CB}(t)}{dt}. \end{aligned} \tag{6.126}$$

Using Eq. (6.121) for $i_n(0,t) - i_n(W_B,t)$ and Eq. (6.122) for $i_p(0,t)$ in Eq. (6.126), we obtain

$$\begin{aligned} i_B(t) &= -\frac{d(Q_B + Q_{BE} + Q_{BC})}{dt} + \frac{dQ_E}{dt} - \frac{Q_B + Q_{BE} + Q_{BC}}{\tau_{nB}} \\ &\quad + \frac{Q_E}{\tau_{pE}} - i_p(-W_E,t) + C_{dBE}\frac{dV'_{BE}(t)}{dt} - C_{dBC}\frac{dV'_{CB}(t)}{dt} \\ &= \frac{dQ_{DE}}{dt} + \frac{-(Q_B + Q_{BE} + Q_{BC})}{\tau_{nB}} + \frac{Q_E}{\tau_{pE}} - i_p(-W_E,t) \\ &\quad + C_{dBE}\frac{dV'_{BE}(t)}{dt} - C_{dBC}\frac{dV'_{CB}(t)}{dt}, \end{aligned} \tag{6.127}$$

where $Q_{DE}$ is the amount of emitter diffusion-capacitance charge given by Eq. (6.112). (Notice that the electron charges $Q_B$, $Q_{BE}$, and $Q_{BC}$ are negative quantities, and the hole charge $Q_E$ is positive. $Q_{DE}$ is always positive.) Equation (6.127) is the charge-control model for the base current. It simply states that the base current feeds the emitter diffusion capacitance, the base–emitter diode depletion-layer capacitance, the base–collector diode depletion-layer capacitance, the recombination current in the base, the recombination current in the emitter, and the hole recombination current at the emitter contact.

- *The transistor equation.* The base-current equation can be reduced to a more useful form by noting the relationship between $V'_{CB}(t)$ and $V'_{BE}(t)$.

We have

$$\begin{aligned} V'_{CB}(t) &= V'_C - V'_B \\ &= -(V'_B - V'_E) + (V'_C - V'_E) \\ &\approx -(V'_B - V'_E) + (V_C - V_E) - i_C(t)(r_e + r_c) \\ &= -V'_{BE}(t) + V_{CC} - i_C(t)(R_L + r_e + r_c), \end{aligned} \tag{6.128}$$

where we have used Eq. (6.108), which relates $V'_{CE}$ to $V_{CE}$, and the fact that $V_{CE} = V_{CC} - i_C R_L$. Therefore,

$$\frac{dV'_{CB}(t)}{dt} = -\frac{dV'_{BE}(t)}{dt} - (R_L + r_e + r_c)\frac{di_C(t)}{dt}. \tag{6.129}$$

Substituting Eq. (6.129) into Eq. (6.127), we have

$$\begin{aligned} i_B(t) &= \frac{dQ_{DE}}{dt} + \frac{-(Q_B + Q_{BE} + Q_{BC})}{\tau_{nB}} + \frac{Q_E}{\tau_{pE}} - i_p(-W_E, t) \\ &\quad + (C_{dBE} + C_{dBC})\frac{dV'_{BE}(t)}{dt} + C_{dBC}(R_L + r_e + r_c)\frac{di_C(t)}{dt} \\ &= \frac{dQ_{DE}}{dt} + \frac{-(Q_B + Q_{BE} + Q_{BC})}{\tau_{nB}} + \frac{Q_E}{\tau_{pE}} - i_p(-W_E, t) \\ &\quad + \frac{(C_{dBE} + C_{dBC})}{g_m}\frac{di_C(t)}{dt} + C_{dBC}(R_L + r_e + r_c)\frac{di_C(t)}{dt}, \end{aligned} \tag{6.130}$$

where we have used Eq. (6.99) for the transconductance $g_m$.

In the steady state, Eq. (6.130) gives

$$\begin{aligned} I_B = \frac{I_C}{\beta_0} &= \left[\frac{-(Q_B + Q_{BE} + Q_{BC})}{\tau_{nB}} + \frac{Q_E}{\tau_{pE}}\right]_{\text{steady state}} \\ &\quad -i_p(-W_E). \end{aligned} \tag{6.131}$$

That is, in the steady state, the base current is simply equal to the sum of the recombination currents in the base, in the emitter, and at the emitter contact. If we assume the time dependence is quasistatic so that the sum of the recombination currents is proportional to the collector current in the nonsteady state as well, i.e.,

$$\frac{i_C(t)}{\beta_0} = \frac{-(Q_B + Q_{BE} + Q_{BC})}{\tau_{nB}} + \frac{Q_E}{\tau_{pE}} - i_p(-W_E, t), \tag{6.132}$$

and that Eq. (6.115) also holds in the nonsteady state, i.e.,

$$Q_{DE}(t) = \tau_F i_C(t), \tag{6.133}$$

then Eq. (6.130) becomes

$$i_B(t) = \left( \tau_F + \frac{C_{dBE} + C_{dBC}}{g_m} + C_{dBC}(R_L + r_e + r_c) \right) \times \frac{di_C(t)}{dt} + \frac{i_C(t)}{\beta_0}. \tag{6.134}$$

This is the differential equation relating the base and collector currents for a transistor with a resistive load $R_L$ (Ghandhi, 1968).

- *Small-signal base-current equation.* If the time-dependent variations from the steady state are small, we can write

$$i_B(t) = I_B + i_b(t) \tag{6.135}$$

and

$$i_C(t) = I_C + i_c(t), \tag{6.136}$$

where $i_b(t)$ represents a small variation of the base current from its static value $I_B$, and $i_c(t)$ is the resultant small variation of the collector current. Substituting these currents into (6.134), we obtain

$$i_b(t) = \left( \tau_F + \frac{C_{dBE} + C_{dBC}}{g_m} + C_{dBC}(R_L + r_e + r_c) \right) \times \frac{di_c(t)}{dt} + \frac{i_c(t)}{\beta_0}. \tag{6.137}$$

Thus, the small-signal base-current equation has the same form as the large-signal one.

If we further assume $i_b$ and $i_c$ to have a time dependence $e^{j\omega t}$, then Eq. (6.137) gives

$$i_b(t) = j\omega \left( \tau_F + \frac{C_{dBE} + C_{dBC}}{g_m} + C_{dBC}(R_L + r_e + r_c) \right) \times i_c(t) + \frac{i_c(t)}{\beta_0}. \tag{6.138}$$

The small-signal frequency-dependent current gain $\beta(\omega)$ is

$$\beta(\omega) = \frac{i_c(t)}{i_b(t)} = \left[ \frac{1}{\beta_0} + j\omega \left( \tau_F + \frac{C_{dBE} + C_{dBC}}{g_m} + C_{dBC}(R_L + r_e + r_c) \right) \right]^{-1} \tag{6.139}$$

Equation (6.139) gives

$$\beta(\omega \to 0) = \beta_0, \tag{6.140}$$

as expected. At the high-frequency limit, Eq. (6.139) gives

$$\beta(\omega) \approx \frac{1}{j\omega\left(\tau_F + \frac{C_{dBE}+C_{dBC}}{g_m} + C_{dBC}(R_L + r_e + r_c)\right)}. \tag{6.141}$$

## 6.5 BREAKDOWN VOLTAGES

The breakdown voltages of a bipolar transistor are often characterized by applying a reverse bias across two of the three device terminals, with the third device terminal left open-circuit. These breakdown voltages are usually denoted by

$BV_{EBO}$ = emitter–base breakdown voltage with the collector open-circuit,
$BV_{CBO}$ = collector–base breakdown voltage with the emitter open-circuit,
$BV_{CEO}$ = collector–emitter breakdown voltage with the base open-circuit.

Since bipolar transistors are usually operated with the emitter-base junction zero-biased or forward biased, their $BV_{EBO}$ values are not important as long as they do not adversely affect the other device parameters. On the other hand, $BV_{CBO}$ and $BV_{CEO}$ must be adequately large for the intended circuit application. $BV_{CBO}$ and $BV_{CEO}$ are often determined, respectively, from the measured common-base and common-emitter current–voltage characteristics. The measurement setups for an n–p–n transistor, and the corresponding $I$–$V$ characteristics, are illustrated schematically in Fig. 6.22.

### 6.5.1 COMMON-BASE CURRENT GAIN IN THE PRESENCE OF BASE-COLLECTOR JUNCTION AVALANCHE

Consider an n–p–n transistor biased in the forward-active mode, as illustrated in Fig. 6.23(a). The corresponding energy-band diagram and the electron and hole current flows inside the transistor are illustrated in Fig. 6.23(b), where the locations of the emitter–base junction and the base–collector space-charge layer, where avalanche multiplication takes place, are also indicated. The emitter current $I_E$ is equal to the sum of the hole current entering the emitter from the base and the electron current entering the base from the emitter, i.e.,

$$I_E = A_E[J_n(0) + J_p(0)], \tag{6.142}$$

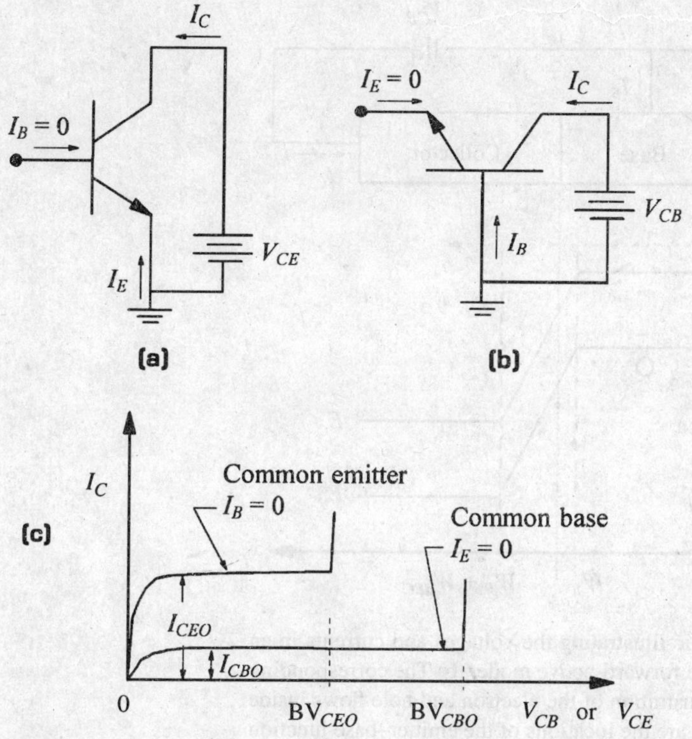


**FIGURE 6.22.** Circuit schematics for measuring (a) $BV_{CEO}$ and (b) $BV_{CBO}$ of an npn transistor. (c) Common-emitter $I_C$–$V_{CE}$ characteristics at $I_B = 0$, and common-base $I_C$–$V_{CB}$ characteristics at $I_E = 0$.

where $A_E$ is the emitter area. It should be noted that $I_E$, defined as the current entering the emitter, is a negative quantity for an n–p–n transistor, since both $J_n$ and $J_p$ are negative.

As the electrons traverse the base layer, some of them can recombine within the base layer. Only those electrons reaching $x = W_B$ contribute to the collector current. In the presence of avalanche multiplication in the reverse-biased base–collector junction, the electron current exiting the base–collector space-charge layer is a factor of $M$ larger than that entering the space-charge layer,. where $M$ is the avalanche multiplication factor (see Section 2.4.1). That is,

$$J_n(W_B + W_{dBC}) = M J_n(W_B). \tag{6.143}$$

The collector current $I_C$ is equal to the electron current exiting the base–collector space-charge layer, i.e.,

$$I_C = -A_E J_n(W_B + W_{dBC}). \tag{6.144}$$

The minus sign in Eq. (6.144) is due to the fact that, as defined, $I_C$ is a current entering the collector. $I_C$ is a positive quantity for an n–p–n transistor.

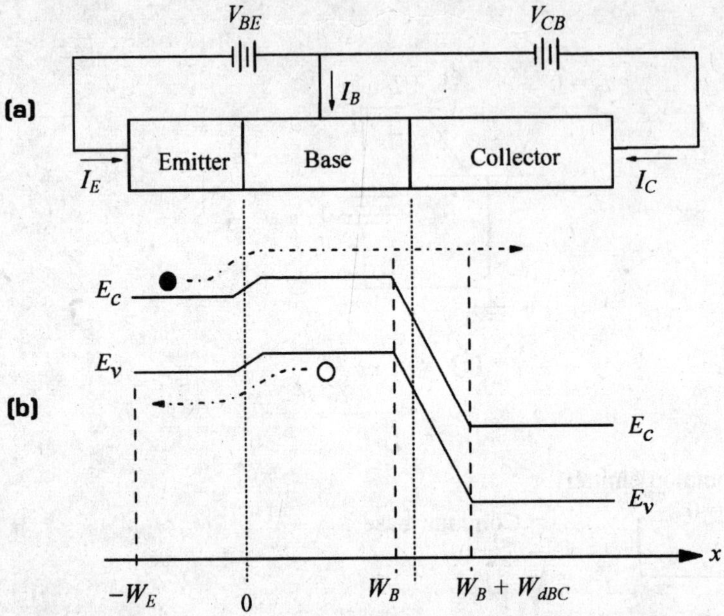


**FIGURE 6.23.** (a) Schematic illustrating the voltages and currents in an n–p–n transistor biased in the forward-active mode. (b) The corresponding energy-band diagram and illustration of the electron and hole flows inside the transistor. Also indicated are the locations of the emitter–base junction and the base–collector space-charge layer.

Using Eqs. (6.142) to (6.144), we can rewrite the static common-base current gain $\alpha_0$ [cf. Eq. (6.54)] as

$$\begin{aligned}\alpha_0 &\equiv \frac{\partial I_C}{\partial(-I_E)} \\ &= \frac{\partial J_n(W_B + W_{dBC})}{\partial[J_n(0) + J_p(0)]} \\ &= \frac{\partial J_n(0)}{\partial[J_n(0) + J_p(0)]} \frac{\partial J_n(W_B)}{\partial J_n(0)} \frac{\partial J_n(W_B + W_{dBC})}{\partial J_n(W_B)} \\ &= \gamma \alpha_T M,\end{aligned} \tag{6.145}$$

where the *emitter injection efficiency* $\gamma$ is defined by

$$\gamma \equiv \frac{\partial J_n(0)}{\partial[J_n(0) + J_p(0)]} = \frac{J_n(0)}{J_n(0) + J_p(0)}, \tag{6.146}$$

and the *base transport factor* $\alpha_T$ is defined by

$$\alpha_T \equiv \frac{\partial J_n(W_B)}{\partial J_n(0)} = \frac{J_n(W_B)}{J_n(0)}. \tag{6.147}$$

When the base–collector junction avalanche effect is negligible, we have $M \approx 1$, and the common-base current gain is

$$\alpha_0 = \gamma\alpha_T \qquad (\text{when} \quad M = 1). \tag{6.148}$$

If we further assume that recombination in the thin base is negligible (see Exercise 6.7), then the common-base current gain is simply

$$\alpha_0 = \gamma = \frac{\partial J_n(0)}{\partial[J_n(0) + J_p(0)]} \qquad (\text{when} \quad M = 1 \text{ and } \alpha_T = 1). \tag{6.149}$$

[*Note:* Throughout this chapter, by equating the collector current to the electron current entering the intrinsic base, i.e., Eq. (6.31), and by equating the base current to the hole current entering the emitter, i.e., Eq. (6.41), we have implicitly made the assumptions that $M = 1$ and $\alpha_T = 1$. That is, we have implicitly assumed that $\alpha_0 = \gamma$.]

### 6.5.2 SATURATION CURRENTS IN A TRANSISTOR

If we define $I_{EBO}$ and $I_{CBO}$ by (Ebers and Moll, 1954)

$$I_{EBO} \equiv I_{F0}(1 - \alpha_R\alpha_F) \tag{6.150}$$

and

$$I_{CBO} \equiv I_{R0}(1 - \alpha_R\alpha_F), \tag{6.151}$$

then Eqs. (6.88) and (6.89) give

$$I_E = -I_{EBO}[\exp(qV_{BE}/kT) - 1] - \alpha_R I_C \tag{6.152}$$

and

$$I_C = -I_{CBO}[\exp(qV_{BC}/kT) - 1] - \alpha_F I_E. \tag{6.153}$$

The physical meaning of $I_{EBO}$ and $I_{CBO}$ is apparent from these equations. $I_{EBO}$ is the saturation current of the emitter–base diode when the collector is open-circuit, i.e., it is the emitter current when the emitter–base diode is reverse biased and $I_C = 0$. This is the current one measures in measuring $BV_{EBO}$. Similarly, $I_{CBO}$ is the saturation current of the collector–base diode when the emitter is open-circuit, i.e., it is the collector current when the base–collector diode is reverse biased and $I_E = 0$. This is the current one measures in measuring $BV_{CBO}$. $I_{CBO}$ is indicated in Fig. 6.22(c).

Let us apply Eq. (6.153) to the $BV_{CEO}$ measurement setup shown in Fig. 6.22(a). We note that when $V_{CE}$ is near $BV_{CEO}$, the collector–base diode is reverse biased.

Also, at $I_B = 0$, $I_C = -I_E$. Therefore, Eq. (6.153) gives, for the common-emitter configuration with the base–collector junction reverse biased and at $I_B = 0$,

$$I_C = \frac{I_{CBO}}{1 - \alpha_F}. \tag{6.154}$$

This is the saturation current in the common-emitter configuration. We shall denote this current by $I_{CEO}$, i.e.,

$$I_{CEO} = \frac{I_{CBO}}{1 - \alpha_0}, \tag{6.155}$$

where we have used the fact that $\alpha_F = \alpha_0$. This current is also indicated in Fig. 6.22(c). It is clear from Eq. (6.155) that $I_{CEO}$ is significantly larger than $I_{CBO}$, since $\alpha_0$ is usually less than but close to unity. This is indicated in Fig. 6.22(c).

### 6.5.3 RELATION BETWEEN $BV_{CEO}$ AND $BV_{CBO}$

As pointed out in Section 2.4.1, the breakdown voltages in VLSI devices are usually determined experimentally, rather than calculated from some model. The avalanche multiplication factor $M$ in a reverse-biased diode is often expressed in terms of its breakdown voltage $BV$ using the *empirical* formula (Miller, 1955)

$$M(V) = \frac{1}{1 - (V/BV)^m}, \tag{6.156}$$

where $V$ is the reverse-bias voltage and $m$ is a number between 3 and 6 depending on the material and its resistivity. Thus, for the reverse-biased collector–base diode, we have

$$M(V_{CB}) = \frac{1}{1 - (V_{CB}/BV_{CBO})^m}. \tag{6.157}$$

Equation (6.155) implies that $I_{CEO}$ becomes infinite when $\alpha_0 = 1$. From Eq. (6.145), this means that when the collector voltage reaches $BV_{CEO}$,

$$\gamma\alpha_T M(V_{CB}) = \gamma\alpha_T M(BV_{CEO}) = 1. \tag{6.158}$$

Equations (6.157) and (6.158) give

$$\frac{BV_{CEO}}{BV_{CBO}} = (1 - \gamma\alpha_T)^{1/m}. \tag{6.159}$$

Since $1 - \gamma\alpha_T \ll 1$, Eq. (6.159) indicates that $BV_{CEO}$ can be substantially smaller than $BV_{CBO}$. This is illustrated in Fig. 6.22(c). Another way of comparing these breakdown voltages is to note that it takes $M$ approaching infinity to cause collector–base breakdown, while it takes $M$ only slightly larger than unity to cause collector–emitter breakdown (see Exercise 6.8). Figure 6.24 is a plot of $BV_{CEO}$ versus $BV_{CBO}$

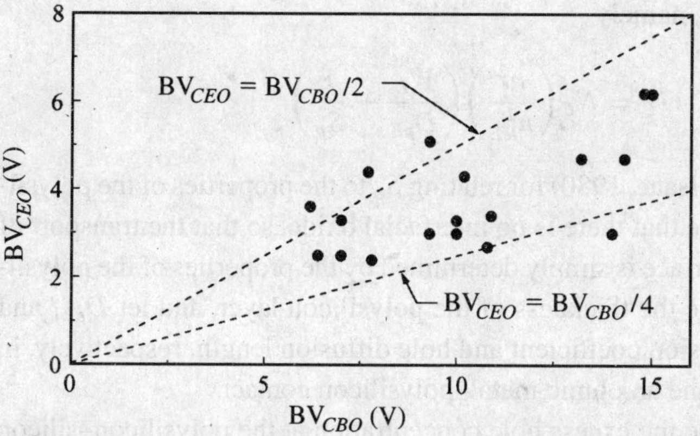


**FIGURE 6.24.** Reported $BV_{CEO}$ versus $BV_{CBO}$ data for recently published n–p–n transistors.

based on data reported in recent literature for modern n–p–n transistors. It shows that for modern n–p–n transistors, $BV_{CEO}$ is typically a factor of 2 to 4 smaller than $BV_{CBO}$.

From Eq. (6.148), $\gamma\alpha_T = \alpha_0(M = 1) = \beta_0/(1+\beta_0)$, where we have used Eq. (6.55) and $\beta_0$ is the current gain at negligible collector–base junction avalanche. Thus, Eq. (6.159) can also be written as

$$\frac{BV_{CEO}}{BV_{CBO}} = \left(\frac{1}{1+\beta_0}\right)^{1/m} \approx \left(\frac{1}{\beta_0}\right)^{1/m} \tag{6.160}$$

Equation (6.160) shows that ***there is a tradeoff between the collector–emitter breakdown voltage and the current gain*** of a transistor.

## EXERCISES

**6.1** The electric field in an n-type semiconductor is given by Eq. (6.20), i.e.,

$$\mathscr{E}(\text{n-region}) = -\frac{kT}{q}\left(\frac{1}{n_n}\frac{dn_n}{dx} - \frac{1}{n_{ie}^2}\frac{dn_{ie}^2}{dx}\right).$$

Derive this equation, stating clearly the approximations made in the derivation.

**6.2** The hole current density in the n-side of a p–n diode is given by Eq. (6.26), i.e.,

$$J_p(x) = -qD_p\frac{n_{ie}^2}{n_n}\frac{d}{dx}\left(\frac{n_n p_n}{n_{ie}^2}\right).$$

Derive this equation, stating clearly the approximations made in the derivation.

**6.3** For a polysilicon emitter with the emitter–base junction located at $x = 0$ and the silicon–polysilicon interface located at $x = -W_E$, the emitter Gummel number is given by Eq. (6.46), namely

$$G_E = N_E\left(\frac{n_i^2}{n_{ieE}^2}\right)\left(\frac{W_E}{D_{pE}} + \frac{1}{S_p}\right).$$

One model (Ning and Isaac, 1980) for relating $S_p$ to the properties of the polysilicon layer is to assume that there is no interfacial oxide, so that the transport of holes through the interface is simply determined by the properties of the polysilicon layer. Let $W_{E1}$ be the thickness of the polysilicon layer, and let $D_{pE1}$ and $L_{pE1}$ be the hole diffusion coefficient and hole diffusion length, respectively, in the polysilicon. Assume an ohmic metal–polysilicon contact.

**(a)** Let $\Delta p_n(-W_E)$ be the excess hole concentration at the polysilicon–silicon interface, and let $x'$ denote the distance from the polysilicon–silicon interface, i.e., $x' = -(x + W_E)$. Show that the excess hole distribution in the polysilicon layer, $\Delta p_n(x')$, is given by [cf. Eq. (2.101)]

$$\Delta p_n(x') = \Delta p_n(-W_E)\frac{\sinh[(W_{E1} - x')/L_{pE1}]}{\sinh(W_{E1}/L_{pE1})}.$$

**(b)** The relationship between $S_p$ and the hole current density entering the polysilicon layer is given by Eq. (6.36). Show that

$$S_p = \frac{D_{pE1}}{L_{pE1}\tanh(W_{E1}/L_{pE1})}.$$

**6.4** From time-dependent charge-control analysis, the small-signal current gain is given by Eq. (6.139), namely

$$\beta(\omega) = \left[\frac{1}{\beta_0} + j\omega\left(\tau_F + \frac{C_{dBE} + C_{dBC}}{g_m} + C_{dBC}(R_L + r_e + r_c)\right)\right]^{-1}$$

Show that in the high-frequency limit, the short-circuit ($R_L = 0$) current gain falls to unity at a frequency $f_T$ given by

$$\frac{1}{2\pi f_T} = \tau_F + \frac{kT}{qI_C}(C_{dBE} + C_{dBC}) + C_{dBC}(r_e + r_c).$$

($f_T$ is called the cutoff frequency, and is usually derived using a small-signal equivalent-circuit model; see Section 8.1.1.)

**6.5** Consider an n–p–n transistor with negligible parasitic resistances (which will be included in Exercise 6.6). Equations (6.152) and (6.153) give

$$I_E = -I_{EBO}\left[\exp\left(qV'_{BE}/kT\right) - 1\right] - \alpha_R I_C$$

and

$$I_C = -I_{CBO}\left[\exp\left(qV'_{BC}/kT\right) - 1\right] - \alpha_F I_E,$$

where $V'_{BE}$ and $V'_{BC}$ are the internal base–emitter and base–collector junction bias voltages. If the transistor is *operated in saturation*, i.e., both $V'_{BE}$ and $V'_{BC}$ are positive, show that the internal collector–emitter voltage, $V'_{CE} = V'_C - V'_E$, is related to the currents by

$$V'_{CE} = \frac{kT}{q}\ln\left[\frac{\alpha_F(I_{EBO} - I_E - \alpha_R I_C)}{\alpha_R(I_{CBO} - I_C - \alpha_F I_E)}\right].$$

[*Hint:* Use Eqs. (6.90), (6.150), and (6.151) to show that $I_{CBO}/I_{EBO} = \alpha_F/\alpha_R$.]

**6.6** From the expression for $V'_{CE}$ in Exercise 6.5, show that if the emitter and collector series resistances $r_e$ and $r_c$ are included, and if the saturation currents $I_{EBO}$ and $I_{CBO}$ are negligible, the voltage drop across the collector and emitter terminals, $V_{CE} = V_C - V_E$, is given by

$$V_{CE} = \frac{kT}{q}\ln\left[\frac{I_B + I_C(1-\alpha_R)}{\alpha_R[I_B - I_C(1-\alpha_F)/\alpha_F]}\right] + r_e(I_B + I_C) + r_c I_C$$

and that for open-circuit collector

$$V_{CE}(I_C = 0) = \frac{kT}{q}\ln\left(\frac{1}{\alpha_R}\right) + r_e I_B.$$

[The emitter resistance $r_e$ is often determined from a plot of the saturation open-collector voltage, $V_{CE}(I_C = 0)$, as a function of $I_B$ (Ebers and Moll, 1954; Filensky and Beneking, 1981). The collector resistance $r_c$ can be determined in a similar way by interchanging the emitter and collector connections.]

**6.7** For an n–p–n transistor, the base transport factor $\alpha_T$ is given in Eq. (6.147), i.e.,

$$\alpha_T \equiv \frac{J_n(x = W_B)}{J_n(x = 0)},$$

where the intrinsic-base layer is located between $x = 0$ and $x = W_B$. For a uniformly doped base, the excess-electron distribution is given by Eq. (2.101), namely

$$n_p - n_{p0} = n_{p0}[\exp(qV_{BE}/kT) - 1]\frac{\sinh[(W_B - x)/L_{nB}]}{\sinh(W_B/L_{nB})}.$$

If the electron current in the base is due to diffusion current only, show that

$$\alpha_T = \left(\cosh\frac{W_B}{L_{nB}}\right)^{-1}.$$

Use Fig. 2.18(c) to estimate $\alpha_T$ for a uniformly doped base with $N_B = 1 \times 10^{18}$ cm$^{-3}$ and $W_B = 100$ nm, and show that our assumption of negligible recombination in the intrinsic base is justified.

**6.8** If $M$ is the avalanche multiplication factor for the base–collector junction, and $\beta_0$ is the common-emitter current gain at negligible base–collector junction avalanche, show that the collector–emitter breakdown occurs when

$$M - 1 = \frac{1}{\beta_0}.$$

[*Hint:* Use Eqs. (6.148) and (6.158).] (It is interesting to note that since $\beta_0$ is typically about 100, collector–emitter breakdown occurs when $M$ is only slightly larger than unity. That is, it does not take much base–collector junction avalanche to cause collector–emitter breakdown.)

# 7 BIPOLAR DEVICE DESIGN

Bipolar device design can be considered in two parts. The first part deals with designing bipolar transistors in general, independent of their intended application. In this case, the goal is to reduce as much as possible, consistent with the start-of-the-art fabrication technology, all the internal resistance and capacitance components of the transistor. The second part deals with designing a bipolar transistor for a specific circuit application. In this case, the optimal device design point depends on the application. The design of a bipolar transistor in general is covered in this chapter, and the optimization of a transistor for a specific application is discussed in Chapter 8.

## 7.1 DESIGN OF THE EMITTER REGION

It was shown in Section 6.2 that the emitter parameters affect only the base current, and have no effect on the collector current. In theory, a device designer can vary the emitter design to vary the base current. In practice, this is rarely done, for two reasons. First, for digital-circuit applications, as long as the current gain is not unusually low or the base current unusually high, the performance of a bipolar transistor is rather insensitive to its base current (Ning *et al.*, 1981). For many analog-circuit applications, once the current gain is adequate, the reproducibility of the base current is more important than its magnitude. Therefore, there is really no particular reason to tune the base current of a bipolar device by tuning the emitter design, once a low and reproducible base current is obtained. Second, as can be seen in Appendix 2, the emitter is formed towards the end of the device fabrication process. Any change to the emitter process to tune the base current could affect the doping profile of the other device regions and hence could affect the other device parameters. As a result, once a bipolar technology is ready for manufacturing, its emitter fabrication process is usually fixed. All that a device designer can do to alter the device and circuit characteristics in this bipolar technology is to change the base and the collector designs, which often can be accomplished independently of the emitter process and hence has no effect on the base current.

The objective in designing the emitter of a bipolar transistor is then to achieve a low but reproducible base current while at the same time minimizing the emitter series resistance. As illustrated in Fig. 6.2, the commonly used bipolar transistors have either a diffused (or implanted-and-diffused) emitter or a polysilicon emitter. The design of both types of emitters is discussed in this section.

### 7.1.1 DIFFUSED OR IMPLANTED-AND-DIFFUSED EMITTER

A diffused or implanted-and-diffused emitter is formed by predoping a surface region of the silicon above the intrinsic base and then thermally diffusing the dopant to a desired depth. As shown in Eq. (6.48), for a diffused emitter, the base current is inversely proportional to the emitter doping concentration. Therefore, to minimize both the base current and the emitter series resistance, a diffused emitter is usually doped as heavily as possible. For n–p–n transistors, arsenic, instead of phosphorus, is usually used as the dopant, because arsenic gives a more abrupt doping profile than phosphorus. A more abrupt emitter doping profile leads to a shallower emitter junction, and, as we shall see later, a shallow emitter junction is needed for achieving a thin intrinsic base. A diffused emitter typically has a peak doping concentration of about $2 \times 10^{20}\ \text{cm}^{-3}$, as indicated in Fig. 6.2(a).

A diffused emitter is contacted either directly by a metal, or by a metal via a metal silicide layer. Commonly used silicides for emitter contact include platinum silicide and titanium silicide. If the fabrication process leaves negligible residual oxide on the emitter prior to contact formation, the resultant contact resistivity, as indicated in Eq. (5.11), is a function of the metal or metal silicide used, as well as a function of the emitter doping concentration at the contact (Yu, 1970). For a doping concentration of $2 \times 10^{20}\ \text{cm}^{-3}$ at the contact, a specific contact resistivity of about $(1\text{–}2) \times 10^{-7}\ \Omega\text{-cm}^2$ should be achievable.

Using the resistivity values of silicon shown in Fig. 2.8, the specific series resistivity of a 0.5-μm-deep silicon region, with an averaged doping concentration of $1 \times 10^{20}\ \text{cm}^{-3}$, is about $4 \times 10^{-8}\ \Omega\text{-cm}^2$. Therefore, the series resistance of a diffused emitter is dominated by its metal–silicon contact resistance; the series resistance of the doped-silicon region itself is negligible in comparison. For a diffused emitter of 1 $\mu\text{m}^2$ in area, the emitter series resistance is typically about 10–20 Ω.

It can be inferred from Fig. 6.1(b) that the intrinsic-base width $W_B$ is related to the emitter junction depth $x_{jE}$ and the base junction depth $x_{jB}$ by

$$W_B = x_{jB} - x_{jE}. \tag{7.1}$$

As we shall see in Section 7.2, one of the objectives in the design of the intrinsic base is to minimize its width. For $W_B$ to be well controlled, reproducible, and thin, $x_{jE}$ should be as small as possible. If $x_{jE}$ is much larger than $W_B$, then $W_B$ is given by the difference of two large numbers and hence will have large fluctuation.

Commonly used metal silicides are formed by depositing a layer of the appropriate metal on the silicon surface and then reacting the metal with the underlying silicon to form silicide. The emitter width $W_E$ is therefore reduced when metal silicide is used for emitter contact, because silicon in the emitter is consumed in the metal silicide formation process. As shown in Section 6.2.2, once $W_E$ is less than the minority-carrier diffusion length, the base current increases as $1/W_E$. As a result, the base current, and hence the current gain, of a bipolar transistor with a shallow diffused emitter varies with the emitter contact process (Ning and Isaac, 1980).

Referring to the minority-carrier diffusion lengths shown in Fig. 2.18(c), we see that the junction depth of a diffused n-type emitter should be larger than 0.3 μm in order to have adequately controllable and reproducible base-current characteristics. ***Diffused emitters are therefore not suitable for base widths of less than 100 nm.***

### 7.1.2 POLYSILICON EMITTER

Practically all modern high-performance bipolar transistors with base widths of 100 nm or smaller employ a polysilicon emitter. In this case, the emitter is formed by doping a polysilicon layer heavily and then activating the doped polysilicon layer just sufficiently to obtain reproducible base current and low emitter series resistance. The emitter junction depth, measured from the silicon–polysilicon interface, can be as small as 25 nm (Warnock, 1995). Consequently, with polysilicon-emitter technology, base widths of 50 nm or less can be obtained. The polysilicon-emitter process recipes are usually considered proprietary. However, there is a vast amount of literature on the physics of polysilicon-emitter devices (Ashburn, 1988; Kapoor and Roulston, 1989). Interested readers are referred to these publications.

The base current of a polysilicon-emitter transistor is given by Eqs. (6.42) to (6.44), with a surface recombination velocity, $S_p$, appropriate for the particular process used for forming the polysilicon emitter. The $S_p$ is usually used as a fitting parameter to the measured base current. In general, the base current of a polysilicon-emitter transistor is sufficiently low so that current gains in excess of 100 are readily achievable.

As will be shown in Chapter 8, the maximum speed of a modern bipolar transistor is determined primarily by its diffusion capacitance. In Section 2.2.6, the diffusion capacitance due to minority-carrier storage in the emitter was shown to be small compared to that due to minority-carrier storage in the base. Therefore, the maximum speed of a bipolar transistor is relatively insensitive to the emitter component of its diffusion capacitance. In other words, as long as the desired base-region characteristics are obtained, the details of the emitter region have relatively little effect on the maximum speed of a bipolar transistor (Ning *et al.*, 1981). Nonetheless, a polysilicon-emitter fabrication process should be designed to give low emitter series resistance and adequate emitter–base breakdown voltage, as well as the desired base-region characteristics.

The series resistance of a polysilicon emitter includes the polysilicon–silicon contact resistance, resistance of the polysilicon layer, and resistance of the metal–polysilicon contact. The specific resistivity of a metal–polysilicon contact is about the same as that of a metal–silicon contact. For arsenic-doped polysilicon

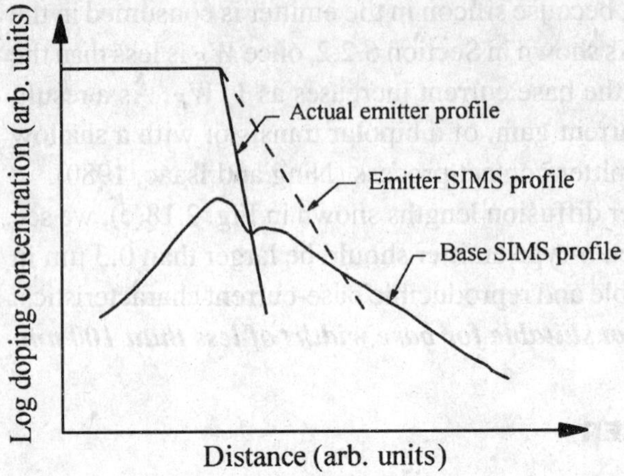


**FIGURE 7.1.** Schematic illustrating the measured SIMS doping profiles of the emitter and base of a modern n–p–n transistor. The measured emitter SIMS profile is usually less abrupt than the real one.

emitters, the reported specific silicide–polysilicon contact resistivity is typically (2–6) × $10^{-7}$ Ω-cm², depending on the arsenic concentration (Iinuma *et al.*, 1995). It is large compared to the series resistance of the polysilicon layer itself. The polysilicon–silicon contact resistance, on the other hand, is a strong function of the polysilicon-emitter fabrication process and can vary by large amounts (Chor *et al.*, 1985). In fact, polysilicon-emitter technology is still an area of active development. The recently published data (Iinuma *et al.*, 1995; Uchino *et al.*, 1995; Kondo *et al.*, 1995; Shiba *et al.*, 1996) suggest that a total emitter specific resistivity, which includes contributions from both the polysilicon–silicon interface and the metal–polysilicon contact, of 7–50 Ω-μm² should be obtainable (see Exercise 7.7).

The small junction depth of a polysilicon emitter implies a relatively small-perimeter, or vertical, extrinsic-base–emitter junction area. The total emitter–base junction capacitance of a polysilicon emitter is therefore much smaller than that of a diffused emitter. For a 0.3-μm emitter stripe, the total emitter–base junction capacitance of a polysilicon emitter can be less than $\frac{1}{3}$ of that of a diffused emitter.

It should be pointed out that the junction of a polysilicon emitter is so shallow that the commonly used secondary-ion mass spectroscopy (SIMS) technique for measuring dopant concentration profiles often indicates an emitter junction deeper than it really is. The real emitter junction depth can be obtained from the p-type base SIMS profile, which shows a dip where the n-type and p-type doping concentrations are equal (Hu and Schmidt, 1968). This is illustrated schematically in Fig. 7.1.

## 7.2 DESIGN OF THE BASE REGION

It was shown in Section 6.2 that the base-region parameters affect only the collector current, not the base current. The base current is determined by the emitter parameters. It has been demonstrated experimentally (Ning *et al.*, 1981), and will be discussed in Chapter 8, that the performance of a bipolar circuit is determined primarily by the collector current, not the base current, at least for circuits where the bipolar transistors do not saturate. Therefore, as long as current gain is adequate, which is the case with a polysilicon emitter, the focus in designing or optimizing a bipolar transistor should be on the collector current, and not on the base current. In other words, the focus should be on the intrinsic base when there is negligible base widening, and on both the intrinsic base and the collector when base widening is not negligible.

The design of the base of a bipolar transistor can be very complex, because of the tradeoffs that must be made between the ac and dc characteristics, which depend on the intended application, and because of the tradeoffs that must be made between the desired device characteristics and the complexity of the fabrication process for realizing the design. In this section, the relationship between the physical and electrical parameters of the base is derived, and the design tradeoffs are discussed. Optimization of the base design for various circuit applications will be covered in Chapter 8.

Referring to Fig. 6.12, we can divide the base region into two parts. The part directly underneath the emitter is the intrinsic base, and the part connecting the intrinsic base to the base terminal is the extrinsic base. As a first-order but good approximation, the intrinsic base is what determines the collector current characteristics, and hence the intrinsic performance of a transistor. The discussions and the collector current characteristics derived in Chapter 6 are all for the intrinsic base. Effects of the extrinsic base were ignored.

The extrinsic base is an integral part of any bipolar transistor. It is a parasitic component in that it does not contribute appreciably to the collector current, at least for properly designed transistors. In general, designing the extrinsic base is very simple: the extrinsic-base area and its associated capacitance and series resistance should all be as small as possible. How this is accomplished depends on the fabrication process used. A major focus in bipolar-technology research and development has been to minimize the parasitic resistance and capacitance associated with the extrinsic base. The interested reader is referred to the vast literature on the subject (Warnock, 1995; Nakamura and Nishizawa, 1995; and the references therein), and to Appendix 2, which outlines the fabrication process for one of the most widely used modern bipolar transistors.

Any adverse effect of the extrinsic base on the breakdown voltages of the emitter–base and base–collector diodes should be minimized. This is accomplished by having the dopant distribution of the extrinsic base not extending appreciably into the intrinsic base. If the extrinsic base encroaches appreciably on the intrinsic base, the encroached-on intrinsic-base region will appear to be wider, as well as more heavily doped, than the rest of the intrinsic base. Extrinsic-base encroachment on the intrinsic base, therefore, will lead to a smaller collector current as well as degraded dc and ac characteristics (Lu *et al.*, 1987; Li *et al.*, 1987).

For an optimally designed bipolar process, extrinsic-base encroachment is usually negligible. As a result, the extrinsic base usually has little effect on the collector current. Therefore, only the design of the intrinsic base will be discussed further in this section.

### 7.2.1 RELATIONSHIP BETWEEN BASE SHEET RESISTIVITY AND COLLECTOR CURRENT DENSITY

As shown in Fig. 6.5, the collector current of a typical bipolar transistor is ideal, i.e., varying as $\exp(qV_{BE}/kT)$, for $V_{BE}$ less than about 0.8 V. For this ideal region, the saturated collector current density for an n–p–n transistor is given by Eq. (6.33), which is repeated here:

$$J_{C0} = \frac{q}{\int_0^{W_B} \frac{p_p(x)}{D_{nB}(x) n_{ieB}^2(x)}\, dx}, \tag{7.2}$$

where $p_p$, $D_{nB}$, and $n_{ieB}$ are the hole density, the electron diffusion coefficient, and the effective intrinsic-carrier concentration, respectively, in the p-type base region. The effective intrinsic-carrier concentration is given by Eq. (6.15). It can be used to allow for the heavy-doping effect as well as any bandgap-engineering effect by properly adjusting the bandgap-narrowing parameter $\Delta E_g$. We shall first consider the case where $n_{ieB}$ is used to allow for heavy-doping effect in the base. The case of using $n_{ieB}$ to allow for base-bandgap engineering will be covered in Section 7.2.5.

For device design purposes, it is often convenient to assume that both $D_{nB}$ and $n_{ieB}$ are slowly varying functions of $x$ and hence can be approximated by some average values. That is, Eq. (7.2) is often written as

$$J_{C0} \approx \frac{q\bar{D}_{nB}\bar{n}_{ieB}^2}{\int_0^{W_B} p_p(x)\, dx}. \tag{7.3}$$

At low currents, the hole concentration in the base is equal to the base doping concentration $N_B(x)$, and Eq. (7.3) can be further simplified to

$$J_{C0} \approx \frac{q\bar{D}_{nB}\bar{n}_{ieB}^2}{\int_0^{W_B} N_B(x)\, dx}. \tag{7.4}$$

The integral in the denominator of Eq. (7.4) is simply the total *integrated base dose*. [In the literature, the denominator in Eq. (7.4) is sometimes referred to as the base Gummel number (Gummel, 1961). However, in this book we follow the convention of de Graaff (de Graaff *et al.*, 1977), where the base Gummel number $G_B$ is defined by Eq. (6.34).] Thus, the collector current density at low currents is approximately inversely proportional to the total integrated base dose. Using ion-implantation techniques for doping the intrinsic base, the integrated base dose, and hence the collector current density, can be controlled quite precisely and reproducibly.

The sheet resistivity of the intrinsic base, $R_{Sbi}$, is

$$R_{Sbi} = \left( q \int_0^{W_B} p_p(x)\mu_p(x)\,dx \right)^{-1}. \tag{7.5}$$

Again, for device design purposes, it is convenient to assume an average mobility and rewrite Eq. (7.5) as

$$R_{Sbi} \approx \left( q\bar{\mu}_p \int_0^{W_B} p_p(x)\,dx \right)^{-1}. \tag{7.6}$$

Substituting Eq. (7.6) into Eq. (7.3), we obtain

$$J_{C0} \approx q^2 \bar{D}_{nB}\bar{\mu}_p \bar{n}_{ieB}^2 R_{Sbi}. \tag{7.7}$$

That is, ***the collector current density is approximately proportional to the intrinsic-base sheet resistivity***. This direct correlation is valid for $R_{Sbi}$ between 500 and $20 \times 10^3\ \Omega/\square$, which is the range of interest in most bipolar device designs (Tang, 1980).

### 7.2.2 INTRINSIC-BASE DOPANT DISTRIBUTION

For a desired intrinsic-base sheet resistivity, the detailed intrinsic-base dopant distribution depends on the fabrication process used. Most modern bipolar-transistor processes employ ion implantation, followed by thermal annealing and/or thermal diffusion, to form the intrinsic base. In this case, the intrinsic-base doping profile is approximately a *Gaussian* distribution, often with an exponentially decreasing tail. As will be shown in Section 7.3, the collector doping concentration of a modern bipolar transistor is relatively high, often in excess of $1 \times 10^{17}\ \text{cm}^{-3}$. This concentration is usually high compared with the tail of the base dopant distribution. As a result, the lightly doped tail is often clipped off by the collector doping profile and has little effect on the collector current. Therefore, for simplicity of discussion and analysis, we shall ignore the tail of the base dopant distribution.

If the Gaussian base dopant distribution peaks at the emitter–base junction located at $x = 0$, then the base doping concentration can be described by

$$N_B(x) = N_{B\,\max} \exp\left( -\frac{x^2}{2\sigma^2} \right), \tag{7.8}$$

where $\sigma$ and $N_{B\,\max}$ are the standard deviation and peak concentration, respectively, of the distribution. For most bipolar device designs, the peak doping concentration

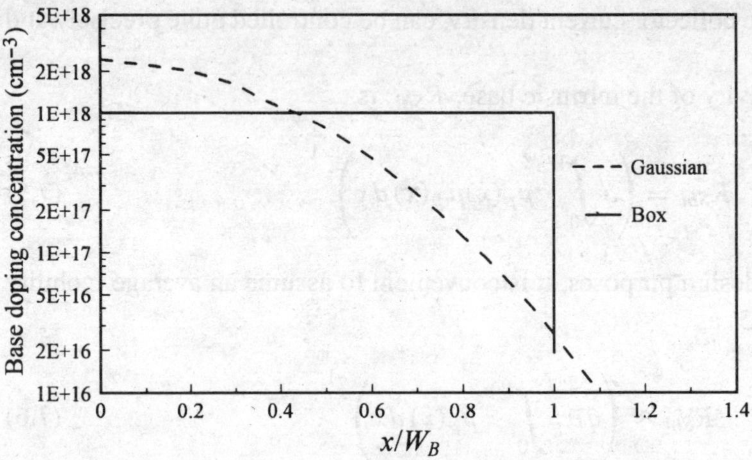


**FIGURE 7.2.** Schematic illustration of a boxlike doping profile and a Gaussian doping profile for the same base width and the same integrated base dose. The peak doping concentration of the Gaussian profile is approximately 2.4 times that of the box profile, and the base width is approximately equal to $3\sigma$.

in the base is approximately 10–100 times that in the collector. Here, for purposes of discussion, we assume $N_{B\,\max}/N_C = 100$. This implies a base width of

$$W_B \approx 3\sigma \tag{7.9}$$

for the Gaussian base dopant distribution.

With the advent of silicon epitaxy processes, instead of implanting dopant ions into silicon, the intrinsic base can be formed by epitaxial growth of a thin, *in situ* doped silicon layer on top of the collector. In this case, the base dopant distribution depends on the *in situ* doping process used. The simplest distribution is an approximately uniform, or *boxlike*, distribution.

Figure 7.2 illustrates a box profile of $N_B = 1 \times 10^{18}$ cm$^{-3}$, and a Gaussian profile of the same integrated base dose and the same base width. It shows that the peak concentration of the Gaussian profile is more than twice that of the box profile. The emitter–base depletion-layer capacitance of a Gaussian base doping profile is therefore larger than that of a boxlike base doping profile. Also, with a higher base doping concentration at the emitter–base junction, high-field effects at the emitter–base junction are also more severe for a Gaussian-profile base than for a box-profile base.

In general, for the same base width and integrated base dose, a base doping profile with a higher peak concentration, at or close to the emitter–base junction, will lead to a larger emitter–base junction capacitance. However, this does not imply that a box-profile base is necessarily preferred over a base with a peak concentration located at or near the emitter–base junction, for there are many other factors or parameters, such as base transit time and ease of fabrication, that must also be considered. (A boxlike base doping profile will certainly lead to a larger Early voltage, as will be shown in Section 8.5.1.) Before we discuss the dependence of the base transit time on the physical parameters of the base, we need to discuss the electric field in a quasineutral base region, since the transport of minority carriers in the base depends on the electric field in it.

### 7.2.3 ELECTRIC FIELD IN THE QUASINEUTRAL INTRINSIC BASE

The electron current density in the base region of an n–p–n transistor is given by Eq. (6.25). Since this is also the collector current density, we can write

$$J_C = qD_{nB}\frac{n_{ieB}^2}{p_p}\frac{d}{dx}\left(\frac{n_p p_p}{n_{ieB}^2}\right), \tag{7.10}$$

where the subscript $B$ has been added to indicate that the parameters are for the base region. It is valid for arbitrary base doping profile and arbitrary bandgap variation in the base. The dependence on bandgap variation is implicit in the effective intrinsic-carrier concentration, which is given by Eq. (6.15), i.e.,

$$n_{ieB}^2 = n_i^2 \exp(\Delta E_{gB}/kT), \tag{7.11}$$

where $\Delta E_{gB}$ is the bandgap-narrowing parameter in the base.

The electric field in the base region can be derived by decomposing Eq. (7.10) into the more familiar drift and diffusion components. To this end, Eq. (7.10) can be rewritten as

$$J_C = qD_{nB}n_p\left(\frac{1}{p_p}\frac{dp_p}{dx} - \frac{1}{n_{ieB}^2}\frac{dn_{ieB}^2}{dx}\right) + qD_{nB}\frac{dn_p}{dx}. \tag{7.12}$$

From Eq. (6.5), the collector current density can also be written in its usual form of

$$J_C = qn_p\mu_{nB}\mathscr{E} + qD_{nB}\frac{dn_p}{dx}. \tag{7.13}$$

Comparison of Eqs. (7.12) and (7.13) shows that the electric field for minority-carrier electrons in the p-type base region is given by

$$\mathscr{E}(\text{p-base}) = \frac{kT}{q}\left(\frac{1}{p_p}\frac{dp_p}{dx} - \frac{1}{n_{ieB}^2}\frac{dn_{ieB}^2}{dx}\right). \tag{7.14}$$

This dependence of the electric field on majority-carrier concentration and effective intrinsic-carrier concentration was stated without derivation in Eq. (6.19). Using Eq. (7.11), Eq. (7.14) can also be written as

$$\mathscr{E}(\text{p-base}) = \frac{kT}{q}\frac{1}{p_p}\frac{dp_p}{dx} - \frac{1}{q}\frac{d\Delta E_{gB}}{dx}, \tag{7.15}$$

which relates the electric field to the majority-carrier concentration and bandgap narrowing in the base region.

#### 7.2.3.1 ELECTRIC FIELD IN THE QUASINEUTRAL BASE REGION AT LOW CURRENTS

At low injection currents, $P_p \approx N_B$, and Eq. (7.14), or Eq. (7.15), gives the built-in electric field in the base caused by the base doping profile and base-bandgap variation. It should be noted that as $N_B$ increases, $\Delta E_{gB}$ increases, and hence $dN_B/dx$ and $d\Delta E_{gB}/dx$ have the same sign. Equation (7.15) shows that ***the electric field due to the heavy-doping effect always tends to offset the electric field due to the dopant distribution.***

When transistors are designed with base widths much larger than 100 nm, the peak base doping concentration is usually about $10^{17}$ cm$^{-3}$ or smaller. For such low concentrations, the effect of heavy doping is negligible. In this case, a graded base doping profile can result in a substantial electric field in the base, which enhances the drift component of the collector current traversing the base layer. These are so-called *drift transistors.* They have higher cutoff frequencies than transistors with a more uniform base doping profile (Sze, 1981; Ghandhi, 1968).

Modern bipolar transistors, however, have peak base doping concentrations larger than $10^{18}$ cm$^{-3}$, as indicated in Fig. 6.2. For these devices, the electric field due to the $d\Delta E_{gB}/dx$ term must be included, which could substantially cancel the electric field due to the $dN_B/dx$ term. As a result, the net electric field in the quasineutral intrinsic base of a modern bipolar transistor can be relatively small. In other words, ***the drift-transistor concept is less important in modern thin-base bipolar device design than in the design of wide-base bipolar transistors***. (This will be demonstrated below for a Gaussian-profile base design.)

- *Electric field in an intrinsic base with a box profile.* For a boxlike base doping profile, both $dN_B/dx$ and $d\Delta E_{gB}/dx$ are equal to zero. (Here we assume $\Delta E_{gB}$ is due to heavy doping alone. $\Delta E_{gB}$ due to base bandgap grading will be covered in Section 7.2.5.) There is no electric field in a box-profile base region at low injection currents.

- *Electric field in an intrinsic base with a Gaussian doping profile.* For the Gaussian base profile given by Eq. (7.8), the electric field at low currents is

$$\mathscr{E}(\text{Gaussian-base}) = \left(-\frac{kT}{q}\frac{x}{\sigma^2}\right) + \left(-\frac{1}{q}\frac{d\Delta E_{gB}}{dx}\right). \tag{7.16}$$

The first term in Eq. (7.16) is negative. The Gaussian base doping profile, with its concentration larger near the emitter–base junction and lower towards the base–collector junction, has a *graded-base* electric field in a direction to drive the electrons across the base layer. However, the second term in Eq. (7.16) is positive, since $\Delta E_{gB}$ is larger near the emitter–base junction, where the base doping concentration is large, than near the base–collector junction, where the base doping concentration is very small.

For the Gaussian base profile shown in Fig. 7.2, the electric field components as well as the total electric field at low injection levels, i.e., for $p_p \approx N_B$, are shown in Fig. 7.3. The bandgap-narrowing parameter given by Eq. (6.18) is used, and the base width is assumed to be 100 nm. Figure 7.3 shows that, for this specific Gaussian base doping profile, the effect of heavy doping almost completely offsets the effect of nonuniform dopant distribution, except for the region near the base–collector junction, where the base doping concentration is relatively small and hence the effect of heavy doping is negligible. This lightly doped base region near the base–collector junction is most likely depleted in normal device operation and hence does not form part of the quasineutral base. Therefore, the net electric field in the entire quasineutral base region is quite negligible.

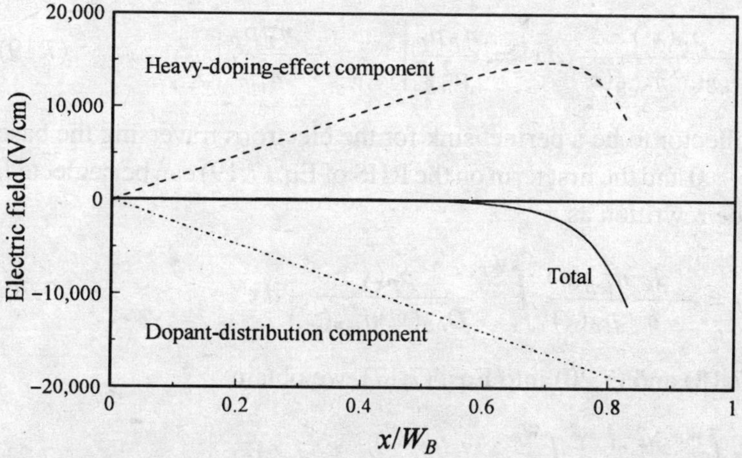


**FIGURE 7.3.** Electric fields in the quasineutral intrinsic-base region with a Gaussian doping profile. The total electric field is the sum of the dopant-distribution and the bandgap-narrowing components. The Gaussian-profile parameters are $\sigma = W_B/3$, $W_B = 100$ nm, and $N_{B\,max} = 2.4 \times 10^{18}$ cm$^{-3}$.

For non-Gaussian base doping profiles, the cancellation of the electric field components may not be as complete as suggested in Fig. 7.3. Nonetheless, the cancellation is substantial. The reader is referred to the literature (Suzuki, 1991) for more examples of similar calculations.

### 7.2.4 BASE TRANSIT TIME

The base transit time $t_B$ is an important and often used figure of merit for bipolar transistors. It was shown in Section 2.2.5 that for a thin-base diode, $t_B$ is the time needed to fill the base region of the diode with minority carriers, and it is also the average time for minority carriers injected from the emitter to traverse the thin base. For a bipolar transistor, the charging current for the base is the current flowing from the emitter into the base, i.e., the collector current. The base transit time for a bipolar transistor is therefore defined by

$$t_B \equiv |Q_B|/|J_C|, \tag{7.17}$$

where $J_C$ is the collector current density and $Q_B$ is the minority-carrier charge per unit area stored in the base region. For an n–p–n transistor, $Q_B$ is given by

$$Q_B = -q \int_0^{W_B} n_p(x)\, dx. \tag{7.18}$$

It should be noted that $Q_B$ is negative for an n–p–n transistor, since the minority carriers in the p-type base are electrons.

Since recombination in a thin base is negligible, $J_C$ is independent of $x$. Therefore, Eq. (7.10) can be integrated to give

$$J_C \int_x^{W_B} \frac{p_p(x')}{q D_{nB}(x') n_{ieB}^2(x')}\, dx' = \left.\frac{n_p p_p}{n_{ieB}^2}\right|_{x'=W_B} - \left.\frac{n_p p_p}{n_{ieB}^2}\right|_{x'=x} \tag{7.19}$$

If we assume the collector to be a perfect sink for the electrons traversing the base layer, then $n_p(W_B) \approx 0$ and the first term on the RHS of Eq. (7.19) can be neglected, and Eq. (7.19) can be rewritten as

$$n_p(x) = -\frac{J_C}{q} \frac{n_{ieB}^2(x)}{p_p(x)} \int_x^{W_B} \frac{p_p(x')}{D_{nB}(x') n_{ieB}^2(x')}\, dx'. \tag{7.20}$$

Substituting Eqs. (7.18) and (7.20) into Eqs. (7.17), we obtain

$$t_B = \int_0^{W_B} \frac{n_{ieB}^2(x)}{p_p(x)} \int_x^{W_B} \frac{p_p(x')}{D_{nB}(x') n_{ieB}^2(x')}\, dx' dx. \tag{7.21}$$

Equation (7.21) is the general expression for the base transit time (Kroemer, 1985). It includes high-current effects, through the dependence of $p_p$ and $W_B$ on minority-carrier concentration, as well as nonuniform-bandgap effects, through the parameter $n_{ieB}$.

#### 7.2.4.1 BASE TRANSIT TIME AT LOW CURRENTS

At low currents, the hole concentration is approximately equal to the base doping concentration, and Eq. (7.21) reduces to

$$t_B \approx \int_0^{W_B} \frac{n_{ieB}^2(x)}{N_B(x)} \int_x^{W_B} \frac{N_B(x')}{D_{nB}(x')n_{ieB}^2(x')} dx'\, dx. \tag{7.22}$$

If the dopant distribution and the dependencies of mobility and bandgap narrowing on doping concentration are known, Eq. (7.22) can be used to calculate the base transit time at low currents (Suzuki, 1991).

In the *uniform-bandgap approximation* for the base transit time at low currents, $n_{ieB}$ is independent of $x$ and Eq. (7.22) reduces to (Moll and Ross, 1956)

$$t_B \approx \int_0^{W_B} \frac{1}{N_B(x)} \int_x^{W_B} \frac{N_B(x')}{D_{nB}(x')} dx'\, dx. \tag{7.23}$$

For a box profile, both $N_B$ and $D_{nB}$ are constant, and Eq. (7.23) reduces further to

$$t_B \approx \frac{W_B^2}{2D_{nB}}, \tag{7.24}$$

which, as expected, is the same as Eq. (2.122) for the transit time for a uniformly doped thin-base diode.

Equation (7.24) suggests that ***an effective way to reduce the base transit time is to reduce the intrinsic-base width***. However, as the base width is reduced, the base doping concentration must be increased appropriately to avoid emitter–collector punch-through or the Early voltage becoming unacceptably small.

### 7.2.5 SiGe BASE

The bandgap of Ge (≈0.66 eV) is significantly smaller than that of Si (≈1.12 eV). By incorporating Ge into the base region of a Si bipolar transistor, the bandgap of the base region, and hence the accompanied device characteristics, can be modified (Iyer *et al.*, 1987). Of all the methods for incorporating Ge into the base layer of a Si bipolar transistor, the ultrahigh-vacuum chemical-vapor-deposition epitaxial growth technique (Meyerson, 1986), where Ge is introduced during epitaxial growth of the intrinsic-base layer, has been most successful so far.

In designing the Ge profile in the base region, it is desirable to have the largest amount of Ge near the base–collector junction and the smallest amount near the emitter–base junction. In this way, the base bandgap is smallest near the base–collector junction and largest near the emitter–base junction, thus maximizing the drift field for electrons in the base (Patton *et al.*, 1990; Harame *et al.*, 1995a, b).

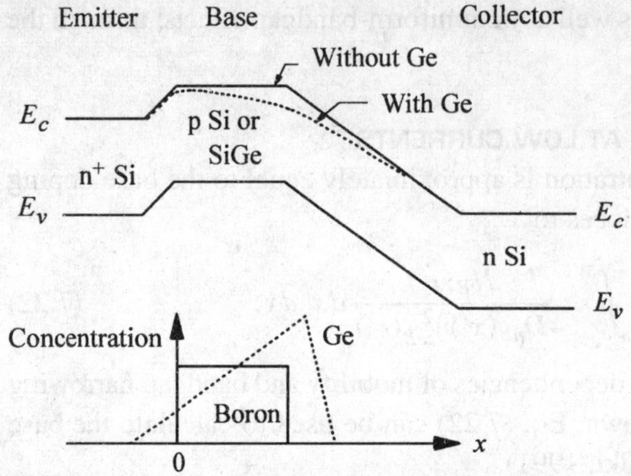


**FIGURE 7.4.** Schematic illustration of the energy bands of a SiGe-base n–p–n transistor (dotted) and a Si-base n–p–n transistor (solid). Both transistors are assumed to have the same base doping profile. The base bandgaps of the two transistors are about the same near the base–emitter junction. The base bandgap of the SiGe-base transistor narrows gradually towards the base–collector junction.

The emitter of a typical SiGe-base bipolar transistor is the same as that of a regular Si-base bipolar transistor. In both transistors, it is simply a polysilicon emitter. With a relatively small Ge concentration near the emitter-base junction, the bandgap within the intrinsic-base layer but near the emitter–base junction of a typical SiGe-base transistor is about the same as the base bandgap of a regular Si-base transistor. Therefore, the commonly made SiGe-base bipolar transistors are really not heterojunction bipolar transistors (HBTs), since a heterojunction bipolar transistor usually refers to one whose emitter is engineered to have a significantly larger bandgap, at the emitter–base junction, than the base. ***The commonly made SiGe-base bipolar transistors are graded-base-bandgap transistors.***

The energy bands of a common SiGe-base n–p–n transistor are illustrated and compared with those of a Si-base transistor in Fig. 7.4. The base bandgaps of the two transistors near the emitter are about the same. The base bandgap of the SiGe-base transistor narrows gradually towards the base–collector junction.

As shown in Section 6.2.2, the base current is determined by the emitter parameters only and is independent of the base parameters. Therefore, since the emitter of a SiGe-base transistor is the same as that of a Si-base transistor, the base current of a SiGe-base transistor should be the same as that of a Si-base transistor. This is indeed the case for the commonly used SiGe-base transistors (Harame *et al.*, 1995a, b). This observation is also consistent with our comment above that the commonly practiced SiGe-base bipolar transistors are not HBTs. ***Only the collector current, which depends on the base parameters, is affected by the Ge profile in the intrinsic base.***

#### 7.2.5.1 SiGe BASE WITH LINEARLY GRADED BANDGAP AND BOXLIKE DOPING PROFILE

The effect of base-bandgap grading can be included in the device model simply by extending the bandgap narrowing parameter in Eq. (7.11) to include base-bandgap grading. The resultant collector current density can be derived in closed form if the base-bandgap grading is assumed to be linear and the base doping profile is assumed to be boxlike (Kroemer, 1985). As it turns out, a linearly graded base bandgap is readily achievable with a triangular Ge profile such as that illustrated in Fig. 7.4, where the Ge concentration increases linearly from the emitter–base junction towards the base–collector junction. Such a linearly graded base bandgap is used in the commonly practiced SiGe-base transistors (Harame *et al.*, 1995a, b).

For a linearly graded base bandgap, the effective intrinsic-carrier concentration in the base can be written as (Kroemer, 1985)

$$n_{ieB}^2(\text{SiGe}) = n_{ieB}^2(\text{Si}) \exp\left(\frac{\Delta E_{g,SiGe}}{kT}\frac{x}{W_B}\right), \tag{7.25}$$

where $n_{ieB}(\text{Si})$ is the effective intrinsic-carrier concentration in a Si-base transistor and is given by Eq. (7.11), and $\Delta E_{g,SiGe}$ is the total bandgap narrowing across the intrinsic-base layer due to the incorporation of Ge.

- *Effect of base-bandgap grading on collector current and current gain.* The collector current density of a SiGe-base transistor can be obtained simply by substituting $n_{ieB}(\text{SiGe})$ for $n_{ieB}$ in all the equations derived earlier for the collector current density. Thus, the saturated collector current density given in Eq. (6.33) becomes

$$J_{C0}(\text{SiGe}) = \frac{q}{\int_0^{W_B} \frac{p_p(x)}{D_{nB}(x) n_{ieB}^2(\text{SiGe},x)}\, dx}. \tag{7.26}$$

For a boxlike base doping profile and for low currents, $p_p \approx N_B$ and is independent of $x$. $D_{nB}$ and $n_{ieB}(\text{Si})$ are also independent of $x$. Therefore, Eq. (7.26) can be rewritten as

$$\begin{aligned} J_{C0}(\text{SiGe}) &\approx \frac{q D_{nB} n_{ieB}^2(\text{Si})}{N_B} \frac{1}{\int_0^{W_B} \exp\left(\frac{-\Delta E_{g,SiGe}}{kT}\frac{x}{W_B}\right) dx} \\ &= \frac{q D_{nB} n_{ieB}^2(\text{Si})}{N_B W_B}\left(\frac{\Delta E_{g,SiGe}/kT}{1-\exp(-\Delta E_{g,SiGe}/kT)}\right). \end{aligned} \tag{7.27}$$

The saturated collector current density for a Si-base transistor with the same doping profile is

$$J_{C0}(\text{Si}) = \frac{q D_{nB} n_{ieB}^2(\text{Si})}{N_B W_B}. \tag{7.28}$$

Therefore, for the same emitter–base forward bias and the same base doping profile, the incorporation of a linearly graded Ge profile in the intrinsic base increases the collector current by a factor of

$$\frac{J_{C0}(\text{SiGe})}{J_{C0}(\text{Si})} = \frac{\Delta E_{g,SiGe}/kT}{1 - \exp(-\Delta E_{g,SiGe}/kT)}. \tag{7.29}$$

As discussed earlier, the base current of a SiGe-base transistor is approximately the same as that of a Si-base transistor; the ratio of the current gains is therefore the same as the ratio of the collector currents, namely

$$\frac{\beta(\text{SiGe})}{\beta(\text{Si})} = \frac{J_{C0}(\text{SiGe})}{J_{C0}(\text{Si})} = \frac{\Delta E_{g,SiGe}/kT}{1 - \exp(-\Delta E_{g,SiGe}/kT)}. \tag{7.30}$$

The readily achievable total base bandgap narrowing is in the range of 100–150 meV (Harame *et al.*, 1995a, b), which means a SiGe-base transistor typically can have a collector current and current gain that are 4–6 times those of a Si-base transistor with the same base doping profile. As shown in Eq. (7.7), the collector current density, and hence the current gain, is proportional to the intrinsic-base sheet resistivity. The enhanced current gain of a SiGe-base transistor can be and is often used to trade off for a smaller intrinsic-base resistance. The smaller intrinsic-base sheet resistivity improves the Early voltage as well, as discussed below.

- *Effect of linearly graded base bandgap on the Early voltage.* The effect of base-bandgap grading on the Early voltage can be obtained from Eq. (6.73) by replacing $n_{ieB}$ by $n_{ieB}(\text{SiGe})$, i.e.,

$$V_A(\text{SiGe}) \approx \frac{qD_{nB}(W_B)n_{ieB}^2(\text{SiGe}, W_B)}{C_{dBC}} \times \int_0^{W_B} \frac{N_B(x)}{D_{nB}(x)n_{ieB}^2(\text{SiGe}, x)}\,dx. \tag{7.31}$$

Again, for a boxlike base profile, $N_B$, $D_{nB}$, and $n_{ieB}(\text{Si})$ are all independent of $x$. Also

$$n_{ieB}^2(\text{SiGe}, W_B) = n_{ieB}^2(\text{Si}, W_B)\exp(\Delta E_{g,SiGe}/kT). \tag{7.32}$$

Therefore, Eq. (7.31) can be rewritten as

$$V_A(\text{SiGe}) \approx \frac{qN_B\exp(\Delta E_{g,SiGe}/kT)}{C_{dBC}} \times \int_0^{W_B} \exp\left(\frac{-\Delta E_{g,SiGe}}{kT}\frac{x}{W_B}\right)dx = \frac{qN_BW_B}{C_{dBC}}\frac{kT}{\Delta E_{g,SiGe}}\left[\exp\left(\frac{\Delta E_{g,SiGe}}{kT}\right) - 1\right]. \tag{7.33}$$

From Eq. (6.72), the Early voltage for a Si-base transistor with a boxlike base profile is

$$V_A(\text{Si}) = \frac{qN_B W_B}{C_{dBC}}. \tag{7.34}$$

The ratio of the Early voltage of a SiGe-base transistor to that of a Si-base transistor of the same base doping profile is therefore

$$\frac{V_A(\text{SiGe})}{V_A(\text{Si})} = \frac{kT}{\Delta E_{g,SiGe}}\left[\exp\left(\frac{\Delta E_{g,SiGe}}{kT}\right) - 1\right]. \tag{7.35}$$

That is, the Early voltage increases almost exponentially with the total base-bandgap narrowing. For a typical value of $\Delta E_{g,SiGe} = 100$ meV, the Early voltage is increased by a factor of 12. Incorporation of Ge into the intrinsic base thus can increase the Early voltage by a large factor. Combining Eqs. (7.30) and (7.35), the ratio of the $\beta V_A$ products is increased by a factor of

$$\frac{\beta(\text{SiGe})V_A(\text{SiGe})}{\beta(\text{Si})V_A(\text{Si})} = \exp\left(\frac{\Delta E_{g,SiGe}}{kT}\right). \tag{7.36}$$

The same result could have been obtained from Eq. (6.74) by using Eq. (7.32) for $n_{ieB}(\text{SiGe}, W_B)$. For $\Delta E_{g,SiGe} = 100$ meV, the product $\beta V_A$ is increased by almost a factor of 50. As mentioned earlier, the larger collector current density of a SiGe-base transistor is often traded off for a smaller intrinsic-base sheet resistivity. A smaller intrinsic-base sheet resistivity implies an increase in the majority-carrier base charge, or $qN_B W_B$, which in turn will increase the Early voltage of the SiGe-base transistor according to Eq. (7.33). Therefore, if desired, a SiGe-base bipolar transistor can be designed to have a very large Early voltage.

- *Effect of linearly graded base bandgap on base transit time.* The graded base bandgap introduces an electric field which helps to drive the electrons across the p-type base layer. For a total base-bandgap narrowing of 100 meV across a base layer of 100 nm, a SiGe-base transistor would have a built-in electric field of $10^4$ V/cm in the base due to the presence of Ge alone. This field is in addition to the electric fields due to the base dopant distribution and due to the heavy-doping effect, which have been discussed earlier in Section 7.2.3. As can be seen from comparing this field with the fields plotted in Fig. 7.3, the electric field due to the presence of a graded Ge profile can be comparable to the maximum fields due to dopant distribution and heavy-doping effect. Consequently, the base transit time of a SiGe-base transistor can be significantly smaller than that of a Si-base transistor of the same base doping profile. The base transit time at low currents can be derived by substituting

$n_{ieB}(\text{SiGe})$ for $n_{ieB}$ in Eq. (7.22), i.e.,

$$t_B(\text{SiGe}) \approx \int_0^{W_B} \frac{n_{ieB}^2(\text{SiGe}, x)}{N_B(x)} \times \int_x^{W_B} \frac{N_B(x')}{D_{nB}(x') n_{ieB}^2(\text{SiGe}, x')} \, dx' \, dx. \tag{7.37}$$

Again, for a boxlike base doping profile, $N_B$, $D_{nB}$, and $n_{ieB}(\text{Si})$ are all independent of $x$, and Eq. (7.37) simplifies to

$$\begin{aligned} t_B(\text{SiGe}) &\approx \frac{1}{D_{nB}} \int_0^{W_B} \exp\left(\frac{\Delta E_{g,SiGe}}{kT} \frac{x}{W_B}\right) \times \int_x^{W_B} \exp\left(\frac{-\Delta E_{g,SiGe}}{kT} \frac{x'}{W_B}\right) dx' \, dx \\ &= \frac{W_B^2}{2D_{nB}} \frac{2kT}{\Delta E_{g,SiGe}} \left\{1 - \frac{kT}{\Delta E_{g,SiGe}} \left[1 - \exp\left(\frac{-\Delta E_{g,SiGe}}{kT}\right)\right]\right\}. \end{aligned} \tag{7.38}$$

The base transit time of a Si-base transistor with boxlike base doping profile is given by Eq. (7.24), i.e.,

$$t_B(\text{Si}) \approx \frac{W_B^2}{2D_{nB}}. \tag{7.39}$$

Therefore, the ratio of the base transit times is

$$\frac{t_B(\text{SiGe})}{t_B(\text{Si})} = \frac{2kT}{\Delta E_{g,SiGe}} \left\{1 - \frac{kT}{\Delta E_{g,SiGe}} \left[1 - \exp\left(\frac{-\Delta E_{g,SiGe}}{kT}\right)\right]\right\}. \tag{7.40}$$

For a total base bandgap narrowing of 100 meV, the low-current base transit time of a SiGe-base transistor is about 2.5 times smaller than that of a Si-base transistor with the same base doping profile.

- *Effect of linearly graded base bandgap on emitter delay time.* It will be shown in Chapter 8 that the cutoff frequency $f_T$ of a bipolar transistor is limited by the sum of the transit times, of which the base transit time, $t_B$, and the emitter delay time, $t_E$, are two of the components. It will also be shown in Chapter 8 [see Eq. (8.16)] that $t_E$ is inversely proportional to the current gain. Thus, with a significantly larger current gain [cf. Eq. (7.30)], a SiGe-base transistor has a much smaller emitter delay time than a Si-base transistor of the same emitter design. The combination of small $t_B$ and small $t_E$ allows a SiGe-base transistor to have a cutoff frequency that is substantially higher than a Si-base transistor (Patton *et al.*, 1990).

## 7.3 DESIGN OF THE COLLECTOR REGION

The cross section of the physical structure of a modern n–p–n bipolar transistor is illustrated schematically in Fig. 7.5. The collector includes all the n-type regions underneath and surrounding the p-type base. It can be subdivided into four parts. The part directly underneath the emitter and intrinsic base (shaded in Fig. 7.5) is the active region of the collector. This region is usually referred to simply as the *collector*. It is the region referred to in all the transport and current equations in this book. The horizontal heavily doped ($n^+$) region underneath the collector is called the *subcollector*, and the heavily doped ($n^+$) vertical region connecting the subcollector to the collector terminal on the silicon surface is called the *reach-through*. The remaining n-type regions make up the parasitic collector, which is usually relatively lightly doped ($n^-$) in order to minimize the total base–collector junction capacitance.

To first order, both the subcollector and the reach-through are there only to reduce the series resistance between the collector terminal and the active collector. However, as will be discussed later, the proximity of the subcollector to the intrinsic base, that is, the thickness of the active collector layer, has a strong effect on the base–collector breakdown voltage and on the collector current characteristics at high current densities.

Just like the extrinsic base, the parasitic collector is an unavoidable part of a bipolar transistor structure. In general, designing the parasitic collector is very simple: the parasitic-collector area and its associated capacitance should be as small as possible. As can be seen from Fig. 7.5, the parasitic collector and the extrinsic base form a p–n diode. Therefore, a bipolar technology that gives a small extrinsic-base area will have a small parasitic-collector area as well. For a given extrinsic-base area, the parasitic-collector doping concentration should be as low as possible, in order to achieve the smallest capacitance and the largest breakdown

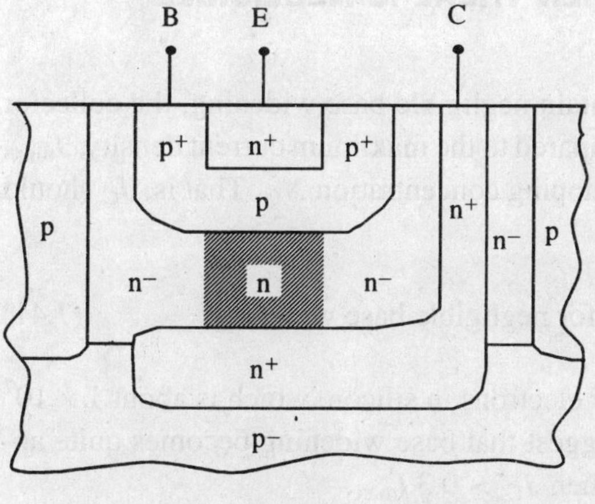


**FIGURE 7.5.** Schematic cross section of a modern n–p–n bipolar transistor, illustrating the various doped regions. This transistor employs p-type regions for isolation.

voltage for the extrinsic-base–collector diode. Interested readers are referred to the vast literature on the research and development of bipolar technology, which describes methods for reducing the extrinsic-base area and reducing the parasitic-collector doping concentration (Warnock, 1995; Nakamura and Nishizawa, 1995), and to Appendix 2 for the outline of a process for making a commonly used modern n–p–n transistor.

The collector (we shall use the terms collector and active collector interchangeably whenever there is no confusion) and the subcollector are the only regions that affect the intrinsic characteristics of a bipolar transistor. Therefore, these are the only regions that will be discussed further in this section.

For bipolar transistors operated in the forward-active mode, i.e., with the base–collector junction reverse-biased at all times, the collector acts simply as a sink for the carriers injected from the emitter and traversing the base layer. As discussed in Sections 6.3.2 and 6.3.3, how the collector and subcollector affect the collector current depends on whether or not the collector current density is large enough to cause significant base widening. Thus, the design of the collector will be discussed in two parts, one when base widening is negligible, and the other when base widening is significant. (The collector parameters have no effect on the base current, which, as shown in Section 6.2.2, depends only on the emitter parameters.)

It should be pointed out that one of the most important trends in VLSI technology development is the continued miniaturization of devices while simultaneously increasing their operating current densities. For bipolar technology, emitter areas of much less than 1 μm$^2$ can be fabricated readily today, while device currents of 0.5 mA and larger are desired in many bipolar circuits. That is, the collector current densities in many modern bipolar transistors can easily exceed 0.5 mA/μm$^2$. At these high current densities, ***base widening can easily occur***, and special attention should be paid to the design of the collector and subcollector to minimize its effects.

### 7.3.1 COLLECTOR DESIGN WHEN THERE IS NEGLIGIBLE BASE WIDENING

As discussed in Section 6.3.3, to maintain negligible base widening, the collector current density $J_C$ should be small compared to the maximum current density, $J_{\max}$, that can be supported by the collector doping concentration $N_C$. That is, $J_C$ should satisfy the condition that

$$|J_C| \ll J_{\max} = q v_{sat} N_C \qquad \text{for negligible base widening,} \tag{7.41}$$

where $v_{sat}$ is the saturated velocity for electrons in silicon, which is about $1 \times 10^7$ cm/s (see Fig. 2.9). Published data suggest that base widening becomes quite appreciable in modern npn transistors when $J_C > 0.3 J_{\max}$.

For $N_C = 1 \times 10^{16}$ cm$^{-3}$, one has $J_{max} = 0.16$ mA/μm$^2$, and the allowed $J_C$ is only about 0.05 mA/μm$^2$, which is much too small for the modern bipolar devices. To increase the collector current density without increasing base-widening effect, $N_C$ must be increased proportionately. However, as $N_C$ is increased, the base–collector junction capacitance is increased, and other device characteristics, such as base–collector junction avalanche, can be adversely affected. Therefore, ***tradeoffs have to be made in the design of the collector***. These design tradeoffs are discussed below.

#### 7.3.1.1 TRADEOFF IN EARLY VOLTAGE

The Early voltage of a bipolar transistor is inversely proportional to the base–collector junction depletion-layer capacitance, $C_{dBC}$ [cf. Eq. (6.71)]. As $N_C$ is increased to allow a larger collector current density, $C_{dBC}$ is increased and $V_A$ will decrease. Therefore, there is a tradeoff between the current-density capability of a transistor and its Early voltage.

#### 7.3.1.2 TRADEOFF IN BASE–COLLECTOR JUNCTION AVALANCHE EFFECT

As discussed in Section 6.3.2, base–collector junction avalanche occurs when the electric field in the junction space-charge region becomes too large. Excessive base–collector junction avalanche can cause the base and collector currents to increase out of control and hence can affect the functionality of the circuits using these transistors. Indeed, when base–collector junction avalanche runs away, device breakdown occurs. Modern VLSI circuits typically operate with a power supply voltage of 3.3 or 5 V. These voltages are sufficiently high that significant base–collector junction avalanche can easily occur in modern bipolar devices, unless care has been taken in the collector design to minimize it (Lu and Chen, 1989).

There are several ways to reduce avalanche multiplication in the base–collector junction. The most straightforward way is to reduce $N_C$, but that will proportionately reduce the allowed collector current density. Alternatively, the base and/or the collector doping profiles, at or near the base–collector junction, can be designed to reduce the electric field in the junction.

Referring to Fig. 7.2, the Gaussian base doping profile, with its graded dopant distribution near the base–collector junction, has a lower electric field in the base–collector junction than the boxlike base doping profile. In practice, ion implantation of boron usually results in an exponential tail in the base doping profile, as can be seen from Fig. 6.2. This tail is caused by a combination of channeling effect during ion implantation and defect-induced enhanced-diffusion effect during postimplantation thermal annealing. The ion-implanted base profile is therefore always graded.

If the intrinsic base is formed by epitaxial growth and is doped *in situ*, its doping profile can be much more boxlike. For the same collector doping profile, such a base doping profile will result in a larger electric field in the base–collector junction. However, this does not imply that a graded base doping profile is preferred over a boxlike profile. This point will be discussed further in Chapter 8 in connection with the optimization of a device design.

The collector doping profile can also be retrograded (i.e., graded with its concentration increasing with distance into the silicon) to reduce the electric field in the base–collector junction (Lee *et al.,* 1996). Retrograding of the collector doping profile can be achieved readily by high-energy ion implantation. The transistor doping profiles illustrated in Fig. 6.2 show collectors with retrograded doping profiles.

Qualitatively, grading the base doping profile, and/or retrograding the collector doping profile, is similar to sandwiching an i-layer between the base and collector doped regions. Introducing a thin i-layer between the p- and n-regions of a diode is quite effective in reducing the electric field in the junction, as discussed in Section 2.2.2.

Reducing base–collector junction avalanche, either by reducing the collector doping concentration or by grading the base doping profile and/or retrograding the collector doping profile, reduces the base–collector junction depletion-layer capacitance as well. This should help to improve the device and circuit performance (Lee *et al.*, 1996). However, as can be seen from Eqs. (6.81) and (6.82), these techniques for reducing the base–collector junction capacitance also lead to more base widening, or to base widening occurring at a lower collector current density. Thus, reducing base–collector junction avalanche can reduce the current-density capability, and hence the maximum speed, of a bipolar transistor (Lu and Chen, 1989). The tradeoff between base–collector junction avalanche effect and device and circuit speed will be discussed further in Chapter 8.

### 7.3.2 COLLECTOR DESIGN WHEN THERE IS APPRECIABLE BASE WIDENING

As mentioned earlier, the operating current densities of a modern bipolar transistor could easily be in excess of 0.5 mA/$\mu$m$^2$, if the base-widening effect were not a concern. Unfortunately, at these high current densities, base widening does occur. The challenge in designing the collector when base widening is unavoidable is to minimize the deleterious effects of base widening.

As shown in Section 6.3.3, when base widening occurs, there are excess minority carriers stored in the collector, and, as shown in Section 6.4.4, these excess minority carriers contribute to the emitter diffusion capacitance. As will be shown in Chapter 8, when a bipolar transistor is operated with significant base widening, it is its emitter diffusion capacitance that limits its circuit speed and cutoff frequency. To minimize emitter diffusion capacitance, the total excess minority carriers stored in the collector should be minimized. To accomplish this goal, in addition to retrograding the collector doping profile as discussed in the previous subsection, the total collector volume available for minority-carrier storage should also be minimized. That is, the thickness of the collector layer should be minimized. This is easily accomplished by reducing the thickness of the epitaxial layer grown after the subcollector region is formed (see Appendix 2).

However, thinning the collector can lead to an increase in the base–collector junction depletion-layer capacitance, if the collector thickness is comparable to the base–collector depletion-layer width. Thus, when operated at low current densities, where base widening is negligible, a circuit using thin-collector transistors could run slower than a circuit using thick-collector transistors. However, at high current densities, circuits with thin-collector transistors often run faster than circuits with thick-collector transistors (Tang *et al.*, 1983). Also, when the collector–base space-charge layer extends all the way to the subcollector, base–collector junction avalanche will increase, and the base–collector junction breakdown voltage will decrease.

Designing the collector of a modern bipolar transistor is therefore a complex tradeoff process. The important point to remember is that ***base widening occurs readily in modern bipolar devices, and optimizing the tradeoff in the collector design is key to realizing the maximum performance of these devices.***

## 7.4 MODERN BIPOLAR TRANSISTOR STRUCTURES

After a large research and development effort worldwide in the 1970s and 1980s, bipolar technology has become fairly mature. Techniques for implementing the device design concepts discussed in this chapter have been developed and in many cases implemented. The most widely used bipolar technology today is probably the deep-trench-isolated, double-polysilicon, self-aligned bipolar technology (Ning *et al.*, 1981; Chen *et al.*, 1989) and variations of it. This device structure is illustrated schematically in Fig. 7.6. The process flow for fabricating this transistor structure

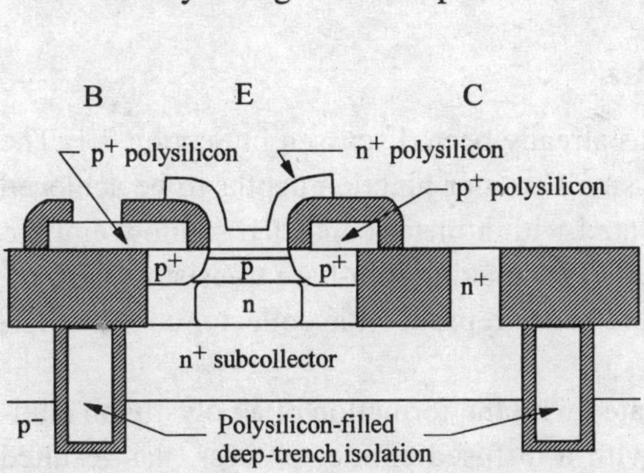


**FIGURE 7.6.** Schematic illustrating the structure of the commonly used bipolar transistor and its salient features.

is outlined in Appendix 2. The salient features of this device are: (a) deep-trench isolation between adjacent transistors, (b) a polysilicon emitter, (c) a polysilicon base contact, which is self-aligned to the emitter, and (d) a pedestal collector, i.e., the collector region directly underneath the emitter is more heavily doped than its surrounding regions. Also, for analog circuit applications, the SiGe-base transistor is particularly advantageous (Iyer *et al.*, 1987; Patton *et al.*, 1990). The basic design concept for each of these features is discussed below.

### 7.4.1 DEEP-TRENCH ISOLATION

The isolation region must be deep enough to isolate the subcollectors of adjacent transistors. Prior to the advent of deep-trench isolation, p-type diffusion regions were used to isolate subcollectors, as illustrated in Fig. 7.5. A p-type diffusion-isolation region is typically as wide as it is deep. This is because as the p-type impurities diffuse downward, they also diffuse laterally. Furthermore, to minimize junction capacitance and to avoid excessively low collector–substrate junction breakdown voltage, the p-type diffusion-isolation regions should not butt against the heavily doped n-type subcollector regions. There is usually an $n^-$ region between a p-type isolation and a subcollector or reach-through, as illustrated in Fig. 7.5. As a result, the total silicon area taken up by the diffusion-isolation regions is very large. The area of a diffusion-isolated bipolar transistor is completely dominated by its isolation.

Replacing a diffusion isolation by deep-trench isolation reduces very significantly the area taken up by isolation. The horizontal dimension of the deep trenches is usually defined by lithography. Using oxide and polysilicon filling, deep trenches less than 1.0 μm wide can be made readily. Furthermore, the deep trenches can cut right through the subcollector layer, resulting in much smaller collector–substrate capacitance than with p-type diffusion isolation.

The diffusion-isolation process is less complex, and hence costs less, than the trench-isolation process. For reasons of cost, diffusion isolation is still used in many bipolar products.

### 7.4.2 POLYSILICON EMITTER

The benefit of polysilicon emitters has already been discussed in Section 7.1. The polysilicon emitter allows extremely small emitter-junction depths to be achieved without the large base current associated with a metal-contacted shallow emitter. Small emitter-junction depths are needed for making thin-base transistors with reproducible base-region parameters, and hence reproducible collector current characteristics.

The very low thermal cycle associated with the formation of a polysilicon emitter, compared with that associated with a diffused-emitter process, has resulted in a drastic reduction in the density of defects called *pipes*, which are localized emitter–collector shorts formed in the intrinsic-base layer. Practically all modern bipolar transistors employ polysilicon-emitter technology.

### 7.4.3 SELF-ALIGNED POLYSILICON BASE CONTACT

As illustrated in Fig. 7.6, instead of being contacted directly by metal, the extrinsic base is contacted indirectly via a layer of p-type polysilicon. Metal contact to the p-type polysilicon is made on top of the field-oxide region. In this way, the extrinsic-base area does not have to accommodate the base metal contact, and hence can be made quite small, resulting in a very small extrinsic-base–collector junction capacitance.

Furthermore, by separating the extrinsic-base polysilicon layer and the emitter polysilicon layer by a thin vertical insulator layer, which is typically 0.1–0.2 μm thick, formed in some nonlithographic self-aligned manner, the extrinsic-base area is further reduced. The ratio of the total collector–base (extrinsic base + intrinsic base) junction area to the emitter–base junction area is typically only 3 : 1 (Ning *et al.*, 1981).

Perhaps more important is the fact that the polysilicon-base technology allows the extrinsic base to be formed independently of the intrinsic base. This decoupling of the intrinsic and the extrinsic base greatly enlarges the intrinsic-base design and process window. It allows thin-base transistors to be made readily. Practically all modern digital bipolar transistors employ polysilicon-base technology, although not necessarily with self-alignment to minimize the extrinsic-base area.

### 7.4.4 PEDESTAL COLLECTOR

The pedestal collector has a higher doping concentration in the active collector directly underneath the intrinsic base than its surrounding regions (Yu, 1971), as illustrated in Fig. 7.5. The high collector doping concentration minimizes the base-widening effect, while the low parasitic-collector doping concentration reduces the total base–collector junction capacitance. A pedestal collector can be achieved quite simply by high-energy ion implantation when the emitter opening is defined in the device fabrication process (see Appendix 2).

### 7.4.5 SiGe-BASE BIPOLAR TRANSISTOR

The benefit of incorporating Ge into the intrinsic base layer has been discussed in Section 7.2.5. The most successful SiGe-base bipolar technology to date uses an ultrahigh-vacuum chemical-vapor-deposition process to form the epitaxially grown, *in situ* doped, intrinsic base (Harame *et al.*, 1995a, b). For maximum device performance, the other device features, such as trench isolation and polysilicon emitter, are usually the same as in a regular Si-base transistor.

## EXERCISES

**7.1** This exercise is designed to show the sensitivity of the current gain to the polysilicon thickness of a polysilicon-emitter transistor. For the polysilicon-emitter model described in Exercise 6.3, the emitter Gummel number is a function of the emitter junction depth $W_E$ and the polysilicon thickness $W_{E1}$, i.e.,

$$G_E(W_E, W_{E1}) = N_E\left(\frac{n_i^2}{n_{ieE}^2}\right)\left(\frac{W_E}{D_{pE}} + \frac{L_{pE1}\tanh(W_{E1}/L_{pE1})}{D_{pE1}}\right).$$

It is reasonable to assume the lifetimes in heavily doped silicon and heavily doped polysilicon to be the same for the same doping concentration, since both are determined by Auger recombination. It is also reasonable to assume the mobility in polysilicon to be smaller than that in silicon, since there is additional grain-boundary scattering in polysilicon.

**(a)** A typical nonpolysilicon emitter has $N_E = 10^{20}$ cm$^{-3}$ and $W_E = 300$ nm, and $W_{E1} = 0$. Estimate $G_E(300 \text{ nm}, 0)$ using the hole mobility and lifetime values in Fig. 2.18(a) and (b).

**(b)** A typical polysilicon emitter has $N_E = 10^{20}$ cm$^{-3}$ and $W_E = 30$ nm. Let us assume the hole mobility in polysilicon to be $\frac{1}{3}$ that in silicon, and assume the hole lifetimes in silicon and polysilicon to be the same. Estimate $G_E(30 \text{ nm}, W_{E1})$ for $W_{E1} = 50$, 100, 200, and 300 nm.

**(c)** Graph the ratio $G_E(30 \text{ nm}, W_{E1})/G_E(300 \text{ nm}, 0)$ as a function of $W_{E1}$.

**7.2** The intrinsic-base sheet resistivity is given by Eq. (7.5), namely

$$R_{Sbi} = \left(q\int_0^{W_B} p_p(x)\mu_p(x)\,dx\right)^{-1}.$$

Most n–p–n transistors have a low-current $R_{Sbi}$ value of about $10^4$ Ω/□. Assuming a boxlike base doping profile, graph $N_B$ as a function of $W_B$ for $W_B$ between 50 and 300 nm for $R_{Sbi} = 10^4$ Ω/□. This graph illustrates how the base doping concentration varies with the intrinsic-base width in scaling in many practical device designs.

**7.3** The base transit time for a box profile is given by

$$t_B \approx \frac{W_B^2}{2D_{nB}}.$$

For a fixed intrinsic-base sheet resistivity of $10^4$ Ω/□, calculate and plot $t_B$ as a function of $W_B$ for $W_B$ between 50 and 300 nm. Use the mobility and lifetime values in Fig. 2.18(a) and (b) to estimate $D_{nB}$. (See Exercise 7.2 for the relation between $W_B$ and $N_B$.)

**7.4** This exercise illustrates the tradeoff between the collector current density and the Early voltage in an optimized bipolar device design. The Early voltage for a boxlike base doping profile is given by $V_A = qN_BW_B/C_{dBC}$, where $C_{dBC}$ is the base–collector junction depletion-layer capacitance per unit area [cf. Eq. (6.72)]. To maintain negligible base widening in scaling, we assume the collector current density is maintained at $J_C = 0.3qv_{sat}N_C$, where $v_{sat} = 10^7$ cm/s is the electron saturated velocity. Thus, as $N_C$ is increased to increase $J_C$, $C_{dBC}$ is increased and $V_A$ is decreased.

**(a)** Use the one-sided junction approximation for the base–collector diode, and assume $V_{CB} = 2$ V (for purposes of calculating $C_{dBC}$). Plot $C_{dBC}$ as a function of $J_C$ for $J_C$ between 0.1 and 5 mA/μm$^2$.

**(b)** For a base design with $qN_BW_B = 1.6 \times 10^{-6}$ C/cm$^2$ (e.g., $N_B = 10^{18}$ cm$^{-3}$ and $W_B = 100$ nm), estimate and plot $V_A$ as a function of $J_C$ for $J_C$ between 0.1 and 5 mA/μm$^2$.

**7.5** Consider an npn transistor with a wide base of $W_B = 500$ nm. Suppose the base doping concentration is linearly graded, i.e., $N_B$ has the form $N_B(x) = A - \alpha x$, with $N_B(0) = 2 \times 10^{17}$ cm$^{-3}$ and $N_B(W_B) = 2 \times 10^{16}$ cm$^{-3}$. For such light doping concentrations, the effect of heavy doping is negligible. Plot the built-in electric field due to the dopant distribution as a function of distance between $x = 0$ and $x = W_B$.

**7.6** Consider an npn transistor with a linearly graded base doping profile (cf. Exercise 7.5). The doping concentration at the emitter–base junction is $N_B(0) = 5 \times 10^{18}$ cm$^{-3}$. The doping concentration at the base–collector junction is $N_B(W_B) = 5 \times 10^{17}$ cm$^{-3}$, and $W_B = 100$ nm. For such high doping concentrations, the heavy-doping effect cannot be ignored. Plot the electric fields due to the dopant distribution and due to the heavy-doping effect, as well as the total electric field, as a function of distance from $x = 0$ to $x = W_B$. [Use the bandgap-narrowing parameter in Eq. (6.18).]

**7.7** The emitter series resistance $r_e$ of a polysilicon-emitter n–p–n transistor, with negligible polysilicon–silicon interface oxide, has three components, namely, the resistance due to the single-crystal emitter region, the resistance due to the polysilicon layer, and the resistance due to the metal–polysilicon contact. Consider a polysilicon emitter with $N_E = 10^{20}$ cm$^{-3}$, a single-crystal region of width $W_E = 30$ nm, a polysilicon layer of thickness $W_{E1} = 200$ nm, and a metal–polysilicon contact resistivity of $2 \times 10^{-7}$ Ω-cm$^2$. Assume that, for the same doping concentration, the resistivity of polysilicon is 3 times that of single-crystal silicon. Calculate the series resistance components, as well as the total series resistance $r_e$, for an emitter 1 μm$^2$ in area. (Use the resistivity for n-type silicon shown in Fig. 2.8.)

**7.8** The incorporation of a linearly retrograded amount of Ge into the intrinsic base increases the current gain and the Early voltage and reduces the base transit time, according to [see Eqs. (7.30), (7.35), and (7.40)]

$$\frac{\beta(\text{SiGe})}{\beta(\text{Si})} = \frac{\Delta E_{g,SiGe}/kT}{1 - \exp(-\Delta E_{g,SiGe}/kT)},$$

$$\frac{V_A(\text{SiGe})}{V_A(\text{Si})} = \frac{kT}{\Delta E_{g,SiGe}}\left[\exp\left(\frac{\Delta E_{g,SiGe}}{kT}\right) - 1\right],$$

and

$$\frac{t_B(\text{SiGe})}{t_B(\text{Si})} = \frac{2kT}{\Delta E_{g,SiGe}}\left\{1 - \frac{kT}{\Delta E_{g,SiGe}}\left[1 - \exp\left(\frac{-\Delta E_{g,SiGe}}{kT}\right)\right]\right\}.$$

Plot these ratios as a function of $\Delta E_{g,SiGe}/kT$ for values of $\Delta E_{g,SiGe}/kT$ between 0 and 10.

**7.9** In the literature, the heavy-doping effect in the emitter is well recognized, but in the base it is often ignored. The saturated collector current density for an n–p–n transistor is [see Eq. (6.33)]

$$J_{C0} = \frac{q}{\int_0^{W_B} \frac{p_p(x)}{D_{nB}(x) n_{ieB}^2(x)}\, dx}.$$

Assume a uniformly doped base with $N_B = 10^{18}$ cm$^{-3}$ and $W_B = 100$ nm. Also assume low current levels so that $p_p = N_B$. Estimate $J_{C0}$ for the following two cases: (a) the heavy-doping effect in the base is neglected, i.e., $n_{ieB} = n_i$, and (b) the heavy-doping effect in the base is included. (This exercise demonstrates that heavy doping in the intrinsic base of modern bipolar transistors cannot be ignored.)

# 8 BIPOLAR PERFORMANCE FACTORS

In Chapter 7, the design of the individual regions and parameters of a bipolar transistor was discussed. It was noted that, during device operation, an individual device region is not isolated from and independent of the other device regions. Optimization of one device parameter often adversely affects the other device parameters. Thus, ***optimization of the design of a bipolar transistor is a tradeoff process***. This design tradeoff should be done at the circuit and/or chip level, for the optimum design of a transistor is a function of its application and environment. In this chapter, we will first discuss soine figures of merit for evaluating a bipolar transistor for typical digital and analog circuit applications, and then discuss the tradeoffs in the design of a bipolar transistor for these applications.

When we consider the performance of a circuit, the wires connecting the transistors and elements that make up the circuit and connecting the output of the circuit to the input of another circuit must be included. The resistance and capacitance as well as the signal propagation delays associated with the interconnect wires have been discussed in Section 5.2.4 in connection with CMOS circuits. The reader is referred to that subsection for details. In this chapter, the wire capacitance which acts as a load on a bipolar circuit is included when we consider the performance and optimization of bipolar transistors and circuits.

In practice, the choice of a particular device design point is often dictated by many nontechnical factors. These factors include cost, time to market, production volume, etc. They will not be considered here.

## 8.1 FIGURES OF MERIT OF A BIPOLAR TRANSISTOR

It is often desirable to consider the merit of a transistor in terms of some simple, and preferably readily measurable, parameters. However, it is important to note that the relevance or significance of a particular figure of merit depends on the application. Some of the commonly used figures of merit are discussed here.

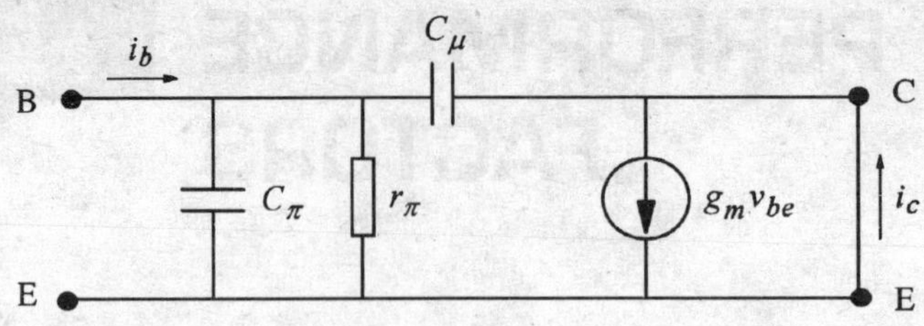


**FIGURE 8.1.** Small-signal equivalent circuit for determining the cutoff frequency of a bipolar transistor. Parasitic resistances are neglected.

### 8.1.1 CUTOFF FREQUENCY

For small-signal applications, the *cutoff frequency*, or *transition frequency*, or *unity-current-gain frequency*, $f_T$, is probably the most often used figure of merit of a bipolar transistor. It is defined as the transition frequency at which the common-emitter, short-circuit-load, small-signal current gain drops to unity. It is a measure of the maximum useful frequency of a transistor when it is used as an amplifier. For simplicity, we shall first neglect the internal parasitic resistances and use the equivalent circuit shown in Fig. 6.19 to determine the cutoff frequency. With the output shorted, $r_0$ and $C_{dCS}$ have no influence, and the resulting equivalent circuit is shown in Fig. 8.1. From this equivalent circuit, the small-signal collector and base currents can be written as

$$i_c = g_m v_{be} - j\omega C_\mu v_{be} \tag{8.1}$$

and

$$i_b = \left(\frac{1}{r_\pi} + j\omega C_\pi + j\omega C_\mu\right) v_{be}, \tag{8.2}$$

where $g_m$ is the transconductance, $r_\pi$ is the input resistance, $C_\mu = C_{dBC}$ is the base–collector junction depletion-layer capacitance, $C_\pi = C_{dBE} + C_{DE}$ is the sum of the base–emitter junction depletion-layer capacitance and the emitter diffusion capacitance, $v_{be}$ is the applied small-signal input voltage, and $i_b$ and $i_c$ are the small-signal base and collector currents due to $v_{be}$. (The reader is referred to Section 6.4.3 for the derivation the of the small-signal equivalent-circuit model of a bipolar transistor.) The small-signal frequency-dependent common-emitter current gain is

$$\beta(\omega) = \frac{i_c}{i_b} = \frac{g_m - j\omega C_\mu}{(1/r_\pi) + j\omega(C_\pi + C_\mu)}. \tag{8.3}$$

In the low-frequency limit, Eq. (8.3) gives $\beta(\omega \to 0) = g_m r_\pi = \beta_0$, which is simply the static common-emitter current gain.

It is instructive to compare the magnitudes of the terms in the numerator of Eq. (8.3). For a typical modern bipolar transistor with a collector–base junction area of 10 $\mu m^2$, a collector doping concentration of $1 \times 10^{16}$ cm$^{-3}$, and a base–collector bias of 5 V, the value of $C_\mu$ is about 1 fF. At a frequency of 10 GHz, $\omega C_\mu \approx 6 \times 10^{-5}\ \Omega^{-1}$. The transconductance is proportional to $I_C$. For a collector current of 1 mA, $g_m \approx 4 \times 10^{-2}\ \Omega^{-1}$. That is, for most frequencies and currents of interest, $\omega C_\mu \ll g_m$, and hence Eq. (8.3) reduces to

$$\beta(\omega) \approx \frac{g_m r_\pi}{1 + j\omega r_\pi (C_\pi + C_\mu)}. \tag{8.4}$$

At high frequencies, the imaginary part in the denominator dominates, and Eq. (8.4) gives

$$\beta(\omega) \approx \frac{g_m}{j\omega(C_\pi + C_\mu)} \qquad \text{(high frequencies)}. \tag{8.5}$$

Therefore, the common-emitter current gain drops to unity at a frequency $f_T$ given by

$$1 = \frac{g_m}{2\pi f_T (C_\pi + C_\mu)}, \tag{8.6}$$

or

$$2\pi f_T = \frac{g_m}{C_\pi + C_\mu}. \tag{8.7}$$

Substituting Eqs. (6.99), (6.102), (6.103), and (6.117) into Eq. (8.7), we obtain

$$\frac{1}{2\pi f_T} = \tau_F + \frac{kT}{qI_C}(C_{dBE} + C_{dBC}) \qquad \text{(resistances neglected)}, \tag{8.8}$$

where $\tau_F$ is the forward transit time given by Eq. (6.114), and $I_C$ is the collector current. This is the commonly used expression for the cutoff frequency. It is often used to determine the low-current value of $\tau_F$. This is done by plotting the measured values of $1/f_T$ as a function of $1/I_C$. Figure 8.2 is an illustration of such a plot. At low currents, $1/f_T$ varies linearly with $1/I_C$. Equation (8.8) suggests that extrapolation of the linear portion of $1/f_T$ to $1/I_C = 0$ gives $2\pi\tau_F$. However, at large currents, the measured $1/f_T$ increases very rapidly as the current is increased. This rapid rise is due to base-widening effects and will be discussed in Section 8.5.1. This is also illustrated in Fig. 8.2.

If the internal parasitic resistances are included, then the equivalent circuit shown in Fig. 6.20 should be used. In this case, the same analysis gives (see

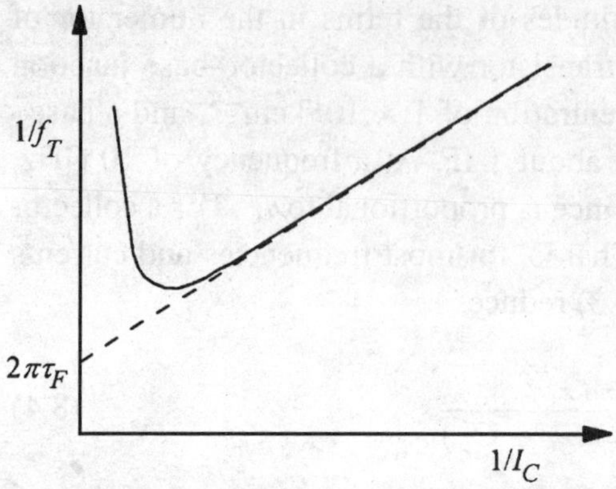


**FIGURE 8.2.** Schematic illustration of a $1/f_T$-versus-$1/I_C$ plot. The extrapolated intercept at $1/I_C = 0$ can be used to determine $\tau_F$.

Exercise 8.1)

$$\frac{1}{2\pi f_T} = \tau_F + \frac{kT}{qI_C}(C_{dBE} + C_{dBC}) + C_{dBC}(r_e + r_c), \tag{8.9}$$

where $r_c$ is the collector series resistance and $r_e$ is the emitter series resistance. In this case, extrapolation of the linear portion of a $1/f_T$-versus-$1/I_C$ plot to $1/I_C = 0$ gives $2\pi[\tau_F + C_{dBC}(r_e + r_c)]$.

In the literature, the peak $f_T$ is often also quoted for a transistor designed for large-signal digital circuit applications. The sensitivity of the performance of a digital bipolar circuit on the peak $f_T$-value of its transistors will be covered in Section 8.3.3.

### 8.1.2 MAXIMUM OSCILLATION FREQUENCY

The cutoff frequency is certainly a good indicator of the low-current forward transit time. However, as a figure of merit, it does not include the effects of base resistance, which are very important in determining the transient response of a bipolar transistor. Consequently, other figures of merit have been proposed and discussed in the literature (Taylor and Simmons, 1986). One that is relatively simple and commonly used is the *maximum oscillation frequency*, $f_{max}$, which is the frequency at which the unilateral power gain becomes unity. It is given by (Pritchard, 1955; Thornton *et al.*, 1966; Roulston, 1990)

$$f_{max} = \left(\frac{f_T}{8\pi r_b C_{dBC}}\right)^{1/2}, \tag{8.10}$$

where $r_b$ is the base resistance.

The important point is that both $f_T$ and $f_{max}$ should be considered only as qualitative indicators of the frequency response of a transistor. There are many other elements that can affect the performance of a transistor, and the magnitude of the effect depends on the circuit application and on the design point of the transistor, which will be discussed further later.

### 8.1.3 RING OSCILLATOR AND GATE DELAY

For large-signal digital- or logic-circuit applications, neither $f_T$ nor $f_{\max}$ is really a good indicator of device performance (Taylor and Simmons, 1986). For a digital circuit, the gate delay itself is often used as a figure of merit for the transistors in the circuit. (For digital circuits, the terms "circuit" and "gate" are used interchangeably.)

Since the merit of a transistor is reflected in the switching speed of the circuit in which the transistor is used, the merit of a transistor therefore depends on the circuit and its design point. That is, ***a transistor optimized for one circuit and its design point may not be optimum for another design point, and certainly not for another circuit***. For high-performance logic applications, the most commonly used bipolar circuit is *the emitter-coupled logic* or *ECL* circuit. Most publications on digital bipolar transistor technology quote the measured or modeled ECL gate delay, often with negligible external capacitance loading, as a figure of merit. In this chapter, an ECL gate is used to illustrate the optimization of a bipolar transistor for digital-circuit applications.

As explained in Section 5.3.1, the switching speed of a logic gate can be measured very easily from a ring-oscillator arrangement of the circuit. For almost all logic circuits, a ring oscillator consists of an odd number of stages of the logic gate connected with the output of one stage feeding the input of the next stage, and the output of the last stage feeding the input of the first stage, thus forming a ring configuration. The average switching delay of a circuit can be measured directly by measuring the period of its ring-oscillator waveform. This average delay is equal to $P/2n$, where $P$ is the period, and $n$ the number of stages, of the ring oscillator. However, for some circuits, such as a bipolar ECL circuit (see next section), which have both an inverted output and a noninverted output, a ring oscillator can be formed by using an even number of stages. In this case, the inverted outputs from one half of the stages and the noninverted outputs from the other half of the stages are used to form the ring. The average stage delay is still given by $P/2n$.

## 8.2 DIGITAL BIPOLAR CIRCUITS

An ECL gate with fan-in of 1 and fan-out of 1 is shown in Fig. 8.3. Both the inverted output $V_{out}$ and the noninverted output $\bar{V}_{out}$ are shown. In this circuit configuration, the voltage $V_S$ and the resistor $R_S$ together set the *switch current* $I_S$ of the ECL gate. This current is constant, i.e., it does not change when the circuit switches. The two resistors $R_L$ are the load resistors of the gate. The capacitor $C_L$ represents the total external load capacitance connected to the output of the gate. The two resistors $R_E$ together with transistors $Q_3$ and $Q_5$ form the two emitter followers. (In an emitter follower, the emitter voltage follows the base voltage. Thus, if the

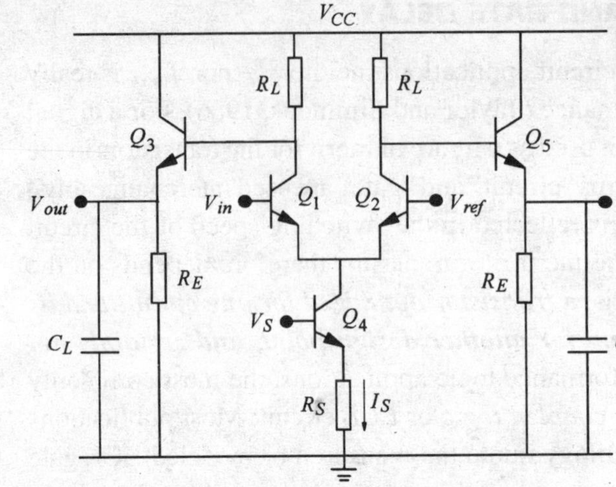


**FIGURE 8.3.** Schematic of an emitter-coupled logic gate for fan-in = fan-out = 1 and an output capacitance loading of $C_L$. Both the inverting and the noninverting outputs are shown.

base voltage of $Q_3$ goes up, the emitter voltage of $Q_3$ also goes up by the same amount.) The input voltage $V_{in}$ and the output voltages $V_{out}$ and $\bar{V}_{out}$ swing above and below a fixed reference voltage $V_{ref}$, usually approximately symmetrically, by one-half of the logic swing $\Delta V$.

When $V_{in}$ is high, i.e., when $V_{in} = V_{ref} + \Delta V/2$, transistor $Q_1$ is turned on much harder than the reference transistor $Q_2$. As a result, the switch current $I_S$ flows mainly through transistor $Q_1$ and its load resistor $R_L$. The *IR* drop across this load resistor in turn lowers the base voltage of transistor $Q_3$. The output voltage $V_{out}$ follows the base voltage of $Q_3$ and hence becomes low. At the same time, with negligible current flowing through transistor $Q_2$ and its load resistor $R_L$, the base voltage of transistor $Q_5$ is pulled up to a high voltage. The output voltage $\bar{V}_{out}$ follows the base voltage of $Q_5$ to high. Thus, $V_{out}$ is inverting, while $\bar{V}_{out}$ is noninverting. A similar analysis shows that when $V_{in}$ is switched to low, i.e., when $V_{in} = V_{ref} - \Delta V/2$, $V_{out}$ is switched to high and $\bar{V}_{out}$ is switched to low.

When the gate switches, the switch current $I_S$ is steered, or switched, from one load resistor to the other. (For its current-switching characteristics, an ECL gate is sometimes called a *current-switch emitter-follower circuit*.) The logic swing $\Delta V$ is equal to the *IR* drop in one of the load resistors, i.e., $\Delta V = I_S R_L$.

### 8.2.1 DELAY COMPONENTS OF A LOGIC GATE

It has been shown (Tang and Solomon, 1979; Chor *et al.*, 1988) that the switching delay of a bipolar logic gate can be expressed as a linear combination of all the time constants of the circuit, with each time constant weighed by a factor that is determined by the detailed arrangement of the circuit. For the ECL gate depicted in Fig. 8.3, the switching delay can therefore be written as

$$T_{delay} = \sum_i K_i R_i C_i + \sum_j K_j \tau_j, \qquad (8.11)$$

where the first sum is over all the resistances and capacitances of the transistors in the circuit, the second sum includes the forward and reverse transit times of the transistors, and $K_i$ and $K_j$ are the corresponding weighing factors. Since the transistors in an ECL circuit are all biased in the forward-active mode (i.e., the emitter–base diodes are zero-biased or forward biased, and the base–collector diodes are zero-biased or reverse biased), the reverse transit times, which are associated with forward-biased base–collector diodes, are zero. Only the forward transit times need to be included in Eq. (8.11).

For simplicity of discussion, it is often assumed that all the transistors in the circuit are the same. In this case, Eq. (8.11) is reduced to (Chor *et al.*, 1988)

$$\begin{aligned} T_{delay} = {} & K_1\tau_F + r_{bi}(K_2C_{dBCi} + K_3C_{dBCx} + K_4C_{dBE}) \\ & + r_{bx}(K_6C_{dBCi} + K_7C_{dBCx} + K_8C_{dBE}) \\ & + R_L(K_{10}C_{dBCi} + K_{11}C_{dBCx} + K_{12}C_{dBE} + K_{13}C_{dCS} + K_{14}C_L) \\ & + r_c(K_{15}C_{dBCi} + K_{16}C_{dBCx} + K_{18}C_{dCS}) \\ & + r_e(K_{19}C_{dBCi} + K_{20}C_{dBCx} + K_{21}C_{dBE} + K_{23}C_{dCS} + K_{24}C_L) \\ & + C_{DE}(K_5r_{bi} + K_9r_{bx} + K_{17}r_c + K_{22}r_e), \end{aligned} \qquad (8.12)$$

where the same numbering system for the $K$-factors as in Chor *et al.* is followed. The internal resistances and capacitances of the transistors are illustrated in Fig. 6.18. The circuit resistances and capacitances are shown in Fig. 8.3. It should be noted that transistor $Q_4$ and resistor $R_S$, functioning only to set the switch current, are not involved in the switching of the circuit and hence do not enter into Eq. (8.12).

In practice, the performance of a bipolar logic gate is often characterized as a function of the operating current of its transistors. Since the power dissipation of a logic gate is proportional to the total current passing through the transistors, the performance of a logic gate can also be characterized as a function of its power dissipation. That is, in principle, the delay-versus-current and the delay-versus-power dissipation characteristics contain the same information, and either one can be used to describe the behavior of the transistors in the circuit. The circuit delay-versus-current or delay-versus-power dissipation characteristics are usually obtained by varying the resistor values in the circuit, keeping the transistor geometries and parameters fixed. In so doing, the collector current density becomes proportional to the collector current. In the published literature, sometimes the circuit delay is plotted as a function of the collector current density, and sometimes simply as a function of

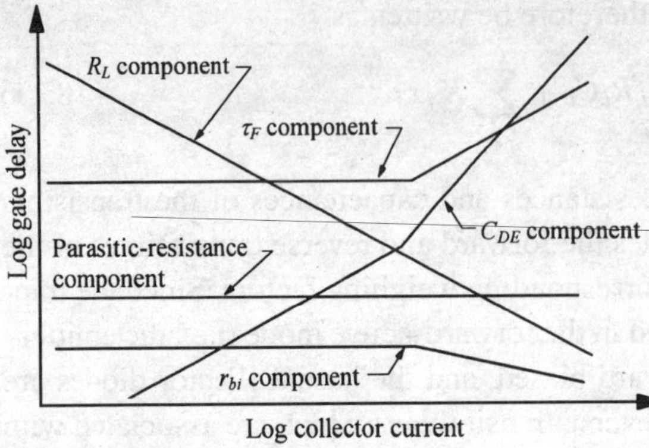


**FIGURE 8.4.** Schematic illustration of the relative magnitudes of the gate delay components of an ECL gate and their dependence on collector current or power dissipation.

the collector current. In any event, the delay-versus-current or delay-versus-power characteristics reflects the performance of a *fixed* transistor design as a function of its collector current density.

The relative magnitudes of the delay components represented in Eq. (8.12) have been evaluated for the modern bipolar transistor shown in Fig. 7.6 (Tang and Solomon, 1979; Chor *et al.*, 1988). The results are illustrated schematically in Fig. 8.4. Each delay component depends on a key device or circuit parameter. In the remainder of this subsection, we analyze each of the delay components qualitatively. Such an analysis is very helpful as a guide to optimizing the device design.

#### 8.2.1.1 TRANSIT-TIME DELAY COMPONENT

The first term in Eq. (8.12) is proportional to the forward transit time $\tau_F$. As shown in Section 6.4.4, at low collector currents (hence low collector current densities), where base widening is negligible, $\tau_F$ is a constant, independent of $I_C$. However, once base widening becomes appreciable, $\tau_F$ increases with $I_C$. Therefore, the transit-time delay component of an ECL gate is expected to be independent of current at low collector currents, but to increase with current once the current density exceeds the base-widening threshold. This is illustrated in Fig. 8.4. Most high-speed digital circuits are designed to have the transit time as one of the dominant delay components (Tang and Solomon, 1979; Chor *et al.*, 1988).

#### 8.2.1.2 INTRINSIC-BASE-RESISTANCE DELAY COMPONENT

The second term in Eq. (8.12) is due to the *RC* time constants associated with the intrinsic-base resistance. At low collector current densities, the intrinsic-base resistance is a constant, independent of current. However, as can be seen from Eq. (7.5) and Appendix 8, once base widening occurs at high collector current densities, the intrinsic-base resistance decreases rapidly with further increase in collector current density. Therefore, the intrinsic-base-resistance delay component of an ECL gate is independent of current as long as base widening is negligible. Once base widening becomes appreciable, this delay component decreases with further increase in current, as illustrated in Fig. 8.4. The base-resistance delay component is usually quite small (Tang and Solomon, 1979; Chor *et al.*, 1988).

#### 8.2.1.3 PARASITIC-RESISTANCE DELAY COMPONENTS

The third, fifth, and sixth terms in Eq. (8.12) are due to the $RC$ time constants associated with the extrinsic-base resistance, the collector resistance, and the emitter resistance, respectively. Since these parasitic resistors, to first order, are all independent of the operating current of a transistor, these delay components are independent of current, as illustrated in Fig. 8.4. The parasitic-resistance delay components are also quite small (Tang and Solomon, 1979; Chor *et al.*, 1988).

#### 8.2.1.4 LOAD-RESISTANCE DELAY COMPONENT

The fourth term in Eq. (8.12) is due to all the $RC$ time constants associated with the load resistors $R_L$. For circuits with a large load capacitance $C_L$, the load-resistance delay component is often dominated by the $R_L C_L$ term. For this reason, the load-resistance delay component is also referred to as the load-capacitance delay component.

Referring to Fig. 8.3, it can be seen that the logic swing $\Delta V$, the switch current $I_S$, and the load resistor $R_L$ are interrelated by

$$R_L = \frac{\Delta V}{I_S}. \tag{8.13}$$

Since the logic swing is fixed, the load-resistance delay component of an ECL circuit is inversely proportional to the switch current. This is illustrated in Fig. 8.4. Most ECL circuits are designed to operate at large currents in order to minimize the load-resistance delay component.

#### 8.2.1.5 DIFFUSION-CAPACITANCE DELAY COMPONENT

The last term in Eq. (8.12) is associated with the emitter diffusion capacitance $C_{DE}$. As shown in Eq. (6.115), the stored charge associated with forward-biasing the emitter–base diode can be written as $Q_{DE} = \tau_F I_C = \tau_F I_S$. For modeling purposes, the emitter diffusion capacitance is often approximated by (stored minority-carrier charge)/(average change in input voltage when the gate changes state) $= 2Q_{DE}/\Delta V$ (Tang and Solomon, 1979; Chor *et al.*, 1988), i.e.,

$$C_{DE} \approx \frac{2\tau_F I_S}{\Delta V}. \tag{8.14}$$

Thus, the diffusion-capacitance delay component of an ECL gate is proportional to the switch current at low currents, where base widening is negligible and $\tau_F$ is independent of current. At high currents where base widening is appreciable, however, $\tau_F$ itself increases with $I_S$, and the diffusion-capacitance delay component increases in proportion to the product $\tau_F I_S$. This is illustrated in Fig. 8.4.

It should be noted that as long as $C_{DE}$ is proportional to $Q_{DE}$, the total gate delay obtained from Eq. (8.12) is independent of the exact approximation used for $C_{DE}$. This is due to the fact that the weighing factors in Eq. (8.12) are obtained from a sensitivity analysis (Tang and Solomon, 1979; Chor *et al.*, 1988). Any "inaccuracy" in the coefficient of proportionality, which is $2/\Delta V$ in Eq. (8.14), is compensated by the corresponding weighing factor obtained from the sensitivity-analysis procedure.

### 8.2.2 DEVICE STRUCTURE AND LAYOUT FOR DIGITAL CIRCUITS

Referring to Eq. (8.12), we see that the $RC$ delay components can be grouped into two categories, namely, delay components associated with the intrinsic-device parameters, and delay components associated with the extrinsic-device and circuit parameters. The intrinsic-device parameters are $\tau_F$, $r_{bi}$, $C_{dBCi}$, $C_{dBE}$, and $C_{DE}$. These parameters determine, or are closely related to, the intrinsic properties of the transistors. Designing the intrinsic parts of a transistor and how the intrinsic-device parameters relate to the device characteristics have already been discussed in Chapter 7.

The extrinsic-device and circuit parameters are just as important as the intrinsic-device parameters in determining the measured device and circuit characteristics. Furthermore, the parasitic emitter and base resistances affect the measured current–voltage characteristics directly, as noted in Section 6.3.1. Thus, the optimal design of the intrinsic-device parameters becomes a function of the extrinsic-device and circuit parameters. For instance, if $C_{dBCx}$ is large compared to $C_{dBCi}$ or $C_{dBE}$, then reducing $C_{dBCi}$ or $C_{dBE}$ is not going to have much effect in improving the speed of a circuit. In general, the extrinsic-device resistance and capacitance of a transistor depend on its physical structure and fabrication process. Therefore, in optimizing the intrinsic-device parameters, we need to specify the device structure and process being used.

It should be noted that the physical structure of a transistor includes its physical layout as well. For the same design of the intrinsic-device parameters, the resultant device characteristics depend on how the transistor is laid out. As an illustration, the plan views of the base–collector diode portion of a non-self-aligned transistor are shown in Fig. 8.5 for two commonly used layouts. Both layouts have the same emitter area and hence have the same intrinsic device characteristics when operated at the same current. The one in (b) with two base contacts has a much larger extrinsic-base–collector junction area, and hence much larger $C_{dBCx}$, than the one in (a) with only one base contact. However, the base current for layout (b) can flow laterally in two directions, while that for layout (a) can flow in only one direction. As shown in Appendix 9, the intrinsic-base resistance for layout (b) is $\frac{1}{4}$ that of layout (a).

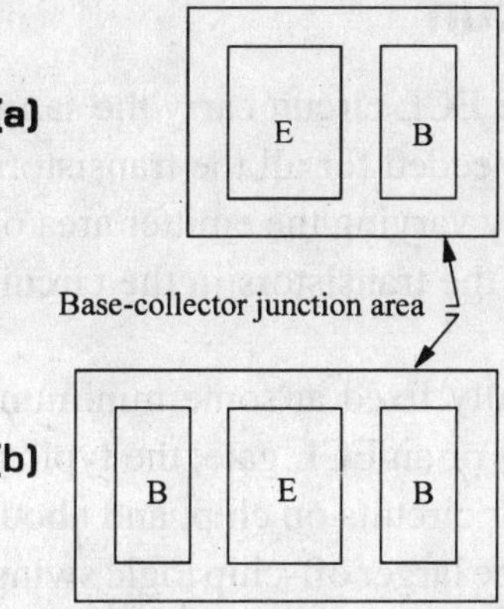


**FIGURE 8.5.** Schematics illustrating the layouts of the base–collector diode region for two bipolar transistors. The transistors are of the usual non-self-aligned type, and both transistors have the same emitter area. Layout (a) has base contact on only one side of the emitter, and layout (b) has base contact on both sides of the emitter.

In general, if a circuit is designed for low power dissipation, or for operation at low collector current densities, layouts such as that shown in Fig. 8.5(b) result in slower circuits because the reduction in base resistance is not enough to compensate for the increase in collector capacitance. For circuits designed to operate at large power dissipation, or for operation at large collector current densities, however, the reduction in base resistance can more than compensate for the increase in collector capacitance. In this case, layout (b) in Fig. 8.5 gives higher circuit speeds than layout (a) (Ranfft and Rein, 1982).

In order to reduce power dissipation and/or delay, most high-speed bipolar transistors now employ a structure such as the one described in Section 7.4, which has near-minimum parasitic resistance and capacitance. To further improve circuit speed and/or reduce power dissipation, the intrinsic-device parameters need to be optimized, as discussed in the next section.

## 8.3 BIPOLAR DEVICE OPTIMIZATION FOR DIGITAL CIRCUITS

In the literature, the switching delay of an ECL gate is often plotted as a function of its power dissipation, or as a function of its switch or collector current. Both the delay-versus-current and the delay-versus-power-dissipation characteristics contain really the same information, since power dissipation is proportional to current. However, it is shown in Chapter 7 that ***the detailed design of a bipolar transistor is closely coupled to its collector current density***. Therefore, in considering the design of a bipolar transistor, we need to translate the delay-versus-power-dissipation, or delay-versus-current, characteristics to delay-versus-collector-current-density characteristics. In this section, we discuss the optimization of a bipolar transistor for ECL circuits by examining the dependence of the dominant ECL delay components on collector current density.

### 8.3.1 DESIGN POINTS FOR A DIGITAL CIRCUIT

For simplicity, we assume that all the transistors in the ECL circuit carry the same current density, and hence only one device design is needed for all the transistors in the circuit. In practice, this can be achieved easily by varying the emitter area of each transistor in proportion to its current, so that all the transistors in the circuit carry the same current density.

The logic swing of a bipolar digital circuit is usually fixed at some minimum value consistent with the noise-margin requirements. For an ECL gate, the typical logic swing is about 400 mV for a circuit driving other circuits on chip, and about 800 to 1000 mV for a circuit driving a signal off chip. The larger off-chip logic swing is due to the noisier off-chip environment. Referring to the circuit configuration in Fig. 8.3, the switch current $I_S$ can be varied readily by adjusting the voltage $V_S$ and/or the resistor $R_S$. To maintain the same logic swing, the load resistors $R_L$ are also varied according to Eq. (8.13).

It should be pointed out that the 400-mV logic swing of a typical ECL gate is much smaller than the logic swing of a CMOS circuit, which is the same as its power-supply voltage $V_{dd}$ (see Section 5.1). $V_{dd}$ for CMOS has been decreasing with device scaling from 5 V towards about 1 V. However, even for CMOS devices with 0.1-μm channel length, $V_{dd}$ is still about 1.2–1.5 V, and is still much larger than the logic swing of an ECL gate. To first order, it is their small logic swings that give bipolar circuits the speed advantage over CMOS circuits for driving heavy load capacitances, such as long wires.

For a given set of transistors in a circuit, as the operating current is changed by changing the resistors, the collector current density of the transistors is changed correspondingly. The expected gate delay as a function of collector current, or current density, is illustrated in Fig. 8.6. For simplicity, only the three dominant delay components in Fig. 8.4, namely the $R_L$ component, the $C_{DE}$ component, and the transit-time component, are shown in this illustration. Three possible design points, $A$, $B$, and $C$, are indicated in Fig 8.6. These design points are discussed further next.

It is clear that design $C$ is to be avoided. At $C$, the gate delay is completely dominated by base widening, which causes both the transit-time and the diffusion-capacitance components to increase rapidly with collector current density. The circuit actually runs slower as additional power is dissipated.

The minimum-delay point is $B$. Here the circuit is running at its maximum speed. For applications where speed is the most important consideration and power dissipation is not a factor at all, this is a reasonable design point.

However, if power dissipation is an important factor, then a low-power design point such as $A$ is preferred. Design $A$ can have a much smaller power–delay product than design $B$. Moving from design $B$ to design $A$, a lot of power can be saved for a small increase in circuit delay. The device designs for points $A$ and $B$ are discussed further next.

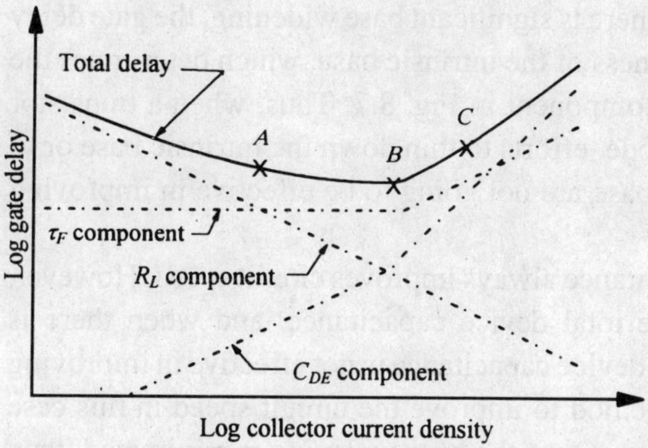


**FIGURE 8.6.** Schematic illustration of the dominant delay components, as well as their sum total, of an ECL gate as a function of collector current density. The design points $A$, $B$, and $C$ are discussed in the text.

## 8.3.2 DEVICE OPTIMIZATION WHEN THERE IS SIGNIFICANT BASE WIDENING

If the load-resistance delay component is large, the gate delay may not reach minimum until the collector current density is so large that there is significant base widening. This is illustrated in Fig. 8.7. To increase the circuit speed further, the transistors should be optimized to reduce base widening.

As discussed in Section 7.3.2, base-widening effects can be reduced by increasing the collector doping concentration and/or reducing the collector layer thickness, provided that the level of base–collector junction avalanche remains acceptable. Alternatively, or concurrently, the transistors can be designed with larger emitter areas to reduce their collector current density. Reducing the collector current density is effective in reducing base widening, as illustrated in Fig. 6.11. Of course, suppressing base widening by increasing the collector doping concentration and/or by increasing the emitter area increases the device capacitance and hence will increase the $R_L$ component of the gate delay. However, often the net result is an increase in circuit speed (Tang and Solomon, 1979).

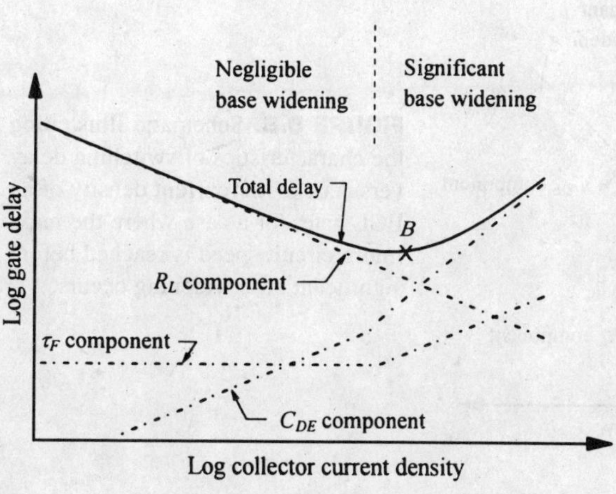


**FIGURE 8.7.** Schematic illustrating the characteristics of switching delay versus collector current density of an ECL gate, for a case where the load capacitance is large. The maximum speed occurs at point $B$, where there is significant base widening.

It should be noted that as long as there is significant base widening, the gate delay is not sensitive to the physical thickness of the intrinsic base, which determines the low-current value of the $\tau_F$ delay component in Fig. 8.7. Thus, when a transistor is operated in the base-widening mode, efforts to thin down the intrinsic base or to incorporate SiGe into the intrinsic base are not going to be effective in improving circuit speed.

In general, reducing device capacitance always improves circuit speed. However, when $C_L$ is large compared to the total device capacitance, and when there is significant base widening, reducing device capacitance is not effective in improving circuit speed. The only effective method to improve the circuit speed in this case is to first minimize the base widening. Once the base widening is minimized, then efforts to reduce the device capacitance may be effective.

### 8.3.3 DEVICE OPTIMIZATION WHEN THERE IS NEGLIGIBLE BASE WIDENING

If the $R_L$ delay component is relatively small, the gate delay can reach its minimum value at a collector current density where base widening is still negligible. This design point is illustrated in Fig. 8.8. To maximize the circuit speed, both the transit-time and the diffusion-capacitance components should be minimized.

Of course, reducing device capacitance also improves circuit speed. In particular, the collector doping concentration can be reduced to be just large enough to maintain negligible base widening, thereby reducing the collector–base junction capacitance. In so doing, the total delay, particularly in the region to the left side of the design point $B$ in Fig. 8.8, can be reduced appreciably (Tang and Solomon, 1979). With the $R_L$ delay component reduced by reducing the device capacitance, the design point $B$ can be moved to a lower collector current density, thus improving the speed of the circuit and/or reducing its power dissipation.

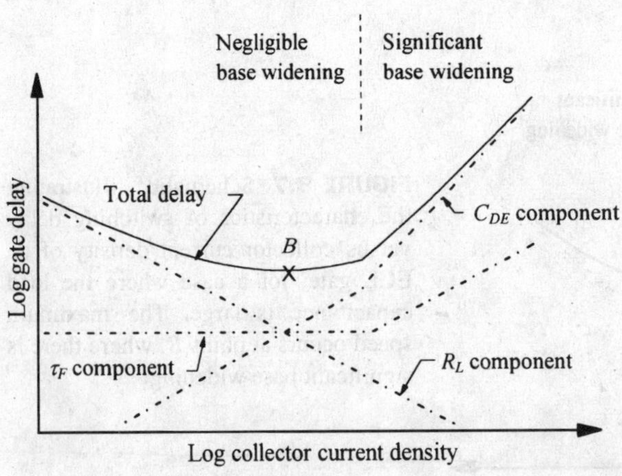


**FIGURE 8.8.** Schematic illustrating the characteristics of switching delay versus collector current density of an ECL gate, for a case where the maximum circuit speed is reached before significant base widening occurs.

The maximum speed, however, is still limited by the transit-time and diffusion-capacitance components. Since both of these components are proportional to $\tau_F$, minimizing $\tau_F$ will increase the maximum circuit speed. In the rest of this subsection, we discuss how to minimize $\tau_F$ and to what extent it can be minimized.

#### 8.3.3.1 MINIMIZING THE FORWARD TRANSIT TIME

The forward transit time $\tau_F$ is given by Eq. (6.114), which is repeated here:

$$\tau_F = t_E + t_B + t_{BE} + t_{BC}, \tag{8.15}$$

where $t_E$ is the emitter delay time, $t_B$ is the base transit time, $t_{BE}$ is the base–emitter depletion-region transit time, and $t_{BC}$ is the base–collector depletion-region transit time. The relative contributions of these components to $\tau_F$ for a typical self-aligned polysilicon-emitter bipolar transistor have been evaluated (Ashburn, 1988). No one component dominates the total transit time when base widening is negligible, although $t_B$ and $t_{BC}$ are often the largest components.

As shown in Section 7.2.5, a SiGe-base transistor can have a significantly smaller $t_E$ and $t_B$ than a Si-base transistor. As a result, a SiGe-base transistor has a significantly smaller $\tau_F$, and hence a higher $f_T$ [see Eqs. (8.8) and (8.9)], than a Si-base transistor. The effect of a higher $f_T$ on circuit speed has been studied (Chuang *et al.*, 1992). The benefit of a higher $f_T$ (70 versus 48 GHz) depends on the circuit family and its design point, as expected. For an unloaded ECL gate the benefit is only about 7% for a low-power design point, and about 20% for a high-power design point. For a loaded (FI = FO = 3, $C_L = 0.3$ pF) ECL gate, the benefit is reduced to less than 3% for the low-power design point, and only 10% for the high-power design point. The reader is referred to the literature (Chuang *et al.*, 1992) for more details and for a discussion on the benefit of higher $f_T$ for other bipolar circuit families.

In the rest of this subsection, we examine the characteristics of each of the components of $\tau_F$. We also discuss how they can be reduced.

- *Emitter delay time.* From Eqs. (6.112) and (6.113), we have $t_E = |Q_E|/I_C$, where $Q_E$ is the minority-carrier charge stored in the emitter. If we assume the transistor to have a deep emitter, then $Q_E$ can be obtained from the equation for the minority-carrier charge stored in a wide-base diode. Therefore, for an n–p–n transistor, we have $Q_E = I_B \tau_{pE}$ from Eq. (2.120), where $\tau_{pE}$ is the hole diffusion length in the emitter, and $I_B$ is the base current. ($Q_E$ is positive for an n–p–n transistor, and the current supplying $Q_E$ is $I_B$). The emitter delay time is

$$t_E(\text{deep-emitter}) = \frac{Q_E}{I_C} = \frac{\tau_{pE}}{\beta_0}, \tag{8.16}$$

where $\beta_0 = I_C/I_B$ is the current gain. It can be seen readily from Fig. 2.18(b) that $\tau_{pE}$ can be reduced by increasing the emitter doping concentration. In practice, the emitter is always doped as heavily as possible already. For a typical emitter doping concentration of $2 \times 10^{20}$ cm$^{-3}$, Fig. 2.18(b) gives $\tau_{pE} \sim 100$ ps. Therefore, $t_E$(deep-emitter) $\sim 1$ ps if $\beta_0 = 100$. The emitter delay time should be smaller for a shallow emitter than for a deep emitter, since the transit time for a shallow emitter is smaller than $\tau_{pE}$. Also, for a polysilicon emitter, the emitter transit time should be smaller than $t_E$(deep-emitter), since the current gain can be increased significantly with a polysilicon emitter. For a 1-μm self-aligned polysilicon-emitter n–p–n transistor, $t_E \approx 0.6$ ps (Ashburn, 1988). Equation (8.16) indicates that $t_E$ can be reduced by increasing the device current gain. For a given emitter process used to fabricate the transistor, the current gain can be increased by tailoring the base parameters to increase the collector current. As discussed in Chapter 7, this can be accomplished by increasing the base sheet resistivity and/or by using SiGe-base technology (see Exercise 8.5). However, in practice, a device designer almost never increases the current gain of a bipolar transistor in order to decrease its emitter delay time, since the emitter delay time is not a dominant component of $\tau_F$. Even with SiGe-base technology, the larger current gain associated with a SiGe-base transistor is usually traded off for a lower base resistance and/or for a larger Early voltage, instead of being used to reduce the emitter delay time.

- *Base–collector depletion-layer transit time.* The base–collector depletion-layer transit time $t_{BC}$ is given approximately by (Meyer and Muller, 1987; see also Exercise 8.6)

$$t_{BC} = \frac{W_{dBC}}{2v_{sat}} \tag{8.17}$$

where $W_{dBC}$ is the width of the base–collector junction depletion layer, and $v_{sat}$ is the saturated velocity of electrons. For a collector with $3 \times 10^{16}$ cm$^{-3}$ doping concentration and reverse biased with respect to the base by 3 V, $W_{dBC}$ is about 0.4 μm, and hence $t_{BC} \approx 2$ ps if $v_{sat} \approx 10^7$ cm/s is assumed. For a 1-μm self-aligned polysilicon-emitter bipolar device, the value for $t_{BC}$ calculated by numerical simulation is about 1.7 ps (Ashburn, 1988). As the collector doping concentration is increased to minimize base widening, $W_{dBC}$, and hence $t_{BC}$, will be reduced. However, in so doing, the base–collector depletion-layer capacitance will be increased, and the base–collector junction breakdown voltage will be reduced. In practice, the collector doping profile is usually designed to minimize base widening and to provide adequate breakdown voltage, instead of to reduce $t_{BC}$.

- *Base–emitter depletion-layer transit time.* The base–emitter depletion-region transit time $t_{BE}$ is smaller than $t_{BC}$, since the width of the base–emitter junction depletion layer, $W_{dBE}$, is much smaller than $W_{dBC}$. This is due to the fact that, for a transistor biased in the forward-active mode, the potential drop across the base–emitter junction is much smaller than that across the base–collector junction. Furthermore, the base is typically about 10 times more heavily doped than the collector. For a 1-μm self-aligned polysilicon-emitter bipolar device, the value of $t_{BE}$ calculated by numerical simulation is about 0.7 ps (Ashburn, 1988).

As will be shown in the next section, the base doping concentration is increased in bipolar device scaling. Therefore, $W_{dBE}$ and $t_{BE}$ are reduced in bipolar device scaling.

- *Base transit time.* The base transit time $t_B$ has been derived and discussed in detail in Section 7.2.4. For a boxlike base doping profile, $t_B$ is given by Eq. (7.24), namely

$$t_B \approx \frac{W_B^2}{2D_{nB}}, \tag{8.18}$$

where $W_B$ is the intrinsic-base width, and $D_{nB}$ is the electron diffusion coefficient in the base. Base doping profiles obtained by ion implantation are seldom boxlike, but often like a Gaussian distribution. The graded dopant distribution in a Gaussian profile results in a built-in electric field in the intrinsic base. However, as discussed in Section 7.2.3 and illustrated in Fig. 7.3, heavy doping in the intrinsic base of modern bipolar transistors also results in a built-in electric field which compensates substantially the electric field due to the graded dopant distribution. As a result, for modern bipolar transistors, the built-in electric fields in the base due to its nonuniform distribution of dopants have relatively little effect on the base transit time. However, this does not mean that the base transit time is totally independent of the base doping profile. In Eq. (8.18), $W_B$ represents the width of the quasineutral region of the intrinsic base, which is always smaller than the physical base width, which is defined by the separation between where the emitter doping concentration equals the base doping concentration in the emitter–base junction and where the base doping concentration equals the collector doping concentration in the base–collector junction. For the same physical base width, $W_B$ for a boxlike doping profile is larger than $W_B$ for a Gaussian or graded base doping profile. This is due to the fact that for the same collector doping profile, $W_{dBC}$ for a boxlike base doping profile is smaller than $W_{dBC}$ for a Gaussian or graded base doping profile. In any event, Eq. (8.18) can be used to obtain a first-order estimate of $t_B$. For a transistor with $W_B = 90$ nm and average $N_B = 3 \times 10^{17}$ cm$^{-3}$, Eq. (8.18) gives $t_B = 3.1$ ps, where an electron mobility of 500 cm$^2$/V-s, from Fig. 2.18(a), is assumed. Numerical simulation of a comparable Gaussian-like doping profile gives a value of 2.9 ps for $t_B$ (Ashburn, 1988). It is clear from Eq. (8.18) that reducing $W_B$ is an effective way of reducing the base transit time. However, when $W_B$ is reduced, the base doping concentration must be increased appropriately to maintain adequately large emitter–collector punch-through voltage and adequately large Early voltage.

As the base doping concentration is increased, the emitter–base junction capacitance will increase, and the emitter–base junction breakdown voltage will decrease. If the breakdown voltage becomes unacceptably low, then it may be necessary to design the base doping profile so that it peaks slightly away from the emitter–base junction. This can be accomplished by inserting an i-layer between the intrinsic-base region and the emitter region (Lu *et al.*, 1990), or by tailoring the base dopant implantation energy. A much more detailed discussion of the dependence of $t_B$ on the base doping profile can be found in the literature (Suzuki, 1991). Also, using SiGe technology for the intrinsic-base layer can reduce the base transit time by a factor of 2 to 3, as shown in Section 7.2.5.

### 8.3.4 DEVICE OPTIMIZATION FOR SMALL POWER–DELAY PRODUCT

For circuits designed to operate at low current densities, the gate delay is dominated by the $R_L$-component. This is illustrated by the design point $A$ in Fig. 8.9. This design point is not meant for maximum circuit speed, but for optimum power–delay tradeoff. To improve circuit speed, the capacitances in the $R_L$ component should be minimized. Referring to Eq. (8.12), we see that these capacitances are the base–collector junction depletion-layer capacitances $C_{dBCi}$ and $C_{dBCx}$, the base–emitter junction depletion-layer capacitance $C_{dBE}$, the collector–substrate junction depletion-layer capacitance $C_{dCS}$, and the load capacitance $C_L$.

The base–collector junction capacitance can be reduced by reducing the collector doping concentration. However, the intrinsic-collector doping concentration must be kept high enough to maintain negligible base widening. This can be achieved with the pedestal-collector structure discussed in Section 7.4.

The base–emitter junction capacitance can be reduced by reducing the intrinsic base doping concentration. However, as the base doping concentration is reduced, the base width may have to be increased to avoid emitter–collector punch-through and to avoid the Early voltage becoming unacceptably low. As the base width

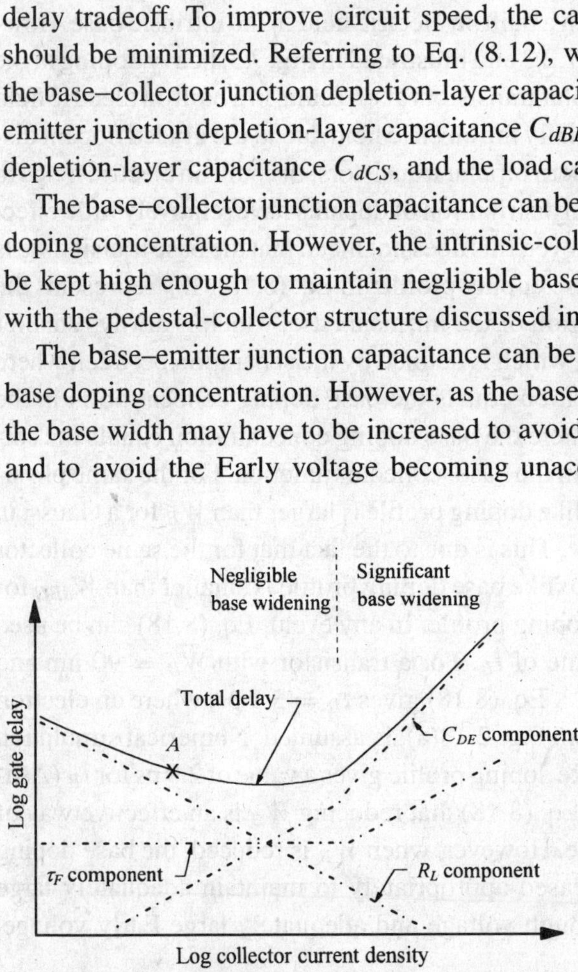


**FIGURE 8.9.** Schematic illustrating a design, point $A$, where the gate delay is dominated by the $R_L$ component.

increases, the transit-time delay component increases. Therefore, there will be an optimum base doping concentration for this design point. Alternatively, the base–emitter junction capacitance can be reduced by sandwiching an i-layer between the emitter and the base, as discussed in connection with the base transit time in the previous subsection. In practice, most bipolar transistors are not optimized specially for reducing the base–emitter junction capacitance, except by using as small an emitter area as possible, consistent with the process technology and the intended collector current density.

The collector–substrate junction capacitance can be reduced by reducing the substrate doping concentration. Consequently, most modern bipolar transistors use lightly doped substrates, typically with a doping concentration of about $1 \times 10^{15}$ cm$^{-3}$. In addition, as discussed in Section 7.4, the use of deep-trench isolation, instead of p-diffusion isolation, reduces the collector–substrate and collector–isolation capacitances very significantly.

The load capacitance consists of two components, namely the interconnect wire capacitance and the input device capacitance. Only the input device capacitance is a function of the device design. It is reduced as the base–emitter and base–collector junction capacitances are reduced. The modern device structure described in Section 7.4 has a base–collector junction area, and hence a base–collector junction capacitance, that is close to minimum.

### 8.3.5 BIPOLAR DEVICE OPTIMIZATION – AN EXAMPLE

It is clear from the above discussion that ***a device optimized for one design point is likely to be nonoptimum for a different design point***. This is demonstrated by the experimental results (Tang *et al.*, 1983) shown in Fig. 8.10. Figure 8.10 shows

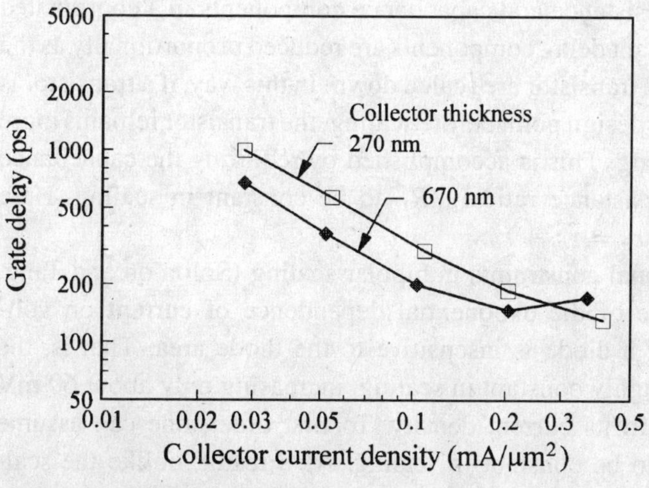


**FIGURE 8.10.** Typical switching delay of a bipolar circuit as a function of collector current density, with collector thickness as a parameter. (After Tang *et al.*, 1983.)

the inverter gate delays for two bipolar transistors, one with a collector thickness (i.e., the distance between the intrinsic base and the subcollector) of 270 nm, and one with a collector thickness of 670 nm. The thin-collector device has a larger base–collector junction capacitance than the thick-collector device. Therefore, at low collector current densities, where there is negligible base widening, the thin-collector device leads to a larger gate delay.

However, as the collector current density increases, base widening, and hence emitter diffusion capacitance, increases. At sufficiently large collector current densities, it is the emitter diffusion capacitance that determines the gate delay. When that happens, the thin-collector device, with less collector volume than the thick-collector device for minority-carrier charge storage, leads to faster circuits.

Figure 8.10 also demonstrates clearly one very important point about modern bipolar transistors for digital-circuit applications, i.e., base-widening effect limits the maximum speed of modern bipolar devices. ***Optimization of the collector doping profile and collector thickness is key to realizing the maximum performance of modern bipolar transistors.***

## 8.4 BIPOLAR DEVICE SCALING FOR ECL CIRCUITS

Since the details of a device design depend on its circuit application, scaling of a bipolar device should be discussed in the context of its circuit application as well. A theory for scaling bipolar transistors for high-performance ECL circuits has been developed by Solomon and Tang (1979). The basic concept in this scaling theory is to reduce the dominant resistance and capacitance components in a coordinated manner so that all the dominant delay components are reduced proportionally as the horizontal dimensions of the transistor are scaled down. In this way, if a transistor is optimized for a given circuit design point before scaling, the transistor remains more or less optimized after scaling. This is accomplished by requiring the capacitance ratio $C_{DE}/C_{dBC}$ and the resistance ratio $r_b/R_L$ to be constant in scaling. Here $C_{dBC} = C_{dBCi} + C_{dBCx}$ and $r_b = r_{bi} + r_{bx}$.

There are several additional constraints in bipolar scaling (Solomon and Tang, 1979). First of all, because of the exponential dependence of current on voltage, the turn-on voltage of a diode is insensitive to the diode area. That is, the diode turn-on voltage is roughly constant in scaling, increasing only about 60 mV for every tenfold increase in its current density. To first order, one can assume the diode turn-on voltage to be constant in scaling. As a result, unlike the scaling of MOSFETs (see Section 4.1), the voltages in a bipolar circuit, including the logic swing $\Delta V$, cannot be reduced in scaling. If the voltages are already optimally small to begin with, then they should remain constant in scaling. Secondly, as explained in Section 6.3.3, the collector doping concentration $N_C$ should be varied in proportion to the collector current density $J_C$ in order to maintain the same degree of base widening in scaling. Thirdly, to avoid emitter–collector punch-through as the base width is reduced in scaling, the base doping concentration must be increased. The base is depleted on the emitter side as well as on the collector side, but the depletion on the emitter side is usually more severe than on the collector side because the emitter is more heavily doped than the base, while the collector is more lightly doped than the base. To avoid excessive base-region depletion near the emitter–base junction, the emitter–base junction depletion-layer width $W_{dBE}$ should remain the same fraction of the base width $W_B$ as $W_B$ is reduced. From the dependence of $W_{dBE}$ on $N_B$ [see Eq. (2.70)], we see that this requirement is met if the base doping concentration $N_B$ is increased so that $N_B \propto W_B^{-2}$.

As shown in Eq. (8.13), for a constant logic swing, $R_L$ is inversely proportional to the switch current $I_S$. The requirement of $r_b/R_L$ being constant means that $r_b$ should be varied inversely proportional to $I_S$ as well, which would greatly complicate the device layout and design, and would also greatly narrow the device design window. It is much more practical to drop this resistance-ratio requirement and only keep the capacitance-ratio requirement in scaling. This approximation is quite reasonable, since, as discussed in Section 8.2.1, the $r_b$-component of the gate delay is relatively small to start with. Furthermore, as shown in Appendix 9, the base resistance can be reduced readily, if desired, by modifying the physical layout of the transistor.

Actually, as will be shown in the next subsection, $I_S$ is often kept constant in scaling in order to achieve circuit delay reduction in proportion to the emitter-stripe width. (This scaling objective is analogous to the constant-field scaling of MOSFETs, the results of which are shown in Table 4.1). In this case, to maintain a constant logic swing in scaling, $R_L$ is also kept constant. Therefore, the ratio $r_b/R_L$ is constant if $r_b$ is kept constant. As shown in Appendix 9, for a given emitter geometry, $r_b$ is constant if the intrinsic-base sheet resistivity is constant. The intrinsic-base sheet resistivity is indeed often maintained around 10 kΩ/□ partly to maintain a current gain of about 100 and partly to maintain a sufficiently large Early voltage. That is, without requiring special effort, $r_b/R_L$ is more or less constant in the scaling of bipolar devices for high-speed digital applications.

## 8.4.1 DEVICE SCALING RULES

The scaling constraints, together with the requirement on the capacitance ratio in scaling, are summarized in Table 8.1. The resulting scaling rules for the device design and circuit delay are summarized in Table 8.2. These rules are for the case where the ECL gate delay is reduced in proportion to the emitter-stripe width of

**TABLE 8.1** Constraints and Requirements in ECL Scaling

| Parameter | Constraint or Requirement |
|---|---|
| Voltage | $V, \Delta V =$ constant |
| Capacitance | $C_{DE}/C_{dBC} =$ constant |
| Base doping concentration | $N_B \propto W_B^{-2}$ |
| Collector doping concentration | $N_C \propto J_C$ |

**TABLE 8.2** Scaling Rules for ECL Circuits

| Parameter | Scaling Rule* |
|---|---|
| Feature size or emitter-stripe width | $1/\kappa$ |
| Base width $W_B$ | $1/\kappa^{0.8}$ |
| Collector current density $J_C$ | $\kappa^2$ |
| Circuit delay | $1/\kappa$ |

*Scaling factor $\kappa > 1$.

the transistors, or in proportion to the minimum lithographic feature size (Solomon and Tang, 1979).

#### 8.4.1.1 A QUALITATIVE DERIVATION OF THE ECL SCALING RULES

The scaling rules shown in Tables 8.1 and 8.2 can also be "derived" qualitatively as follows. The emitter–base and base–collector junctions can be approximated by one-sided diodes. With the power-supply voltage $V$ and the logic swing $\Delta V$ held constant, the voltages across these junctions do not change in scaling. As a result, the capacitance per unit area for the emitter–base junction, as given by Eqs. (2.68) and (2.70), is simply proportional to $N_B^{1/2}$. The device areas decrease as $1/\kappa^2$. Therefore, if $N_B$ is increased in proportion to $\kappa^2$ to avoid emitter–collector punch-through, then the emitter–base junction capacitance decreases as $1/\kappa$. (The more accurate scaling rules of Solomon and Tang, shown in Tables 8.1 and 8.2, suggest that $N_B$ increases as $\kappa^{1.6}$ for the case where $W_B$ decreases as $1/\kappa^{0.8}$.) Similarly, as $N_C$ is increased in proportion to $\kappa^2$ in scaling to maintain the same degree of base widening, the base-collector junction capacitance also decreases as $1/\kappa$ in scaling.

It is shown in Section 5.2.4 that the wiring capacitance per unit length is approximately constant, independent of the wire physical dimensions. Therefore, the capacitance due to wire loading decreases as $1/\kappa$ in scaling. This fact, together with the scaling properties of the device capacitance components discussed above, suggests that the total capacitance $C$ in a bipolar circuit decreases as $1/\kappa$ in scaling.

The gate delay scales as $C\Delta V/I$, where $I$ is the device current charging and discharging the capacitance $C$. Since $C$ scales as $1/\kappa$ and $\Delta V$ is constant, the gate

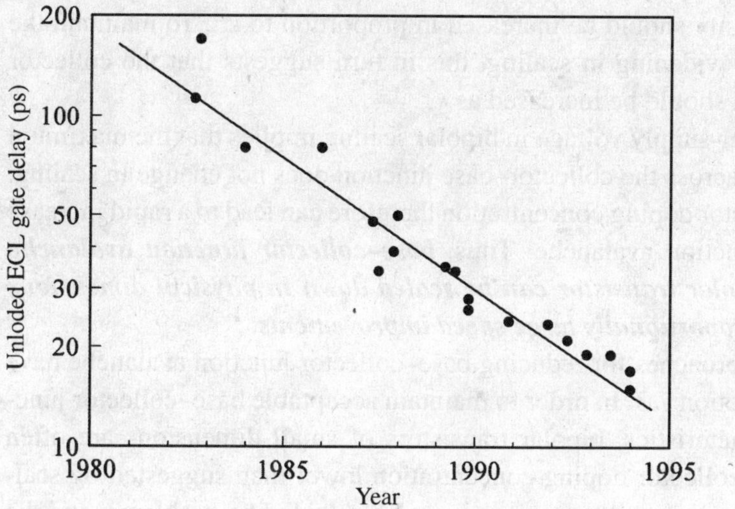


**FIGURE 8.11.** Reported ECL gate delays over time. (After Warnock, 1995.)

delay scales as $1/\kappa$ if $I$ is held constant. In other words, in order to obtain a gate delay decreasing as $1/\kappa$ in scaling, $I$ should be kept constant, and hence the current density should increase as $\kappa^2$, as indicated in Table 8.2.

#### 8.4.1.2 ECL CIRCUIT SCALING IN PRACTICE

In practice, device designers do not follow any scaling rules exactly in designing transistors for product applications. Nonetheless, the Solomon–Tang ECL scaling rules have provided a guide to identifying the important delay components in bipolar device scaling. Once these important delay components have been identified, device designers can then focus on minimizing them. As lithographic dimensions were reduced over time, the reported best-of-breed ECL delays were also reduced steadily over time, as expected from scaling. This is illustrated in Fig. 8.11, which was compiled from published data in the literature (Warnock, 1995).

### 8.4.2 LIMITS IN BIPOLAR DEVICE SCALING FOR ECL CIRCUITS

In this subsection, we examine the limits in scaling bipolar devices for ECL circuits. We do this by examining the scaling constraints in Table 8.1 and the scaling rules in Table 8.2, and by examining the implications of these constraints and rules for the design of small-dimension bipolar transistors.

#### 8.4.2.1 COLLECTOR-CURRENT-DENSITY LIMIT

The scaling rules in Table 8.2 suggest that in order to reduce the delay of an ECL circuit in scaling in proportion to the minimum lithographic feature size, the collector current density should be increased in proportion to $\kappa^2$. To maintain the same degree of base widening in scaling, this in turn suggests that the collector doping concentration should be increased as $\kappa^2$.

The constant power-supply voltage in bipolar scaling implies that the maximum reverse-bias voltage across the collector–base junction does not change in scaling. The increasing collector doping concentration therefore can lead to a rapid increase in base–collector junction avalanche. Thus, ***base–collector junction avalanche limits how far a bipolar transistor can be scaled down in physical dimensions and can still yield proportionally large speed improvements***.

Several design approaches for reducing base–collector junction avalanche have been discussed in Section 7.3. In order to maintain acceptable base–collector junction avalanche characteristics, bipolar transistors of small dimensions are often designed with their collector doping concentration lower than suggested by scaling. As a result, these devices have more severe base-widening problems, and the circuit delays tend to saturate with collector current density. This is illustrated in Fig. 8.12, where the reported delays of many ECL circuits are plotted as a function of the quoted collector current densities (Warnock, 1995).

#### 8.4.2.2 LIMITATION DUE TO DEVICE BREAKDOWN

With the power-supply voltage held constant, the minimum breakdown voltage required for the proper operation of a bipolar transistor does not change in scaling. As discussed in Section 6.5, the most important breakdown voltage to consider is $BV_{CEO}$, which is related to but is much smaller than $BV_{CBO}$. In the literature, there are many reports of high-speed bipolar devices with $BV_{CEO}$ of only slightly larger

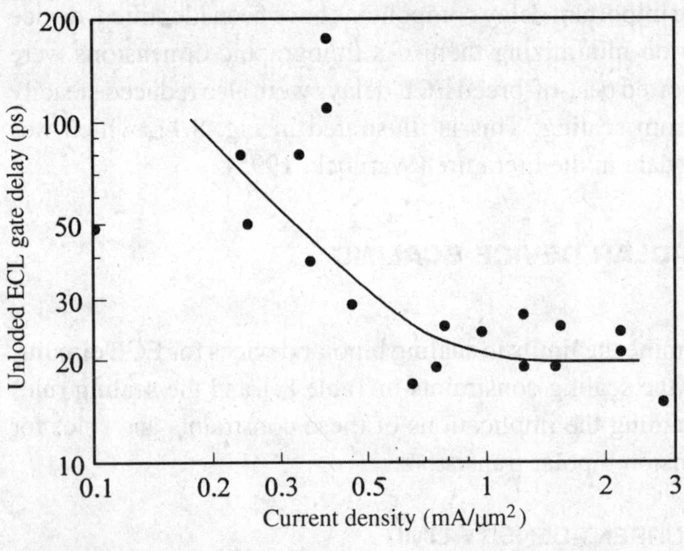


**FIGURE 8.12.** Reported ECL circuit delays plotted as a function of the quoted collector current densities. (After Warnock, 1995.)

than 2.0 V (see Fig. 6.24). If these devices were optimally designed already, then they cannot be scaled much further, if at all.

It should be pointed out that there are many reports of ECL circuits running at "record" speeds in the literature. Unfortunately, many of these reports do not state clearly if the bipolar transistors also have acceptable breakdown voltages. Without such information, it is impossible to judge the significance of the record speeds claimed. However, one point for sure is that transistors for circuits that operate with a smaller power-supply voltage can be scaled to larger collector current densities than those for circuits that operate with a larger power supply voltage.

#### 8.4.2.3 LIMITATION DUE TO POWER DENSITY

With the current density increasing as $\kappa^2$, the device currents are in effect kept constant in scaling. This implies that the power dissipation of the circuit, which is simply the total current times the power-supply voltage, is approximately constant. That is, the power density increases as $\kappa^2$ in scaling. Thus, if a 1000-circuit bipolar chip at 1-μm design rules dissipates 3 W, the same-size chip at 0.25-μm design rules could hold 16,000 circuits, with each circuit running 4 times as fast, but the chip would dissipate 48 W.

Unlike a CMOS gate, which dissipates negligible power during standby (see Section 5.1.1), an ECL gate dissipates about the same power whether it is in standby or switching. It is this large standby power dissipation of an ECL gate that makes the averaged power dissipation of an ECL chip much larger than that of a CMOS chip with the same number of gates. Thus while a bipolar gate can run much faster than a CMOS gate, the very high averaged power dissipation of bipolar circuits has limited bipolar chips to circuit integration levels that are very small compared to CMOS chips. As a result, bipolar technology has lost out to CMOS for digital VLSI applications, where the system-level speed is determined not just by transistor speed, but by chip-level integration and package-level integration as well. For a few applications, e.g., high-end mainframe computers, where circuit speed is more important than power dissipation, bipolar circuits are still used. In this case, the circuits are usually designed to operate at less than maximum speed, such as the design point *A* in Fig. 8.9, in order to keep the chip-level power dissipation at an acceptable level.

It was discussed in Section 5.2.4 that to minimize the *RC* delay due to scaling of the interconnect wires, a wiring hierarchy (Fig. 5.24) is preferred where only the wires for local interconnection are scaled down in size while the wires for global interconnection are unscaled or even scaled up. For designs where the chip area is wire-limited, which is the case for most high-speed processors, such a wiring hierarchy usually results in an increased chip area, which actually facilitates heat removal from the chip, which in turn allows higher power and faster circuits to be used. Suffice it to say that the performance of a bipolar process increases with the heat-removal capability of the package used (Sai-Halasz, 1995).

#### 8.4.2.4 LIMITATION DUE TO EMITTER SERIES RESISTANCE

It was shown in Section 6.3.1 that the emitter series resistance $r_e$ introduces an emitter–base diode voltage drop of $-I_E r_e = (I_C + I_B)\, r_e \approx I_C r_e$ and causes the $I_C$-versus-$V_{BE}$ characteristics to saturate rapidly once this voltage drop is larger than about 60 mV. As an emitter is scaled down in size, the emitter series resistance increases in inverse proportion to its area. Therefore, the emitter–base diode voltage drop due to emitter series resistance increases rapidly in scaling and could severely limit the current-carrying capability of small-emitter-area transistors.

Fortunately, for an ECL gate, the switching delay attributable to emitter series resistance is relatively small. For a properly optimized 0.5-μm self-aligned bipolar transistor, with an emitter series resistance of 70 Ω, the ECL gate delay attributable to $r_e$ is only about 10% of the total gate delay (Chor *et al.*, 1988).

Also helping to alleviate the emitter-series-resistance problem is the fact that the collector-current-density limit discussed earlier forces designers to use emitters that are significantly larger than the minimum allowed by the lithography, in order to minimize the much more detrimental effects of base widening. A relatively narrow emitter stripe is often employed, with the total emitter area determined by the desired operating current and the current-density limit. The stripe-emitter geometry is chosen to reduce base resistance (see Appendix 9).

## 8.5 BIPOLAR DEVICE OPTIMIZATION AND SCALING FOR ANALOG CIRCUITS

In general, the techniques used to optimize a bipolar transistor for digital circuits, such as minimizing the base resistance and collector capacitance, are also applicable to optimizing a bipolar transistor for analog circuits. Also, the current-density limit due to base widening applies to transistors for digital as well as analog circuits. However, there are some device design differences between analog and digital circuits.

For a digital circuit, the overall circuit switching speed and power dissipation are the important factors governing the device design, provided that the design meets the breakdown voltage requirements of the circuit. For an analog circuit, perhaps the most important device parameters or figures of merit are the cutoff frequency, the maximum oscillation frequency, the base resistance, and the Early voltage. The cutoff frequency and the maximum oscillation frequency should be high, the base resistance should be small, and the Early voltage should be large. The relative importance of these parameters or figures of merit depends on the specific application. In this section, we discuss how these parameters or figures of merit can be optimized in general, and how they can be traded off among one another. We also discuss how the technology of growing the intrinsic base epitaxially can lead to superior device characteristics for analog circuit applications.

### 8.5.1 OPTIMIZING THE INDIVIDUAL PARAMETERS

The frequency response, the base resistance, and the Early voltage are all closely coupled. In a practical device design, they are and should be considered together. Here we describe how each of them can be optimized and discuss how the optimization of one may adversely affect the others.

#### 8.5.1.1 MAXIMIZING THE CUTOFF FREQUENCY

The cutoff frequency $f_T$ is given by Eq. (8.8) or (8.9). To maximize $f_T$, the capacitances $C_{dBE}$ and $C_{dBC}$ and the forward transit time $\tau_F$ should all be minimized. The simplest way to minimize these capacitances is to use advanced device structures that have small parasitic capacitance, the same as for digital circuits (see Section 7.4). The techniques for minimizing $\tau_F$ have already been discussed in Section 8.3.3, in connection with the device optimization for digital circuits. The most important component of $\tau_F$ is the base transit time, which can be reduced effectively by reducing the intrinsic-base width $W_B$. However, reducing $W_B$ alone will lead to a larger intrinsic-base resistance and a smaller Early voltage.

It should be noted that $\tau_F$ is a function of both the intrinsic-device doping profiles and the collector current density. It is independent of the horizontal device dimensions and geometry. That is, for a given vertical device doping profile, a large-emitter device has the same $\tau_F$ as a small-emitter device, if both devices are operated at the same collector current density. Equation (8.8) shows that the maximum $f_T$ is determined by $\tau_F$. Therefore, the maximum $f_T$ of a transistor is independent of its emitter area but is a function of its vertical doping profile.

At sufficiently large collector current densities, base widening becomes important, and $\tau_F$ increases rapidly, and hence $f_T$ decreases rapidly, with further increase in current density. This is illustrated in Fig. 8.13, where the measured $f_T$ as a function of collector current is shown for two transistors of different collector doping concentrations (Crabbé *et al.*, 1993). It clearly shows that the $f_T$ roll-off characteristics shift towards higher current in proportion to the increased collector doping concentration, as expected from base-widening effects. It also shows that a simple way to increase the peak $f_T$ of a transistor is to increase its collector doping concentration, provided that the device breakdown voltages remain acceptable.

#### 8.5.1.2 MINIMIZING THE INTRINSIC-BASE RESISTANCE

As shown in Appendix 9, the intrinsic-base resistance $r_{bi}$ for an emitter stripe of width $W$ and length $L$ is proportional to $(W/L)R_{Sbi}$, where $R_{Sbi}$ is the sheet resistivity of the intrinsic-base layer. Thus the intrinsic-base resistance can be reduced by making the emitter stripe as narrow as possible and using long emitter stripes. Emitter stripes of less than 0.2 μm wide can now be made readily (Shiba *et al.*,

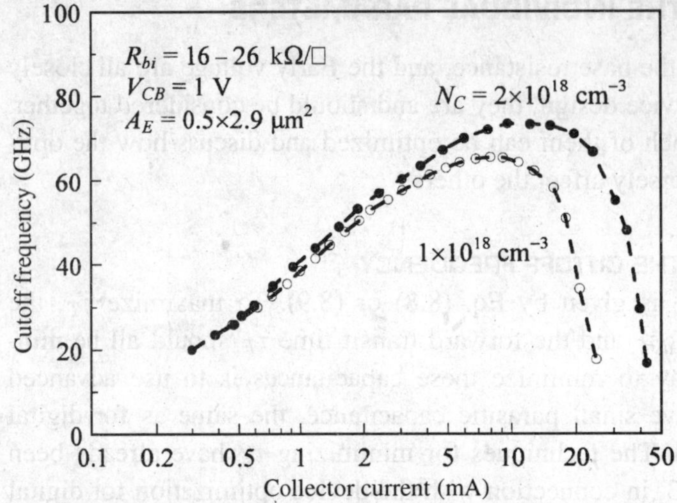


**FIGURE 8.13.** Typical measured cutoff frequency as a function of collector current, with collector doping concentration as a parameter. (After Crabbé *et al.*, 1993.)

1996). Furthermore, contacting the intrinsic base from both sides of an emitter stripe, instead of simply from one side, reduces $r_{bi}$ by a factor of four. In practice, all high-performance bipolar transistors have base contacts on both sides of their emitter stripes.

With a polysilicon emitter, current gain is usually not an issue. As a result, the intrinsic-base layer can be doped rather heavily to reduce $R_{Sbi}$, which in turn reduces $r_{bi}$. However, if the base doping concentration is too high, emitter–base band-to-band tunneling current can degrade the current gain at low and moderate currents (see Section 6.3.4). In practice, the peak base doping concentration is usually kept below $5 \times 10^{18}$ cm$^{-3}$ to avoid excessive emitter–base tunneling current.

The design window in the tradeoff between $f_T$ and $r_{bi}$ can be enlarged by using a boxlike instead of a graded intrinsic-base doping profile. It was shown in Section 7.2 that, as a result of the heavy-doping effect, the built-in electric field in the intrinsic base of a modern bipolar transistor is not significant, whether the intrinsic-base doping profile is graded or boxlike. Thus, for the same $W_B$ and peak base doping concentration, a boxlike intrinsic-base doping profile and a graded intrinsic-base doping profile should give about the same maximum $f_T$. The boxlike base profile, however, has a larger majority-carrier hole charge, and hence a smaller $R_{Sbi}$ and $r_{bi}$, than a graded base profile.

#### 8.5.1.3 MAXIMIZING THE MAXIMUM OSCILLATION FREQUENCY

The maximum oscillation frequency is given by Eq. (8.10). It is a function of $f_T$, $r_b$, and $C_{dBC}$. Therefore, designs that increase $f_T$ at the expense of increasing $r_b$ can result in a decrease in $f_{max}$. In fact, if a slightly reduced $f_T$ allows a significantly reduced $r_b$, the net result can be a larger $f_{max}$.

Reducing $C_{dBC}$ increases $f_{max}$. However, if $C_{dBC}$ is reduced by reducing the collector doping concentration, base widening will set in at a lower collector current density, which in turn can reduce the maximum $f_T$ of the transistor (see Fig. 8.13). Thus, maximizing $f_{max}$ is a complex tradeoff process. This point will be discussed further in the next subsection.

#### 8.5.1.4 MAXIMIZING THE EARLY VOLTAGE

The Early voltage for a uniformly doped intrinsic base is give by Eq. (6.72). It is the ratio of the majority-carrier hole charge per unit area, $Q_{pB}$, to the intrinsic-base–collector depletion-layer capacitance per unit area, $C_{dBCi}$. An obvious way to increase the Early voltage is to increase the majority-carrier hole charge. If this is done at a fixed $W_B$ by increasing the base doping concentration, then it has relatively little effect on $\tau_F$, and it will reduce $r_{bi}$. It will have relatively little effect on $f_T$, but it will lead to a higher $f_{max}$. However, it will also reduce the current gain, since, as shown in Eq. (7.7), the current gain is inversely proportional to the intrinsic-base sheet resistivity. On the other hand, if the majority-carrier hole charge is increased by increasing $W_B$, it will decrease $f_T$ and may adversely affect $f_{max}$ as well. It will definitely reduce the current gain.

The Early voltage can also be increased by reducing $C_{dBCi}$. As long as base widening is not an issue, this capacitance can be reduced by reducing the collector doping concentration and/or increasing the collector layer thickness.

For a given collector doping profile, a boxlike intrinsic-base doping profile has a larger $C_{dBCi}$ as well as a larger $Q_{pB}$ than a graded base doping profile of the same width and same peak doping concentration. However, the increase in $C_{dBCi}$ is not enough to offset the increase in $Q_{pB}$, and the net effect is that the Early voltage, $V_A = Q_{pB}/C_{dBCi}$, is larger for a boxlike base doping profile than for a graded base doping profile (see Exercise 8.8).

### 8.5.2 TECHNOLOGY FOR ANALOG BIPOLAR DEVICES

It is apparent from the above discussion that ***a boxlike intrinsic-base doping profile is preferred for analog circuit applications***. It provides a larger device design window, and hence allows more optimized tradeoffs to be made among the various device parameters. In practice, a boxlike intrinsic-base doping profile can be obtained readily by using an epitaxially grown intrinsic-base layer. The intrinsic base is doped *in situ* during growth, instead of doped subsequently by ion implantation or diffusion. Using such technology, a tradeoff study for analog device design has been made (Yoshino *et al.*, 1995). The results are shown in Figs. 8.14 and 8.15. These results clearly show that $f_T$ can be traded off for a larger $f_{max}$ and/or a larger $V_A$.

By incorporating Ge during epitaxial growth of the intrinsic-base layer, a SiGe-base transistor can be made readily. It is shown in Section 7.2.5 that, compared to the

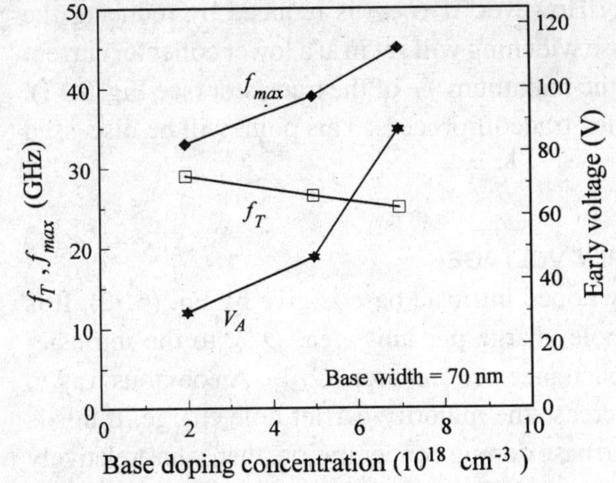


**FIGURE 8.14.** Experimental results showing the dependence of $f_T$, $f_{\max}$, and $V_A$ on base doping concentration $N_B$. The base has a boxlike doping profile, formed by epitaxial growth of the intrinsic-base layer. (After Yoshino *et al.*, 1995.)

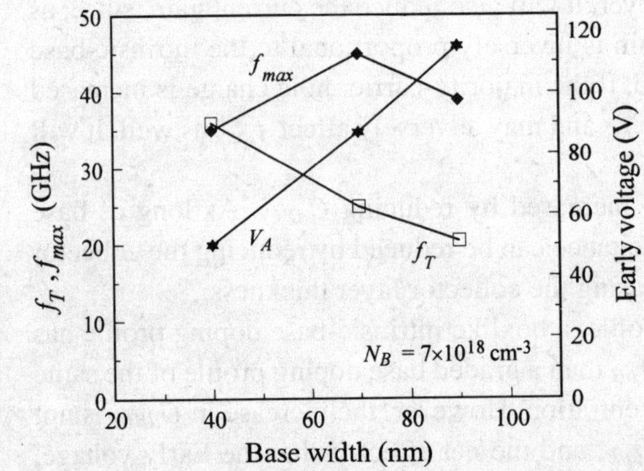


**FIGURE 8.15.** Experimental results showing the dependence of $f_T$, $f_{\max}$, and $V_A$ on intrinsic-base layer thickness. The intrinsic base is formed by epitaxial growth of silicon. (After Yoshino *et al.*, 1995.)

regular Si-base transistor, a SiGe-base transistor has significantly larger collector current for the same base–emitter forward bias, significantly smaller base transit time, and significantly larger Early voltage. The larger collector current can be traded off for a smaller base resistance. Also, as shown in Section 8.3.3, a SiGe-base transistor has a smaller emitter delay time than a Si-base transistor because of its larger current gain. Thus, SiGe-base technology is superior to Si-base technology for analog circuit applications. The interested reader is referred to the vast literature on SiGe-base transistors for such applications (Patton *et al.*, 1990; Harame *et al.*, 1995a, b; and the references therein).

### 8.5.3 LIMITS IN SCALING ANALOG BIPOLAR TRANSISTORS

Although quantitatively the effects are different, the physical mechanisms that limit the scaling of bipolar transistors for digital circuits also limit the scaling of bipolar

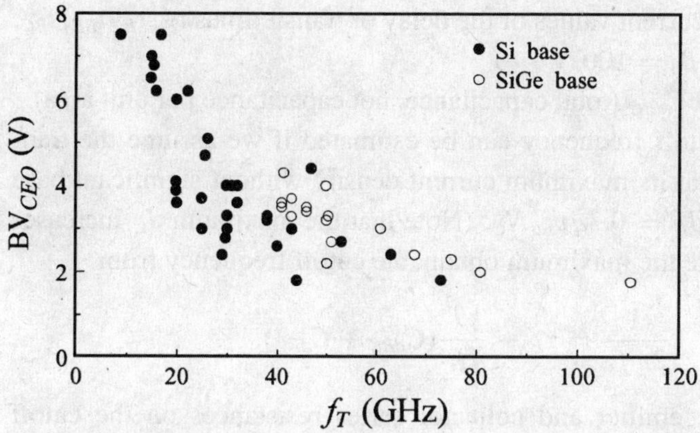


**FIGURE 8.16.** $BV_{CEO}$ versus $f_T$ for many recently reported n–p–n transistors.

transistors for analog circuits. Thus base widening limits the maximum collector current density, and collector–emitter breakdown limits how heavily the collector can be doped to minimize base widening and hence limits the maximum obtainable cutoff frequency.

Figure 8.16 is a plot of $BV_{CEO}$ versus $f_T$ for many n–p–n transistors reported in recent years. Data for both Si-base and SiGe-base transistors are included. It shows that there is definitely a tradeoff between $BV_{CEO}$ and $f_T$. As expected, for the same $BV_{CEO}$ value, the SiGe-base transistors can reach a higher $f_T$. However, it is also apparent that if a minimum $BV_{CEO}$ value of 3.0 V is desired, even with SiGe base, the maximum $f_T$ obtainable is only about 60 GHz.

## EXERCISES

**8.1** The small-signal equivalent circuit for a bipolar transistor, including its internal series resistances, is shown in Fig. 6.20. Ignore the collector–substrate capacitance $C_{dCS}$, which is quite small in modern bipolar transistors. Show that the cutoff frequency is given by Eq. (8.9), i.e.,

$$\frac{1}{2\pi f_T} = \tau_F + \frac{kT}{qI_C}(C_{dBE} + C_{dBC}) + C_{dBC}(r_e + r_c).$$

State the assumptions used in the derivation.

**8.2** Consider an n–p–n bipolar transistor, with an emitter area of $A_E = 1\times 2\ \mu m^2$ and a base–collector junction area of $A_{BC} = 10\ \mu m^2$, a deep emitter with $N_E = 10^{20}$ cm$^{-3}$, a boxlike intrinsic base with $N_B = 1 \times 10^{18}$ cm$^{-3}$ and $W_B = 100$ nm, and a uniformly doped collector of $N_C = 5 \times 10^{16}$ cm$^{-3}$. Assume a one-sided junction approximation for all the junctions, and that the transistor is biased with $V_{BE} = 0.8$ V (for purposes of $t_{BE}$ and $C_{dBE}$ calculations) and $V_{CB} = 2$ V (for purposes of $t_{BC}$ and $C_{dBC}$ calculations).

**(a)** Estimate the low-current values of the delay or transit times $t_E$, $t_B$, $t_{BE}$, $t_{BC}$, and $\tau_F$, assuming $\beta_0 = 100$.

**(b)** Estimate $C_{dBE}$ and $C_{dBC}$ (total capacitance, not capacitance per unit area).

**(c)** The maximum cutoff frequency can be estimated if we assume the transistor is operated at its maximum current density without significant base widening, i.e., at $J_C = 0.3qv_{sat}N_C$. (Note that the maximum $J_C$ increases with $N_C$.) Estimate the maximum obtainable cutoff frequency from

$$\frac{1}{2\pi f_T} = \tau_F + \frac{kT}{qI_C}(C_{dBE} + C_{dBC}).$$

**(d)** The effect of the emitter and collector series resistances on the cutoff frequency can be estimated from

$$\frac{1}{2\pi f_T} = \tau_F + \frac{kT}{qI_C}(C_{dBE} + C_{dBC}) + C_{dBC}(r_e + r_c).$$

Assume $r_e + r_c = 50\ \Omega$. Estimate the maximum cutoff frequency.

**8.3** This exercise is designed to illustrate the advantage of the pedestal-collector design. Consider the n–p–n transistor of Exercise 8.2. Let $N_C(\text{int}) = 5 \times 10^{16}\ \text{cm}^{-3}$ be the collector doping concentration directly underneath the emitter and intrinsic base, and $N_C(\text{ext}) = 5 \times 10^{15}\ \text{cm}^{-3}$ be the doping concentration of the extrinsic part of the collector. $C_{dBC}$ of the pedestal-collector design is smaller than that of the non-pedestal-collector design. Repeat Exercise 8.2 for this pedestal-collector transistor.

**8.4** This exercise is designed to illustrate the sensitivity of the maximum cutoff frequency to the pedestal-collector doping concentration. Consider the pedestal-collector transistor of Exercise 8.3. The maximum collector current density without significant base widening can be increased by increasing $N_C(\text{int})$. Repeat Exercise 8.3 for the case of $N_C(\text{int}) = 1 \times 10^{17}\ \text{cm}^{-3}$.

**8.5** The incorporation of Ge into the intrinsic base will reduce the base transit time and the emitter delay time. For a linearly retrograded Ge profile, the transit time is reduced by a factor [see Eq. (7.40)]

$$\frac{t_B(\text{SiGe})}{t_B(\text{Si})} = \frac{2kT}{\Delta E_{g,SiGe}}\left\{1 - \frac{kT}{\Delta E_{g,SiGe}}\left[1 - \exp\left(-\frac{\Delta E_{g.SiGe}}{kT}\right)\right]\right\},$$

and the emitter delay time is reduced by a factor [see Eq. (8.16)]

$$\frac{t_E(\text{SiGe})}{t_E(\text{Si})} = \frac{\beta_0(\text{Si})}{\beta_0(\text{SiGe})},$$

where $\Delta E_{g,SiGe}$ is the total bandgap narrowing due to Ge. The current gain ratio is given by Eq. (7.30), namely

$$\frac{\beta(\text{SiGe})}{\beta(\text{Si})} = \frac{\Delta E_{g.SiGe}/kT}{1 - \exp(-\Delta E_{g,SiGe}/kT)}.$$

Repeat Exercise 8.4 for a SiGe-base transistor with $\Delta E_{g,SiGe} = 100$ meV. (Assume the same $V_{BE}$ of 0.8 V.)

(Comparison of the results from Exercises 8.4 and 8.5 illustrates the effect of SiGe-base technology on $f_T$.)

**8.6** The depletion-layer transit time is given by Eq. (8.17). Here we want to show that it can also be derived by considering the average transit time of the stored charge after the charging current is switched off. Consider a depletion layer of width $W$, located between $x = 0$ and $x = W$. Assume there is a constant current density $J$ flowing through this depletion layer, and all the charges contributing to this current flow are traveling at the same velocity $v$, so that $J = qnv$, where $qn$ is the mobile charge density. We further assume that $qn$ is small enough so that the presence of the current flow does not affect the depletion-layer thickness. Imagine at time $t = 0$ the current is suddenly switched off, and the mobile charge carriers stored in the depletion layer continue to drain out at the same velocity $v$. The time needed to drain out the last mobile charge carrier from the depletion layer is $W/v$. Show that the average transit time $t_{avg}$, i.e., the transit time averaged over all the stored charge carriers, is only $W/2v$. [*Hint:* The transit time for a charge carrier located at $x$ is $(W - x)/v$.]

**8.7** This exercise is designed to evaluate the tradeoff between performance and base–collector junction avalanche in an n–p–n transistor. In Exercise 8.4, as $N_C(\text{int})$ is increased from $5 \times 10^{16}$ to $1 \times 10^{17}$ cm$^{-3}$, the maximum electric field $\mathscr{E}_m$ in the base–collector junction depletion layer is increased, which could lead to significantly increased avalanche multiplication in the junction. One way to minimize this problem is to sandwich an i-layer between the base and the collector doped regions.

**(a)** Assume the base–collector junction is reverse-biased at 2 V, calculate the i-layer thickness $d$ needed so that the maximum base–collector junction fields for the following two designs are the same: (i) $N_C(\text{int}) = 1 \times 10^{17}$ cm$^{-3}$ with an i-layer, and (ii) $N_C(\text{int}) = 5 \times 10^{16}$ cm$^{-3}$ without an i-layer.

**(b)** Assume the emitter and base designs are the same as the transistor in Exercise 8.2. Calculate the delay or transit times $t_E$, $t_B$, $t_{BE}$, $t_{BC}$, and $\tau_F$, assuming $\beta_0 = 100$, for the two designs in (a).

**(c)** Calculate $C_{dBE}$ and $C_{dBC}$ (total capacitance, not capacitance per unit area) for the two pedestal-collector designs in (a).

**(d)** Estimate the maximum cutoff frequency for the two designs from

$$\frac{1}{2\pi f_T} = \tau_F + \frac{kT}{qI_C}(C_{dBE} + C_{dBC}),$$

assuming that in each case the transistor is operated at its maximum collector current density of $J_C = 0.3qv_{sat}N_C(\text{int})$ to avoid significant base widening. [*Hint:* The depletion-layer width, maximum electric field, and capacitance for a p–i–n diode are derived in Section 2.2.2. Use the one-sided

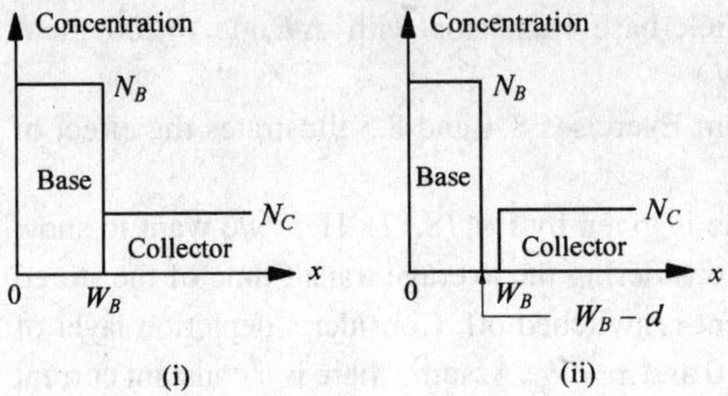


**FIGURE 8.17.** Base and collector doping profiles for Exercise 8.8.

junction approximation for both the emitter–base and the base–collector diodes.)

**8.8** This exercise is designed to show that for the same peak base doping concentration and base width, a boxlike doping profile gives a larger Early voltage than a graded base doping profile. The Early voltage $V_A$ is given approximately by Eq. (6.72), namely

$$V_A \approx \frac{Q_{pB}}{C_{dBC}},$$

where $Q_{pB}$ is the majority-carrier hole charge per unit area in the base [see Eq. (6.66)] and $C_{dBC}$ is the base–collector junction depletion-layer capacitance per unit area. Consider the following two intrinsic-base profiles (see Fig. 8.17):

**(i)** a boxlike base of doping concentration $N_B$ and width $W_B$, and

**(ii)** a steplike base doping profile (as an approximation for a graded doping profile) such that the doping concentration between $x = 0$ (emitter–base junction) and $x = W_B - d$ is $N_B$, and the doping concentration between $x = W_B - d$ and $x = W_B$ (base–collector junction) is zero.

[Thus, profile (i) is like profile (ii) with $d = 0$.] If we assume the collector doping concentration is $N_C$ for both transistors, then it is clear that $Q_{pB}$ is larger for profile (i) than for profile (ii), i.e., $Q_{pB}(d = 0) > Q_{pB}(d)$, and $C_{dBC}$ is also larger for profile (i) than for profile (ii), i.e., $C_{dBC}(d = 0) > C_{dBC}(d)$. We want to show that $V_A(d) < V_A(d = 0)$.

**(a)** For simplicity, we assume $W_B$ to be large enough so that the quasineutral base width is also $W_B$. Show that

$$\frac{V_A(d)}{V_A(d = 0)} = \left(1 - \frac{d}{W_B}\right)\sqrt{1 + \frac{d^2}{W_{dBC}^2(d = 0)}},$$

where $W_{dBC}(d = 0)$ is the base–collector junction depletion-layer width for profile (i). [*Hint:* Use Eq. (2.82) for the capacitance ratio $C_{dBC}(d)/C_{dBC}(d = 0)$].

**(b)** The results in (a) show that the ratio $V_A(d)/V_A(d = 0)$ is less than 1 unless the ratio $d/W_{dBC}(d = 0)$ is quite large. Estimate the largest value of

$d/W_{dBC}(d=0)$ for which the ratio $V_A(d)/V_A(d=0)$ is still less than 1, for $d/W_B = 0.1$ and for $d/W_B = 0.2$.

**(c)** Let us put in some real numbers. Estimate the ratio $V_A(d)/V_A(d=0)$ for $d =$ 10 nm and for $d = 20$ nm, assuming $N_B = 1 \times 10^{18}$ cm$^{-3}$, $N_C = 5 \times 10^{16}$ cm$^{-3}$, $W_B = 100$ nm. Assume the one-sided junction approximation and $V_{BC} = 0$ in calculating $W_{dBC}$.

APPENDIX

# 1 CMOS PROCESS FLOW

A more or less generic CMOS process flow is described below. It features shallow trench isolation (STI) (Davari *et al.*, 1988b, 1989), dual $n^+/p^+$ polysilicon gates (Wong *et al.*, 1988; Sun *et al.*, 1989), and self-aligned silicide (Ting *et al.*, 1982). The front-end-of-the-line process consists of six or seven masking levels and is suitable for sub-0.5-μm generations of VLSI logic and SRAM technology.

- Starting material p-type substrate or $p^-$ epi on $p^+$ substrate for latch-up prevention (Taur *et al.*, 1984).
- Grow pad oxide. Deposit CVD (Chemical Vapor Deposition) nitride. (See Fig. A1.1.)
- Lithography to cover the active region with photoresist.
- Reactive ion etching (RIE) nitride and oxide in the field region.
- RIE shallow trench in silicon. (See Fig. A1.2.)
- Grow pad oxide. Deposit thick CVD oxide. (See Fig. A1.3)
- Chemical–mechanical polishing planarization. (See Fig. A1.4)
- n-well lithography and implant (also channel doping).
- p-well lithography and implant (also channel doping). (See Fig. A1.5.)
- Grow gate oxide.
- Deposit polysilicon film. (See Fig. A1.6.)
- Gate lithography.
- RIE polysilicon gate. (See Fig. A1.7.)
- Sidewall reoxidation.
- $n^+$ source–drain lithography and implant (also dope $n^+$ polysilicon gate).
- $p^+$ source–drain lithography and implant (also dope $p^+$ polysilicon gate). (See Fig. A1.8.)
- Oxide (or nitride) spacer formation by CVD and RIE.
- Source–drain anneal. (See Fig. A1.9.)
- Self-aligned silicide process. (See Fig. A1.10.)
- Back-end-of-the-line process.

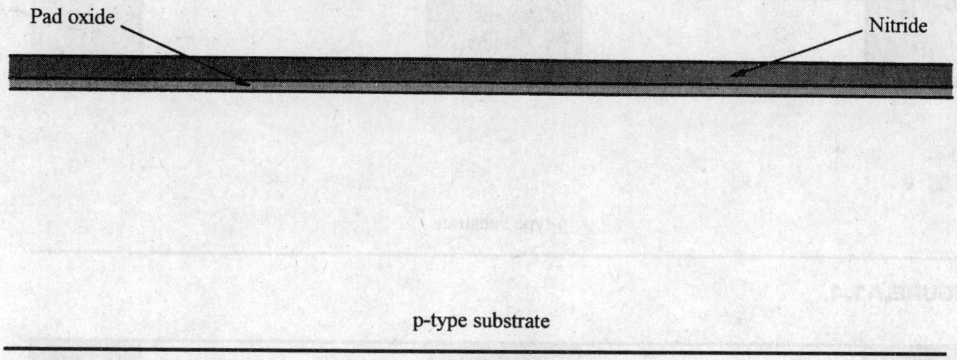


**FIGURE A1.1.**

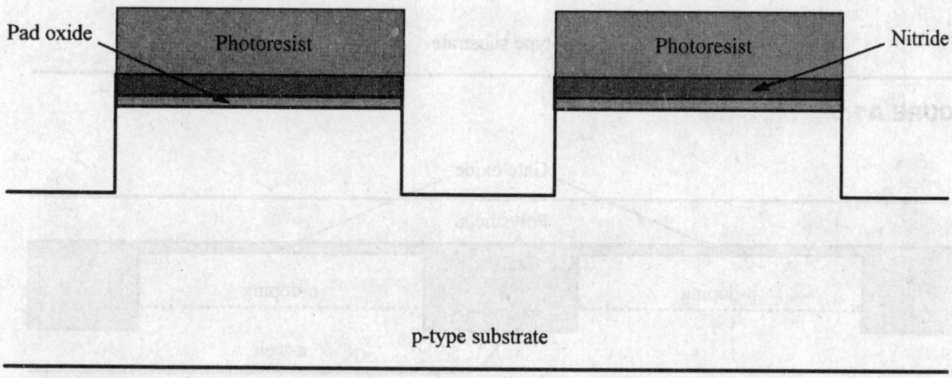


**FIGURE A1.2.**

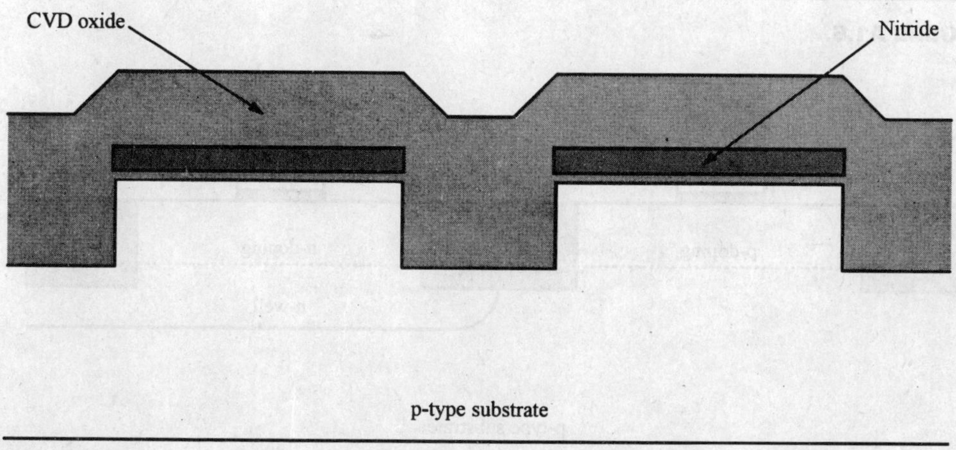


**FIGURE A1.3.**

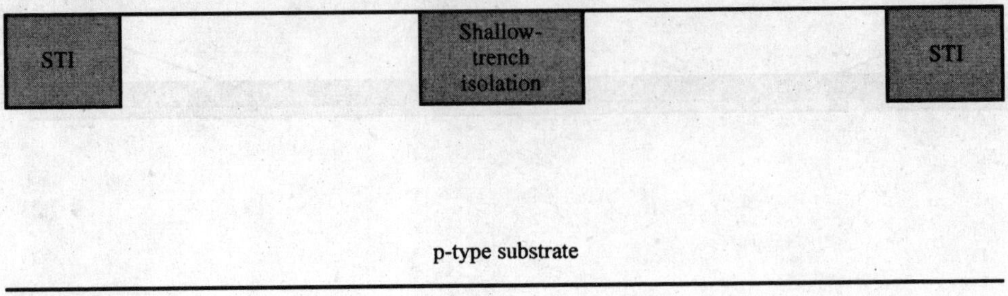


**FIGURE A1.4.**

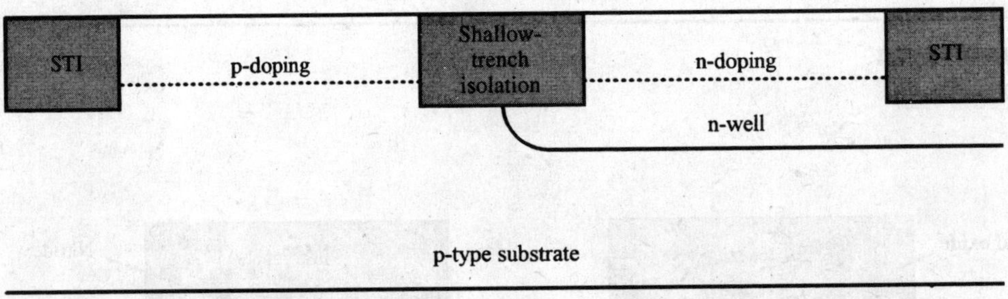


**FIGURE A1.5.**

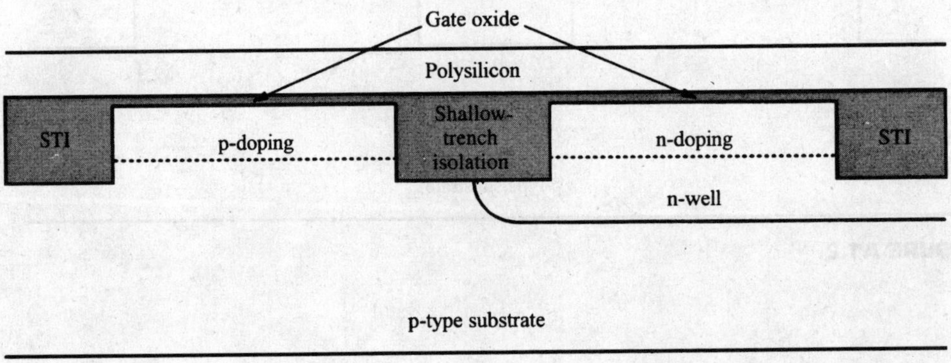


**FIGURE A1.6.**

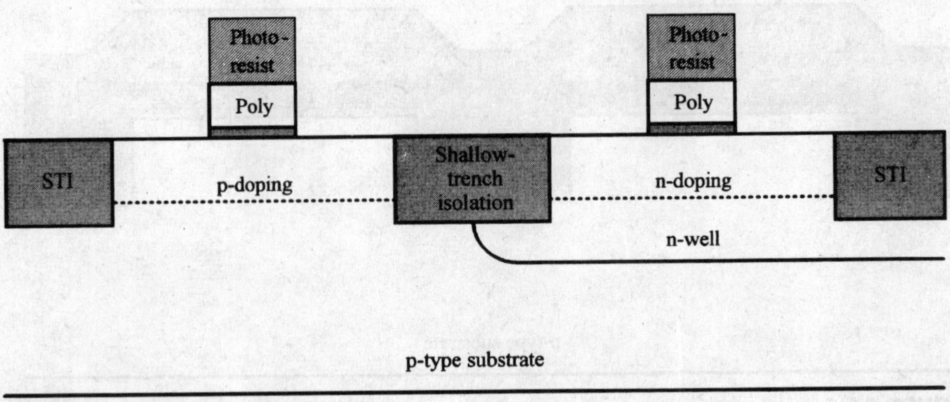


**FIGURE A1.7.**

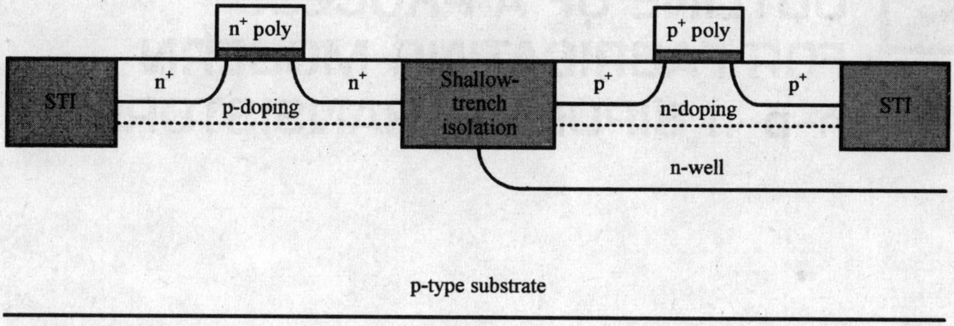


**FIGURE A1.8.**

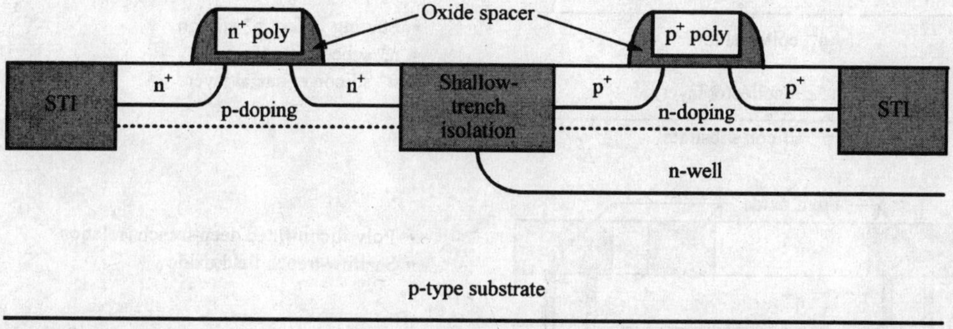


**FIGURE A1.9.**

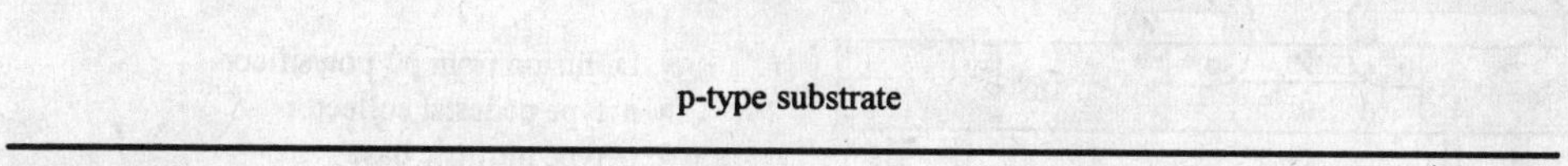


**FIGURE A1.10.**

APPENDIX 2

# OUTLINE OF A PROCESS FOR FABRICATING MODERN n–p–n BIPOLAR TRANSISTORS

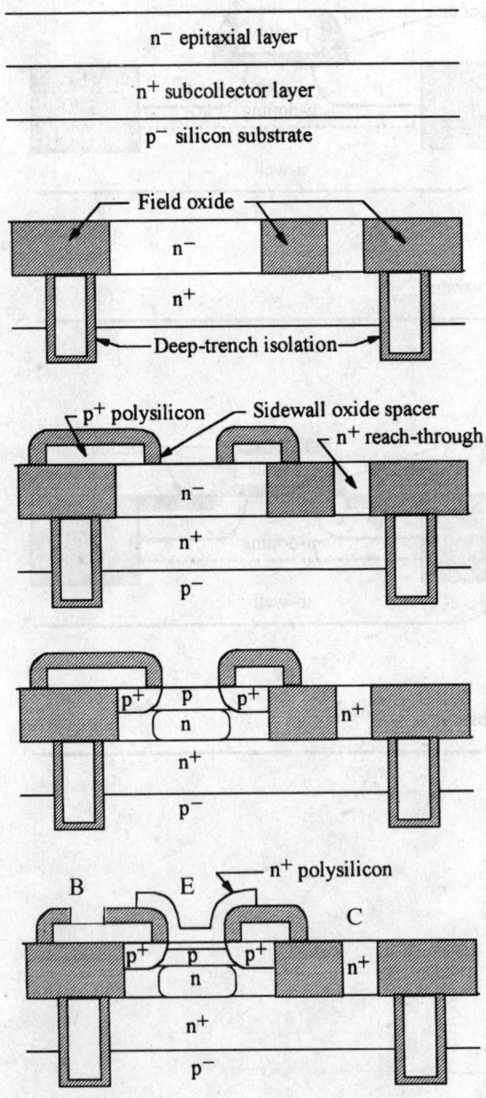


- Starting wafer: $p^-$ silicon
- $n^+$ subcollector layer
- $n^-$ silicon epitaxial layer

- Polysilicon-filled deep-trench isolation
- Shallow-trench field oxide

- $n^+$ reach-through
- $p^+$ polysilicon layer for base contact
- Sidewall oxide spacer

- Diffusion from $p^+$ polysilicon
- n-type pedestal collector
- p-type intrinsic base

- $n^+$ polysilicon layer for emitter
- Contact windows
- Metal (not shown)

**FIGURE A2.1.**

APPENDIX 3

# EFFECTIVE DENSITY OF STATES

The effective densities of states, $N_c$ and $N_v$, in Eqs. (2.4) and (2.5) relate the electron concentration in the conduction band and the hole concentration in the valence band to the Fermi level in the bandgap. In this appendix, we derive an expression for the effective density of states, using electrons in the conduction band as an example.

To determine the density of electronic states within a certain energy range, it is first necessary to determine the number of different momentum values that can be acquired by electrons in this energy range. Based on quantum mechanics, there is one allowed state in a phase space of volume $(\Delta x\, \Delta p_x)(\Delta y\, \Delta p_y)(\Delta z\, \Delta p_z) = h^3$, where $p_x$, $p_y$, $p_z$ are the $x$-, $y$-, $z$- components of the electron momentum and $h$ is Planck's constant. If we let $N(E)\, dE$ be the number of electronic states per unit volume with an energy between $E$ and $E + dE$ in the conduction band, then

$$N(E)\, dE = 2g \frac{dp_x\, dp_y\, dp_z}{h^3}, \tag{A3.1}$$

where $dp_x\, dp_y\, dp_z$ is the volume in the momentum space within which the electron energy lies between $E$ and $E + dE$, $g$ is the number of equivalent minima in the conduction band, and the factor of two arises from the two possible directions of electron spin. The conduction band of silicon has a sixfold degeneracy, so $g = 6$. Note that MKS units are used throughout this appendix (i.e., length must be in meters, not centimeters).

If the electron kinetic energy is not too high, one can consider the energy–momentum relationship near the conduction-band minima as being parabolic and write

$$E - E_c = \frac{p_x^2}{2m_x} + \frac{p_y^2}{2m_y} + \frac{p_z^2}{2m_z}, \tag{A3.2}$$

where $E - E_c$ is the electron kinetic energy, and $m_x, m_y, m_z$ are the effective masses, inversely proportional to the curvatures of the band. For the silicon conduction band in the $\langle 100 \rangle$ direction, the longitudinal mass is $m_x = 0.92m_0$ and the transverse masses are $m_y = m_z = 0.19m_0$, where $m_0$ is the free-electron mass. To obtain an expression for $N(E)$, we change the variables in Eq. (A3.2) to convert the energy–momentum relationship to a spherical one: $p'_x = p_x/(2m_x)^{1/2}$, $p'_y = p_y/(2m_y)^{1/2}$,

$p'_z = p_z/(2m_z)^{1/2}$, and

$$E - E_c = p'^2_x + p'^2_y + p'^2_z = p'^2, \tag{A3.3}$$

where $p'$ is the magnitude of the new "momentum" vector. The RHS of Eq. (A3.1) can be expressed in terms of a small volume element $dp'_x\, dp'_y\, dp'_z$, which can be replaced by the volume of a spherical shell of radius $p'$ and thickness $dp'$: $4\pi p'^2\, dp'$. Using Eq. (A3.3) and $dE = 2p'\, dp'$, one obtains

$$N(E)\, dE = \frac{8\pi g\sqrt{2m_x m_y m_z}}{h^3}\sqrt{E - E_c}\, dE. \tag{A3.4}$$

The electron density of states in an energy diagram is therefore a parabolic function with its downward apex at the conduction-band edge, and vice versa for the hole density of states in the valence band. These are shown schematically in Fig. A3.1 (Sze, 1981), together with the Fermi–Dirac distribution function $f(E)$, given by Eq. (2.1). Since $f(E)$ is the probability that an electronic state at energy $E$ is occupied by an electron, the total number of electrons per unit volume in the conduction band is given by

$$n = \int_{E_c}^{\infty} N(E) f(E)\, dE. \tag{A3.5}$$

Here the upper limit of integration (the top of the conduction band) is taken as infinity. In general, Eq. (A3.5) is a Fermi integral of the order $\frac{1}{2}$ and must be evaluated numerically (Ghandhi, 1968). For nondegenerate silicon with a Fermi level at least $3kT/q$ below the edge of the conduction band, the Fermi–Dirac

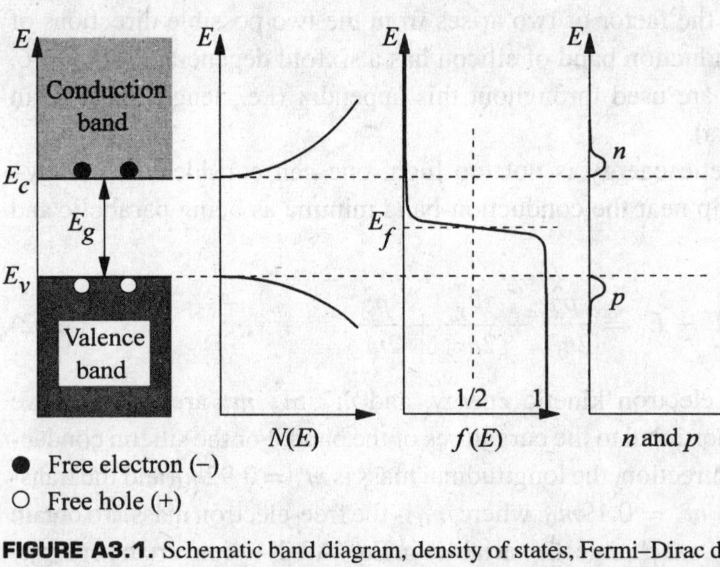


**FIGURE A3.1.** Schematic band diagram, density of states, Fermi–Dirac distribution function, and carrier concentrations versus energy. (After Sze, 1981.)

distribution function can be approximated by the Maxwell–Boltzmann distribution, Eq. (2.2). Equation (A3.5) then becomes

$$n = \frac{8\pi g\sqrt{2m_x m_y m_z}}{h^3}\int_{E_c}^{\infty}\sqrt{E - E_c}e^{-(E-E_f)/kT}\,dE. \qquad \text{(A3.6)}$$

With a change of variable, Eq. (A3.6) is expressed as

$$n = \frac{8\pi g\sqrt{2m_x m_y m_z}}{h^3}(kT)^{3/2}e^{-(E_c-E_f)/kT}\int_0^{\infty}\sqrt{x}e^{-x}\,dx. \qquad \text{(A3.7)}$$

The integral on the RHS of Eq. (A3.7) is of the standard form of a gamma function, $\Gamma(\frac{3}{2})$, which equals $\pi^{1/2}/2$. Now we obtain

$$n = 2g\sqrt{m_x m_y m_z}\left(\frac{2\pi kT}{h^2}\right)^{3/2} e^{-(E_c-E_f)/kT}, \qquad \text{(A3.8)}$$

which is of the same form as Eq. (2.4) with the effective density of states $N_c$ equal to the preexponential factor, i.e.,

$$N_c = 2g\sqrt{m_x m_y m_z}\left(\frac{2\pi kT}{h^2}\right)^{3/2}. \qquad \text{(A3.9)}$$

A similar expression can be derived for the effective density of states $N_v$ in Eq. (2.5) in terms of the hole effective mass and the valence-band degeneracy.

APPENDIX 4

# EINSTEIN RELATIONS

It was stated in Section 2.1.3 that carrier motion in silicon consists of *drift* in the presence of an electric field and *diffusion* in the presence of a concentration gradient. Drift is characterized by the *mobility* defined in Eq. (2.18), and diffusion by the *diffusion coefficient* defined in Eqs. (2.28) and (2.29). Since both mechanisms are closely tied to the random thermal motion of electrons (or holes), the diffusion coefficient and the mobility are related by the Einstein relations, Eqs. (2.30) and (2.31). In this appendix, we briefly describe the physical picture of the drift and diffusion processes, leading to the basic concept behind the Einstein relations. Note that MKS units are used throughout this appendix (i.e., length must be in meters, not centimeters).

## A4.1 DRIFT

Under thermal equilibrium, electrons possess an average kinetic energy proportional to $kT$. They move in random directions through the silicon crystal with an average thermal velocity $v_{th}$. At room temperature, $v_{th}$ is of the order of $10^7$ cm/s. Electrons scatter frequently with the lattice (phonons) and ionized impurity atoms. The average distance electrons travel between collisions is called the *mean free path* $l$, and the average time between collisions is called the *mean free time* $\tau = l/v_{th}$. Typically, $l \approx 100$ Å and $\tau \approx 0.1$ ps.

In the absence of electric field, the net velocity of electrons in any particular direction is zero, since the thermal motion is completely random. When a small electric field $\mathscr{E}$ is applied, however, electrons are accelerated in the direction opposite to the field during the time between collisions. If the effective mass of electrons is $m^*$, the acceleration is given by $-q\mathscr{E}/m^*$. The drift velocity the electrons gain during a mean free time $\tau$ between collisions is thus

$$v_d = -q\mathscr{E}\tau/m^*. \tag{A4.1}$$

After a collision event, electron velocities become randomized again, which means that the average drift velocity is reset to zero and the process starts over again. Since the mobility $\mu$ is defined by Eq. (2.18) as the proportionality of the drift velocity to the electric field, one obtains

$$\mu = \frac{q\tau}{m^*} = \frac{ql}{m^* v_{th}}. \quad \text{(A4.2)}$$

The last step follows from $l = v_{th}\tau$. The drift current density is given by

$$J_{drift} = qnv_d = qn\mu\mathscr{E}, \quad \text{(A4.3)}$$

as indicated by Eq. (2.20).

## A4.2 DIFFUSION

Diffusion current in silicon arises when there is a spatial variation of carrier concentration in the material, that is, the carriers tend to move from a region of high concentration to a region of low concentration. To illustrate the diffusion process, let us assume a one-dimensional case shown in Fig. A4.1, in which the electron density $n$ varies in the $x$-direction (Muller and Kamins, 1977). We consider the number of electrons crossing the plane at $x$ per unit area per unit time. The electrons are moving at thermal velocity $v_{th}$ either in the left or in the right direction and are scattered each time after they travel, on the average, a distance equal to the mean free path $l$. Therefore, the electrons crossing the plane at $x$ from the left in each collision time start at approximately $x - l$, i.e., one mean free path away on the left side of $x$. Since electrons have equal chances of moving left or right, half of the electrons at $x - l$ will move across the plane at $x$ before the next collision takes place. The current density per unit area resulting from the motion of these carriers is then

$$J_{\rightharpoonup} = \frac{1}{2} qn(x - l)v_{th}. \quad \text{(A4.4)}$$

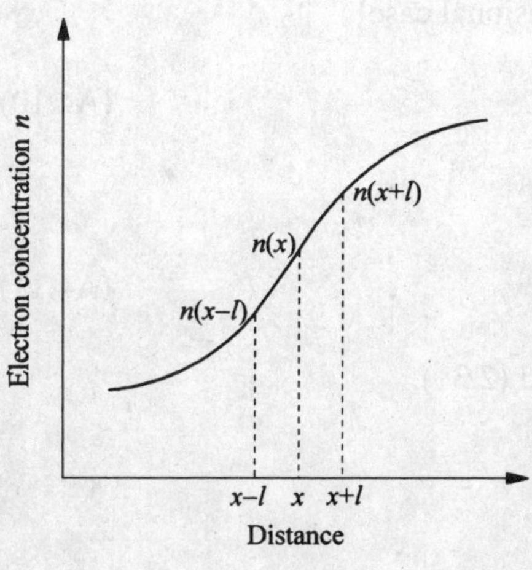


**FIGURE A4.1.** Schematic diagram of electron concentration versus distance in a one-dimensional case. (After Muller and Kamins, 1977.)

This current is negative, since it consists of negative charge moving in the positive direction. Similarly, half of the electrons at $x + l$, one mean free path away on the right side of $x$, will move across the plane $x$ from the right, resulting in a current density in the positive direction:

$$J_+ = \frac{1}{2} q n(x + l) v_{th}. \tag{A4.5}$$

The net diffusion current density at $x$ is then

$$J_{diff} = J_+ - J_- = \frac{1}{2} q v_{th}[n(x + l) - n(x - l)]. \tag{A4.6}$$

Using a Taylor series expansion of $n(x + l)$ and $n(x - l)$, and keeping only the first-order terms, one obtains

$$J_{diff} = \frac{1}{2} q v_{th}\left[\left(n(x) + l\frac{dn}{dx}\right) - \left(n(x) - l\frac{dn}{dx}\right)\right] = q v_{th} l \frac{dn}{dx}. \tag{A4.7}$$

We thus see that the diffusion current density is proportional to the spatial derivative of the electron concentration. In other words, diffusion current arises because of the random thermal motion of charged carriers in a concentration gradient. Equation (A4.7) can be written in the form of Eq. (2.28) with the diffusion coefficient defined as

$$D \equiv v_{th} l. \tag{A4.8}$$

It is now straightforward to derive the Einstein relations. Taking the ratio of Eq. (A4.8) to Eq. (A4.2), one obtains

$$\frac{D}{\mu} = \frac{m^* v_{th}^2}{q}. \tag{A4.9}$$

From the theorem for the equipartition of energy in a one-dimensional case [see Exercise 2.3(a) for the proof of a three-dimensional case],

$$\frac{1}{2} m^* v_{th}^2 = \frac{1}{2} kT. \tag{A4.10}$$

Therefore,

$$\frac{D}{\mu} = \frac{kT}{q}, \tag{A4.11}$$

which is the Einstein relations Eqs. (2.30) and (2.31).

APPENDIX 5

# ELECTRON-INITIATED AND HOLE-INITIATED AVALANCHE BREAKDOWN

In Section 2.4.1, we showed that the multiplication factor for hole-initiated impact ionization is

$$\frac{1}{M_p} = \exp\left(-\int_0^W (\alpha_p - \alpha_n)\,dx\right) - \int_0^W \alpha_n \exp\left(-\int_0^x (\alpha_p - \alpha_n)\,dx'\right)dx, \tag{A5.1}$$

and that for electron-initiated impact ionization is

$$\frac{1}{M_n} = \exp\left(-\int_0^W (\alpha_n - \alpha_p)\,dx\right) - \int_0^W \alpha_p \exp\left(-\int_x^W (\alpha_n - \alpha_p)\,dx'\right)dx. \tag{A5.2}$$

It can be shown (see Exercise 2.12) that in general

$$\int_0^W f(x) \exp\left(-\int_0^x f(x')\,dx'\right)dx = 1 - \exp\left(-\int_0^W f(x)\,dx\right) \tag{A5.3}$$

and

$$\int_0^W f(x) \exp\left(-\int_x^W f(x')\,dx'\right)dx = 1 - \exp\left(-\int_0^W f(x)\,dx\right). \tag{A5.4}$$

Applying Eq. (A5.3) to Eq. (A5.1), we obtain

$$1 - \frac{1}{M_p} = \int_0^W \alpha_p \exp\left(-\int_0^x (\alpha_p - \alpha_n)\,dx'\right)dx. \tag{A5.5}$$

Similarly, applying Eq. (A5.4) to Eq. (A5.2), we have

$$1 - \frac{1}{M_n} = \int_0^W \alpha_n \exp\left(-\int_x^W (\alpha_n - \alpha_p)\,dx'\right)dx. \tag{A5.6}$$

Avalanche breakdown occurs when carrier multiplication by impact ionization runs away, i.e., when the multiplication factor becomes infinite. Thus, Eq. (A5.5) gives the hole-initiated breakdown condition as

$$\int_0^W \alpha_p \exp\left(-\int_0^x (\alpha_p - \alpha_n)\, dx'\right) dx = 1. \tag{A5.7}$$

Using Eq. (A5.3), the left-hand side of Eq. (A5.7) can be rearranged to give

$$\begin{aligned}\int_0^W \alpha_p \exp\left(-\int_0^x (\alpha_p - \alpha_n)\, dx'\right) dx &\\ = 1 - \exp\left(-\int_0^W (\alpha_p - \alpha_n)\, dx'\right)&\\ + \int_0^W \alpha_n \exp\left(-\int_0^x (\alpha_p - \alpha_n)\, dx'\right) dx.&\end{aligned} \tag{A5.8}$$

Substituting Eq. (A5.7) into Eq. (A5.8) gives

$$\begin{aligned}\int_0^W \alpha_n \exp\left(-\int_0^x (\alpha_p - \alpha_n)\, dx'\right) dx &\\ = \exp\left(-\int_0^W (\alpha_p - \alpha_n)\, dx\right).&\end{aligned} \tag{A5.9}$$

Dividing both sides of Eq. (A5.9) by its RHS gives

$$\int_0^W \alpha_n \exp\left(-\int_x^W (\alpha_n - \alpha_p)\, dx'\right) dx = 1, \tag{A5.10}$$

which, according to Eq. (A5.6), is simply the condition for electron-initiated breakdown. Thus ***the condition for avalanche breakdown is the same whether the breakdown process is initiated by electrons or by holes.***

APPENDIX 6

# AN ANALYTICAL SOLUTION FOR THE SHORT-CHANNEL EFFECT IN SUBTHRESHOLD

In this appendix, we outline the mathematical approach that leads to the analytical expression, Eq. (3.66) in Section 3.2.1, for the short-channel threshold roll-off. The short-channel effect (SCE) is a very complex mathematical problem involving the solution of an irregular 2-D boundary-value problem. It is impractical to derive an exact analytical solution applicable to all general cases. Numerical simulations running on a finite-element program should be used to obtain accurate solutions for specific device geometries and doping conditions. Nevertheless, an analytical expression, even an approximate one, goes a long way in providing valuable insights into the fundamentals of the short-channel effect and its controlling parameters. The approach here essentially follows the Ph.D. thesis of Thao N. Nguyen published in 1984.

## A6.1 DEFINING THE PROBLEM WITH SIMPLIFIED BOUNDARY CONDITIONS

To simplify the 2-D boundary-value problem to a manageable level, we make a number of approximations so as to retain only the most basic aspects of the short-channel effect. A simplified short-channel MOSFET geometry is shown in Fig. A6.1 (Nguyen, 1984). The $x$-axis is along the vertical direction, the $y$-axis along the horizontal direction, and the origin at point $A$. As in Section 2.3.2, $\psi(x, y) = \psi_i(x, y) - \psi_i(x = \infty)$ is defined as the intrinsic potential at a point $(x, y)$ with respect to the intrinsic potential of the p-type substrate. The substrate is assumed to be uniformly doped with a concentration $N_a$. In the oxide region $AFGH$, Poisson's equation becomes a homogeneous (Laplace) equation,

$$\frac{\partial^2 \psi}{\partial x^2} + \frac{\partial^2 \psi}{\partial y^2} = 0. \tag{A6.1}$$

In the depletion region in silicon, the concentrations of both types of mobile carriers are negligible under subthreshold conditions. Therefore, in $ABEF$, Poisson's equation is approximated by

$$\frac{\partial^2\psi}{\partial x^2} + \frac{\partial^2\psi}{\partial y^2} = \frac{qN_a}{\varepsilon_{si}}. \tag{A6.2}$$

The length of the silicon region is equal to the channel length $L$. The depth is given by the depletion layer width, $W_d$, to be determined later.

As described by Eq. (2.146), the normal component of the electric field changes by a factor of $\varepsilon_{si}/\varepsilon_{ox} \approx 3$ across the silicon–oxide boundary $AF$. To eliminate this boundary condition so that $\psi$ and its derivatives are continuous, the oxide is replaced by an equivalent region of the same dielectric constant as silicon, but with a thickness equal to $3t_{ox}$. This preserves the capacitance and allows the entire rectangular region to be treated as a homogeneous material of dielectric constant $\varepsilon_{si}$. The drawback is that it may cause some errors in the tangential field, whose magnitude does not change across the silicon–oxide boundary. In the equivalent structure, the tangential field apparently experiences a thicker-than-actual oxide. The errors are expected to be small when the gate oxide is thin compared to the silicon depletion depth $W_d$ so that the oxide field is dominated by its normal component.

If we assume that the source and drain junctions are abrupt and deeper than $W_d$, we can write down the following set of simplified boundary conditions:

$$\psi(-3t_{ox}, y) = V_g - V_{fb} \quad \text{along } GH, \tag{A6.3}$$

$$\psi(x, 0) = \psi_{bi} \quad \text{along } AB, \tag{A6.4}$$

$$\psi(x, L) = \psi_{bi} + V_{ds} \quad \text{along } EF, \tag{A6.5}$$

$$\psi(W_d, y) = 0 \quad \text{along } CD, \tag{A6.6}$$

where $V_g$ and $V_{ds}$ are the gate and source–drain voltages defined in Section 3.1.1, $V_{fb}$ is the flat-band voltage of the gate electrode, and $\psi_{bi}$ is the built-in potential of the source- or drain-to-substrate junction. For an abrupt n$^+$–p junction, $\psi_{bi} = E_g/2q + \psi_B$, where $\psi_B$ is given by Eq. (2.37). If there is a substrate bias $-V_{bs}$, then $\psi_{bi}$ should be replaced by $\psi_{bi} + V_{bs}$ in Eqs. (A6.4) and (A6.5), and $V_g$ by $V_g + V_{bs}$ in Eq. (A6.3). The bottom boundary is actually a movable one, as $W_d$ will change with the gate voltage $V_g$. The distance $BC$ is approximately given by the source junction depletion width,

$$W_S = \sqrt{\frac{2\varepsilon_{si}\psi_{bi}}{qN_a}}. \tag{A6.7}$$

Similarly, $DE$ is given by the drain junction depletion width,

$$W_D = \sqrt{\frac{2\varepsilon_{si}(\psi_{bi} + V_{ds})}{qN_a}}. \tag{A6.8}$$

The boundary conditions along $FG$ and $HA$ are assumed to vary linearly between the end-point values, while those along $BC$ and $DE$ are assumed to vary parabolically between the end points.

## A6.2 SOLUTION TECHNIQUES

The solution technique makes use of the superposition principle and breaks the electrostatic potential into the following terms:

$$\psi(x, y) = v(x, y) + u_L(x, y) + u_R(x, y) + u_B(x, y). \tag{A6.9}$$

Here $v(x, y)$ is a solution to the inhomogeneous (Poisson's) equation and satisfies the top boundary condition, Eq. (A6.3). $u_L, u_R, u_B$ are all solutions to the homogeneous (Laplace) equation, and are chosen in order for $\psi(x, y)$ to satisfy the rest of the boundary conditions, namely, on the left, the right, and the bottom of the rectangular box in Fig. A6.1. For example, $u_L$ is zero on the top, bottom, and right boundaries, but $v + u_L$ satisfies the left boundary condition, Eq. (A6.4). Likewise, $u_R$ is zero on the top, bottom, and left boundaries, but $v + u_R$ satisfies the right boundary condition, Eq. (A6.5), and so on.

A natural choice for $v(x, y)$ [actually $v(x)$] is the long-channel, 1-D MOS solution employing the depletion approximation discussed in Section 2.3.2:

$$v(x, y) = \psi_s^0 - \frac{V_g - V_{fb} - \psi_s^0}{3t_{ox}} x \tag{A6.10}$$

for the oxide region, $\quad -3t_{ox} \leq x \leq 0,$

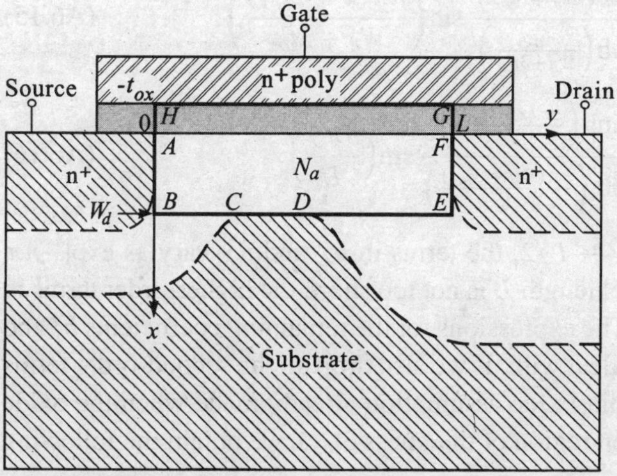


**FIGURE A6.1.** Simplified geometry for analytically solving Poisson's equation in a short-channel MOSFET. (After Nguyen, 1984.)

and

$$v(x,y) = \psi_s^0 \left(1 - \sqrt{\frac{qN_a}{2\varepsilon_{si}\psi_s^0}}\,x\right)^2 \tag{A6.11}$$

for the silicon region, $0 \le x \le W_d$.

Here the long-channel surface potential $\psi_s^0$ is related to $V_g$ by the requirement that $\partial v/\partial x$ be continuous at $x = 0$ (both $v$ and $\partial v/\partial y$ are already continuous in the above equations):

$$\frac{V_g - V_{fb} - \psi_s^0}{3t_{ox}} = \sqrt{\frac{2qN_a\psi_s^0}{\varepsilon_{si}}}. \tag{A6.12}$$

Note that both $v$ and $\partial v/\partial x$ are zero at $x = W_d^0$, where $W_d^0$ is the long-channel depletion width given by

$$W_d^0 = \sqrt{\frac{2\varepsilon_{si}\psi_s^0}{qN_a}}. \tag{A6.13}$$

As we shall see later, the depletion width of a short-channel device can differ from $W_d^0$.

The rest of the solutions are of the following forms (Nguyen, 1984):

$$u_L(x,y) = \sum_{n=1}^{\infty} b_n^* \frac{\sinh\left(\frac{n\pi(L-y)}{W_d+3t_{ox}}\right)}{\sinh\left(\frac{n\pi L}{W_d+3t_{ox}}\right)} \sin\left(\frac{n\pi(x+3t_{ox})}{W_d+3t_{ox}}\right), \tag{A6.14}$$

$$u_R(x,y) = \sum_{n=1}^{\infty} c_n^* \frac{\sinh\left(\frac{n\pi y}{W_d+3t_{ox}}\right)}{\sinh\left(\frac{n\pi L}{W_d+3t_{ox}}\right)} \sin\left(\frac{n\pi(x+3t_{ox})}{W_d+3t_{ox}}\right), \tag{A6.15}$$

$$u_B(x,y) = \sum_{n=1}^{\infty} d_n^* \frac{\sinh\left(\frac{n\pi(x+3t_{ox})}{L}\right)}{\sinh\left(\frac{n\pi(W_d+3t_{ox})}{L}\right)} \sin\left(\frac{n\pi y}{L}\right). \tag{A6.16}$$

In the middle of the device, $y \approx L/2$, the terms in $u_L$ and $u_R$ vary as $\exp[-n\pi L/2(W_d + 3t_{ox})]$. If the channel length $L$ is not too short, the higher-order terms in both series may be neglected. The expressions for the remaining coefficients $b_1^*$ and $c_1^*$ are still too cumbersome to deal with. If we also drop all the second-order terms in $3t_{ox}/W_d$ (similarly to the thin-oxide assumption already made when the oxide region was replaced by a silicon region of thickness $3t_{ox}$), we obtain the following simplified expressions:

$$b_1^* = \frac{4}{\pi}\psi_{bi} - \frac{2}{\pi}\left(1 - \frac{4}{\pi^2}\right)\left(1 + \frac{6t_{ox}}{W_d}\right)\psi_s^0, \tag{A6.17}$$

and

$$c_1^* = \frac{4}{\pi}(\psi_{bi} + V_{ds}) - \frac{2}{\pi}\left(1 - \frac{4}{\pi^2}\right)\left(1 + \frac{6t_{ox}}{W_d}\right)\psi_s^0. \tag{A6.18}$$

The third series, $u_B$, however, cannot be treated similarly, since $\exp[-n\pi(W_d + 3t_{ox})/L]$ decreases much more slowly with $n$. Fortunately, the coefficients $d_n^*$ are at least an order of magnitude less than $b_1^*$ and $c_1^*$ combined. The entire $u_B$ series can therefore be neglected altogether.

An approximate analytical solution under subthreshold conditions is then

$$\psi(x, y) = \psi_s^0 \left(1 - \sqrt{\frac{qN_a}{2\varepsilon_{si}\psi_s^0}}\,x\right)^2 + \frac{b_1^* \sinh\left(\frac{\pi(L-y)}{W_d+3t_{ox}}\right) + c_1^* \sinh\left(\frac{\pi y}{W_d+3t_{ox}}\right)}{\sinh\left(\frac{\pi L}{W_d+3t_{ox}}\right)} \times \sin\left(\frac{\pi(x + 3t_{ox})}{W_d + 3t_{ox}}\right) \tag{A6.19}$$

for the silicon region, $0 \le x \le W_d$. It is straightforward to evaluate $\mathscr{E}_y = -\partial\psi/\partial y$ from the above equation and show that it behaves as depicted in Fig. 3.22(a) and (b). The characteristic length of the exponential decay is $(W_d + 3t_{ox})/\pi$, which scales with the vertical depth of the rectangular region in Fig. A6.1.

## A6.3 SHORT-CHANNEL THRESHOLD VOLTAGE

To find the threshold voltage of a short-channel device, we consider the potential at the silicon surface, $\psi(0, y)$. It has a minimum value at $y = y_c$, determined by solving $\partial\psi(0, y)/\partial y = 0$ with the approximation $\sinh z \approx e^z/2$ for $z > 1$ (Nguyen, 1984):

$$y_c = \frac{L}{2} - \frac{W_d + 3t_{ox}}{2\pi}\ln\left(\frac{c_1^*}{b_1^*}\right) \approx \frac{L}{2} - \frac{W_d + 3t_{ox}}{2\pi}\ln\left(1 + \frac{V_{ds}}{\psi_{bi}}\right). \tag{A6.20}$$

This point corresponds to the point of maximum barrier height in Fig. 3.20. It is close to the midpoint of the channel when the drain voltage is low. When the drain voltage is high, the point of highest barrier moves closer to the source, as depicted in Fig. 3.18(b). Knowing $y_c$ allows one to determine the short-channel depletion width $W_d$ from the self-consistency condition $\partial\psi(W_d, y_c)/\partial x = 0$. As was discussed in Section 3.2.1, $W_d$ is deeper than the long-channel value $W_d^0$ given by Eq. (A6.13). Solving the above condition, one obtains an implicit equation for $W_d$ in terms of $W_d^0$:

$$\frac{W_d}{W_d^0} = 1 + 4\left(\frac{W_d^0}{W_d + 3t_{ox}}\right)\left(\frac{\sqrt{\psi_{bi}(\psi_{bi} + V_{ds})}}{\psi_s^0}\right) e^{-\pi L/2(W_d+3t_{ox})}, \qquad \text{(A6.21)}$$

where only the leading terms in $b_1^*$ and $c_1^*$ were kept. Equation (A6.21) must be solved iteratively. It is clear that for long-channel devices, $W_d = W_d^0$. When $L$ is shorter than about $3W_d^0$ or so, however, the second term on the RHS of Eq. (A6.21) becomes appreciable and $W_d$ can be significantly greater than $W_d^0$. Once this happens, the short-channel effect deteriorates rapidly.

Under subthreshold conditions, current conduction is dominated by diffusion and is controlled by the highest potential barrier along the channel. The threshold voltage of a short-channel device is then determined by the potential at $y = y_c$. Substituting Eq. (A6.20) into Eq. (A6.19) and letting $x = 0$, we obtain

$$\begin{aligned}\psi(0, y_c) &= \psi_s^0 + \frac{6\pi t_{ox}}{W_d + 3t_{ox}} \sqrt{b_1^* c_1^* e^{-\pi L/2(W_d+3t_{ox})}} \\ &\approx \psi_s^0 + \frac{24t_{ox}}{W_d + 3t_{ox}} \sqrt{\psi_{bi}(\psi_{bi} + V_{ds})e^{-\pi L/2(W_d+3t_{ox})}} \\ &\quad - \frac{11t_{ox}}{W_d} \psi_s^0 e^{-\pi L/2(W_d+3t_{ox})}.\end{aligned} \qquad \text{(A6.22)}$$

The first term is the long-channel surface potential. The second term stems from the source–drain fields. It raises the surface potential and helps turn on a short-channel device. The third term is much smaller than the second term but is responsible for the degradation of subthreshold slope in a short-channel device. Because of the second term in Eq. (A6.22), the threshold criterion, $\psi(0, y_c) = 2\psi_B$, is met with a $\psi_s^0$ less than the long-channel threshold value, $\psi_s^0 = 2\psi_B$. Such a $\psi_s^0$ reduction, $\Delta\psi_s^0$, can be translated into a $V_g$ reduction, $\Delta V_g$, by differentiating Eq. (A6.12):

$$\frac{\Delta V_g}{\Delta \psi_s^0} = 1 + \frac{3t_{ox}}{\sqrt{2\varepsilon_{si}\psi_s^0/qN_a}} \approx 1 + \frac{3t_{ox}}{W_{dm}^0} = m, \qquad \text{(A6.23)}$$

where $W_{dm}^0$ is the long-channel maximum depletion width given by Eq. (2.162), and $m$ is the long-channel body-effect coefficient defined by Eq. (3.22). The approximation in Eq. (A6.23) is valid if the short-channel effect is not too severe and $\psi_s^0$ in the square-root expression can be replaced by its long-channel value $2\psi_B$. The threshold voltage lowering in a short-channel device is therefore

$$\Delta V_t = \frac{24mt_{ox}}{W_{dm} + 3t_{ox}} \sqrt{\psi_{bi}(\psi_{bi} + V_{ds})e^{-\pi L/2(W_{dm}+3t_{ox})}}, \qquad \text{(A6.24)}$$

where $W_{dm}$ is the maximum gate depletion width in a short-channel device at threshold. It can be solved for in Eq. (A6.21) with $W_d^0$ replaced by $W_{dm}^0$.

For very short channel lengths, the maximum depletion width is wider than the long-channel value, as can be seen from Figs. 3.18 and 3.23. If $L$ is not too short, the maximum depletion width can be approximated by the long-channel value, and $W_{dm}$ and $W_{dm}^0$ are interchangeable in the definition of $m$. In other words, $m = 1 + (3t_{ox}/W_{dm}^0) \approx 1 + (3t_{ox}/W_{dm})$. Equation (A6.24) can then be expressed alternatively as

$$\Delta V_t = \frac{24t_{ox}}{W_{dm}}\sqrt{\psi_{bi}(\psi_{bi} + V_{ds})}e^{-\pi L/2(W_{dm}+3t_{ox})}, \tag{A6.25}$$

or as

$$\Delta V_t = 8(m-1)\sqrt{\psi_{bi}(\psi_{bi} + V_{ds})}e^{-\pi L/2mW_{dm}}, \tag{A6.26}$$

using Eq. (A6.23).

## A6.4 SHORT-CHANNEL SUBTHRESHOLD SLOPE AND SUBSTRATE SENSITIVITY

The third term in Eq. (A6.22) degrades the subthreshold slope of a short-channel device, since it reduces the sensitivity of $\psi(0, y_c)$ to variations in $\psi_s^0$ and therefore to variations in gate voltage. Following Eq. (3.37), one can write the inverse subthreshold slope of a short-channel device as

$$S \approx 2.3\frac{mkT}{q}\left(1 + \frac{11t_{ox}}{W_{dm}}e^{-\pi L/2(W_{dm}+3t_{ox})}\right), \tag{A6.27}$$

where $m$ is the reciprocal of the rate of change of $\psi_s^0$ with respect to $V_g$. Significant degradation of the subthreshold slope from its long-channel value is usually accompanied by severe short-channel effects. A weak dependence of $S$ on the drain voltage $V_{ds}$ was neglected in the steps leading to Eq. (A6.22).

When there is a substrate bias $-V_{bs}$ present, $\psi_{bi}$ and $V_g$ are replaced by $\psi_{bi} + V_{bs}$ and $V_g + V_{bs}$, and the threshold condition becomes $\psi(0, y_c) = 2\psi_B + V_{bs}$. Following similar steps, one obtains the threshold voltage as a function of the substrate bias, which can be used to evaluate the substrate sensitivity $dV_t/dV_{bs}$ of a short-channel device. At $V_{bs} = 0$, the result is

$$\frac{dV_t}{dV_{bs}} \approx (m-1) - \frac{30t_{ox}}{W_{dm}}e^{-\pi L/2(W_{dm}+3t_{ox})}, \tag{A6.28}$$

which is slightly lower (better) than the long-channel value, $m - 1$. This is because some of the substrate depletion charge terminates on the source and drain instead of on the gate in a short-channel device (Fig. 3.19).

APPENDIX

# 7 QUANTUM-MECHANICAL SOLUTION IN WEAK INVERSION

In this appendix, we derive the expressions for the 2DEG (two-dimensional electron gas) density of states and inversion charge density used in Section 4.2.4. A quantum-mechanical treatment of inversion layer is necessary because of the confinement of electron motion in the direction normal to the surface (Stern and Howard, 1967). Note that MKS units are used throughout this appendix (i.e., length must be in meters, not centimeters).

## A7.1 2-D DENSITY OF STATES

Following an approach in parallel with that of Appendix 3 for the 3-D density of states, one can derive an expression for the density of states of a 2-D gas. Based on quantum mechanics, there is one allowed state in a phase space of volume $(\Delta y\,\Delta p_y)(\Delta z\,\Delta p_z) = h^2$, where $p_y$ and $p_z$ are the $y$- and $z$-components of the electron momentum and $h$ is Planck's constant. If we let $N(E)\,dE$ be the number of electronic states per unit area with an energy between $E$ and $E + dE$, then

$$N(E)\,dE = 2g\frac{dp_y\,dp_z}{h^2}, \tag{A7.1}$$

where $dp_y\,dp_z$ is the area in the momentum space within which the electron energy lies between $E$ and $E + dE$, $g$ is the degeneracy of the subband, and the factor of two arises from the two possible directions of electron spin.

If $E_{min}$ is the ground-state energy of a particular subband, the energy–momentum relationship near the bottom of that subband is

$$E - E_{min} = \frac{p_y^2}{2m_y} + \frac{p_z^2}{2m_z}, \tag{A7.2}$$

where $E - E_{min}$ is the electron kinetic energy, and $m_y$, $m_z$ are the effective masses. Following a change of variables like that in Appendix 3, except replacing spheres and spherical shells with circles and circular rings for the 2-D case at hand, one obtains an expression for $N(E)$:

$$N(E)\,dE = \frac{4\pi g\sqrt{m_y m_z}}{h^2}\,dE. \tag{A7.3}$$

The number of electrons per unit area in this subband is then given by

$$n = \int_{E_{\min}}^{\infty} N(E) f(E)\,dE, \tag{A7.4}$$

where $f(E)$ is the Fermi–Dirac distribution function, Eq. (2.1). Since $N(E)$ is a constant and can be taken out of the integral, Eq. (A7.4) can be easily integrated to yield

$$n = \frac{4\pi g kT\sqrt{m_y m_z}}{h^2}\ln\left[1 + e^{(E_f - E_{\min})/kT}\right]. \tag{A7.5}$$

## A7.2 QM INVERSION CHARGE DENSITY

Figure A7.1 shows the energy-band diagram of a quantized inversion layer at the silicon surface, where the bottom of the conduction band at an energy $E_c'$ is lower than the conduction band $E_c$ in the bulk by an amount $q\psi_s$ due to the applied field from the gate. Here $\psi_s$ is the surface potential or band bending described in Section 2.3.2. In the subthreshold region, the potential function can be approximated by a triangular well of slope $\mathscr{E}_s$ equal to the surface electric field the electrons experience. The quantized subband energies are then represented by $E_j$ ($g = 2$) and $E'_j$ ($g' = 4$) above the conduction band at the surface, where $E_j$ and $E'_j$ are given by Eqs. (4.49) and (4.51) in Section 4.2.4. For the $j$th subband, the minimum

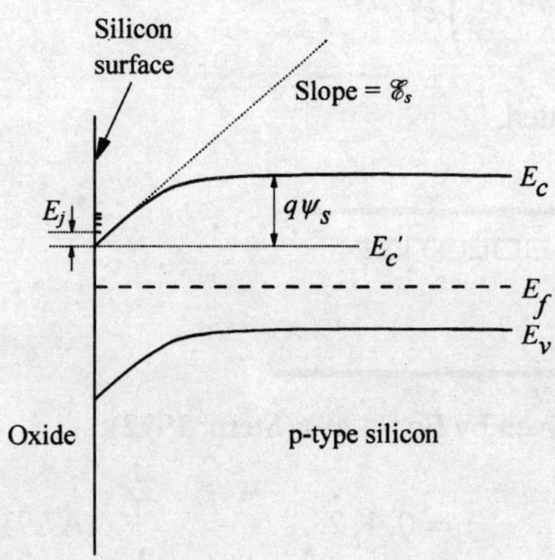


**FIGURE A7.1.** Schematic band diagram showing band bending in the subthreshold region and quantized electron energy levels in the inversion layer at the silicon surface.

energy is

$$E_{\min} = E'_c + E_j = E_c - q\psi_s + E_j. \tag{A7.6}$$

Summing over all the subbands in both valleys using Eq. (A7.5), one obtains the total inversion charge density per unit area (Stern and Howard, 1967):

$$Q_i^{QM} = \frac{4\pi qkT}{h^2}\Bigg\{ g\sqrt{m_y m_z}\sum_j \ln\left[1 + e^{(E_f - E_c - E_j + q\psi_s)/kT}\right] + g'\sqrt{m_x m_y}\sum_j \ln\left[1 + e^{(E_f - E_c - E'_j + q\psi_s)/kT}\right]\Bigg\}, \tag{A7.7}$$

which is the same as Eq. (4.52). Note that for silicon conduction band in the $\langle 100 \rangle$ direction, the longitudinal mass is $m_x = 0.92m_0$ and the transverse masses are $m_y = m_z = 0.19m_0$, where $m_0$ is the free-electron mass. For the first group of subbands in Eq. (A7.7), the longitudinal direction is perpendicular to the surface. For the second group of subbands in Eq. (A7.7), the longitudinal direction is parallel to the surface. In the subthreshold region, the Fermi level is at least a few $kT$ below the lowest subband of energy $E_c - q\psi_s + E_0$, and the factors $\ln(1 + e^x)$ can be approximated by $e^x$ in both terms of Eq. (A7.7). Furthermore, from Eq. (2.4), $E_f - E_c$ in the bulk silicon is equal to $kT\ \ln(n/N_c)$ or $kT\ \ln(n_i^2/N_a N_c)$, where $N_c$ is the effective density of states of the conduction band, and $n = n_i^2/N_a$ is the equilibrium electron concentration in the bulk silicon. For a nonuniformly doped channel, $N_a$ refers to the p-type concentration at the edge of the depletion layer. Equation (A7.7) is then simplified to

$$Q_i^{QM} = \frac{4\pi qkTn_i^2}{h^2 N_c N_a}\left(2\sqrt{m_y m_z}\sum_j e^{-E_j/kT} + 4\sqrt{m_x m_y}\sum_j e^{-E'_j/kT}\right) e^{q\psi_s/kT}, \tag{A7.8}$$

where $g = 2$ and $g' = 4$ have been substituted.

## A7.3 CONVERGENCE OF THE QM SOLUTION AT LOW FIELDS TO THE 3-D CONTINUUM CASE

The energy levels of the lower valley are given by Eq. (4.49) (Stern, 1972):

$$E_j = \left[\frac{3hq\mathscr{E}_s}{4\sqrt{2m_x}}\left(j + \frac{3}{4}\right)\right]^{2/3}, \qquad j = 0, 1, 2, \ldots. \tag{A7.9}$$

When the surface field is low ($\mathscr{E}_s < 10^4$ V/cm at room temperature), the spacings between the quantized energy levels are small compared with $kT$ and the 2-D quantum effect is weak. In this case, the serial summations in Eq. (A7.8) can be replaced by integrals using the identity

$$\sum_n e^{-(n\Delta y)^{2/3}} \Delta y = \int_0^\infty e^{-y^{2/3}} dy \tag{A7.10}$$

in the limit of $\Delta y \to 0$. By a simple change of variable ($u = y^{1/3}$), the integral on the RHS of Eq. (A7.10) can be easily converted to a gamma function, whose value is $3\pi^{1/2}/4$. Therefore,

$$\sum_j e^{-E_j/kT} \to \frac{3\sqrt{\pi}}{4}\left(\frac{4\sqrt{2m_x}(kT)^{3/2}}{3hq\mathscr{E}_s}\right) \tag{A7.11}$$

in the low-field limit. Now Eq. (A7.8) can be evaluated as

$$Q_i^{QM} = \frac{4\pi qkTn_i^2}{h^2N_cN_a}\left[2\sqrt{m_ym_z}\frac{\sqrt{2\pi m_x}(kT)^{3/2}}{hq\mathscr{E}_s} + 4\sqrt{m_xm_y}\frac{\sqrt{2\pi m_z}(kT)^{3/2}}{hq\mathscr{E}_s}\right]e^{q\psi_s/kT}. \tag{A7.12}$$

Substituting $N_c$ from Eq. (A3.9) with the 3-D degeneracy factor $g = 6$ into Eq. (A7.12) yields

$$Q_i^{QM} = \frac{kTn_i^2}{\mathscr{E}_sN_a}e^{q\psi_s/kT}. \tag{A7.13}$$

This is the same as Eq. (3.31) for the 3-D inversion charge density per unit area in the subthreshold region. In other words, when the surface field is low and/or the temperature is high, the 2-D quantum solution converges to the 3-D continuum case. At high fields, however, $Q_i^{QM}$ is lower than the 3-D inversion charge density for the same band bending, which results in a higher threshold voltage than predicted by the classical model, as discussed in Section 4.2.4.

APPENDIX

# 8 DETERMINATION OF EMITTER AND BASE SERIES RESISTANCES

The ideal base current $I_{B0}$ is proportional to $\exp(qV_{BE}/kT)$, where $V_{BE}$ is the forward-bias voltage applied to the emitter and base terminals. In Section 6.3.1, it is shown that the effect of the parasitic emitter and base series resistances is to reduce the voltage appearing across the immediate emitter–base junction by an amount $\Delta V_{BE}$ given by

$$\Delta V_{BE} = I_C r_e + I_B(r_e + r_{bi} + r_{bx}), \tag{A8.1}$$

where $r_e$ is the emitter series resistance, $r_{bi}$ is the intrinsic-base series resistance, $r_{bx}$ is the extrinsic-base series resistance, $I_B$ is the measured base current, and $I_C$ is the measured collector current. As a result of this voltage reduction, the measured base current is reduced from its ideal value by the ratio

$$\frac{I_B}{I_{B0}} = \exp\left(-\frac{q\Delta V_{BE}}{kT}\right). \tag{A8.2}$$

Using Eq. (A8.1), Eq. (A8.2) can be rearranged (Ning and Tang, 1984) to give

$$\frac{kT}{qI_C}\ln\left(\frac{I_{B0}}{I_B}\right) = \left(r_e + \frac{r_{bi}}{\beta_0}\right) + \frac{r_e + r_{bx}}{\beta_0}, \tag{A8.3}$$

where $\beta_0 = I_C/I_B$ is the measured static common-emitter current gain.

## A8.1 THE CASE WHERE THE EMITTER SERIES RESISTANCE IS CONSTANT, INDEPENDENT OF $V_{BE}$

For most bipolar device fabrication processes, the emitter series resistance is a constant, independent of $V_{BE}$ or the device operating current. In this case, $r_e$ and $r_{bx}$ are constants. Also, it is shown in Section 7.2.1 that the collector current density is proportional to the intrinsic-base sheet resistivity. Therefore, $r_{bi}$ is

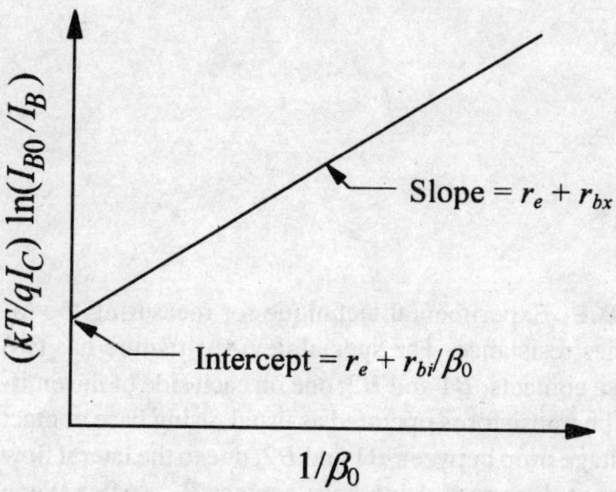


**FIGURE A8.1.** Schematic illustrating the determination of the emitter and base series resistances from Eq. (A8.3). This method works for devices where $r_e$ is a constant, independent of $V_{BE}$.

proportional to $\beta_0$, and the ratio $r_{bi}/\beta_0$ is constant, independent of current. Thus, if $(kT/qI_C)\ln(I_{B0}/I_B)$ is plotted as a function of $1/\beta_0$, the slope gives $r_e + r_{bx}$ and the intercept gives $r_e + r_{bi}/\beta_0$. This is illustrated in Fig. A8.1.

The ratio $r_{bi}/\beta_0$ can be obtained at low currents where the individual values of $r_{bi}$ and $\beta_0$ are relatively independent of current. $\beta_0$ is directly measurable from $I_C$ and $I_B$, and $r_{bi}$ can be calculated as described in Appendix 9, or measured as described below. Once the ratio $r_{bi}/\beta_0$ is determined, $r_e$ and $r_{bx}$ can be extracted from the intercept and the slope of the plot described above. Thus, all three resistance components can be obtained.

## A8.2 THE CASE WHERE THE EMITTER SERIES RESISTANCE IS A FUNCTION OF $V_{BE}$

Sometimes, a particular bipolar device fabrication process produces devices with an emitter series resistance that appears to depend on $V_{BE}$ or on the device operating current. This is often the case with polysilicon-emitter processes, and certainly the case when the transport in and/or across the emitter is governed by tunneling (Yu *et al.*, 1984). In this case, a plot like Fig. A8.1 does not yield a straight line (Riccó *et al.*, 1984; Dubois *et al.*, 1994), and hence cannot be used to extract the emitter and base series resistances.

## A8.3 DIRECT MEASUREMENT OF BASE RESISTANCES

The total base resistance can be measured directly by using a test device structure with two separate base contacts, one on each side of the emitter stripe (Weng *et al.*, 1992). This device structure and the measurement technique are illustrated in Fig. A8.2. Here the mask dimension of the emitter stripe is $W_{mask}$, and the electrical emitter stripe width is $W$. Base contact B1 is used to operate the transistor as usual,

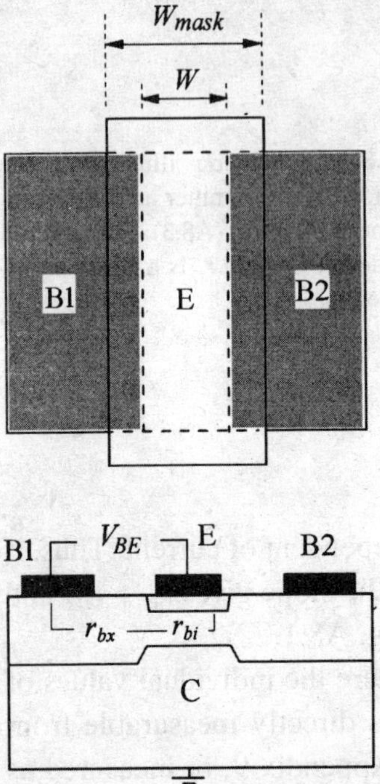


**FIGURE A8.2.** Experimental technique for measuring the total base series resistance. The special device structure has two separate base contacts, $B1$ and $B2$, one on each side of the emitter stripe. The transistor is operated as usual, using base contact $B1$. The voltage drop between $B1$ and $B2$, due to the lateral flow of base current, is sensed using base contact $B2$. (After Weng *et al.*, 1992).

while base contact B2 is used only to sense the potential difference that develops between B1 and B2 when a base current flows from B1 into the emitter terminal. As a terminal for voltage sensing only, B2 draws negligible base current. Thus, the device operates like a transistor with base contact on only side of the emitter stripe. The total base series resistance is simply the measured potential difference between B1 and B2, divided by the measured base current.

Let $r_b$ be the total base resistance measured as described above. That is,

$$r_b = r_{bx} + r_{bi}. \tag{A8.4}$$

It is reasonable to assume that, for a given device fabrication process, $W$ is related to its mask dimension $W_{mask}$ by

$$W = W_{mask} - 2\,\Delta W, \tag{A8.5}$$

where $\Delta W$ represents the overlap per edge between the emitter mask and the extrinsic-base region. Also, it is shown in Appendix 9 that $r_{bi}$ is proportional to the emitter stripe width. Therefore,

$$r_{bi} \propto W_{mask} - 2\,\Delta W. \tag{A8.6}$$

By using specially designed device structures of various $W_{mask}$ dimensions and measuring $r_b$ as a function of $W_{mask}$, $r_{bx}$ can be determined (Weng *et al.*, 1992). This is illustrated in Fig. A8.3. Once $r_{bx}$ is known, $r_{bi}$ for a given $W_{mask}$ can be determined.

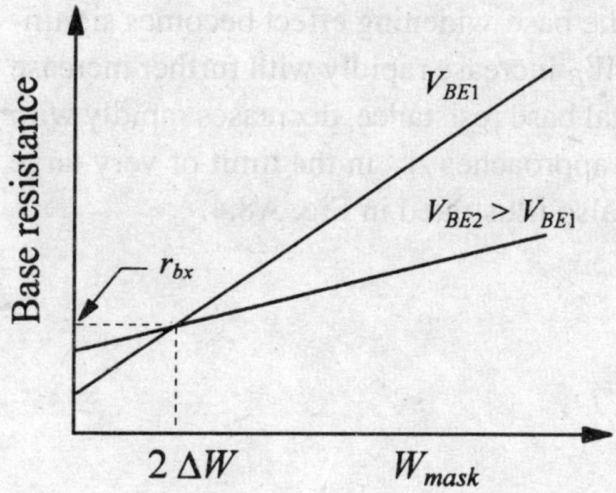


**FIGURE A8.3.** Schematic illustrating the determination of the total base resistance $r_b$ and its extrinsic-base component $r_{bx}$ using the device structure shown in Fig. A8.2.

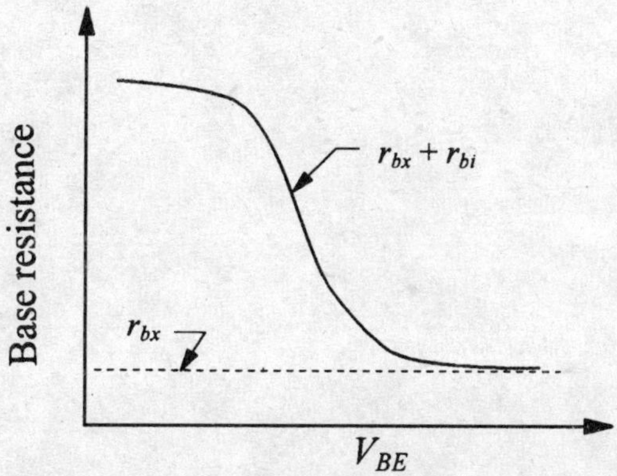


**FIGURE A8.4.** Schematic illustrating the dependence of the total base resistance on forward $V_{BE}$.

## A8.4 DEPENDENCE OF BASE RESISTANCE ON $V_{BE}$

The intrinsic-base sheet resistivity is given by Eq. (7.5), i.e.,

$$R_{Sbi} = \left( q \int_0^{W_B} p_p(x)\mu_p(x)\,dx \right)^{-1} \tag{A8.7}$$

As the emitter–base diode is forward-biased, the emitter–base junction depletion-layer width decreases, and hence the intrinsic-base width $W_B$ increases, as can be seen readily from Eq. (2.70). Also, as shown in Eq. (6.82), the electrons injected from the emitter and traversing the base–collector junction depletion layer causes the base width to increase. The amount of base-width increase increases with the injected electron concentration. As a result, at small emitter–base forward biases, $R_{Sbi}$ decreases slowly as $W_B$ increases with increasing $V_{BE}$. The total base resistance therefore decreases slowly with increasing $V_{BE}$. This is illustrated in Fig. A8.4.

At large emitter–base forward biases, the base-widening effect becomes significant. When that happens, both $p_p(x)$ and $W_B$ increase rapidly with further increase in $V_{BE}$. As a result, $R_{Sbi}$, and hence the total base resistance, decreases rapidly with increasing $V_{BE}$. The total base resistance approaches $r_{bx}$ in the limit of very large forward $V_{BE}$ (Weng *et al.*, 1992). This is also illustrated in Fig. A8.4.

APPENDIX 9

# INTRINSIC-BASE RESISTANCE

Consider the intrinsic part of a bipolar transistor, the cross section of which is shown schematically in Fig. A9.1. The base current $I_B$ enters the intrinsic-base region at the base contact and then spreads out, turns upward, and enters the emitter. Thus, the effective intrinsic-base resistance $r_{bi}$ as seen by the base current depends on how the base current spreads out inside the intrinsic-base layer. One commonly used method for evaluating $r_{bi}$ is to consider the power dissipation $P$ in the intrinsic base (Hauser, 1968), and define $r_{bi}$ by

$$P = I_B^2 r_{bi}. \tag{A9.1}$$

## A9.1 THE CASE OF NEGLIGIBLE CURRENT CROWDING

The power dissipation can be evaluated readily for low-current situations where there is no current crowding and the base-current density $J_B$ entering the emitter can be assumed to be uniform. That this is a good assumption for modern bipolar transistors will be shown later.

Let us assume the emitter stripe has a width $W$ and a length $L$, and the base is contacted on one side of the emitter only, as shown in Fig. A9.1. Consider a slice of the intrinsic base between points $y$ and $y + \Delta y$. The resistance of this slice as seen by the base current is

$$\Delta R = \frac{R_{Sbi}}{L} \Delta y, \tag{A9.2}$$

where $R_{Sbi}$ is the intrinsic-base sheet resistivity given by Eq. (7.5). The base current passing through this slice is

$$i_B(y) = J_B L(W - y), \tag{A9.3}$$

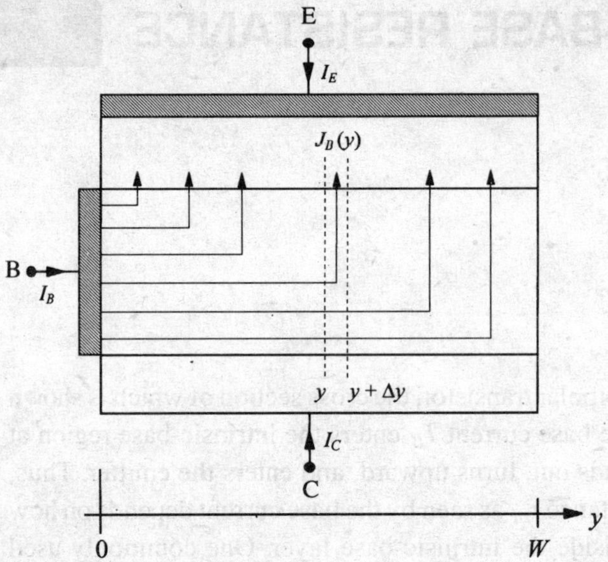


**FIGURE A9.1.** Schematic of the intrinsic part of a bipolar transistor illustrating the flow of base current. The transistor has an emitter stripe of width $W$ and a base contact on only one side.

and the power dissipated in this slice is

$$\Delta P = i_B^2(y)\,\Delta R = \frac{R_{Sbi}}{L} i_B^2(y)\Delta y. \tag{A9.4}$$

$R_{Sbi}$ is a function of collector current density, which in turn is a function of the base–emitter diode voltage. The assumption of negligible current crowding implies that there is negligible lateral voltage drop along the intrinsic base, and $R_{Sbi}$ is independent of $y$. Therefore, the total power dissipated by the base current in the intrinsic base is

$$\begin{aligned} P &= \frac{R_{Sbi}}{L} \int_0^W i_B^2(y)\,dy \qquad\qquad (A9.5) \\ &= R_{Sbi} J_B^2 L \int_0^W (W-y)^2\,dy \\ &= \frac{1}{3} R_{Sbi} J_B^2 L W^3 \\ &= \frac{1}{3}\left(\frac{W}{L}\right) R_{Sbi} I_B^2. \end{aligned}$$

Comparison of Eq. (A9.1) with Eq. (A9.5) gives

$$r_{bi} = \frac{1}{3}\left(\frac{W}{L}\right) R_{Sbi}. \tag{A9.6}$$

It should be noted that, as illustrated in Fig. A8.4, ***$R_{Sbi}$, and hence $r_{bi}$, rolls off with collector current density***. The roll-off can be very rapid once the collector current density is large enough to cause appreciable base widening.

## A9.2 OTHER EMITTER GEOMETRIES

The above calculations can be applied to other emitter geometries and/or base contact schemes. For the same emitter-stripe geometry, but with base contact on both sides of the emitter, the base resistance is reduced by a factor of 4, namely

$$r_{bi} = \frac{1}{12}\left(\frac{W}{L}\right) R_{Sbi} \quad \text{for two-sided base contact.} \tag{A9.7}$$

The base resistance for a square emitter with base contact on all four sides is

$$r_{bi} = \frac{1}{32} R_{Sbi} \quad \text{for four-sided base contact.} \tag{A9.8}$$

For a round emitter, it is

$$r_{bi} = \frac{1}{8\pi} R_{Sbi} \quad \text{for round emitter.} \tag{A9.9}$$

## A9.3 ESTIMATION OF EMITTER CURRENT-CROWDING EFFECT

The assumption of uniform base current density used above is valid when the maximum lateral voltage drop in the intrinsic-base layer, caused by the lateral flow of the base current, is small compared to $kT/q$. If the lateral voltage drop is not negligible, then the base current density becomes a function of $y$, with

$$J_B(y) = J_B(0) \exp\left(\frac{qV_{BE}(y)}{kT} - \frac{qV_{BE}(0)}{kT}\right). \tag{A9.10}$$

Since $V_{BE}(y) \le V_{BE}(0)$, we have $J_B(y) \le J_B(0)$. That is, the emitter current density is larger at the emitter edge than towards the middle of the emitter. This is known as the emitter current-crowding effect. The general expressions for $J_B(y)$ and $V_{BE}(y)$

have been derived by Hauser (1964, 1968). The derivation is rather involved, and the interested reader is referred to the references for details. Here we simply give an upper-bound estimate of the emitter current-crowding effect.

When the base current density is a function of distance from the emitter edge, then instead of Eq. (A9.3), the base current passing through the intrinsic-base slice at point $y$ is

$$i_B(y) = L \int_y^W J_B(y')\,dy'. \tag{A9.11}$$

The voltage drop across an intrinsic-base slice between $y$ and $y + \Delta y$ is

$$\begin{aligned} \Delta V_{BE}(y) &= V_{BE}(y + \Delta y) - V_{BE}(y) \\ &= -i_B(y)\,\Delta R(y) = -\Delta y\, R_{Sbi}(J_C(y)) \int_y^W J_B(y')\,dy', \end{aligned} \tag{A9.12}$$

where Eq. (A9.2) has been used for $\Delta R$. In Eq. (A9.12), $R_{Sbi}$ is denoted explicitly as a function of $J_C$, which in turn is a function of $y$. The gradient of the base–emitter voltage along the intrinsic-base layer is therefore

$$\begin{aligned} \frac{dV_{BE}(y)}{dy} &= -R_{Sbi}(J_C(y)) \int_y^W J_B(y')\,dy' \\ &= -R_{Sbi}(J_C(y))J_B(0) \int_y^W \exp\left(\frac{qV_{BE}(y')}{kT} - \frac{qV_{BE}(0)}{kT}\right) dy', \end{aligned} \tag{A9.13}$$

where we have used Eq. (A9.10) for $J_B(y)$. Notice that this voltage gradient is negative. An upperbound of the magnitude of this voltage gradient can be obtained by replacing the exponential inside the integral, which is smaller than unity because $V_{BE}(y) \le V_{BE}(0)$, by unity, and replacing $R_{Sbi}$ by its low-current value. That is,

$$\begin{aligned} \left|\frac{dV_{BE}(y)}{dy}\right| &\le R_{Sbi}(J_C \to 0)J_B(0) \int_y^W dy' \\ &= R_{Sbi}(J_c \to 0)J_B(0)(W - y). \end{aligned} \tag{A9.14}$$

Integrating Eq. (A9.14) gives an upper-bound estimate of the maximum voltage drop along the intrinsic base due to the lateral flow of base current. This estimate is

$$\begin{aligned} V_{BE}(0) - V_{BE}(W) &< R_{Sbi}(J_C \to 0)J_B(0) \int_0^W (W - y)\,dy \\ &= R_{Sbi}(J_C \to 0)J_B(0)\frac{W^2}{2}. \end{aligned} \tag{A9.15}$$

Consider a typical modern bipolar transistor. To avoid excessive base-widening effect, the collector current density is typically under 1 mA/µm$^2$. For a typical current gain of 100, $J_B$ is therefore typically under 0.01 mA/μm$^2$. The low-current value of $R_{Sbi}$ is typically 10 kΩ/□. If the transistor has an emitter-stripe width of 0.5 μm, with base contact on both sides of the stripe, then $W = 0.25$ μm. For this typical transistor, Eq. (A9.15) suggests that the maximum lateral voltage drop in the intrinsic base is less than 3 mV, much smaller than $kT/q$, which is about 26 mV at room temperature. As the intrinsic-base sheet resistivity rolls off with emitter–base forward bias, this lateral voltage drop becomes even smaller. Therefore, ***emitter current-crowding effect is negligible for modern bipolar transistors***. Of course, for transistors of earlier generations, where the emitter stripes were much wider, emitter current crowding could be significant. Our estimates are consistent with the results obtained from a more exact calculation of the internal base voltage distribution (Chiu *et al.*, 1992).

# REFERENCES


N. Arora (1993). *MOSFET Models for VLSI Circuit Simulation*, Springer-Verlag, Wien.

N. D. Arora and G. S. Gildenblat (1987). "A semi-empirical model of the MOSFET inversion layer mobility for low-temperature operation," *IEEE Trans. Electron Devices* **ED-34**, p. 89.

P. Ashburn (1988). *Design and Realization of Bipolar Transistors*, Wiley, Chichester.

G. Baccarani and G. A. Sai-Halasz (1983). "Spreading resistance in submicron MOSFETs," *IEEE Electron Device Lett.* **EDL-4**, p. 27.

G. Baccarani and M. R. Wordeman (1983). "Transconductance degradation in thin-oxide MOSFETs," *IEEE Trans. Electron Devices* **ED-30**, p. 1295.

G. Baccarani, M. R. Wordeman, and R. H. Dennard (1984). "Generalized scaling theory and its application to a 1/4 micrometer MOSFET design," *IEEE Trans. Electron Devices* **ED-31**, p. 452.

H. B. Bakoglu (1990). *Circuits, Interconnections, and Packaging for VLSI*, Addison-Wesley.

P. Balk, P. G. Burkhardt, and L. V. Gregor (1965). "Orientation dependence of built-in surface charge on thermally oxidized silicon," *IEEE Proc.* **53**, p. 2133.

B. Balland and G. Barbottin (1989). "The trapping and detrapping of carriers injected into $SiO_2$," Chapter 10 in *Instabilities in Silicon Devices*, Vol. 2, edited by G. Barbottin and A. Vapaille, North-Holland.

H. H. Berger (1972). "Models for contacts to planar devices," *Solid-State Electron.* **15**, p. 145.

J. C. Bourgoin (1989). "Radiation-induced defects in the $Si–SiO_2$ structure," Chapter 17 in *Instabilities in Silicon Devices*, Vol. 2, edited by G. Barbottin and A. Vapaille, North-Holland.

J. R. Brews (1978). "A charge sheet model of the MOSFET," *Solid-State Electron.* **21**, p. 345.

J. R. Brews (1979). "Threshold shifts due to nonuniform doping profiles in surface channel MOSFETs," *IEEE Trans. Electron Devices* **ED-26**, p. 1696.

J. R. Brews (1981). "Physics of the MOS transistor," *Applied Solid State Science*, Suppl. 2A, edited by D. Kahng, Academic, New York.

J. R. Burns (1964). "Switching response of complementary-symmetry MOS transistor logic circuits," *RCA Rev.* **25**, p. 627.

D. M. Caughey and R. E. Thomas (1967). "Carrier mobilities in silicon empirically related to doping and field," *Proc. IEEE* **55**, p. 2192.

T. Y. Chan, J. Chen, P. K. Ko, and C. Hu (1987a). "The impact of gate-induced drain leakage current on MOSFET scaling," *1987 IEDM Technical Digest*, pp. 718–721.

T. Y. Chan, A. T. Wu, P. K. Ko, and C. Hu (1987b). "Effects of the gate-to-drain/source overlap on MOSFET characteristics," *IEEE Electron Device Lett.* **EDL-8**, p. 326.

L. L. Chang, P. J. Stiles, and L. Esaki (1967). "Electron tunneling between a metal and a semiconductor: characteristics of Al–$Al_2O_3$–SnTe and –GeTe junctions," *J. Appl. Phys.* **38**, pp. 4440–4445.

I. C. Chen, S. Holland, and C. Hu (1986). "Oxide breakdown dependence on thickness and hole current-enhanced reliability of ultra-thin oxides," *1986 IEDM Technical Digest*, pp. 660–663.

T. C. Chen, J. D. Cressler, K. Y. Toh, J. Warnock, P. F. Lu, K. A. Jenkins, S. Basavaiah, M. P. Manny, H. Y. Ng, D. D. Tang, G. P. Li, C. T. Chuang, M. R. Polcari, M. B. Ketchen, and T. H. Ning (1989). "A submicron high performance bipolar technology," *1989 Symp. VLSI Technology Digest of Tech. Papers*, pp. 87–88.

W. Chen, Y. Taur, D. Sadana, K. A. Jenkins, J. Y.-C. Sun, and S. Cohen (1996). "Suppression of the SOI floating-body effects by linked-body structure," *1996 VLSI Technology Symp. Technical Digest*, IEEE, p. 92.

J. G. J. Chern, P. Chang, R. F. Motta, and N. Godinho (1980). "A new method to determine MOSFET channel length," *IEEE Electron Device Lett.* **ED-1**, p. 170.

T.-Y. Chiu, P. K. Tien, J. Sung, and T.-Y. M. Liu (1992). "A new analytical model and the impact of base charge storage on base potential distribution, emitter current crowding and base resistance," *1992 IEDM Technical Digest*, pp. 573–576.

E. F. Chor, P. Ashburn, and A. Brunnschweiler (1985). "Emitter resistance of arsenic- and phosphorus-doped polysilicon emitter transistors," *IEEE Electron Device Lett.* **EDL-6**, pp. 516–518.

E. F. Chor, A. Brunnschweiler, and P. Ashburn (1988). "A propagation-delay expression and its application to the optimization of polysilicon emitter ECL processes," *IEEE J. Solid-State Circuits* **23**, pp. 251–259.

C. T. Chuang, K. Chin, J. M. C. Stork, G. L. Patton, E. F. Crabbé, and J. H. Comfort (1992). "On the leverage of high-$f_T$ transistors for advanced high-speed bipolar circuits," *IEEE J. Solid-State Circuits* **27**, pp. 225–228.

A. G. Chynoweth (1957). "Ionization rates for electrons and holes in silicon," *Phys. Rev.* **109**, pp. 1537–1540.

A. G. Chynoweth, W. L. Feldmann, C. A. Lee, R. A. Logan, and G. L. Pearson (1960). "Internal field emission at narrow silicon and germanium p–n junctions," *Phys. Rev.* **118**, pp. 425–434.

R. W. Coen and R. S. Muller (1980). "Velocity of surface carriers in inversion layers on silicon," *Solid-State Electron.* **23**, p. 35.

E. F. Crabbé, B. S. Meyerson, J. M. C. Stork, and D. L. Harame (1993). "Vertical profile optimization of very high frequency epitaxial Si- and SiGe-base bipolar transistors," *1993 IEDM Technical Digest*, pp. 83–86.

C. R. Crowell and S. M. Sze (1966). "Temperature dependence of avalanche multiplication in semiconductors," *Appl. Phys. Lett.* **9**, pp. 242–244.

B. Davari, C. Y. Ting, K. Y. Ahn, S. Basavaiah, C. K. Hu, Y. Taur, M. R. Wordeman, O. Aboelfotoh, L. Krusin-Elbaum, R. V. Joshi, and M. R. Polcari (1987). "Submicron tungsten-gate MOSFET with 10-nm gate oxide," *1987 VLSI Technology Symp. Technical Digest*, IEEE, pp. 61–62.

B. Davari, W. H. Chang, M. R. Wordeman, C. S. Oh, Y. Taur, K. E. Petrillo, D. Moy, J. Bucchignano, H. Y. Ng, M. Rosenfield, F. J. Hohn, and M. Rodriguez (1988a). "A high-performance 0.25-μm CMOS technology," *1988 IEDM Technical Digest*, pp. 56–59.

B. Davari, C. Koburger, T. Furukawa, Y. Taur, W. Noble, A. Megdanis, J. Warnock, and J. Mauer (1988b). "A variable-size shallow trench isolation technology with diffused sidewall doping for submicron CMOS," *1988 IEDM Technical Digest*, pp. 92–95.

B. Davari, C. W. Koburger, R. Schulz, J. D. Warnock, T. Furukawa, M. Jost, Y. Taur, W. G. Schwittek, J. K. DeBrosse, M. L. Kerbaugh, and J. L. Mauer (1989). "A new planarization technique using a combination of RIE and chemical mechanical polish," *1989 IEDM Technical Digest*, pp. 61–64.

B. E. Deal (1980). "Standardized terminology for oxide charges associated with thermally oxidized silicon," *IEEE Trans. Electron Devices* **27**, pp. 606–608.

B. E. Deal, M. Sklar, A. S. Grove, and E. H. Snow (1967). "Characteristics of the surface-state charge of thermally oxidized silicon," *J. Electrochem. Soc.* **114**, pp. 266–274.

H. C. de Graaff, J. W. Slotboom, and A. Schmitz (1977). "The emitter efficiency of bipolar transistors," *Solid-State Electron.* **20**, pp. 515–521.

J. del Alamo, S. Swirhun, and R. M. Swanson (1985a). "Measuring and modeling minority carrier transport in heavily doped silicon," *Solid-State Electron.* **28**, pp. 47–54.

J. del Alamo, S. Swirhun, and R. M. Swanson (1985b). "Simultaneous measurement of hole lifetime, hole mobility and bandgap narrowing in heavily doped n-type silicon," *1985 IEDM Technical Digest*, pp. 290–293.

R. H. Dennard (1968). "Field-effect transistor memory," U. S. Patent 3,387,286 issued June 4.

R. H. Dennard (1984). "Evolution of the MOSFET dynamic RAM – a personal view," *IEEE Trans. Electron Devices* **ED-31**, p. 1549.

R. H. Dennard (1986). "Scaling limits of silicon VLSI technology," in *The Physics and Fabrication of Microstructures and Microdevices*, edited by M. J. Kelly and C. Weisbuch, Springer-Verlag, Berlin.

R. H. Dennard, F. H. Gaensslen, H. N. Yu, V. L. Rideout, E. Bassous, and A. R. LeBlanc (1974). "Design of ion-implanted MOSFETs with very small physical dimensions," *IEEE J. Solid-State Circuits* **SC-9**, p. 256.

D. J. DiMaria and J. H. Stathis (1997). "Explanation for the oxide thickness dependence of breakdown characteristics of metal–oxide semiconductor structures," *Appl. Phys. Lett.* **70**, pp. 2708–2710.

D. J. DiMaria, E. Cartier, and D. Arnold (1993). "Impact ionization, trap creation, degradation, and breakdown in silicon dioxide films on silicon," *J. Appl. Phys.* **73**, pp. 3367–3384.

T. H. DiStefano and M. Shatzkes (1974). "Impact ionization model for dielectric instability and breakdown," *Appl. Phys. Lett.* **25**, pp. 685–687.

E. Dubois, P.-H. Bricout, and E. Robilliart (1994). "Accuracy of series resistances extraction schemes for polysilicon bipolar transistors," *Proc. Bipolar/BiCMOS Circuit & Technology Meetings, IEEE*, pp. 148–151.

J. Dziewior and W. Schmid (1977). "Auger coefficients for highly doped and highly excited silicon," *Appl. Phys. Lett.* **31**, pp. 346–348.

J. Dziewior and D. Silber (1979). "Minority-carrier diffusion coefficients in highly doped silicon," *Appl. Phys. Lett.* **35**, pp. 170–172.

J. M. Early (1952). "Effects of space-charge layer widening in junction transistors," *Proc. IRE* **40**, pp. 1401–1406.

J. J. Ebers and J. L. Moll (1954). "Large-signal behavior of junction transistors," *Proc. IRE* **42**, pp. 1761–1772.

Y. A. El-Mansy and A. R. Boothroyd (1977). "A simple two-dimensional model for IGFET," *IEEE Trans. Electron Devices* **ED-24**, p. 254.

EMIS Datareviews Series No. 4 (1988). *Properties of Silicon*, INSPEC, The Institute of Electrical Engineers, London.

R. B. Fair and H. W. Wivell (1976). "Zener and avalanche breakdown in As-implanted low voltage Si n–p junctions," *IEEE Trans. Electron Devices* **ED-23**, pp. 512–518.

W. Filensky and H. Beneking (1981). "New technique for determination of static emitter and collector series resistances of bipolar transistors," *Electron. Lett.* **17**, pp. 503–504.

M. V. Fischetti and S. E. Laux (1996). "Band structure, deformation potentials, and carrier mobility in strained Si, Ge, and SiGe alloys," *J. Appl. Phys.* **80**(4), p. 2234.

M. V. Fischetti, S. E. Laux, and E. Crabbé (1995). "Understanding hot-electron transport in silicon devices: is there a shortcut?," *J. Appl. Phys.* **78**, pp. 1058–1087.

F. H. Gaensslen, V. L. Rideout, E. J. Walker, and J. J. Walker (1977). "Very small MOSFETs for low temperature operation," *IEEE Trans. Electron Devices* **ED-24**, p. 218.

J. Gautier, K. A. Jenkins, and J. Y.-C. Sun (1995). "Body charge related transient effects in floating-body SOI nMOSFETs," *1995 IEDM Technical Digest*, p. 623.

S. K. Ghandhi (1968). *The Theory and Practice of Microelectronics*, Wiley, New York.

R. H. Good, Jr. and E. W. Müller (1956). In *Handbuch der Physik*, Vol. XXI, pp. 176–231, Springer-Verlag, Berlin.

A. M. Goodman (1966). "Photoemission of holes from silicon into silicon dioxide," *Phys. Rev.* **152**, pp. 780–784.

W. H. Grant (1973). "Electron and hole ionization rates in epitaxial silicon at high electric fields," *Solid-State Electron.* **16**, pp. 1189–1203.

P. E. Gray, D. DeWitt, A. R. Boothroyd, and J. F. Gibbons (1964). *Physical Electronics and Circuit Models of Transistors*, Vol. 2, Semiconductor Electronics Education Committee, Wiley, New York.

A. S. Grove (1967). *Physics and Technology of Semiconductor Devices*, Wiley, New York.

A. S. Grove and D. J. Fitzgerald (1966). "Surface effects on p–n junctions – characteristics of surface space-charge regions under non-equilibrium conditions," *Solid-State Electron.* **9**, pp. 783–806.

H. K. Gummel (1961). "Measurement of the number of impurities in the base layer of a transistor," *Proc. IRE* **49**, 834.

H. K. Gummel (1967). "Hole–electron product of p–n junctions," *Solid-State Electron.* **10**, pp. 209–212.

J. C. Guo, S. S. Chung, and C. C. Hsu (1994). "A new approach to determine the effective channel length and the drain-and-source series resistance of miniaturized MOSFETs," *IEEE Trans. Electron Devices* **ED-41**, p. 1811.

D. L. Harame, J. H. Comfort, J. D. Cressler, E. F. Crabbé, J. Y.-C. Sun, B. S. Meyerson, and T. Tice (1995a). "Si/SiGe epitaxial-base transistors–part I: materials, physics, and circuits," *IEEE Trans. Electron Devices* **42**, pp. 455–468.

D. L. Harame, J. H. Comfort, J. D. Cressler, E. F. Crabbé, J. Y.-C. Sun, B. S. Meyerson, and T. Tice (1995b). "Si/SiGe epitaxial-base transistors–part II: process integration and analog applications," *IEEE Trans. Electron Devices* **42**, pp. 469–482.

E. Harari (1978). "Dielectric breakdown in electrically stressed thin films of thermal $SiO_2$," *J. Appl. Phys.* **49**, pp. 2478–2489.

J. R. Hauser (1964). "The effects of distributed base potential on emitter current injection density and effective base resistance for stripe transistor geometries," *IEEE Trans. Electron Devices* **ED-11**, pp. 238–242.

J. R. Hauser (1968). "Bipolar transistors," *in Fundamentals of Silicon Integrated Device Technology,* Vol. II, edited by R. M. Burger and R. P. Donovan, Prentice-Hall, Englewood Cliffs, N. J.

N. Hedenstierna and K. O. Jeppson (1987). "CMOS circuit speed and buffer optimization," *IEEE Trans. Computer-Aided Design* **CAD-6**, p. 270.

M. W. Hillen and J. F. Verwey (1986). "Mobile ions in $SiO_2$ layers on Si," Chapter 8 in *Instabilities in Silicon Devices*, Vol. 1, edited by G. Barbottin and A. Vapaille, North-Holland.

C. C.-H. Hsu and T. H. Ning (1991). "Voltage and temperature dependence of interface trap generation by hot electrons in p- and n-poly gated MOSFETs," *1991 Symp. VLSI Technology Digest of Tech. Papers, IEEE,* pp. 17–18.

S. M. Hu and S. Schmidt (1968). "Interactions in sequential diffusion processes in semiconductors," *J. Appl. Phys.* **39,** pp. 4272–4283.

G. J. Hu, C. Chang, and Y. T. Chia (1987). "Gate-voltage dependent effective channel length and series resistance of LDD MOSFETs," *IEEE Trans. Electron Devices* **ED-34**, p. 2469.

J. Hui, S. Wong, and J. Moll (1985). "Specific contact resistivity of $TiSi_2$ to $p^+$ and $n^+$ junctions," *IEEE Electron Device Lett.* **EDL-6**, p. 479.

T. Iinuma, N. Itoh, H. Nakajima, K. Inou, S. Matsuda, C. Yoshino, Y. Tsuboi, Y. Katsumata, and H. Iwai (1995). "Sub-20 ps high-speed ECL bipolar transistor with low parasitic architecture," *IEEE Trans. Electron Devices* **42**, pp. 399–405.

K. Ismail (1995). "Si/SiGe high-speed field-effect transistors," *1995 IEDM Technical Digest,* p. 509.

S. S. Iyer, G. L. Patton, S. S. Delage, S. Tiwari, and J. M. C. Stork (1987). "Silicon-germanium base heterojunction bipolar transistors by molecular beam epitaxy," *1987 IEDM Technical Digest*, pp. 874–876.

C. Jund and R. Poirier (1966). "Carrier concentration and minority carrier lifetime measurement in semiconductor epitaxial layers by the MOS capacitance method," *Solid-State Electron.* **9**, p. 315.

D. Kahng and M. M. Atalla (1960). "Silicon dioxide field surface devices," presented at Device Research Conf. IEEE, Pittsburgh.

E. O. Kane (1961). "Theory of tunneling," *J. Appl. Phys.* **32**, pp. 83–91.

A. K. Kapoor and D. J. Roulston, Editors (1989). *Polysilicon Emitter Bipolar Transistors*, IEEE Press, New York.

R. E. Kerwin, D. L. Klein, and J. C. Sarace (1969). "Method for making MIS structures," U. S. Patent 3,475,234 issued Oct. 28.

V. P. Kesan, S. Subbanna, P. J. Restle, M. J. Tejwani, J. M. Aitken, S. S. Iyer, and J. A. Ott (1991). "High performance 0.25 μm p-MOSFETs with silicon–germanium channels for 300 K and 77 K operation," *1991 IEDM Technical Digest*, pp. 25–28.

C. J. Kircher (1975). "Comparison of leakage currents in ion-implanted and diffused p–n junctions," *J. Appl. Phys.* **46**, pp. 2167–2173.

C. T. Kirk, Jr. (1962). "A theory of transistor cutoff frequency ($f_T$) falloff at high current densities," *IEEE Trans. Electron Devices* **ED-9**, pp. 164–174.

C. Kittel (1976). *Introduction to Solid State Physics*, Wiley, New York.

P. K. Ko (1982). "Hot-electron effects in MOSFET," *Ph.D. Thesis*, University of California, Berkeley.

P. K. Ko (1989). "Approaches to scaling," in *Advanced MOS Device Physics*, edited by N. G. Einspruch and G. Gildenblat, Academic, San Diego.

P. K. Ko, R. S. Muller, and C. Hu (1981). "A unified model for hot-electron currents in MOSFETs," *1981 IEDM Technical Digest*, pp. 600–603.

M. Kondo, T. Kobayashi, and Y. Tamaki (1995). "Hetero-emitter-like characteristics of phosphorus doped polysilicon emitter transistors–part I: band structure in the polysilicon emitter obtained from electrical measurements," *IEEE Trans. Electron Devices* **42**, pp. 419–426.

H. Kroemer (1985). "Two integral relations pertaining to the electron transport through a bipolar transistor with a non-uniform energy gap in the base region," *Solid-State Electron.* **28**, pp. 1101–1103.

H. J. Kuno (1964). "Analysis and characterization of p–n junction diode switching," *IEEE Trans. Devices* **ED-11**, pp. 8–14.

S. E. Laux (1984). "Accuracy of an effective channel Length/external resistance extraction algorithm for MOSFETs," *IEEE Trans. Electron Devices* **ED-31**, p. 1245.

S. E. Laux and M. V. Fischetti (1988). "Monte-Carlo simulation of submicrometer Si n-MOSFET's at 77 and 300 K," *IEEE Electron Device Lett.* **EDL-9**, p. 467.

M. Lax (1960). "Cascade capture of electrons in solids," *Phys. Rev.* **119**, pp. 1502–1523.

J.-H. Lee, W.-G. Kang, J.-S. Lyu, and J. D. Lee (1996). "Modeling of the critical current density of bipolar transistor with retrograde collector doping profile," *IEEE Electron Device Lett.* **17**, pp. 109–111.

M. Lenzlinger and E. H. Snow (1969). "Fowler–Nordheim tunneling into thermally grown $SiO_2$," *J. Appl. Phys.* **40**, pp. 278–283.

G. P. Li, C. T. Chuang, T. C. Chen, and T. H. Ning (1987). "On the narrow-emitter effect of advanced shallow-profile bipolar transistors," *IEEE Trans. Electron Devices* **35**, pp. 1942–1950.

G. P. Li, E. Hackbarth, and T. C. Chen (1988). "Indentification and implication of a perimeter tunneling current component in advanced self-aligned bipolar transistors," *IEEE Trans. Electron Devices* **ED-35**, pp. 89–95.

S.-H. Lo, D. A. Buchanan, Y. Taur, and W. Wang (1997). "Quantum-mechanical modeling of electron tunneling current from the inversion layer of ultra-thin-oxide nMOSFET's," *IEEE Electron Device Lett.* **18**, pp. 209–211.

P.-F. Lu and T.-C. Chen (1989). "Collector–base junction avalanche effects in advanced double-poly self-aligned bipolar transistors," *IEEE Trans. Electron Devices* **ED-36**, pp. 1182–1188.

P. F. Lu, G. P. Li, and D. D. Tang (1987). "Lateral encroachment of extrinsic base dopant in submicron bipolar transistors," *IEEE Electron Device Lett.* **8**, pp. 496–498.

P.-F. Lu, J. H. Comfort, D. D. Tang, B. S. Meyerson, and J. Y.-C. Sun (1990). "The implementation of a reduced-field profile design for high-performance bipolar transistors," *IEEE Electron Device Lett.* **11**, pp. 336–338.

M. Lundstrom (1997). "Elementary scattering theory of the Si MOSFET," *IEEE Electron Device Lett.* **18**, p. 361.

A. Many, Y. Goldstein, and N. B. Grover (1965). *Semiconductor Surfaces*, Wiley, New York.

R. G. Meyer and R. S. Muller (1987). "Charge-control analysis of the collector–base space-charge- region contribution to bipolar-transistor time constant $\tau_T$," *IEEE Trans. Electron Devices* **ED-34**, pp. 450–452.

B. S. Meyerson (1986). "Low-temperature silicon epitaxy by ultra-high vacuum/chemical vapor deposition," *Appl. Phys. Lett.* **48**, pp. 797–799.

Y. Mii, S. Wind, Y. Taur, Y. Lii, D. Klaus, and J. Bucchignano (1994). "An ultra-low power 0.1 μm CMOS,"*1994 VLSI Technology Symp. Technical Digest*, IEEE, pp. 9–10.

S. L. Miller (1955). "Avalanche breakdown in germanium," *Phys. Rev.* **99**, pp. 1234–1241.

J. L. Moll (1964). *Physics of Semiconductors*, McGraw-Hill, New York.

J. L. Moll, and I. M. Ross (1956). "The dependence of transistor parameters on base resistivity," *Proc. IRE* **44**, pp. 72–78.

R. S. Muller and T. I. Kamins (1977). *Device Electronics for Integrated Circuits*, Wiley, New York.

T. Nakamura and H. Nishizawa (1995). "Recent progress in bipolar transistor technology," *IEEE Trans. Electron Devices* **ED-42**, pp. 390–398.

K. K. Ng and W. T. Lynch (1986). "Analysis of the gate-voltage dependent series resistance of MOSFETs," *IEEE Trans. Electron Devices* **ED-33**, p. 965.

T. N. Nguyen (1984). "Small-geometry MOS transistors: physics and modeling of surface- and buried-channel MOSFETs," *Ph.D. Thesis*, Stanford University.

T. N. Nguyen and J. D. Plummer (1981). "Physical mechanisms responsible for short-channel effects in MOS devices," *1981 IEDM Technical Digest*, pp. 596–599.

E. H. Nicollian, C. N. Berglund, P. F. Schmidt, and J. M. Andrews (1971). "Electrochemical charging of thermal $SiO_2$ films by injected electron currents," *J. Appl. Phys.* **42**, pp. 5654–5664.

E. H. Nicollian and J. R. Brews (1982). *MOS Physics and Technology*, Wiley, New York.

T. H. Ning (1976a). "Capture cross section and trap concentration of holes in silicon dioxide," *J. Appl. Phys.* **47**, pp. 1079–1081.

T. H. Ning (1976b). "High-field capture of electrons by Coulomb-attractive centers in silicon dioxide," *J. Appl. Phys.* **47**, pp. 3203–3208.

T. H. Ning (1978). "Electron trapping in $SiO_2$ due to electron-beam deposition of aluminum," *J. Appl. Phys.* **49**, pp. 4077–4082.

T. H. Ning and R. D. Isaac (1980). "Effect of emitter contact on current gain of silicon bipolar devices," *IEEE Trans. Electron Devices* **ED-27**, pp. 2051–2055.

T. H. Ning and D. D. Tang (1984). "Method for determining the emitter and base series resistances of bipolar transistors," *IEEE Trans. Electron Devices* **ED-31**, pp. 409–412.

T. H. Ning, C. M. Osburn, and H. N. Yu (1975). "Electron trapping at positively charged centers in $SiO_2$," *Appl. Phys. Lett.* **26**, pp. 248–250.

T. H. Ning, C. M. Osburn, and H. N. Yu (1977). "Emission probability of hot electrons from silicon into silicon dioxide," *J. Appl. Phys.* **48**, pp. 286–293.

T. H. Ning, R. D. Isaac, P. M. Solomon, D. D. Tang, H. N. Yu, G. C. Feth, and S. K. Wiedmann (1981). "Self-aligned bipolar transistors for high-performance and low-power-delay VLSI," *IEEE Trans. Electron Devices* **ED-28**, pp. 1010–1013.

W. P. Noble, S. H. Voldman, and A. Bryant (1989). "The effects of gate field on the leakage characteristics of heavily doped junctions," *IEEE Trans. Electron Devices* **ED-36**, pp. 720–726.

S. Ogura, C. F. Codella, N. Rovedo, J. F. Shepard, and J. Riseman (1982). "A half-micron MOSFET using double-implanted LDD," *1982 IEDM Technical Digest*, p. 718.

C. S. Oh, W. H. Chang, B. Davari, and Y. Taur (1990). "Voltage dependence of the MOSFET gate-to-source/drain overlap," *Solid-State Electron.* **33**, p. 1650.

Y. Ohkura (1990). "Quantum effects in Si n-MOS inversion layer at high substrate concentration," *Solid-State Electron.* **33**, p. 1581.

H. C. Pao and C. T. Sah (1966). "Effects of diffusion current on characteristics of metal–oxide (insulator)–semiconductor transistors," *Solid-State Electron.* **9**, p. 927.

G. L. Patton, J. M. C. Stork, J. H. Comfort, E. F. Crabbé, B. S. Meyerson, D. L. Harame, and J. Y.-C. Sun (1990). "SiGe-base heterojunction bipolar transistors: physics and design issues," *1990 IEDM Technical Digest*, pp. 13–16.

R. People (1986). "Physics and applications of $Ge_xSi_{1-x}$/Si strained-layer heterostructures," *IEEE J. Quantum Electron.* **QE-22**, pp. 1696–1710.

H. C. Poon, H. K. Gummel, and D. L. Scharfetter (1969). "High injection in epitaxial transistors," *IEEE Trans. Electron Devices* **ED-16**, pp. 455–457.

E. J. Prinz and J. C. Sturm (1991). "Current gain–Early voltage products in heterojunction bipolar transistors with non-uniform base bandgaps," *IEEE Electron Device Lett.* **12**, pp. 661–663.

R. L. Pritchard (1955). "High-frequency power gain of junction transistors," *Proc. IRE* **43**, pp. 1075–1085.

R. Ranfft and H.-M. Rein (1982). "A simple optimization procedure for bipolar subnanosecond ICs with low power dissipation," *Microelectron. J.* **13**, pp. 23–28.

B. Razavi, R.-H. Yan, and K. F. Lee (1994). "Impact of distributed gate resistance on the performance of MOS devices," *IEEE Trans. Circuits & Systems* **41**, p. 750.

R. R. Razouk and B. E. Deal (1979). "Dependence of interface state density on silicon thermal oxidation process variables," *J. Electrochem. Soc.* **126**, pp. 1573–1581.

V. L. Rideout, F. H. Gaensslen, and A. LeBlanc (1975). "Device design consideration for ion-implanted n-channel MOSFETs," *IBM J. Res. and Devel.* **19**, p. 50.

B. Riccó, J. M. C. Stork, and M. Arienzo (1984). "Characterization of non-ohmic behavior of emitter contacts of bipolar transistors," *IEEE Electron Device Lett.* **EDL-5**, pp. 221–223.

R. Rios and N. D. Arora (1994). "Determination of ultra-thin gate oxide thickness for CMOS structures using quantum effects," *1994 IEDM Technical Digest*, pp. 613–616.

D. J. Roulston (1990). *Bipolar Semiconductor Devices*, McGraw-Hill, New York.

A. G. Sabnis and J. T. Clemens (1979). "Characterization of the electron mobility in the inverted ⟨100⟩ Si surface," *1979 IEDM Technical Digest*, pp. 18–21.

C. T. Sah (1988). "Evolution of the MOS transistor – from concept to VLSI," *Proc. IEEE* **76**, pp. 1280–1326.

C. T. Sah, R. N. Noyce, and W. Shockley (1957). "Carrier generation and recombination in p–n junction and p–n junction characteristics," *Proc. IRE* **45**, pp. 1228–1243.

C. T. Sah, T. H. Ning, and L. L. Tschopp (1972). "The scattering of electrons by surface oxide charges and by lattice vibrations at the silicon–silicon dioxide interface," *Surface Sci.* **32**, pp. 561–575.

G. A. Sai-Halasz (1995). "Performance trends in high-end processors," *IEEE Proc.* **83**, 20–36.

G. A. Sai-Halasz, M. R. Wordeman, D. P. Kern, S. Rishton, and E. Ganin (1988). "High transconductance and velocity overshoot in NMOS devices at the 0.1 μm gate-length level," *IEEE Electron Device Lett.* **EDL-9**, p. 464.

G. A. Sai-Halasz, M. R. Wordeman, D. P. Kern, S. A. Rishton, E. Ganin, T. H. P. Chang, and R. H. Dennard (1990). "Experimental technology and performance of 0.1 μm gate length FETs operated at liquid-nitrogen temperature," *IBM J. Res. and Devel.* **34**, p. 452.

T. Sakurai (1983). "Approximation of wiring delay in MOSFET LSI," *IEEE J. Solid-State Circuits* **SC-18**, p. 418.

L. W. Schaper and D. I. Amey (1983). "Improved electrical performance required for future MOS packaging," *IEEE Trans. Components, Hybrids, and Manufacturing Tech.* **CHMT-6**, p. 283.

D. K. Schroder (1990). *Semiconductor Material and Device Characterization*, Wiley, New York.

K. F. Schuegraf and C. Hu (1994). "Metal–oxide–semiconductor field-effect-transistor substrate current during Fowler–Nordheim tunneling stress and silicon dioxide reliability," *J. Appl. Phys.* **76**, pp. 3695–3700.

K. F. Schuegraf, C. C. King, and C. Hu (1992). "Ultra-thin dioxide leakage current and scaling limit," *1992 Symp. VLSI Technology Digest of Tech. Papers, IEEE*, pp. 18–19.

G. G. Shahidi, D. A. Antoniadis, and H. I. Smith (1989). "Indium channel implants for improved MOSFET behavior at the 100 nm channel length regime," DRC Abstract, *IEEE Trans. Electron Devices* **ED-36**, p. 2605.

M. J. Sherony, A. Wei, and D. A. Antoniadis (1996). "Effect of body charge on fully- and partially-depleted SOI MOSFET design," *1996 IEDM Technical Digest*, p. 125.

B. J. Sheu and P. K. Ko (1984). "A capacitive method to determine channel lengths for conventional and LDD MOSFET's," *IEEE Electron Device Lett.* **EDL-5**, p. 491.

T. Shiba, T. Uchino, K. Ohnishi, and Y. Tamaki (1996). "In-situ phosphorus-doped polysilicon emitter technology for very high-speed, small emitter bipolar transistors," *IEEE Trans. Electron Devices* **43**, pp. 889–897.

M. A. Shibib, F. A. Lindholm, and R. Therez (1979). "Heavily doped transparent-emitter regions in junction solar cells, diodes, and transistors," *IEEE Trans. Electron Devices* **ED-26**, pp. 959–965.

W. Shockley, *Electrons and Holes in Semiconductors* (1950). Van Nostrand, Princeton, N.J.

W. Shockley (1961). "Problems related to p–n junctions in silicon," *Solid-State Electron.* **2**, pp. 35–67.

W. Shockley and W. T. Read (1952). "Statistics of the recombination of holes and electrons," *Phys. Rev.* **87**, p. 835.

R. Shrivastava and K. Fitzpatrick (1982). "A simple model for the overlap capacitance of a VLSI MOS device," *IEEE Trans. Electron Devices* **ED-29**, p. 1870.

J. W. Slotboom and H. D. de Graaff (1976). "Measurements of bandgap narrowing in Si bipolar transistors," *Solid-State Electron.* **19**, pp. 857–862.

C. G. Sodini, T. W. Ekstedt, and J. L. Moll (1982). "Charge accumulation and mobility in thin dielectric MOS transistors," *Solid-State Electron.* **25**, pp. 833–841.

C. G. Sodini, P. K. Ko, and J. L. Moll (1984). "The effect of high fields on MOS device and circuit performance," *IEEE Trans. Electron Devices*. **ED-31**, p. 1386.

P. M. Solomon (1982). "A comparison of semiconductor devices for high speed logic," *IEEE Proc.* **70**, p. 489.

P. M. Solomon and D. D. Tang (1979). "Bipolar circuit scaling," *1979 ISSCC Digest of Technical Papers*, pp. 86–87.

F. Stern (1972). "Self-consistent results for n-type Si inversion layers," *Phys. Rev. B* **5**, p. 4891.

F. Stern (1974). "Quantum properties of surface space-charge layers," *CRC Crit. Rev. Solid-State Sci.* **4**, p. 499.

F. Stern and W. E. Howard (1967). "Properties of semiconductor surface inversion layers in the electric quantum limit," *Phys. Rev.* **163**, p. 816.

J. M. C. Stork and R. D. Isaac (1983). "Tunneling in base-emitter junctions," *IEEE Trans. Electron Devices* **ED-30**, pp. 1527–1534.

L. T. Su, J. B. Jacobs, J. Chung, and D. A. Antoniadis (1994). "Deep-submicrometer channel design in silicon-on-insulator (SOI) MOSFETs," *IEEE Electron Device Lett.* **15**, p. 183.

J. Y.-C. Sun, M. R. Wordeman, and S. E. Laux (1986). "On the accuracy of channel length characterization of LDD MOSFETs," *IEEE Trans. Electron Devices* **ED-33**, p. 1556.

J. Y.-C. Sun, Y. Taur, R. H. Dennard, and S. P. Klepner (1987). "Submicrometer-channel CMOS for low-temperature operation," *IEEE Trans. Electron Devices* **ED-34**, p. 19.

J. Y.-C. Sun, C. Y. Wong, Y. Taur, and C. Hsu (1989). "Study of boron penetration through thin oxide with $p^+$ polysilicon gate," *1989 VLSI Technology Symp. Technical Digest*, IEEE, pp. 17–18.

S. C. Sun and J. D. Plummer (1980). "Electron mobility in inversion and accumulation layers on thermally oxidized silicon surfaces," *IEEE Trans. Electron Devices* **ED-27**, p. 1497.

K. Suzuki (1991). "Optimum base doping profile for minimum base transit time," *IEEE Trans. Electron Devices* **38**, pp. 2128–2133.

R. M. Swanson and J. D. Meindl (1972). "Ion-implanted complementary MOS transistors in low-voltage circuits," *IEEE J. Solid-State Circuits*. **SC-7**, p. 146.

S. E. Swirhun, Y.-H. Kwark, and R. M. Swanson (1986). "Measurement of electron lifetime, electron mobility and band-gap narrowing in heavily doped p-type silicon," *1986 IEDM Technical Digest*, pp. 24–27.

S. M. Sze (1981). *Physics of Semiconductor Devices*, Wiley, New York.

S. Takagi, M. Iwase, and A. Toriumi (1988). "On universality of inversion-layer mobility in n- and p-channel MOSFETs," *1988 IEDM Technical Digest*, pp. 398–401.

D. D. Tang (1980). "Heavy doping effects in pnp bipolar transistors," *IEEE Trans. Electron Devices* **ED-27**, pp. 563–570.

D. D. Tang and P.-F. Lu (1989). "A reduced-field design concept for high-performance bipolar transistors," *IEEE Electron Device Lett.* **10**, pp. 67–69.

D. D. Tang and P. M. Solomon (1979). "Bipolar transistor design for optimized power-delay logic circuits," *IEEE J. Solid-State Circuits* **SC-14**, pp. 679–684.

D. D. Tang, K. P. MacWilliams, and P. M. Solomon (1983). "Effects of collector epitaxial layer on the switching speed of high-performance bipolar transistors," *IEEE Electron Device Lett.* **EDL-4**, pp. 17–19.

Y. Taur and E. J. Nowak (1997). "CMOS devices below 0.1 μm; How high will performance go?" *1997 IEDM Technical Digest*, p. 215.

Y. Taur, W. H. Chang, and R. H. Dennard (1984). "Characterization and modeling of a latchup-free 1-μm CMOS technology," *1984 IEDM Technical Digest*, pp. 398–401.

Y. Taur, G. J. Hu, R. H. Dennard, L. M. Terman, C. Y. Ting, and K. E. Petrillo (1985). "A self-aligned 1 μm channel CMOS technology with retrograde n-well and thin epitaxy," *IEEE Trans. Electron Devices* **ED-32**, p. 203.

Y. Taur, J. Y.-C. Sun, D. Moy, L. K. Wang, B. Davari, S. P. Klepner, and C.Y. Ting (1987). "Source–drain contact resistance in CMOS with self-aligned $TiSi_2$," *IEEE Trans. Electron Devices* **ED-34**, p. 575.

Y. Taur, D. S. Zicherman, D. R. Lombardi, P. J. Restle, C. H. Hsu, H. I. Hanafi, M. R. Wordeman, B. Davari, and G. G. Shahidi (1992). "A new shift and ratio method for MOSFET channel length extraction," *IEEE Electron Device Lett.* **EDL-13**, p. 267.

Y. Taur, C. H. Hsu, B. Wu, R. Kiehl, B. Davari, and G. Shahidi (1993a). "Saturation transconductance of deep-submicron-channel MOSFETs," *Solid-State Electron.* **36**, p. 1085.

Y. Taur, S. Cohen, S. Wind, T. Lii, C. Hsu, D. Quinlan, C. Chang, D. Buchanan, P. Agnello, Y. Mii, C. Reeves, A. Acovic, and V. Kesan (1993b). "Experimental 0.1-μm p-channel MOSFET with $p^+$ polysilicon gate on 35-Å gate oxide," *IEEE Electron Device Lett.* **EDL-14**, p. 304.

Y. Taur, S. Wind, Y. Mii, Y. Lii, D. Moy, K. Jenkins, C. L. Chen, P. J. Coane, D. Klaus, J. Bucchignano, M. Rosenfield, M. Thomson, and M. Polcari (1993c). "High performance 0.1 μm CMOS devices with 1.5 V power supply," *1993 IEDM Technical Digest*, pp. 127–130.

Y. Taur, Y.-J. Mii, D. Frank, H.-S. Wong, D. A. Buchanan, S. Wind, S. Rishton, G. Sai-Halasz, E. Nowak (1995a). "CMOS scaling into the 21st century: 0.1 μm and beyond," *IBM J. Res. and Devel.* **39**, p. 245.

Y. Taur, Y. J. Mii, R. Logan, and H. S. Wong (1995b). "On effective channel length in 0.1 μm MOSFETs," *IEEE Electron Device Lett.* **EDL-16**, p. 136.

Y. Taur, D. A. Buchanan, W. Chen, D. J. Frank, K. E. Ismail, S.-H. Lo, G. A. Sai-Halasz, R. G. Viswanathan, H.-J. C. Wann, S. J. Wind, and H.-S. Wong (1997). "CMOS scaling into the nanometer regime," *IEEE Proc.* **85**, pp. 486–504.

G. W. Taylor (1984). "Velocity saturated characteristics of short-channel MOSFETs," *AT&T Bell Labs. Tech. J.* **63**, p. 1325.

G. W. Taylor and J. G. Simmons (1986). "Figure of merit for integrated bipolar transistors," *Solid-State Electron.* **29**, pp. 941–946.

R. D. Thornton, D. DeWitt, P. E. Gray, and E. R. Chenette (1966). *Characteristics and Limitations of Transistors*, Vol. 4, Semiconductor Electronics Education Committee, Wiley, New York.

C. Y. Ting, S. S. Iyer, C. M. Osburn, G. J. Hu, and A. M. Schweighart (1982). "The use of $TiSi_2$ in a self-aligned silicide technology," Presented at ECS Symposium on VLSI Science and Technology.

R. R. Troutman (1979). "VLSI limitations from drain-induced barrier lowering," *IEEE Trans. Electron Devices*. **ED-26**, p. 461.

F. A. Trumbore (1960). "Solid solubilities of impurity elements in germanium and silicon," *Bell System Tech. J.* **39**, p. 205.

T. Uchino, T. Shiba, T. Kikuchi, Y. Tamaki, A. Watanabe, and Y. Kiyota (1995). "Very-high-speed silicon bipolar transistors with *in-situ* doped polysilicon emitter and rapid vapor-phase doping base," *IEEE Trans. Electron Devices* **42**, pp. 406–412.

M. J. van Dort, P. H. Woerlee, and A. J. Walker (1994). "A simple model for quantization effects in heavily-doped silicon MOSFETs at inversion conditions," *Solid-State Electron.* 37, p. 411.

R. van Overstraeten and H. de Man (1970). "Measurement of the ionization rates in diffused silicon p–n junctions," *Solid-State Electron.* **13**, pp. 583–608.

R. J. van Overstraeten, H. J. de Man, and R. P. Mertens (1973). "Transport equations in heavily doped silicon," *IEEE Trans. Electron Devices* **ED-20**, pp. 290–298.

J. F. Verwey (1972). "Hole currents in thermally grown $SiO_2$," *J. Appl. Phys.* **43**, pp. 2273–2277.

F. Wanlass and C. T. Sah (1963). "Nanowatt logic using field-effect metal–oxide–semiconductor triodes," *Technical Digest of the 1963 Int. Solid-State Circuit Conf.*, IEEE, pp. 32–33.

J. D. Warnock (1995). "Silicon bipolar device structures for digital applications: technology trends and future directions," *IEEE Trans. Electron Devices* **42**, pp. 377–389.

W. M. Webster (1954). "On the variation of junction-transistor current amplification factor with emitter current," *Proc. IRE* **42**, pp. 914–920.

J. Weng, J. Holz, and T. F. Meister (1992). "New method to determine the base resistance of bipolar transistors," *IEEE Electron Device Lett.* **13**, pp. 158–160.

W. M. Werner (1976). "The influence of fixed interface charges on the current-gain falloff of planar n–p–n transistors," *J. Electrochem. Soc.* **123**, pp. 540–543.

H. S. Wong and Y. Taur (1993). "Three-dimensional atomistic simulation of discrete random dopant distribution effects in sub-0.1 μm MOSFETs," *1993 IEDM Technical Digest*, pp. 705–708.

C. Y. Wong, J. Y.-C. Sun, Y. Taur, C. S. Oh, R. Angelucci, and B. Davari (1988). "Doping of $n^+$ and $p^+$ polysilicon in a dual-gate CMOS process," *1988 IEDM Technical Digest*, pp. 238–241.

C. Y. Wong, J. Piccirillo, A. Bhattacharyya, Y. Taur, and H. I. Hanafi (1989). "Sidewall oxidation of polycrystalline-silicon gate," *IEEE Electron Device Lett.* **EDL-10**, p. 420.

M. R. Wordeman (1986). "Design and modeling of miniaturized MOSFETs," *Ph.D. Thesis*, Columbia University.

M. R. Wordeman (1989). Private communications.

M. R. Wordeman, Y. C. Sun, and S. E. Laux (1985). "Geometry effects in MOSFET channel length extraction algorithms," *IEEE Electron Device Lett.* **EDL-6**, p. 186.

R. H. Yan, A. Ourmazd, K. F. Lee, D. Y. Jeon, C. S. Rafferty, and M. R. Pinto (1991). "Scaling the Si metal–oxide–semiconductor field-effect transistor into the 0.1-μm regime using vertical doping engineering," *Appl. Phys. Lett.* **59**, p. 3315.

L. D. Yau (1974). "A simple theory to predict the threshold voltage of short-channel IGFETs," *Solid-State Electron.* **17**, p. 1059.

M. Yoshimi, H. Hazama, M. Takahashi, S. Kambayashi, T. Wada, K. Kato, and H. Tango (1989). "Two-dimensional simulation and measurement of high-performance MOSFETs made on a very thin SOI film," *IEEE Trans. Electron Devices* **ED-36**, p. 493.

C. Yoshino, K. Inou, S. Matsuda, H. Nakajima, Y. Tsuboi, H. Naruse, H. Sugaya, Y. Katsumata, and H. Iwai (1995). "A 62.8-GHz $f_{max}$ LP-CVD epitaxially grown silicon-base bipolar transistor with extremely high Early voltage of 85.7 V," *1995 Symp. VLSI Technology, Digest of Technical Papers*, pp. 131–132.

A. Y. C. Yu (1970). "Electron tunneling and contact resistance of metal–silicon contact barriers," *Solid-State Electron.* **13**, p. 239.

H. N. Yu (1971). "Transistor with limited-area base-collector junction," U.S. Patent Re. 27,045, reissued February 2.

Z. Yu, B. Riccó, and R. W. Dutton (1984). "A comprehensive analytical and numerical model of polysilicon emitter contacts in bipolar transistors," *IEEE Trans. Electron Devices* **ED-31**, pp.773–785.

# INDEX